3 0000 000 328 124

Scientific Computing
An Introduction with
Parallel Computing

Scientific Computing
An Introduction with
Parallel Computing

Gene Golub
Department of Computer Science
Stanford University
Stanford, California

James M. Ortega
School of Engineering & Applied Science
University of Virginia
Charlottesville, Virginia

ACADEMIC PRESS, INC.
Harcourt Brace Jovanovich, Publishers
Boston   San Diego   New York
London   Sydney   Tokyo   Toronto

This book is printed on acid-free paper. ∞

Copyright © 1993 by Academic Press, Inc.

All rights reserved.
No part of this publication may be reproduced or
transmitted in any form or by any means, electronic
or mechanical, including photocopy, recording, or
any information storage and retrieval system, without
permission in writing from the publisher.

ACADEMIC PRESS, INC.
1250 Sixth Avenue, San Diego, CA 92101-4311

United Kingdom Edition published by
ACADEMIC PRESS LIMITED
24–28 Oval Road, London NW1 7DX

ISBN 0-12-289253-4

Printed in the United States of America

93 94 95 96 97 EB 9 8 7 6 5 4 3 2 1

*Dedicated to the memories
of our mothers*

Table of Contents

Preface	ix
Chapter 1. The World of Scientific Computing	1
1.1 What Is Scientific Computing?	1
1.2 Mathematical Modeling	3
1.3 The Process of Numerical Solution	6
1.4 The Computational Environment	11
Chapter 2. Linear Algebra	15
2.1 Matrices and Vectors	15
2.2 Eigenvalues and Canonical Forms	28
2.3 Norms	39
Chapter 3 Parallel and Vector Computing	49
3.1 Parallel and Vector Computers	49
3.2 Basic Concepts of Parallel Computing	62
3.3 Matrix Multiplication	71
Chapter 4. Polynomial Approximation	91
4.1 Taylor Series, Interpolation and Splines	91
4.2 Least Squares Approximation	106
4.3 Application to Root-Finding	117

Chapter 5 Continuous Problems Solved Discretely — 137

5.1 Numerical Integration — 137

5.2 Initial Value Problems — 150

5.3 Boundary Value Problems — 173

5.4 Space and Time — 191

5.5 The Curse of Dimensionality — 202

Chapter 6. Direct Solution of Linear Equations — 215

6.1 Gaussian Elimination — 215

6.2 Errors in Gaussian Elimination — 236

6.3 Other Factorizations — 257

Chapter 7. Parallel Direct Methods — 275

7.1 Basic Methods — 275

7.2 Other Organizations of Factorization — 294

7.3 Banded and Tridiagonal Systems — 302

Chapter 8. Iterative Methods — 321

8.1 Relaxation-Type Methods — 321

8.2 Parallel and Vector Implementations — 333

8.3 The Multigrid Method — 353

Chapter 9. Conjugate Gradient-Type Methods — 371

9.1 The Conjugate Gradient Method — 371

9.2 Preconditioning — 381

9.3 Nonsymmetric and Nonlinear Problems — 397

Bibliography — 413

Author Index — 429

Subject Index — 435

Preface

This book is a modification of "Scientific Computing and Differential Equations: An Introduction to Numerical Methods." The intent of the present book is to introduce the basic ideas of vector and parallel computing as part of the introduction of basic numerical methods. We believe this will be useful for computer science students as well as those in science and engineering. It is clear that supercomputing will make an increasing impact on scientific and engineering endeavors in the years to come and supercomputing is necessarily based on vector and parallel computers. It would be useful for students to have access to such machines during this course, although that is not necessary. Much of the material on parallel and vector computing can still be studied profitably.

The book is meant to be used at the advanced undergraduate or beginning graduate level and it is assumed that such students will have a background in calculus, including some differential equations, and also some linear algebra. Linear algebra is the most important mathematical tool in scientific computing, both for the formulation of problems as well as for analysis of numerical methods for their solution. Even most students who have had a full undergraduate course in linear algebra will not have been exposed to some of the necessary material, such as norms and some aspects of eigenvalues, eigenvectors and canonical forms. Chapter 2 contains a review of such material. It is suggested that it be used primarily as a reference as the corresponding material arises later in the book.

Chapter 1 is an overview of scientific computing, especially the topics of mathematical modeling, the general process of numerical solution of problems, and the computational environment in which these solutions are obtained. Chapter 3 gives an introduction to some of the basic ideas in parallel and vector computing including a review of the architecture of such computers as well as some fundamental concepts. Some of these ideas and techniques are then illustrated by the simple but important problem of matrix-vector multiplication, a topic that arises repeatedly in the remainder of the book. Chapter 4 treats some elementary topics that are covered in any first course on numerical methods: Taylor series approximation, interpolation and splines, least squares approximation, and finding solutions of nonlinear equations.

A large fraction of the book is devoted to the solution of linear systems of equations, since linear systems are at the center of many problems in scientific computing. Chapter 5 shows how linear systems arise by discretizing differential equations. An important aspect of these systems is the non-zero structure of the coefficient matrix. Thus, we may obtain tridiagonal matrices, banded matrices or very large sparse matrices, depending on the type of differential equation and the method of discretization.

Chapter 6 begins the solution of linear systems, concentrating on those systems for which direct methods such as Gaussian elimination are appropriate. There are discussions of rounding error and the effect of ill-conditioning as well as an introduction to other direct methods such as QR factorization. Chapter 7 shows how the basic direct methods must be reorganized to be efficient on parallel and vector computers.

For the very large sparse linear systems that arise from partial differential equations, direct methods are not always appropriate and Chapter 8 begins the study of iterative methods. The classical Jacobi, Gauss-Seidel and SOR methods, and their parallel and vector implementations, are treated leading up to their use in the multigrid method. Then, in Chapter 9, we consider conjugate gradient methods, for both symmetric and nonsymmetric systems. Included in this chapter are discussions of preconditioning, which utilizes some of the methods of Chapter 8, and parallel and vector implementations. This chapter ends with a nonlinear partial differential equation.

Many important topics have not been included; for example, computation of eigenvalues and eigenvectors and solution of linear or nonlinear programming problems. However, such areas also rely heavily on techniques for the solution of linear equations.

We believe that this book can be used successfully at different levels. For an introductory one semester course at the undergraduate level, concentration would be on Chapters 3, 4, 6 and parts of 5 and 7. For a first course at the graduate level for those who have had an undergraduate numerical methods course, much of Chapters 4, 5, and 6 can be covered rapidly in review and emphasis placed on the more advanced topics of Chapters 7, 8, and 9. Or, with some supplementary material from the instructor, the book could form the basis for a full year course.

We owe thanks to many colleagues and students for their comments on our previous book and on the draft of the current one. We are also indebted to Ms. Brenda Lynch for her expert LaTeXing of the manuscript.

Stanford, California
Charlottesville, Virginia

Chapter 1

The World of Scientific Computing

1.1 What Is Scientific Computing?

The many thousands of computers now installed in this country and abroad are used for a bewildering – and increasing – variety of tasks: accounting and inventory control for industry and government, airline and other reservation systems, limited translation of natural languages such as Russian to English, monitoring of process control, and on and on. One of the earliest – and still one of the largest – uses of computers was to solve problems in science and engineering and, more specifically, to obtain solutions of mathematical models that represent some physical situation. The techniques used to obtain such solutions are part of the general area called *scientific computing*, and the use of these techniques to elicit insight into scientific or engineering problems is called *computational science* (or *computational engineering*).

There is now hardly an area of science or engineering that does not use computers for modeling. Trajectories for earth satellites and for planetary missions are routinely computed. Engineers use computers to simulate the flow of air about an aircraft or other aerospace vehicle as it passes through the atmosphere, and to verify the structural integrity of aircraft. Such studies are of crucial importance to the aerospace industry in the design of safe and economical aircraft and spacecraft. Modeling new designs on a computer can save many millions of dollars compared to building a series of prototypes. Similar considerations apply to the design of automobiles and many other products, including new computers.

Civil engineers study the structural characteristics of large buildings, dams, and highways. Meteorologists use large amounts of computer time to predict tomorrow's weather as well as to make much longer range predictions, including the possible change of the earth's climate. Astronomers and astrophysicists have modeled the evolution of stars, and much of our basic knowledge about such phenomena as red giants and pulsating stars has come from such calculations coupled with observations. Ecologists and biologists are increasingly using the computer in such diverse areas as population dynamics (including the study of natural predator and prey relationships), the flow of blood in the human body, and the dispersion of pollutants in the oceans and atmosphere.

The mathematical models of all of these problems are systems of differential equations, either ordinary or partial. Differential equations come in all "sizes and shapes," and even with the largest computers we are nowhere near being able to solve many of the problems posed by scientists and engineers. But there is more to scientific computing, and the scope of the field is changing rapidly. There are many other mathematical models, each with its own challenges. In operations research and economics, large linear or nonlinear optimization problems need to be solved. Data reduction – the condensation of a large number of measurements into usable statistics – has always been an important, if somewhat mundane, part of scientific computing. But now we have tools (such as earth satellites) that have increased our ability to make measurements faster than our ability to assimilate them; fresh insights are needed into ways to preserve and use this irreplaceable information. In more developed areas of engineering, what formerly were difficult problems to solve even once on a computer are today's routine problems that are being solved over and over with changes in design parameters. This has given rise to an increasing number of computer-aided design systems. Similar considerations apply in a variety of other areas.

Although this discussion begins to delimit the area that we call scientific computing, it is difficult to define it exactly, especially the boundaries and overlaps with other areas. We will accept as our working definition that *scientific computing is the collection of tools, techniques, and theories required to solve on a computer mathematical models of problems in science and engineering.*

A majority of these tools, techniques, and theories originally developed in mathematics, many of them having their genesis long before the advent of electronic computers. This set of mathematical theories and techniques is called numerical analysis (or numerical mathematics) and constitutes a major part of scientific computing. The development of the electronic computer, however, signaled a new era in the approach to the solution of scientific problems. Many of the numerical methods that had been developed for the purpose of hand calculation (including the use of desk calculators for the actual arithmetic) had to be revised and sometimes abandoned. Considerations that were irrelevant or unimportant for hand calculation now became of utmost importance for the efficient and correct use of a large computer system. Many of these considerations – programming languages, operating systems, management of large quantities of data, correctness of programs – were subsumed under the discipline of computer science, on which scientific computing now depends heavily. But mathematics itself continues to play a major role in scientific computing: it provides the language of the mathematical models that are to be solved and information about the suitability of a model (Does it have a solution? Is the solution unique?), and it provides the theoretical foundation for the numerical methods and, increasingly, many of the tools from computer science.

In summary, then, scientific computing draws on mathematics and computer science to develop the best ways to use computer systems to solve problems from science and engineering. This relationship is depicted schematically in Figure 1.1.1. In the remainder of this chapter, we will go a little deeper into these various areas.

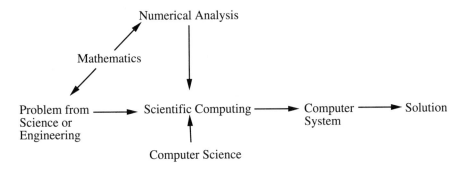

Figure 1.1.1: *Scientific Computing and Related Areas*

1.2 Mathematical Modeling

As was discussed in Section 1.1, we view scientific computing as the discipline that achieves a computer solution of mathematical models of problems from science and engineering. Hence, the first step in the overall solution process is the formulation of a suitable mathematical model of the problem at hand.

Modeling

The formulation of a mathematical model begins with a statement of the factors to be considered. In many physical problems, these factors concern the balance of forces and other conservation laws of physics. For example, in the formulation of a model of a trajectory problem the basic physical law is Newton's second law of motion, which requires that the forces acting on a body equal the rate of change of momentum of the body. This general law must then be specialized to the particular problem by enumerating and quantifying the forces that will be of importance. For example, the gravitational attraction of Jupiter will exert a force on a rocket in Earth's atmosphere, but its effect will be so minute compared to the earth's gravitational force that it can usually be neglected. Other forces may also be small compared to the dominant ones but their effects not so easily dismissed, and the construction of the model will invariably be a compromise between retaining all factors that could likely have a bearing on the validity of the model and keeping the mathematical model sufficiently simple that it is solvable using the tools at hand. Classically, only very simple models of most phenomena were considered since the solutions had to be achieved by hand, either analytically or numerically. As the power of computers and numerical methods has developed, increasingly complicated models have become tractable.

In addition to the basic relations of the model – which in most situations in scientific computing take the form of differential equations – there usually will be a number of initial or boundary conditions. For example, in the predator-prey problem to be discussed in Chapter 5, the initial population of the two species being studied is specified. In studying the flow in a blood vessel, we may require a boundary condition that the flow cannot penetrate the walls of the vessel. In other cases, boundary conditions may not be so physically evident but are still required so that the mathematical problem has a unique solution. Or the mathematical model as first formulated may indeed have many solutions, the one of interest to be selected by some constraint such as a requirement that the solution be positive, or that it be the solution with minimum energy. In any case, it is usually assumed that the final mathematical model with all appropriate initial, boundary, and side conditions indeed has a unique solution. The next step, then, is to find this solution. For problems of current interest, such solutions rarely can be obtained in "closed form." The solution must be approximated by some method, and the methods to be considered in this book are numerical methods suitable for a computer. In the next section we will consider the general steps to be taken to achieve a numerical solution, and the remainder of the book will be devoted to a detailed discussion of these steps for a number of different problems.

Validation

Once we are able to compute solutions of the model, the next step usually is called the *validation of the model*. By this we mean a verification that the solution we compute is sufficiently accurate to serve the purposes for which the model was constructed. There are two main sources of possible error. First, there invariably are errors in the numerical solution. The general nature of these errors will be discussed in the next section, and one of the major themes in the remainder of the book will be a better understanding of the source and control of these numerical errors. But there is also invariably an error in the model itself. As mentioned previously, this is a necessary aspect of modeling:

1.2 Mathematical Modeling

the modeler has attempted to take into account all the factors in the physical problem but then, in order to keep the model tractable, has neglected or approximated those factors that would seem to have a small effect on the solution. The question is whether neglecting these effects was justified. The first test of the validity of the model is whether the solution satisfies obvious physical and mathematical constraints. For example, if the problem is to compute a rocket trajectory where the expected maximum height is 100 kilometers and the computed solution shows heights of 200 kilometers, obviously some blunder has been committed. Or, it may be that we are solving a problem for which we know, mathematically, that the solution must be increasing but the computed solution is not increasing. Once such gross errors are eliminated – which is usually fairly easy – the next phase begins, which is, whenever possible, comparison of the computed results with whatever experimental or observational data are available. Many times this is a subtle undertaking, since even though the experimental results may have been obtained in a controlled setting, the physics of the experiment may differ from the mathematical model. For example, the mathematical model of airflow over an aircraft wing may assume the idealization of an aircraft flying in an infinite atmosphere, whereas the corresponding experimental results will be obtained from a wind tunnel where there will be effects from the walls of the enclosure. (Note that neither the experiment, nor the mathematical model represents the true situation of an aircraft flying in our finite atmosphere.) The experience and intuition of the investigator are required to make a human judgement as to whether the results from the mathematical model are corresponding sufficiently well with observational data.

At the outset of an investigation this is quite often not the case, and the model must be modified. This may mean that additional terms – which were thought negligible but may not be – are added to the model. Sometimes a complete revision of the model is required and the physical situation must be approached from an entirely different point of view. In any case, once the model is modified the cycle begins again: a new numerical solution, revalidation, additional modifications, and so on. This process is depicted schematically in Figure 1.2.1.

Once the model is deemed adequate from the validation and modification process, it is ready to be used for prediction. This, of course, was the whole purpose. We should now be able to answer the questions that gave rise to the modeling effort: How high will the rocket go? Will the wolves eat all the rabbits? Of course, we must always take the answers with a healthy skepticism. Our physical world is simply too complicated and our knowledge of it too meager for us to be able to predict the future perfectly. Nevertheless, we hope that our computer solutions will give increased insight into the problem being studied, be it a physical phenomenon or an engineering design.

6   Chapter 1 The World of Scientific Computing

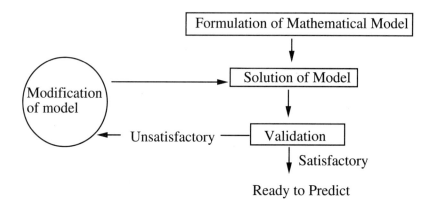

Figure 1.2.1: *The Mathematical Modeling and Solution Process*

1.3 The Process of Numerical Solution

We will discuss in this section the general considerations that arise in the computer solution of a mathematical model, and in the remainder of the book these matters will be discussed in more detail.

Once the mathematical model is given, our first thought typically is to try to obtain an explicit closed-form solution, but such a solution will usually only be possible for certain (perhaps drastic) simplifications of the problem. These simplified problems with known solutions may be of great utility in providing "check cases" for the more general problem.

After realizing that explicit solutions are not possible, we then turn to the task of developing a numerical method for the solution. Implicit in our thinking at the outset – and increasingly explicit as the development proceeds – will be the computing equipment as well as the software environment that is at our disposal. Our approach may be quite different for a microcomputer than for a very large computer. But certain general factors must be considered regardless of the computer to be used.

Rounding Errors

Perhaps the most important factor is that computers deal with a finite number of digits or characters. Because of this we cannot, in general, do arithmetic within the real number system as we do in pure mathematics. That is, the arithmetic done by a computer is restricted to finitely many digits, whereas the numerical representation of most real numbers requires infinitely many. For example, such fundamental constants as π and e require an infinite number of digits for a numerical representation and can *never* be entered exactly in a computer. Moreover, even if we could start with numbers that have

1.3 The Process of Numerical Solution

an exact numerical representation in the computer, the processes of arithmetic require that eventually we make certain errors. For example, the quotient of two numbers may require infinitely many digits for its numerical representation. Therefore, we resign ourselves at the outset to the fact that we cannot do arithmetic exactly on a computer. We shall make small errors, called *rounding errors*, on almost all arithmetic operations, and our task is to insure that these small errors do not accumulate to such an extent as to invalidate the computation.

Computers use the binary number system, and each machine allows a number of binary digits that can be carried in the usual arithmetic, called *single-precision* arithmetic, of the machine. On most scientific computers, this is the equivalent of between 7 and 14 decimal digits. *Higher-precision* arithmetic can also be carried out. On many machines *double-precision* arithmetic, which essentially doubles the number of digits that are carried, is part of the hardware; in this case, programs with double-precision arithmetic usually require only modest, if any, increases in execution time compared to single-precision. On the other hand, some machines implement double precision by software, which may require several times as much time as single precision. Precision higher than double is essentially always carried out by means of software and becomes increasingly inefficient as the precision increases. Higher-precision arithmetic is rarely used on practical problems, but it may be useful for generating "exact" solutions or other information for testing purposes.

Round-off errors can affect the final computed result in different ways. First, during a sequence of millions of operations, each subject to a small error, there is the danger that these small errors will accumulate so as to eliminate much of the accuracy of the computed result. If we round to the nearest digit, the individual errors will tend to cancel out, but the standard deviation of the accumulated error will tend to increase with the number of operations, leaving the possibility of a large final error. If chopping – that is, dropping the trailing digits rather than rounding – is used, there is a bias to errors in one direction, and the possibility of a large final error is increased.

In addition to the possible accumulation of errors over a large number of operations, there is the danger of *catastrophic cancellation*. Suppose that two numbers a and b are equal to within their last digit. Then the difference $c = a - b$ will have only one significant digit of accuracy *even though no round-off error will be made in the subtraction*. Future calculations with c may then limit the final result to one correct digit. Whenever possible, one tries to eliminate the possibility of catastrophic cancellation by rearranging the operations. Catastrophic cancellation is one way in which an algorithm can be *numerically unstable*, although in exact arithmetic it may be a correct algorithm. Indeed, it is possible for the results of a computation to be completely erroneous because of round-off error even though only a small number of arithmetic operations have been performed. Examples of this will be given later.

Detailed round-off error analyses have now been completed for a number of the simpler and more basic algorithms such as those that occur in the solution of linear systems of equations; some of these results will be described in more detail in Chapter 6. A particular type of analysis that has proved to be very powerful is *backward error analysis*. In this approach the round-off errors are shown to have the same effect as that caused by changes in the original problem data. When this analysis is possible, it can be stated that the error in the solution caused by round off is no worse than that caused by certain errors in the original model. The question of errors in the solution is then equivalent to the study of the sensitivity of the solution to perturbations in the model. If the solution is highly sensitive, the problem is said to be *ill-posed* or *ill-conditioned*, and numerical solutions are apt to be meaningless.

Discretization Error

Another way that the finiteness of computers manifests itself in causing errors in numerical computation is due to the need to replace "continuous" problems by "discrete" ones. Chapter 5 is devoted to showing how continuous problems such as differential equations are approximated by discrete problems. As a simple example, discussed further in Section 5.1, the integral of a continuous function requires knowledge of the integrand along the whole interval of integration, whereas a computer approximation to the integral can use values of the integrand at only finitely many points. Hence, even if the subsequent arithmetic were done exactly with no rounding errors, there would still be the error due to the discrete approximation to the integral. This type of error is usually called *discretization error* or *truncation error*, and it affects, except in trivial cases, all numerical solutions of differential equations and other "continuous" problems.

There is one more type of error, which is somewhat akin to discretization error. Many numerical methods are based on the idea of an *iterative process*. In such a process, a sequence of approximations to a solution is generated with the hope that the approximations will converge to the solution; in many cases mathematical proofs can be given that show convergence as the number of iterations tends to infinity. However, only finitely many such approximations can ever be generated on a computer, and, therefore, we must necessarily stop short of mathematical convergence. The error caused by such finite termination of an iterative process is sometimes called *convergence error*, although there is no generally accepted terminology here.

If we rule out trivial problems that are of no interest in scientific computing, we can summarize the situation with respect to computational errors as follows. Every calculation will be subject to rounding error. Whenever the mathematical model of the problem is a differential equation or other "continuous" problem, there also will be discretization error. And if an iterative method is used for the solution, there will be convergence error. These types of

1.3 The Process of Numerical Solution

errors and methods of analyzing and controlling them will be discussed more fully in concrete situations throughout the remainder of the book. But it is important to keep in mind that an acceptable error is very much dependent on the particular problem. Rarely is very high accuracy – say, 14 digits – needed in the final solution; indeed, for many problems arising in industry or other applications two or three digit accuracy is quite acceptable.

Efficiency

The other major consideration besides accuracy in the development of computer methods for the solution of mathematical models is *efficiency*. For most problems, such as solving a system of linear algebraic equations, there are many possible methods, some going back tens or even hundreds of years. Clearly, we would like to choose a method that minimizes the computing time yet retains suitable accuracy in the approximate solution. This turns out to be a difficult problem which involves a number of considerations. Although it is frequently possible to estimate the computing time of an algorithm by counting the required arithmetic operations, the amount of computation necessary to solve a problem to a given tolerance is still an open question except in a few cases. Even if one ignores the effects of round-off error, surprisingly little is known. In the past several years these questions have spawned the subject of *computational complexity*. However, even if such theoretical results were known, they would still give only approximations to the actual computing time, which depends on a number of factors involving the computer system. And these factors change as the result of new systems and architectures. Indeed, the design and analysis of numerical algorithms should provide incentives and directions for such changes.

We give a simple example of the way a very inefficient method can arise. Many elementary textbooks on matrix theory or linear algebra present Cramer's rule for solving systems of linear equations. This rule involves quotients of certain determinants, and the definition of a determinant is usually given as the sum of all possible products (some with minus signs) of elements of the matrix, one element from each row and each column. There are $n!$ such products for an $n \times n$ matrix. Now, if we proceed to carry out the computation of a determinant based on a straightforward implementation of this definition, it would require about $n!$ multiplications and additions. For n very small, say $n = 2$ or $n = 3$, this is a small amount of work. Suppose, however, that we have a 20×20 matrix, a very small size in current scientific computing. If we assume that each arithmetic operation requires 1 microsecond (10^{-6} second), then the time required for this calculation – even ignoring all overhead operations in the computer program – will exceed one million years! On the other hand, the Gaussian elimination method, which will be discussed in Chapter 6, will do the arithmetic operations for the solution of a 20×20 linear system in less than 0.005 second, again assuming 1 microsecond per operation. Although this is an extreme example, it does illustrate the difficulties that can occur by naively following a mathematical prescription in order to solve a problem on a computer.

Good Programs

Even if a method is intrinsically "good," it is extremely important to implement the corresponding computer code in the best way possible, especially if other people are to use it. Some of the criteria for a good code are the following:

1. *Reliability* – The code does not have errors and can be trusted to compute what it is supposed to compute.

2. *Robustness*, which is closely related to reliability – The code has a wide range of applicability as well as the ability to detect bad data, "singular" or other problems that it cannot be expected to handle, and other abnormal situations, and deal with them in a way that is satisfactory to the user.

3. *Portability* – The code can be transferred from one computer to another with a minimum effort and without losing reliability. Usually this means that the code has been written in a general high-level language like FORTRAN and uses no "tricks" that are dependent on the characteristics of a particular computer. Any machine characteristics that must be used are clearly delineated.

4. *Maintainability* – Any code will necessarily need to be changed from time to time, either to make corrections or to add enhancements, and this should be possible with minimum effort.

The code should be written in a clear and straightforward way so that such changes can be made easily and with a minimum likelihood of creating new errors. An important part of maintainability is that there be good *documentation* of the program so that it can be changed efficiently by individuals who did not write the code originally. Good documentation is also important so that the program user will understand not only how to use the code, but also its limitations. Finally, extensive *testing* of the program must be done to ensure that the preceding criteria have been met.

As examples of good software, LINPACK and EISPACK have been two standard packages for the solution of linear systems and eigenvalue problems, respectively. They are now being combined and revised into LAPACK, which is being designed to run on parallel and vector computers (see the next section). Another very useful system is MATLAB, which contains programs for linear systems, eigenvalues and many other mathematical problems and also allows for easy manipulation of matrices.

1.4 The Computational Environment

As indicated in the last section, there is usually a long road from a mathematical model to a successful computer program. Such programs are developed within the overall *computational environment*, which includes the computers to be used, the operating system and other systems software, the languages in which the program is to be written, techniques and software for data management and graphics output of the results, and programs that do symbolic computation. In addition, network facilities allow the use of computers at distant sites as well as the exchange of software and data.

Hardware

The computer hardware itself is of primary importance. Scientific computing is done on computers ranging from small PC's, which execute a few thousand floating point operations per second, to supercomputers capable of billions of such operations per second. Supercomputers that utilize hardware vector instructions are called *vector computers*, while those that incorporate multiple processors are called *parallel computers*. In the latter case, the computer system may contain a few, perhaps very powerful, processors or as many as several tens of thousands of relatively simple processors. Generally, algorithms designed for single processor "serial" computers will not be satisfactory, without modification, for parallel computers. Indeed, a very active area of research in scientific computing is the development of algorithms suitable for vector and parallel computers. Further discussion of vector and parallel computers will be given in Chapter 3.

It is quite common to do program development on a workstation or PC prior to production runs on a larger computer. Unfortunately, a program will not always produce the same answers on two different machines due to different rounding errors. This, of course, will be the case if different precision arithmetic is used. For example, a machine using 48 digit binary arithmetic (14 decimal digits) can be expected to produce less rounding error then one using 24 binary digits (7 decimal digits). However, even when the precision is the same, two machines may produce slightly different results due to different conventions for handling rounding error. This is an unsatisfactory situation that has been addressed by the IEEE standard for floating point arithmetic. Although not all computers currently follow this standard, in the future they probably will, and then machines with the same precision arithmetic will produce identical results on the same problem. On the other hand, algorithms for parallel computers often do the arithmetic operations in a different order than on a serial machine and this causes different errors to occur.

Systems and Languages

In order to be useful, computer hardware must be supplemented by systems software, including operating systems and compilers for high level languages. Although there are many operating systems, UNIX and its variants have increasingly become the standard for scientific computing and essentially all computer manufacturers now offer a version of UNIX for their machines. This is true for vector and parallel computers as well as more conventional ones. The use of a common operating system helps to make programs more portable. The same is true of programming languages. Since its inception in the mid 1950's, Fortran has been the primary programming language for scientific computing. It has been continually modified and extended over the years, and now versions of Fortran also exist for parallel and vector computers. Other languages, especially the systems language C, are sometimes used for scientific computing. However, it is expected that Fortran will continue to evolve and be the standard for the foreseeable future, at least in part because of the large investment in existing Fortran programs.

Data Management

Many of the problems in scientific computing require large amounts of data, both input and output, as well as data generated during the course of the computation. The storing and retrieving of these data in an efficient manner is called *data management*. As an example of this in the area of computer-aided design, a data base containing all information relevant to a particular design application – which might be for an aircraft, an automobile, or a dam – may contain several billion characters. In an aircraft design this information would include everything relevant about the geometry of each part of the aircraft, the material properties of each part, and so on. An engineer may use this data base simply to find all the materials with a certain property. On the other hand, the data base will also be used in doing various analyses of the structural properties of the aircraft, which requires the solution of certain linear or nonlinear systems of equations. It is interesting to note that in many scientific computing programs the number of lines of code to handle data management is far larger than that for the actual computation.

Visualization

The results of a scientific computation are numbers that may represent, for example, the solution of a differential equation at selected points. For large computations, such results may consist of the values of four or five functions at a million or more points. Such a volume of data cannot just be printed. *Scientific visualization* techniques allow the results of computations to be represented pictorially. For example, the output of a fluid flow computation might be a movie that depicts the flow as a function of time in either two or three

1.4 The Computational Environment

dimensions. The results of a calculation of the temperature distribution in a solid might be a color-coded representation in which regions of high temperatures are red and regions of low temperatures are blue, with a gradation of hues between the extremes. Or, a design model may be rotated in three-dimensional space to allow views from any angle. Such visual representations allow a quick understanding of the computation, although more detailed analysis of selected tables of numerical results may be needed for certain purposes, such as error checking.

Symbolic Computation

Another development that is having an increasing impact on scientific computing is *symbolic computation*. Systems such as MACSYMA, REDUCE, MAPLE, and MATHEMATICA allow the symbolic (as opposed to numerical) computation of derivatives, integrals and various algebraic quantities. For example, such systems can add, multiply and divide polynomials or rational expressions; differentiate expressions to obtain the same results that one would obtain using pencil and paper; and integrate expressions that have a "closed form" integral. This capability can relieve the drudgery of manipulating by hand lengthy algebraic expressions, perhaps as a prelude to a subsequent numerical computation. Symbolic computation systems can also solve certain mathematical problems, such as systems of linear equations, without rounding error. However, their use in this regard is limited since the size of the system must be small. In any case, symbolic computation is continuing to develop and can be expected to play an increasing role in scientific computation.

In this section we have discussed briefly some of the major components of the overall computing environment that pertain to scientific computing. In the remainder of the book we will point out in various places where these techniques can be used, although it is beyond the scope of this book to pursue their application in detail.

Supplementary Discussion and References: Chapter 1

For further reading on the computer science areas discussed in this chapter, see Hennesy and Patterson [1990] for computer architecture, Peterson and Silberschatz [1985] for operating systems, Pratt [1984] and Sethi [1989] for programming languages, Aho, Sethi, and Ullman [1988] and Fischer and LeBlanc [1988] for compilers, Elmasri and Navathe [1989] for data management, and Earnshaw and Wiseman [1992], Friedhoff and Benzon [1989] and Mendez [1990] for visualization. A reference for computer graphics, which provides much of the technical foundation for visualization techniques, is Newman and Sproul [1979]. The symbolic computation systems mentioned in the text are covered in Symbolics [1987] for MACSYMA, Rayna [1987] for REDUCE, Char et al. [1992] for MAPLE, and Wolfram [1988] for MATHEMATICA.

The packages EISPACK and LINPACK are discussed in Garbow et al. [1977] and Dongarra and Bunch et al. [1979], respectively, and LAPACK in Anderson et al. [1992]. These and many other software packages are available on NETLIB; see, for example, Dongarra and Duff et al. [1990]. MATLAB can be obtained from The Math Works, Inc., South Natick, MA 01760.

Chapter 2

Linear Algebra

2.1 Matrices and Vectors

An important tool in many areas of scientific computing is linear algebra (also called matrix theory) and we review in this chapter some of the basic results that will be used in the remainder of the book.

A *column vector* is the n-tuple of real or complex numbers

$$\mathbf{x} = \begin{bmatrix} x_1 \\ \vdots \\ x_n \end{bmatrix}, \qquad (2.1.1)$$

and a *row vector* is $(x_1, \ldots, x_n)$. R^n will denote the collection of all column vectors with n real components, and C^n will denote the collection of all column vectors with n complex components. We will sometimes use the term *n-vector* for a vector with n components.

An $m \times n$ *matrix* is the array

$$A = \begin{bmatrix} a_{11} & \cdots & a_{1n} \\ \vdots & & \vdots \\ a_{m1} & \cdots & a_{mn} \end{bmatrix} \qquad (2.1.2)$$

with m rows and n columns, where, again, the components a_{ij} may be real or complex numbers. The (i,j) element of the matrix is a_{ij}. The numbers m and n are the *dimensions* of A. When the dimensions are clear, we will also use the notation $A = (a_{ij})$ to denote an $m \times n$ matrix. If $n = m$, the matrix is *square*; otherwise it is *rectangular*. As indicated above, we will generally denote vectors by lower case boldface letters, and matrices by italicized capital letters.

Submatrices

If $n = 1$ in (2.1.2), then A may be considered as a vector of m components or as an $m \times 1$ matrix. More generally, it is sometimes useful to view the columns of the matrix A as column vectors and the rows as row vectors. In this context we would write A in the form

$$A = (\mathbf{a}_1, \ldots, \mathbf{a}_n), \qquad (2.1.3)$$

where each $\mathbf{a}_i$ is a column vector with m components. Likewise, we may also write A as a collection of row vectors. One should keep in mind that matrices and vectors are conceptually quite different but the identifications and partitionings indicated above are useful for manipulations.

A *submatrix* of A is any matrix obtained from A by deleting rows and columns. The partitioning (2.1.3) into a collection of vectors is a special case of partitioning of a matrix A into submatrices. More generally, we consider partitionings of the form

$$A = \begin{bmatrix} A_{11} & \cdots & A_{1q} \\ \vdots & & \vdots \\ A_{p1} & \cdots & A_{pq} \end{bmatrix}, \qquad (2.1.4)$$

where each A_{ij} is itself a matrix (possibly 1×1), and dimensions of these submatrices must be consistent - that is, all matrices in a given row must have the same number of rows, and all matrices in a given column must have the same number of columns. The partitioned matrix (2.1.4) is sometimes called a *block* matrix. An example is given below for a 5×6 matrix in which the lines indicate the partitioning:

$$A = \begin{bmatrix} 1 & 2 & 3 & 4 & 5 & 6 \\ 1 & 4 & 2 & 1 & 6 & 8 \\ 2 & 4 & 5 & 9 & 1 & 3 \\ 8 & 2 & 4 & 3 & 2 & 1 \\ 3 & 7 & 6 & 4 & 1 & 2 \end{bmatrix} = \begin{bmatrix} A_{11} & A_{12} & A_{13} \\ A_{21} & A_{22} & \backslash A_{23} \end{bmatrix}.$$

A *principle submatrix* of an $n \times n$ matrix is obtained by deleting q rows and the corresponding q columns. A *leading principal submatrix* of order m of an $n \times n$ matrix is obtained by deleting the last $n - m$ rows and columns. In the example below, a 2×2 leading principal submatrix is obtained by deleting the last row and column, and a 2×2 principal submatrix is obtained by deleting the second row and second column:

$$\begin{bmatrix} 1 & 2 & 4 \\ 3 & 2 & 1 \\ 4 & 5 & 6 \end{bmatrix} \quad \begin{bmatrix} 1 & 2 \\ 3 & 2 \end{bmatrix} \quad \begin{bmatrix} 1 & 4 \\ 4 & 6 \end{bmatrix}.$$

2.1 Matrices and Vectors

Special Matrices

We next discuss some important special classes of matrices and the corresponding block matrices. An $n \times n$ matrix A is *diagonal* if $a_{ij} = 0$ whenever $i \neq j$, so that all elements off the main diagonal are zero. A is *upper triangular* if $a_{ij} = 0$ for $i > j$, so that all elements below the main diagonal are zero; similarly, A is *lower triangular* if all elements above the main diagonal are zero. The matrix A is *block upper triangular* or *block lower triangular* if it has the partitioned forms

$$\begin{bmatrix} A_{11} & \cdots & A_{1p} \\ & \ddots & \vdots \\ & & A_{pp} \end{bmatrix}, \quad \begin{bmatrix} A_{11} & & \\ \vdots & \ddots & \\ A_{p1} & \cdots & A_{pp} \end{bmatrix}.$$

Here, the matrices A_{ii} are all square but not necessarily the same size, and only the (possibly) nonzero matrices are shown. In the case that $A_{ij} = 0, i \neq j$, the matrix is *block diagonal*. We will sometimes use the notation $\mathrm{diag}(A_{11}, \ldots, A_{pp})$ to denote a block diagonal matrix, and $\mathrm{diag}(a_{11}, \ldots, a_{nn})$ to denote a diagonal matrix. An extremely important matrix is $I = \mathrm{diag}(1, \ldots, 1)$, called the *identity matrix*.

The *transpose* of the $m \times n$ matrix (2.1.2) is the $n \times m$ matrix

$$A^T = (a_{ji}) = \begin{bmatrix} a_{11} & \cdots & a_{m1} \\ \vdots & & \vdots \\ a_{1n} & \cdots & a_{mn} \end{bmatrix}.$$

If A has complex elements, then the transpose is still well defined, but a more useful matrix is the *conjugate transpose* given by

$$A^* = (\bar{a}_{ji}) = \begin{bmatrix} \bar{a}_{11} & \cdots & \bar{a}_{m1} \\ \vdots & & \vdots \\ \bar{a}_{1n} & \cdots & \bar{a}_{mn} \end{bmatrix},$$

where the bar denotes complex conjugate. An example is

$$A = \begin{bmatrix} 2-i & 3 & 2+2i \\ i & 4i & 1 \end{bmatrix}, \quad A^* = \begin{bmatrix} 2+i & -i \\ 3 & -4i \\ 2-2i & 1 \end{bmatrix},$$

where $i = \sqrt{-1}$. If $\mathbf{x}$ is a column vector, its transpose is the row vector $\mathbf{x}^T = (x_1, \ldots, x_n)$; similarly, the transpose of a row vector is a column vector. The conjugate transpose is $\mathbf{x}^* = (\bar{x}_1, \ldots, \bar{x}_n)$. Another standard notation for the conjugate transpose is A^H or $\mathbf{x}^H$.

A matrix A is *symmetric* if $A = A^T$, and *Hermitian* if $A = A^*$. Examples are

$$A = \begin{bmatrix} 2 & 1 \\ 1 & 2 \end{bmatrix}, \qquad A = \begin{bmatrix} 2 & 1+i \\ 1-i & 2 \end{bmatrix},$$

where the first matrix is real and symmetric, and the second is complex and Hermitian. We will see later that symmetric matrices play a central role in scientific computing. Unless specifically stated to the contrary, we will assume in the remainder of the book that all matrices and vectors are real.

Basic Operations

Addition of matrices or vectors is allowed whenever the dimensions are the same:

$$\begin{bmatrix} x_1 \\ \vdots \\ x_n \end{bmatrix} + \begin{bmatrix} y_1 \\ \vdots \\ y_n \end{bmatrix} = \begin{bmatrix} x_1 + y_1 \\ \vdots \\ x_n + y_n \end{bmatrix},$$

$$\begin{bmatrix} a_{11} & \cdots & a_{1n} \\ \vdots & & \vdots \\ a_{m1} & \cdots & a_{mn} \end{bmatrix} + \begin{bmatrix} b_{11} & \cdots & b_{1n} \\ \vdots & & \vdots \\ b_{m1} & \cdots & b_{mn} \end{bmatrix} = \begin{bmatrix} a_{11} + b_{11} & \cdots & a_{1n} + b_{1n} \\ \vdots & & \vdots \\ a_{m1} + b_{m1} & \cdots & a_{mn} + b_{mn} \end{bmatrix}.$$

Multiplication of a matrix $A = (a_{ij})$ by a scalar α is defined by $\alpha A = (\alpha a_{ij})$ and similarly for vectors. A sum of the form $\sum_{i=1}^{m} \alpha_i \mathbf{x}_i$, where the α_i are scalars, is a *linear combination* of the vectors $\mathbf{x}_1, \ldots, \mathbf{x}_m$, and $\sum_{i=1}^{m} \alpha_i A_i$ is a linear combination of the matrices A_i.

The product AB of two matrices A and B is defined only if the number of columns of A is equal to the number of rows of B. If A is $n \times p$ and B is $p \times m$, the product $C = AB$ is an $n \times m$ matrix C whose ij element is given by

$$c_{ij} = \sum_{k=1}^{p} a_{ik} b_{kj}. \tag{2.1.5}$$

The associative and distributive laws

$$(\mathbf{x} + \mathbf{y}) + \mathbf{z} = \mathbf{x} + (\mathbf{y} + \mathbf{z}), \qquad (AB)C = A(BC),$$

$$A(B + C) = AB + AC,$$

hold for addition and multiplication, as is easily verified. But, whereas addition of matrices is commutative ($A + B = B + A$), multiplication generally is not, even if A and B are both square. Thus, in general, $AB \neq BA$.

The transpose operation satisfies

$$(AB)^T = B^T A^T, \qquad (A + B)^T = A^T + B^T,$$

2.1 Matrices and Vectors

where A and B may be either matrices or vectors of the correct dimensions. These relations are easily proved by direct computation (Exercise 2.1.12).

The definition of matrix multiplication given above arises naturally as a consequence of considering matrices as linear transformations. However, it is also possible to define an element-wise multiplication, analogous to addition, whenever the dimensions of the matrices A and B are the same. This is the *Schur product* (also called the *Hadamard* product), which is defined for two $m \times n$ matrices by

$$A \circ B = \begin{bmatrix} a_{11}b_{11} & \cdots & a_{1n}b_{1n} \\ \vdots & & \vdots \\ a_{m1}b_{m1} & \cdots & a_{mn}b_{mn} \end{bmatrix}.$$

Unless explicitly stated to the contrary, we will always use the matrix multiplication defined by (2.1.5).

The above operations work equally well on block matrices if we assume that all submatrices are of the right dimensions for the indicated operations. For example, for the matrix $A = (A_{ij})$ of (2.1.4), if $B = (B_{ij})$ is a block matrix of proper dimensions, then $C = AB$ is a block matrix $C = (C_{ij})$ with

$$C_{ij} = \sum_{k=1}^{q} A_{ik} B_{kj}.$$

An important special case of matrix multiplication is when A is $1 \times n$ and B is $n \times 1$. Then AB is 1×1, a scalar. On the other hand, if A is $m \times 1$ and B is $1 \times n$, then AB is $m \times n$. These products are given explicitly in terms of vectors by

$$(a_1, \ldots, a_n) \begin{bmatrix} b_1 \\ \vdots \\ b_n \end{bmatrix} = \sum_{i=1}^{n} a_i b_i, \tag{2.1.6}$$

$$\begin{bmatrix} a_1 \\ \vdots \\ a_m \end{bmatrix} (b_1, \ldots, b_n) = \begin{bmatrix} a_1 b_1 & \cdots & a_1 b_n \\ a_2 b_1 & & a_2 b_n \\ \vdots & & \vdots \\ a_m b_1 & \cdots & a_m b_n \end{bmatrix}. \tag{2.1.7}$$

If **a** and **b** are column vectors, the above matrix multiplication rules define the products $\mathbf{a}^T \mathbf{b}$ and $\mathbf{a}\mathbf{b}^T$. The former is called the *inner product* or *dot product* of **a** and **b** and the latter is the *outer product*.

Orthogonality and Linear Independence

Two nonzero column vectors **x** and **y** are *orthogonal* if their inner product, $\mathbf{x}^T \mathbf{y}$, is zero. If, in addition, $\mathbf{x}^T \mathbf{x} = \mathbf{y}^T \mathbf{y} = 1$, the vectors are *orthonormal*.

The nonzero vectors $\mathbf{x}_1, \ldots, \mathbf{x}_m$ are orthogonal if $\mathbf{x}_i^T \mathbf{x}_j = 0$ whenever $i \neq j$, and orthonormal if, in addition, $\mathbf{x}_i^T \mathbf{x}_i = 1, i = 1, \ldots, m$. A set of orthogonal vectors can always be made orthonormal by scaling in the following way. For any nonzero vector $\mathbf{x}, \mathbf{x}^T\mathbf{x} = \sum_{i=1}^n |x_i|^2 > 0$. Then $\mathbf{y} = \mathbf{x}/(\mathbf{x}^T\mathbf{x})^{\frac{1}{2}}$ satisfies $\mathbf{y}^T\mathbf{y} = 1$. Thus, if $\mathbf{x}_1, \ldots, \mathbf{x}_m$ are orthogonal vectors, then the vectors $\mathbf{y}_i = \mathbf{x}_i/(\mathbf{x}_i^T\mathbf{x}_i)^{\frac{1}{2}}, i = 1, \ldots, m$, are orthonormal. For complex vectors, T is replaced by $*$ in all these definitions.

Orthogonal vectors are important special cases of linearly independent vectors, which are defined as follows. A set of n-vectors $\mathbf{x}_1, \ldots, \mathbf{x}_m$ is *linearly dependent* if

$$\sum_{i=1}^m c_i \mathbf{x}_i = 0 \qquad (2.1.8)$$

where the scalars $c_1, \ldots, c_m$ are not all zero. If no linear dependence exists – that is, if no equation of the form (2.1.8) is possible unless all c_i are zero – then the vectors are *linearly independent*. For example, if $c_1\mathbf{x}_1 + c_2\mathbf{x}_2 = 0$ with $c_1 \neq 0$, then $\mathbf{x}_1$ and $\mathbf{x}_2$ are linearly dependent; thus, $\mathbf{x}_1 = -c_1^{-1}c_2\mathbf{x}_2$, so that $\mathbf{x}_1$ is a multiple of $\mathbf{x}_2$. On the other hand, if one vector is not a multiple of the other, then $c_1\mathbf{x}_1 + c_2\mathbf{x}_2 = 0$ implies that $c_1 = c_2 = 0$, and the vectors are linearly independent. Similarly, if $m = n = 3$ and (2.1.8) holds with $c_1 \neq 0$, then

$$\mathbf{x}_1 = -(c_2\mathbf{x}_2 + c_3\mathbf{x}_3)/c_1,$$

which says that $\mathbf{x}_1$ is in the plane containing $\mathbf{x}_2$ and $\mathbf{x}_3$; thus, three 3-vectors are linearly dependent if they lie in a plane; otherwise they are linearly independent.

The set of all linear combinations of the n-vectors $\mathbf{x}_1, \ldots, \mathbf{x}_m$, denoted by span($\mathbf{x}_1, \ldots, \mathbf{x}_m$), is a *subspace* of the linear space of all n-vectors. If $\mathbf{x}_1, \ldots, \mathbf{x}_m$ are linearly independent, then the dimension of span($\mathbf{x}_1, \ldots, \mathbf{x}_m$) is m; and the vectors $\mathbf{x}_1, \ldots, \mathbf{x}_m$ form a *basis* for this subspace. If $\mathbf{x}_1, \ldots, \mathbf{x}_m$ are linearly dependent, they may be so in different ways: They may all be multiples of one of the vectors, say $\mathbf{x}_m$; in this case the set is said to have only one linearly independent vector. Or they may all be linear combinations of two of the vectors, say $\mathbf{x}_m$ and $\mathbf{x}_{m-1}$; in this case, we say there are two linearly independent vectors. More generally, if all of the vectors may be written as a linear combination of r vectors but not $r - 1$ vectors, then we say that the set has r linearly independent vectors. In this case, span($x_1, \ldots, \mathbf{x}_m$) has dimension r.

An important property of a matrix is the number of linearly independent columns, which is the *rank* of the matrix. An important theorem is that the rank is also equal to the number of linearly independent rows. If A is $m \times n$ and p is the minimum of m and n, then rank(A) $\leq p$. If rank(A) $= p$, then A is said to have *full rank*.

2.1 Matrices and Vectors

Orthogonal Matrices

A real $n \times n$ matrix A is *orthogonal* if the matrix equation

$$A^T A = I = A A^T \tag{2.1.9}$$

holds, where $I = \text{diag}(1, \ldots, 1)$ is the identity matrix. This equation may be interpreted as saying that A is orthogonal if its columns (and its rows), viewed as vectors, are orthonormal. (If the matrix A is complex with orthonormal columns, then it is called *unitary* and (2.1.9) becomes $A^* A = I = A A^*$.)

Consider the two matrices

$$\begin{bmatrix} 1 & 1 \\ -1 & 1 \end{bmatrix}, \quad \frac{1}{\sqrt{2}} \begin{bmatrix} 1 & 1 \\ -1 & 1 \end{bmatrix}.$$

The first has orthogonal columns but it is not an orthogonal matrix. The second matrix is the same as the first with a scale factor added so that it is orthogonal.

An important class of orthogonal matrices are *permutation matrices*, which are obtained from the identity matrix by interchange of rows or columns. For example,

$$\begin{bmatrix} 0 & 0 & 1 \\ 0 & 1 & 0 \\ 1 & 0 & 0 \end{bmatrix}$$

is a 3×3 permutation matrix obtained by interchanging the first and third rows of the identity matrix.

Inverses

The identity matrix I in (2.1.9) obviously has the property that $IB = B$ for any matrix or vector B of commensurate dimensions; likewise $BI = B$ whenever the multiplication is defined. We denote the columns of the identity matrix by $\mathbf{e}_1, \ldots, \mathbf{e}_n$, so that $\mathbf{e}_i$ is the vector with a 1 in the ith position and 0's elsewhere. These vectors are a particularly important set of orthogonal (and orthonormal) vectors.

Now let A be an $n \times n$ matrix and suppose that each of the n systems of equations

$$A \mathbf{x}_i = \mathbf{e}_i, \quad i = 1, \ldots, n, \tag{2.1.10}$$

has a unique solution. Let $X = (\mathbf{x}_1, \ldots, \mathbf{x}_n)$ be the matrix with these solutions as its columns. Then (2.1.10) is equivalent to

$$AX = I, \tag{2.1.11}$$

where I is the identity matrix. Then, the matrix X is defined to be the *inverse* of A, denoted by A^{-1}. As an example, consider

$$A = \begin{bmatrix} 2 & 1 \\ 2 & 2 \end{bmatrix}, \quad A\mathbf{x}_1 = \begin{bmatrix} 1 \\ 0 \end{bmatrix}, \quad A\mathbf{x}_2 = \begin{bmatrix} 0 \\ 1 \end{bmatrix}.$$

The solutions of the two systems are $\mathbf{x}_1 = (1, -1)^T$ and $\mathbf{x}_2 = \frac{1}{2}(-1, 2)^T$, so that
$$A^{-1} = X = \frac{1}{2} \begin{bmatrix} 2 & -1 \\ -2 & 2 \end{bmatrix}.$$

We can also multiply the equation $A\mathbf{x} = \mathbf{b}$ by A^{-1} and write the solution as

$$\mathbf{x} = A^{-1}\mathbf{b}, \tag{2.1.12}$$

which is sometimes useful theoretically (but, we stress, not in practice; that is, do *not* solve $A\mathbf{x} = \mathbf{b}$ by finding A^{-1} and then $A^{-1}\mathbf{b}$). Note that in the special case that A is an orthogonal matrix, (2.1.9) shows that $A^{-1} = A^T$.

Determinants

The *determinant* of a $n \times n$ matrix can be defined as

$$\det A = \sum_P (-1)^{S(P)} a_{\sigma(1)1} a_{\sigma(2)2} \cdots a_{\sigma(n)n}. \tag{2.1.13}$$

This rather formidable expression means the following. The summation is taken over all $n!$ possible permutations P of the integers $1, \ldots, n$. Each such permutation is denoted by $\sigma(1), \ldots, \sigma(n)$ so that each of the products in (2.1.13) is made up of exactly one element from each row and each column of A. The sign of each product is positive if $S(P)$ is even (an even permutation) or negative if $S(P)$ is odd (an odd permutation). The formula (2.1.13) is the natural extension to $n \times n$ matrices of the (hopefully) familiar rules for determinants of 2×2 and 3×3 matrices:

$$n = 2: \quad \det A = a_{11}a_{22} - a_{12}a_{21}$$
$$n = 3: \quad \det A = a_{11}a_{22}a_{33} + a_{12}a_{23}a_{31} + a_{13}a_{21}a_{32}$$
$$\qquad\qquad - a_{11}a_{23}a_{32} - a_{13}a_{22}a_{31} - a_{12}a_{21}a_{33}.$$

We next collect, without proof, a number of basic facts about determinants of $n \times n$ matrices A.

THEOREM 2.1.1.

a. If any two rows (or columns) of A are equal, or if A has a zero row or column, then $\det A = 0$.

b. If any two rows (or columns) of A are interchanged, the sign of the determinant changes, but the magnitude remains unchanged.

c. If any row (or column) of A is multiplied by a scalar α, then the determinant is multiplied by α.

d. $\det A^T = \det A$, $\quad \det A^* = \overline{\det A}$.

e. If a scalar multiple of any row (or column) of A is added to another row (or column), the determinant remains unchanged.

2.1 Matrices and Vectors

f. If B is also an $n \times n$ matrix, then $\det(AB) = (\det A)(\det B)$.

g. The determinant of a triangular matrix is the product of the main diagonal elements.

h. Let A_{ij} denote the $(n-1) \times (n-1)$ submatrix of A obtained by deleting the ith row and jth column. Then

$$\det A = \sum_{j=1}^{n} a_{ij}(-1)^{i+j} \det A_{ij}, \quad \text{for any } i, \qquad (2.1.14)$$

$$\det A = \sum_{i=1}^{n} a_{ij}(-1)^{i+j} \det A_{ij}, \quad \text{for any } j. \qquad (2.1.15)$$

The quantity $(-1)^{i+j} \det A_{ij}$ is called the *cofactor* of a_{ij}, and (2.1.14) and (2.1.15) are called *cofactor expansions* of the determinant. For example

$$\det \begin{bmatrix} 1 & 2 & 3 \\ 4 & 5 & 6 \\ 7 & 8 & 9 \end{bmatrix} = \det \begin{bmatrix} 5 & 6 \\ 8 & 9 \end{bmatrix} - 4 \det \begin{bmatrix} 2 & 3 \\ 8 & 9 \end{bmatrix} + 7 \det \begin{bmatrix} 2 & 3 \\ 5 & 6 \end{bmatrix}.$$

Nonsingular Matrices

By Theorem 2.1.1 f and g, we can conclude from (2.1.11) that

$$(\det A)(\det X) = \det I = 1 \qquad (2.1.16)$$

so that $\det A \neq 0$. We will say that an $n \times n$ matrix A is *nonsingular* if the inverse A^{-1} exists and by (2.1.16) this is equivalent to $\det A \neq 0$. The following basic result summarizes various equivalent ways of saying that a matrix is nonsingular.

THEOREM 2.1.2 *For an $n \times n$ matrix A, the following are equivalent:*

a. A^{-1} *exists.*

b. $\det A \neq 0$.

c. *The linear system $A\mathbf{x} = 0$ has only the solution $\mathbf{x} = 0$.*

d. *For any vector $\mathbf{b}$, the linear system $A\mathbf{x} = \mathbf{b}$ has a unique solution.*

e. *The columns (rows) of A are linearly independent.*

f. $Rank(A) = n$.

Positive Definite Matrices

In general, it is difficult to check if a given matrix is nonsingular. One exception is triangular matrices since by Theorem 2.1.1g, a triangular matrix is nonsingular if and only if all its main diagonal elements are non-zero. Another important class of nonsingular matrices are those that are orthogonal. Still another class, which will be extremely important throughout the book, are those that satisfy

$$\mathbf{x}^T A \mathbf{x} > 0, \quad \mathbf{x} \neq 0, \quad \mathbf{x} \text{ real} \qquad (2.1.17)$$

A matrix that satisfies (2.1.17) is *positive definite*. To conclude that A is nonsingular if (2.1.17) holds, suppose that it is singular. Then by c. of Theorem 2.1.2 there is a non-zero $\mathbf{x}$ such that $A\mathbf{x} = 0$. But then (2.1.17) would not hold, which is a contradiction.

An important property of symmetric positive definite matrices is given by the following statement.

THEOREM 2.1.3 *If A is an $n \times n$ positive definite symmetric matrix, then every principal submatrix of A is also symmetric and positive definite. In particular, the diagonal elements of A are positive.*

To see why this is true, let A_p be any $p \times p$ principal submatrix. It is clear that A_p is symmetric, because by deleting corresponding rows and columns of A to obtain A_p we maintain symmetry in the elements of A_p. Now let $\mathbf{x}_p$ be any nonzero p-vector, and let $\mathbf{x}$ be the n-vector obtained from $\mathbf{x}_p$ by inserting zeros in those positions corresponding to the rows deleted from A. Then a direct calculation shows that

$$\mathbf{x}_p^T A_p \mathbf{x}_p = \mathbf{x}^T A \mathbf{x} > 0$$

so that A_p is positive definite.

Reducible Matrices

An $n \times n$ matrix A is *reducible* if there is a permutation matrix P so that

$$PAP^T = \begin{bmatrix} B_{11} & B_{12} \\ 0 & B_{22} \end{bmatrix}, \qquad (2.1.18)$$

where B_{11} and B_{22} are square matrices. For example, if P is the permutation matrix which interchanges the second and third rows, then

$$P \begin{bmatrix} 1 & 0 & 1 \\ 0 & 1 & 0 \\ 1 & 1 & 1 \end{bmatrix} P^T = \begin{bmatrix} 1 & 1 & 0 \\ 1 & 1 & 1 \\ 0 & 0 & 1 \end{bmatrix},$$

so that the original matrix is reducible.

2.1 Matrices and Vectors

If (2.1.18) holds, then we can rewrite the system of equations $A\mathbf{x} = \mathbf{b}$ as $PAP^T\mathbf{y} = \mathbf{c}$, with $\mathbf{y} = P\mathbf{x}$ and $\mathbf{c} = P\mathbf{b}$. Thus, the system may be written as

$$\begin{bmatrix} B_{11} & B_{12} \\ 0 & B_{22} \end{bmatrix} \begin{bmatrix} \mathbf{y}_1 \\ \mathbf{y}_2 \end{bmatrix} = \begin{bmatrix} \mathbf{c}_1 \\ \mathbf{c}_2 \end{bmatrix},$$

and the unknowns in the vector $\mathbf{y}_2$ can be obtained independently of $\mathbf{y}_1$. Thus, in the variables $\mathbf{y}$, the system reduces to two smaller systems.

A matrix A is *irreducible* if it is not reducible. Clearly, any matrix with all non-zero elements is irreducible. If zero elements are present, their location determines whether or not A is irreducible. We state the following criterion without proof.

THEOREM 2.1.4 *An $n \times n$ matrix A is irreducible if and only if for any two distinct indices $1 \leq i, j \leq n$, there is a sequence of nonzero elements of A of the form*

$$\{a_{i,i_1}, a_{i_1 i_2}, \ldots, a_{i_m j}\}. \qquad (2.1.19)$$

We now interpret Theorem 2.1.4 in a geometric way. Let $P_1, \ldots, P_n$ be n distinct points in the plane. Then, for every non-zero element a_{ij} in A we construct a *directed link* (P_i, P_j) from P_i to P_j. As a result, we associate with A a *directed graph*. For example, for the matrix

$$A = \begin{bmatrix} 0 & 0 & 1 & 1 \\ 0 & 0 & 1 & 1 \\ 1 & 0 & 0 & 0 \\ 1 & 1 & 0 & 0 \end{bmatrix} \qquad (2.1.20)$$

the associated directed graph is shown in Figure 2.1.1 (see Exercise 2.1.22).

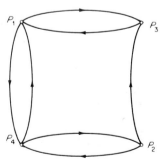

Figure 2.1.1: *A Directed Graph*

A directed graph is *strongly connected* if for any two points P_i and P_j there is a path

$$(P_i, P_{i_1}), (P_{i_1}, P_{i_2}), \ldots, (P_{i_m}, P_j) \qquad (2.1.21)$$

connecting P_i and P_j. The graph in Figure 2.1.1 is strongly connected since there is a path between any two points (to go from P_3 to P_2, go first to P_1, then P_4 and then P_2). The existence of a path (2.1.21) is equivalent to the sequence of non-zero elements (2.1.19). Thus, Theorem 2.1.4 may be stated in terms of graphs as:

THEOREM 2.1.5 *An $n \times n$ matrix is irreducible if and only if its associated directed graph is strongly connected.*

In Chapter 5, we will see examples of the use of this theorem for matrices arising from differential equations.

EXERCISES 2.1

2.1.1. Partition the matrix
$$\begin{bmatrix} 1 & 2 & 3 \\ 4 & 5 & 6 \\ 7 & 8 & 9 \end{bmatrix}$$
into $2 \times 2, 1 \times 2, 2 \times 1$, and 1×1 submatrices in three different ways.

2.1.2. Compute the transpose of the matrix in Exercise 1. Give an example of a 3×3 complex matrix and compute its conjugate transpose.

2.1.3. Do the indicated multiplications:

$$\begin{bmatrix} 1 & 3 \\ 1 & 4 \end{bmatrix} \begin{bmatrix} 2 & 1 & 1 \\ 3 & 2 & 4 \end{bmatrix} \quad (1,3) \begin{bmatrix} 2 & 1 \\ 1 & 2 \end{bmatrix} \quad \begin{bmatrix} 2 \\ 1 \end{bmatrix}(4,2) \quad (4,2) \begin{bmatrix} 2 \\ 1 \end{bmatrix}$$

2.1.4. Ascertain if the following pairs of vectors are orthogonal and orthonormal:

$$\begin{bmatrix} 1 \\ 0 \end{bmatrix} \begin{bmatrix} 0 \\ 1 \end{bmatrix} \quad \begin{bmatrix} 1 \\ 1 \end{bmatrix} \begin{bmatrix} 1 \\ -2 \end{bmatrix} \quad \begin{bmatrix} 1+i \\ i-1 \end{bmatrix} \begin{bmatrix} 1-i \\ 1+i \end{bmatrix}$$

2.1.5. Ascertain if the pairs of vectors in Exercise 4 are linearly independent.

2.1.6. An $n \times n$ Hadamard matrix A has elements that are all ± 1 and satisfies $A^T A = nI$. Show that $|\det A| = n^{n/2}$.

2.1.7. Let $x_1, \ldots, x_m$ be nonzero orthogonal vectors. Show that $x_1, \ldots, x_m$ are linearly independent. *Hint:* Assume that a linear dependence (2.1.8) exists, and show that orthogonality implies that all c_i must be zero.

2.1.8. Show that the following matrix is orthogonal for any real value of θ:

$$\begin{bmatrix} \cos\theta & \sin\theta \\ -\sin\theta & \cos\theta \end{bmatrix}$$

2.1 Matrices and Vectors

2.1.9. Ascertain if the following matrices are orthogonal:
$$\begin{bmatrix} 1 & 0 \\ 0 & 1 \end{bmatrix} \quad \begin{bmatrix} 0 & 1 \\ 1 & 0 \end{bmatrix} \quad \begin{bmatrix} 1 & 1 \\ -1 & 1 \end{bmatrix} \quad \begin{bmatrix} \cos\theta & \sin\theta & 0 \\ -\sin\theta & \cos\theta & 0 \\ 0 & 0 & 1 \end{bmatrix}$$

2.1.10. Show that the product of diagonal matrices is diagonal and that the product of lower (upper) triangular matrices is lower (upper) triangular.

2.1.11. Let A and B be $m \times n$ matrices. Show that $(A+B)^T = A^T + B^T$.

2.1.12. Let A and B be $n \times p$ and $p \times m$ matrices, respectively. Show by direct calculation that $(AB)^T = B^T A^T$.

2.1.13. Show that the transpose of an orthogonal matrix is orthogonal. Show that the inverse of an orthogonal matrix is orthogonal.

2.1.14. Show that the product of $n \times n$ orthogonal matrices is orthogonal.

2.1.15. If A is nonsingular, show that $\det(A^{-1}) = (\det A)^{-1}$.

2.1.16. If A and B are $n \times n$ matrices, show that AB is nonsingular if and only if both A and B are nonsingular. Moreover, $(AB)^{-1} = B^{-1}A^{-1}$.

2.1.17. Let A be a real $n \times n$ symmetric matrix, and let P be a real $n \times m$ matrix. Use the result of Exercise 12 to show that the $m \times m$ matrix $P^T A P$ is symmetric. If P is of full rank with $m \leq n$ and A is positive definite, use Theorem 2.1.2 to show that $P^T A P$ is positive definite.

2.1.18. The *trace*, $Tr(A)$, of an $n \times n$ matrix A is the sum of the main diagonal elements of A. For $n \times n$ matrices A and B, and scalars α and β, show that $\mathrm{Tr}(AB) = \mathrm{Tr}(BA)$ and $\mathrm{Tr}(\alpha A + \beta B) = \alpha \mathrm{Tr}(A) + \beta \mathrm{Tr}(B)$.

2.1.19. Ascertain whether the following matrices are singular or nonsingular:
$$\begin{bmatrix} 2 & 1 \\ 1 & 2 \end{bmatrix} \quad \begin{bmatrix} 1 & 1 & 1 \\ 2 & 2 & 2 \\ 3 & 3 & 3 \end{bmatrix} \quad \begin{bmatrix} 4 & 4 & 4 & 4 \\ 0 & 3 & 3 & 3 \\ 0 & 0 & 2 & 2 \\ 0 & 0 & 0 & 1 \end{bmatrix}$$

2.1.20. Show that m n-vectors cannot be linearly independent if $m > n$.

2.1.21. An $n \times n$ matrix A is *idempotent* if $A^2 = A$. Show that an idempotent matrix is singular unless it is the identity matrix.

2.1.22. Show that Figure 2.1.1 is the directed graph of the matrix A of (2.1.20). Conclude from Theorem 2.1.5 that A is irreducible.

2.1.23. Let $A = P + iQ, B = R + iS$ be complex matrices where P, Q, R and S are real. Show how to construct the product $AB = L + iM$, where L and M are real. How could you compute A if you know the product UV, where $U = P+Q, V = R-S$? How many matrix multiplications are required in each of your approaches?

2.2 Eigenvalues and Canonical Forms

One of the most important parts of matrix theory deals with the eigenvalues, and corresponding eigenvectors, of a matrix. An *eigenvalue* (also called a proper value, characteristic value, or latent root) of an $n \times n$ matrix A is a real or complex scalar λ satisfying the equation

$$A\mathbf{x} = \lambda \mathbf{x} \qquad (2.2.1)$$

for some nonzero vector $\mathbf{x}$, called an *eigenvector*. Note that an eigenvector is only determined up to a scalar multiple, since if $\mathbf{x}$ is an eigenvector and α is a nonzero scalar, then $\alpha \mathbf{x}$ is also an eigenvector. We stress that an eigenvector is nonzero, since (2.2.1) is trivially satisfied for the zero vector for any λ.

We can rewrite (2.2.1) in the form

$$(A - \lambda I)\mathbf{x} = 0. \qquad (2.2.2)$$

Thus, from Theorem 2.1.2, (2.2.2) can have a nonzero solution $\mathbf{x}$ only if $A - \lambda I$ is singular. Therefore, any eigenvalue λ must satisfy the equation

$$\det(A - \lambda I) = 0. \qquad (2.2.3)$$

Suppose that A is a 2×2 matrix. Then

$$\det(A - \lambda I) = \det \begin{bmatrix} a_{11} - \lambda & a_{12} \\ a_{21} & a_{22} - \lambda \end{bmatrix} = (a_{11} - \lambda)(a_{22} - \lambda) - a_{21}a_{12},$$

which is a second-degree polynomial in λ. In general, by isolating the product of the diagonal elements in (2.1.13), we have that

$$\det(A - \lambda I) = \Pi_{i=1}^{n}(a_{ii} - \lambda) + q(\lambda),$$

and it is easy to show that q is a polynomial of degree no more than $n - 1$. Thus, (2.2.3) is a polynomial of degree n in λ and is called the *characteristic polynomial* of A. Its roots are the eigenvalues of A and the set of eigenvalues is called the *spectrum* of A. The collection of both eigenvalues and eigenvectors is called the *eigensystem* of A. By the Fundamental Theorem of Algebra, a polynomial of degree n has exactly n real or complex roots, counting multiplicities. Hence, an $n \times n$ matrix has exactly n eigenvalues, although they are not necessarily distinct. For example, if $A = I$ is the identity matrix, then (2.2.3) reduces to $(1 - \lambda)^n = 0$, which has the root 1 with multiplicity n. Thus, the eigenvalues of the identity matrix are $1, 1, \ldots, 1$ (n times).

Since

$$\det(A^T - \lambda I) = \det(A - \lambda I),$$

2.2 Eigenvalues and Canonical Forms

the eigenvalues of A and A^T are the same. A and A^T do not necessarily have the same eigenvectors, however (see Exercise 2.2.2). If $A^T\mathbf{x} = \lambda\mathbf{x}$, then $\mathbf{x}^T A = \lambda \mathbf{x}^T$, and the row vector $\mathbf{x}^T$ is called a *left eigenvector* of A.

Eigenvalues and Singularity

We next relate the determinant of A to its eigenvalues. We can write the characteristic polynomial of A in terms of the eigenvalues $\lambda_1, \ldots, \lambda_n$ of A as

$$\det(A - \lambda I) = (\lambda_1 - \lambda)(\lambda_2 - \lambda) \cdots (\lambda_n - \lambda). \tag{2.2.4}$$

This is just the factorization of a polynomial in terms of its roots. By evaluating (2.2.4) for $\lambda = 0$, we conclude that

$$\det A = \lambda_1 \cdots \lambda_n, \tag{2.2.5}$$

so that the determinant of A is just the product of its n eigenvalues. Thus, we have another characterization of the nonsingularity of a matrix to complement Theorem 2.1.2.

THEOREM 2.2.1 *An $n \times n$ real or complex matrix A is nonsingular if and only if all of its eigenvalues are nonzero.*

Theorem 2.2.1 allows a simple proof of the assertion of Theorem 2.1.2 that the linear independence of the columns of A implies that $\det A \neq 0$. The linear independence ensures that there is no zero eigenvalue since $A\mathbf{x} = 0$ for a nonzero $\mathbf{x}$ would violate the linear independence of the columns of A.

We note that (2.2.4) can be written in the form

$$\det(A - \lambda I) = (\lambda_1 - \lambda)^{r_1}(\lambda_2 - \lambda)^{r_2} \cdots (\lambda_p - \lambda)^{r_p},$$

where the λ_i are distinct and $\sum_{i=1}^p r_i = n$. Then r_i is the *multiplicity* of the eigenvalue λ_i. An eigenvalue of multiplicity one is called *simple*.

Eigenvalues in Special Cases

If A is a triangular matrix, then by Theorem 2.1.1g

$$\det(A - \lambda I) = (a_{11} - \lambda)(a_{22} - \lambda) \cdots (a_{nn} - \lambda). \tag{2.2.6}$$

Thus, in this case, the eigenvalues are simply the diagonal elements of A. The same also holds, of course, for diagonal matrices. Except for triangular matrices and a few other special cases, eigenvalues are relatively difficult to compute, but a number of excellent algorithms for their computation are now known. (We stress that one does *not* compute the characteristic polynomial (2.2.3) and then find its roots.) However, it is easy to compute at least some of the eigenvalues and eigenvectors of certain functions of a matrix A if the eigenvalues and eigenvectors of A are known. As the simplest example, let λ and $\mathbf{x}$ be any eigenvalue and corresponding eigenvector of A. Then

$$A^2\mathbf{x} = A(A\mathbf{x}) = A(\lambda\mathbf{x}) = \lambda A\mathbf{x} = \lambda^2\mathbf{x},$$

so that λ^2 is an eigenvalue of A^2 and $\mathbf{x}$ is a corresponding eigenvector. Proceeding in the same way for any integer power of A, we conclude the following:

THEOREM 2.2.2 *If the $n \times n$ matrix A has eigenvalues $\lambda_1, \ldots, \lambda_n$, then for any positive integer m, $\lambda_1^m, \ldots, \lambda_n^m$ are eigenvalues of A^m. Moreover, any eigenvector of A is an eigenvector of A^m.*

The converse is not true. For example, if

$$A = \begin{bmatrix} 1 & 0 \\ 0 & -1 \end{bmatrix}, \quad A^2 = \begin{bmatrix} 1 & 0 \\ 0 & 1 \end{bmatrix},$$

then every vector is an eigenvector of A^2, but the only eigenvectors of A are multiples of $(1,0)^T$ and $(0,-1)^T$. Similarly,

$$A = \begin{bmatrix} 0 & 1 \\ 0 & 0 \end{bmatrix}$$

has only one linearly independent eigenvector, but, again, every vector is an eigenvector of A^2. Moreover, all three matrices

$$A = \begin{bmatrix} 1 & 0 \\ 0 & -1 \end{bmatrix}, \quad A = \begin{bmatrix} -1 & 0 \\ 0 & -1 \end{bmatrix}, \quad A = \begin{bmatrix} 1 & 0 \\ 0 & 1 \end{bmatrix}$$

give $A^2 = I$, so that knowing the eigenvalues of A^2 does not allow a precise determination of the eigenvalues of A.

By taking linear combinations of powers of a matrix, we can form a *polynomial* in A defined by

$$p(A) \equiv \alpha_0 I + \alpha_1 A + \alpha_2 A^2 + \cdots + \alpha_m A^m \tag{2.2.7}$$

for given scalars $\alpha_0, \ldots, \alpha_m$. By means of Theorem 2.2.1 it is immediate that if λ and $\mathbf{x}$ are any eigenvalue and corresponding eigenvector of A, then

$$p(A)\mathbf{x} = \alpha_0 \mathbf{x} + \alpha_1 A\mathbf{x} + \cdots + \alpha_m A^m \mathbf{x} = (\alpha_0 + \alpha_1 \lambda + \cdots + \alpha_m \lambda^m)\mathbf{x}.$$

We can summarize this as follows by using the somewhat ambiguous (but standard) notation of p for both the matrix polynomial (2.2.7) as well as the scalar polynomial with the same coefficients.

THEOREM 2.2.3 *If the $n \times n$ matrix A has eigenvalues $\lambda_1, \ldots, \lambda_n$, then for any polynomial p, $p(\lambda_1), \ldots, p(\lambda_n)$ are eigenvalues of the matrix $p(A)$ of (2.2.7). Moreover, any eigenvector of A is an eigenvector of $p(A)$.*

2.2 Eigenvalues and Canonical Forms

A particularly useful special case of this last result is when $p(A) = \alpha I + A$. Thus, the eigenvalues of $\alpha I + A$ are just $\alpha + \lambda_i, i = 1, \ldots, n$, where the λ_i are the eigenvalues of A. An example is

$$A = \begin{bmatrix} 0 & 1 \\ 1 & 0 \end{bmatrix}, \quad B = \begin{bmatrix} \alpha & 1 \\ 1 & \alpha \end{bmatrix}.$$

Here, A has eigenvalues ± 1 so that B has eigenvalues $\alpha + 1$ and $\alpha - 1$.

We can also immediately compute the eigenvalues of A^{-1}, if A is nonsingular, by multiplying the basic relation $A\mathbf{x} = \lambda\mathbf{x}$ by A^{-1}. Thus,

$$\mathbf{x} = \lambda A^{-1}\mathbf{x} \quad \text{or} \quad A^{-1}\mathbf{x} = \lambda^{-1}\mathbf{x},$$

so that the eigenvalues of A^{-1} are the inverses of the eigenvalues of A. (Note that by Theorem 2.2.1 no eigenvalue can be zero if A is nonsingular.) Hence, we can conclude the following theorem, which is a special case of the more general result in Exercise 2.2.18.

THEOREM 2.2.4 *If A is an $n \times n$ nonsingular matrix with eigenvalues $\lambda_1, \ldots, \lambda_n$, then $\lambda_1^{-1}, \ldots, \lambda_n^{-1}$ are eigenvalues of A^{-1}. Moreover, any eigenvector of A is an eigenvector of A^{-1}.*

Eigenvalue calculations also simplify considerably for block triangular matrices. If

$$A = \begin{bmatrix} A_{11} & & \cdots & A_{1p} \\ & A_{22} & & \vdots \\ & & \ddots & \\ & & & A_{pp} \end{bmatrix},$$

where each A_{ii} is a square matrix, then by using the cofactor expansion of Theorem 2.1.1h it can be shown that

$$\det(A - \lambda I) = \det(A_{11} - \lambda I)\det(A_{22} - \lambda I)\cdots\det(A_{pp} - \lambda I). \quad (2.2.8)$$

Thus, the characteristic polynomial of A factors into a product of the characteristic polynomials of the matrices A_{ii}, and to compute the eigenvalues of A we need only compute the eigenvalues of the smaller matrices A_{ii} (see Exercise 2.2.5 for an example).

Symmetric Matrices

The eigenvalues of a matrix A may be either real or complex, even if A itself is real. This is simply a manifestation of the fact that a polynomial with real coefficients may have complex roots. There is an important case, however, in which the eigenvalues must be real. Recall that a real matrix A is symmetric if $A^T = A$. Suppose that λ is an eigenvalue of A and $\mathbf{x}$ is a corresponding eigenvector so that $A\mathbf{x} = \lambda \mathbf{x}$. If λ is complex, then $\mathbf{x}$ will also be complex. Multiplying both sides by the conjugate transpose $\mathbf{x}^*$ gives

$$\mathbf{x}^* A \mathbf{x} = \lambda \mathbf{x}^* \mathbf{x}. \qquad (2.2.9)$$

Now take the complex conjugate of $\mathbf{x}^* A \mathbf{x}$:

$$\overline{\mathbf{x}^* A \mathbf{x}} = (\mathbf{x}^* A \mathbf{x})^* = \mathbf{x}^* A^* \mathbf{x} = \mathbf{x}^* A \mathbf{x}$$

The first step is valid since $\mathbf{x}^* A \mathbf{x}$ is a scalar, the second step is just the product rule for transposes, and the third step uses the fact that A is real and symmetric. Thus, $\mathbf{x}^* A \mathbf{x}$ is real since only a real number can be equal to its complex conjugate. Since $\mathbf{x}^* \mathbf{x} = \sum |x_i|^2 > 0$, it follows from (2.2.9) that λ must be real. Thus, all the eigenvalues of a real symmetric matrix are real, and corresponding eigenvectors may be chosen as real.

A real $n \times n$ matrix is *skew-symmetric* if $A^T = -A$. In this case,

$$\overline{(x^* A x)} = -x^* A x,$$

so that $\mathbf{x}^* A \mathbf{x}$ is imaginary (or zero). Hence, it follows from (2.2.9) that λ is imaginary. Thus, all the eigenvalues of a skew-symmetric matrix are imaginary.

A general real $n \times n$ matrix may be written as

$$A = \frac{1}{2}(A + A^T) + \frac{1}{2}(A - A^T) = A_s + A_{ss}. \qquad (2.2.10)$$

It is easy to verify that A_s is symmetric and A_{ss} is skew-symmetric. Thus, any real $n \times n$ matrix may be written as the sum of a symmetric and a skew-symmetric matrix.

The quantity $\mathbf{x}^T A \mathbf{x}$ is important in many applications, as we shall see later. It is also used to define the concept of definiteness of a matrix. A real matrix A is

(a) *positive definite* if $\mathbf{x}^T A \mathbf{x} > 0$ for all $\mathbf{x} \neq 0$;

(b) *positive semidefinite* if $\mathbf{x}^T A \mathbf{x} \geq 0$ for all $\mathbf{x}$;

(c) *negative definite* if $\mathbf{x}^T A \mathbf{x} < 0$ for all $\mathbf{x} \neq 0$;

(d) *negative semidefinite* if $\mathbf{x}^T A \mathbf{x} \leq 0$ for all $\mathbf{x}$;

(e) *indefinite* if none of the above holds.

An important type of positive definite (or semidefinite) matrix arises as a product $A = B^T B$, where B is an $m \times n$ matrix. For a given $\mathbf{x}$, set $\mathbf{y} = B\mathbf{x}$. Then $\mathbf{x}^T A \mathbf{x} = \mathbf{y}^T \mathbf{y} \geq 0$ and is positive unless $\mathbf{y} = 0$. If B is $n \times n$ and nonsingular, then $\mathbf{y}$ cannot be zero for nonzero $\mathbf{x}$; hence, A is positive definite. If B is singular, then A is positive semidefinite since $\mathbf{y}^T\mathbf{y} \geq 0$.

If A is a symmetric positive definite matrix and λ is an eigenvalue with corresponding eigenvector $\mathbf{x}$, then $\lambda = \mathbf{x}^T A \mathbf{x}/\mathbf{x}^T \mathbf{x} > 0$, which shows that all eigenvalues of a positive definite matrix are positive. In a similar way one may prove (Exercise 2.2.13) the first part of the following statements.

THEOREM 2.2.5 *If A is an $n \times n$ real symmetric matrix, then all of its eigenvalues are:*

a. *positive if and only if A is positive definite;*

b. *nonnegative if and only if A is positive semidefinite;*

c. *negative if and only if A is negative definite;*

d. *nonpositive if and only if A is negative semidefinite.*

We will see shortly how to prove the converse statements in this theorem. Thus, the signs of the eigenvalues of a symmetric matrix characterize its definiteness properties.

Similarity Transformations and Canonical Forms

A *similarity transformation* of an $n \times n$ matrix A is

$$B = PAP^{-1}, \qquad (2.2.11)$$

where P is nonsingular. A and B are then said to be *similar*. Similarity transformations arise from changes of variables. For example, consider the system of equations $A\mathbf{x} = \mathbf{b}$ and make the change of variables $\mathbf{y} = P\mathbf{x}$ and $\mathbf{c} = P\mathbf{b}$, where P is a nonsingular matrix. In the new variables, the system of equations is $AP^{-1}\mathbf{y} = P^{-1}\mathbf{c}$ or, upon multiplying through by P, $PAP^{-1}\mathbf{y} = \mathbf{c}$.

An important property of similarity transformations is that the matrices A and PAP^{-1} have the same eigenvalues. This is easily seen by considering the characteristic polynomial and using the fact (Theorem 2.1.1f) that the determinant of a product of matrices is the product of the determinants. Thus,

$$\det(A - \lambda I) = \det(PP^{-1})\det(A - \lambda I) = \det(P)\det(A - \lambda I)\det(P^{-1})$$
$$= \det(PAP^{-1} - \lambda I),$$

which shows that the characteristic polynomials, and hence the eigenvalues, of A and PAP^{-1} are identical. However, the eigenvectors change under a similarity transformation. Indeed,

$$PAP^{-1}\mathbf{y} = \lambda \mathbf{y} \quad or \quad AP^{-1}\mathbf{y} = \lambda P^{-1}\mathbf{y}$$

shows that the eigenvector $\mathbf{y}$ of PAP^{-1} is related to the eigenvector $\mathbf{x}$ of A by $P^{-1}\mathbf{y} = \mathbf{x}$ or $\mathbf{y} = P\mathbf{x}$.

An important question is how "simple" the matrix A may be made under a similarity transformation. A basic result in this regard is the following:

THEOREM 2.2.6 *A matrix A is similar to a diagonal matrix if and only if A has n linearly independent eigenvectors.*

The proof of this theorem is both simple and illustrative. Let $\mathbf{x}_1, \ldots, \mathbf{x}_n$ be n linearly independent eigenvectors of A with corresponding eigenvalues $\lambda_1, \ldots, \lambda_n$, and let P be the matrix with columns $\mathbf{x}_1, \ldots, \mathbf{x}_n$; then P is nonsingular since its columns are linearly independent. By the basic definition $A\mathbf{x}_i = \lambda_i \mathbf{x}_i$ applied to each column of P, we have

$$AP = A(\mathbf{x}_1, \mathbf{x}_2, \ldots, \mathbf{x}_n) = (\lambda_1 \mathbf{x}_1, \ldots, \lambda_n \mathbf{x}_n) = PD, \qquad (2.2.12)$$

where D is the diagonal matrix $\mathrm{diag}(\lambda_1, \lambda_2, \ldots, \lambda_n)$. Equation (2.2.12) is equivalent to $A = PDP^{-1}$, which shows that A is similar to a diagonal matrix whose diagonal entries are the eigenvalues of A. Conversely, if A is similar to a diagonal matrix, then (2.2.12) shows that the columns of the similarity matrix P must be eigenvectors of A, and they are linearly independent by the nonsingularity of P.

Two important special cases of the preceding result are the following theorems, which we state without proof.

THEOREM 2.2.7 *If A has distinct eigenvalues, then A is similar to a diagonal matrix.*

THEOREM 2.2.8 *If A is a real symmetric matrix, then A is similar to a diagonal matrix, and the similarity matrix may be taken to be orthogonal.*

A useful consequence of Theorem 2.2.8 is the following:

THEOREM 2.2.9 *Let A be an $n \times n$ real symmetric matrix with eigenvalues $\lambda_1 \leq \cdots \leq \lambda_n$. Then for all $\mathbf{x}$*

$$\lambda_1 \mathbf{x}^T \mathbf{x} \leq \mathbf{x}^T A \mathbf{x} \leq \lambda_n \mathbf{x}^T \mathbf{x}. \qquad (2.2.13)$$

This can be proved as follows. By Theorem 2.2.8 there is an orthogonal matrix P such that
$$P^T A P = \mathrm{diag}(\lambda_1, \ldots, \lambda_n).$$
Hence, with $\mathbf{y} = P^T \mathbf{x}$, we have $\mathbf{y}^T \mathbf{y} = \mathbf{x}^T \mathbf{x}$ and

$$\mathbf{x}^T A \mathbf{x} = \mathbf{y}^T P^T A P \mathbf{y} = \sum_{i=1}^{n} \lambda_i y_i^2 \leq \lambda_n \mathbf{y}^T \mathbf{y} = \lambda_n \mathbf{x}^T \mathbf{x}$$

The other inequality is proved analogously.

2.2 Eigenvalues and Canonical Forms

By using Theorem 2.2.9 we can prove the converses of the statements of Theorem 2.2.5. For example, if all eigenvalues of A are positive, then $\lambda_1 > 0$ so that (2.2.13) shows that A is positive definite. Similarly, if $\lambda_1 \geq 0$, then (2.2.13) shows that A is positive semidefinite. On the other hand, if $\lambda_n < 0$, then $\mathbf{x}^T A \mathbf{x} < 0$ for all $\mathbf{x} \neq 0$ so that A is negative definite, and if $\lambda_n \leq 0$, then $\mathbf{x}^T A \mathbf{x} \leq 0$ so that A is negative semidefinite.

The Jordan Canonical Form

Theorems 2.2.6 and 2.2.7 show that if a matrix A does not have n linearly independent eigenvectors, then necessarily it has multiple eigenvalues. (But note that a matrix may have n linearly independent eigenvectors even though it has multiple eigenvalues; the identity matrix is an example.) The matrix

$$A = \begin{bmatrix} 1 & 1 \\ 0 & 1 \end{bmatrix} \tag{2.2.14}$$

is a simple example of a matrix that does not have two linearly independent eigenvectors (see Exercise 2.2.10) and is not similar to a diagonal matrix. More generally, any $n \times n$ matrix may be made similar to a matrix of the form

$$J = \begin{bmatrix} \lambda_1 & \delta_1 & & & \\ & \lambda_2 & \delta_2 & & \\ & & \ddots & \ddots & \\ & & & & \delta_{n-1} \\ & & & & \lambda_n \end{bmatrix},$$

where λ_i are the eigenvalues of A, the δ_i are either 0 or 1, and whenever a δ_i is nonzero, then the eigenvalues λ_i and λ_{i-1} are identical. If q is the number of δ_i that are nonzero, then A has $n - q$ linearly independent eigenvectors. If there are p linearly independent eigenvectors, the matrix J can be partitioned as

$$J = \begin{bmatrix} J_1 & & \\ & \ddots & \\ & & J_p \end{bmatrix}, \tag{2.2.15a}$$

where each J_i is a matrix of the form

$$J_i = \begin{bmatrix} \lambda_i & 1 & & \\ & \ddots & \ddots & \\ & & & 1 \\ & & & \lambda_i \end{bmatrix} \tag{2.2.15b}$$

with identical eigenvalues and all 1's on the first superdiagonal. The matrix J of (2.2.15) is called the *Jordan canonical form* of A. As with (2.2.14), a matrix of the form (2.2.15b) has only one linearly independent eigenvector. If A has n linearly independent eigenvectors, then $p = n$; in this case each J_i reduces to a 1×1 matrix, and J is diagonal.

The Jordan canonical form is useful for theoretical purposes but not very useful in practice. For many computational purposes it is very desirable to work with orthogonal or unitary matrices. We next state without proof two basic results on similarity transformations with unitary or orthogonal matrices.

THEOREM 2.2.10 (Schur's Theorem) *For an arbitrary $n \times n$ matrix A, there is a unitary matrix U such that UAU^* is triangular.*

THEOREM 2.2.11 (Murnaghan-Wintner Theorem) *For a real $n \times n$ matrix A, there is an orthogonal matrix P so that*

$$PAP^T = \begin{bmatrix} T_{11} & & \cdots & T_{1m} \\ & T_{22} & & \\ & & \ddots & \vdots \\ & & & T_{mm} \end{bmatrix}, \qquad (2.2.16)$$

where each T_{ii} is either 2×2 or 1×1.

In the case of Schur's Theorem, the diagonal elements of UAU^* are the eigenvalues of A since UAU^* is a similarity transformation. If A is real but has some complex eigenvalues, then U is necessarily complex. The Murnaghan-Wintner Theorem gives as close to a triangular form as possible with a real orthogonal matrix. In this case, if T_{ii} is 1×1, then it is a real eigenvalue of A, whereas if T_{ii} is 2×2, its two eigenvalues are a complex conjugate pair of eigenvalues of A. The Murnaghan-Wintner form (2.2.16) is also known as the *real Schur form*.

The Singular Value Decomposition (SVD)

If we restrict ourselves to orthogonal or unitary similarity transformations, the previous two theorems given the simplest form, in general, to which A may be reduced. If we do not insist on a similarity transformation we can reduce any $n \times n$ matrix to a diagonal matrix by multiplication on the left and right by orthogonal (or unitary) matrices. We state the following theorem without proof.

THEOREM 2.2.12. (Singular Value Decomposition) *If A is a real $n \times n$ matrix, then there are orthogonal matrices U and V so that*

$$A = UDV, \qquad (2.2.17)$$

2.2 Eigenvalues and Canonical Forms

where
$$D = \text{diag}(\sigma_1, \ldots, \sigma_n). \qquad (2.2.18)$$

The quantities $\sigma_1, \ldots, \sigma_n$ in (2.2.18) are non-negative and are the *singular values* of A. If (2.2.17) holds, then

$$A^T A = V^T D U^T U D V = V^T D^2 V,$$

which is a similarity transformation. Thus, the singular values are the non-negative square roots of the eigenvalues of $A^T A$.

It can be shown that the number of non-zero singular values is equal to the rank of the matrix. For example,

$$A = \begin{bmatrix} 0 & 1 & 0 \\ 0 & 0 & 1 \\ 0 & 0 & 0 \end{bmatrix}, \quad A^T A = \begin{bmatrix} 0 & 0 & 0 \\ 0 & 1 & 0 \\ 0 & 0 & 1 \end{bmatrix},$$

so that the number of non-zero singular values is 2, which is the rank of A. Note that the eigenvalues of A are all zero in this example, and do not give an indication of the rank.

Theorem 2.2.12 extends to complex matrices. In this case U and V are unitary matrices but the singular values are still real. Theorem 2.2.12 also extends to rectangular matrices. If A is $m \times n$, then U is $m \times m$ and V is $n \times n$. D is now $m \times n$ with the singular values again on its main diagonal.

EXERCISES 2.2

2.2.1. Show that the eigenvalues of the matrix

$$\begin{bmatrix} a & b \\ c & d \end{bmatrix}$$

are $\{a + d \pm [(a-d)^2 + 4bc]^{1/2}\}/2$. Then compute the eigenvalues of

(a) $\begin{bmatrix} 1 & 2 \\ 2 & 2 \end{bmatrix}$ (b) $\begin{bmatrix} 1 & 4 \\ 1 & 2 \end{bmatrix}$ (c) $\begin{bmatrix} 0 & 1 \\ -1 & 0 \end{bmatrix}$

2.2.2. Compute the eigenvalues and eigenvectors of A and A^T, where

$$A = \begin{bmatrix} 2 & 2 \\ 1 & 1 \end{bmatrix},$$

and conclude that A and A^T do not have the same eigenvectors.

2.2.3. Compute the eigenvalues and eigenvectors of

$$A = \begin{bmatrix} 2 & 1 \\ 1 & 2 \end{bmatrix}.$$

Then use Theorem 2.2.3 to compute the eigenvalues and eigenvectors of

(a) $\begin{bmatrix} 1 & 1 \\ 1 & 1 \end{bmatrix}$ (b) $\begin{bmatrix} 0 & 1 \\ 1 & 0 \end{bmatrix}$ (c) $\begin{bmatrix} -1 & 1 \\ 1 & -1 \end{bmatrix}$

2.2.4. For the matrix A of Exercise 3, find $\det A^4$ without computing A^4. In addition, find the eigenvalues and eigenvectors of $A^{-1}, A^m, m = 2, 3, 4$ and $I + 2A + 4A^2$, without computing these matrices.

2.2.5. Use (2.2.8) to compute the eigenvalues of the matrix

$$A = \begin{bmatrix} 1 & 2 & 2 & 4 \\ 2 & 2 & 1 & 5 \\ 0 & 0 & 1 & 4 \\ 0 & 0 & 1 & 2 \end{bmatrix}$$

2.2.6. Given the polynomial $p(\lambda) = a_0 + a_1\lambda + \cdots + a_{n-1}\lambda^{n-1} + \lambda^n$, the matrix

$$A = \begin{bmatrix} 0 & 1 & & \\ & & 1 & \\ & & & \ddots & \\ & & & & 1 \\ -a_0 & -a_1 & \cdots & -a_{n-1} \end{bmatrix}$$

is called the *companion matrix* (or *Frobenius matrix*) of p. For $n = 3$, show that $\det(\lambda I - A) = p(\lambda)$.

2.2.7. Ascertain if the matrices (a) and (b) of Exercise 3 are positive definite. If not, what are they?

2.2.8. Show that the determinant of a negative definite $n \times n$ symmetric matrix is positive if n is even and negative if n is odd.

2.2.9. List all leading principle submatrices and all principle submatrices of the matrix of Exercise 5.

2.2.10. Compute an eigenvector of the matrix (2.2.14) and show that there are no other linearly independent eigenvectors.

2.2.11. Assume that a matrix A has two eigenvalues $\lambda_1 = \lambda_2$ and corresponding linearly independent eigenvectors $\mathbf{x}_1, \mathbf{x}_2$. Show that any linear combination $c_1\mathbf{x}_1 + c_2\mathbf{x}_2$ is also an eigenvector.

2.2.12. If A and B are $n \times n$ matrices at least one of which is nonsingular, show that AB and BA have the same eigenvalues.

2.3 Norms

2.2.13. Prove the *if* part of the statements of Theorem 2.2.5.

2.2.14. Let A be a real, symmetric matrix. Show that Schur's Theorem implies that there exists an orthogonal matrix Q such that $Q^T A Q = D$, where D is a diagonal matrix.

2.2.15. Let A be a real, skew-symmetric matrix and $PAP^T = T$, where T is given by the Murnaghan-Wintner Theorem. Describe the structure of T in this case.

2.2.16. Consider a matrix J_i of the form (2.2.15b). Show that there exists a diagonal matrix D so that DJ_iD^{-1} is the same as J_i except that the off-diagonal 1's are replaced by ϵ.

2.2.17. Use just the definition of a positive (semi)definite matrix to show that if A is positive definite and B is positive semidefinite, then $A + B$ is positive definite.

2.2.18. Let $p(A)$ and $q(A)$ be two polynomials in A such that $q(A)$ is non-singular. If λ is an eigenvalue of A show that $p(\lambda)/q(\lambda)$ is an eigenvalue of $q(A)^{-1}p(A)$.

2.2.19. If A is a real skew-symmetric matrix, show that $\mathbf{x}^T A \mathbf{x} = 0$ for any real vector $\mathbf{x}$. Use this to conclude that a general real $n \times n$ matrix is positive definite if and only if its symmetric part in (2.2.10) is positive definite.

2.2.20. Let A be an $n \times n$ matrix and let

$$\tilde{A} = \begin{bmatrix} 0 & A \\ A^T & 0 \end{bmatrix}.$$

Show that the magnitudes of the eigenvalues of $\tilde{A}$ are equal to the singular values of A.

2.2.21. The *polar decomposition* of a real $n \times n$ matrix A is $A = QH$, where Q is orthogonal and H is symmetric. Show how to construct the polar decomposition from the singular value decomposition of A.

2.2.22. Give an example of a real 4×4 orthogonal matrix all of whose eigenvalues are purely imaginary.

2.2.23. Let $\mathbf{u}$ and $\mathbf{v}$ be column vectors and $A = I + \mathbf{u}\mathbf{v}^T$. Find the eigenvalues, eigenvectors and Jordan canonical form of A.

2.3 Norms

The Euclidean length of a vector $\mathbf{x}$ is defined by

$$||\mathbf{x}||_2 = \left(\sum_{i=1}^n x_i^2 \right)^{1/2}. \qquad (2.3.1)$$

This is a special case of a *vector norm*, which is a real-valued function that satisfies the following distance-like properties:

$||\mathbf{x}|| \geq 0$ for any vector $\mathbf{x}$ and $||\mathbf{x}|| = 0$ only if $\mathbf{x} = 0$. (2.3.2a)

$||\alpha \mathbf{x}|| = |\alpha| ||\mathbf{x}||$ for any scalar α. (2.3.2b)

$||\mathbf{x} + \mathbf{y}|| \leq ||\mathbf{x}|| + ||\mathbf{y}||$ for all vectors $\mathbf{x}$ and $\mathbf{y}$. (2.3.2c)

Property (2.3.2c) is known as the *triangle inequality*.

The Euclidean length (2.3.1) satisfies these properties and is usually called the *Euclidean norm*, or l_2 norm. Other commonly used norms are defined by

$$||\mathbf{x}||_1 = \sum_{i=1}^{n} |x_i|, \quad ||\mathbf{x}||_\infty = \max_{1 \leq i \leq n} |x_i|, \quad (2.3.3)$$

which are known as the l_1 norm, and the l_∞ or max norm, respectively. The three norms (2.3.1) and (2.3.3) are special cases of the general class of l_p norms

$$||\mathbf{x}||_p = \left(\sum_{i=1}^{n} |x_i^p| \right)^{1/p}, \quad (2.3.4)$$

defined for any real number $p \geq 1$. The l_∞ norm is the limiting case of (2.3.4) as $p \to \infty$. Another important class of norms consists of the *elliptic norms* defined by

$$||\mathbf{x}||_B = (\mathbf{x}^T B \mathbf{x})^{1/2}$$

for a given symmetric positive-definite matrix B; the Euclidean norm is the special case $B = I$. These various norms can be visualized geometrically in terms of the set of vectors $\{\mathbf{x} : ||\mathbf{x}|| = 1\}$, which is known as the *unit sphere*. (The set $\{\mathbf{x} : ||\mathbf{x}|| \leq 1\}$ is the *unit ball*.) These are shown in Figure 2.3.1 for vectors in the plane. Note that only for the Euclidean norm are the unit vectors on the circle of radius 1.

The elliptic norms play a particularly central role in matrix theory because they arise in terms of an inner product, which in turn defines orthogonality of

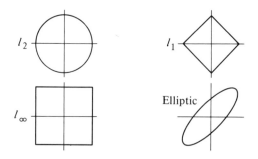

Figure 2.3.1: *Unit Spheres of Several Norms*

2.3 Norms

vectors. An *inner product* is a real-valued function of two vector variables that satisfies the following conditions (stated only for real vectors):

$(\mathbf{x}, \mathbf{x}) \geq 0$ for all vectors $\mathbf{x}$; $(\mathbf{x}, \mathbf{x}) = 0$ only if $\mathbf{x} = 0$. (2.3.5a)

$(\alpha\mathbf{x}, \mathbf{y}) = \alpha(\mathbf{x}, \mathbf{y})$ for all vectors $\mathbf{x}$ and $\mathbf{y}$ and scalars α. (2.3.5b)

$(\mathbf{x}, \mathbf{y}) = (\mathbf{y}, \mathbf{x})$ for all vectors $\mathbf{x}$ and $\mathbf{y}$. (2.3.5c)

$(\mathbf{x} + \mathbf{z}, \mathbf{y}) = (\mathbf{x}, \mathbf{y}) + (\mathbf{z}, \mathbf{y})$ for all vectors $\mathbf{x}, \mathbf{y}$, and $\mathbf{z}$. (2.3.5d)

For any inner product a norm may be defined by

$$||\mathbf{x}|| = (\mathbf{x}, \mathbf{x})^{1/2},$$

and the elliptic norms then derive from the inner product

$$(\mathbf{x}, \mathbf{y}) \equiv \mathbf{x}^T B \mathbf{y}. \qquad (2.3.6)$$

Two nonzero vectors $\mathbf{x}$ and $\mathbf{y}$ are *orthogonal* with respect to some inner product if

$$(\mathbf{x}, \mathbf{y}) = 0.$$

If the inner product is the Euclidean one defined by (2.3.6) with $B = I$, then this gives the usual and intuitive concept of orthogonality used in Section 2.1. A set of nonzero vectors $\mathbf{x}_1, \ldots, \mathbf{x}_m$ is *orthogonal* if

$$(\mathbf{x}_i, \mathbf{x}_j) = 0, \qquad i \neq j.$$

A set of orthogonal vectors is necessarily linearly independent, and a set of n such vectors is an *orthogonal basis*. If, in addition, $||\mathbf{x}_i|| = 1$, $i = 1, \cdots, n$, the vectors are *orthonormal*. As discussed in Section 2.1, if the columns of a matrix A are orthonormal in the inner product $\mathbf{x}^T \mathbf{y}$, then $A^T A = I$ and the matrix is orthogonal. Orthogonal matrices have the important property that they preserve the length of a vector; that is, $||A\mathbf{x}||_2 = ||\mathbf{x}||_2$.

Convergence of a sequence of vectors $\{\mathbf{x}^k\}$ to a limit vector $\mathbf{x}$ is defined in terms of a norm by

$$||\mathbf{x}^k - \mathbf{x}|| \to 0 \quad \text{as} \quad k \to \infty.$$

It is natural to suppose that a sequence might converge in one norm but not in another. Surprisingly, this cannot happen.

THEOREM 2.3.1: *The following are equivalent:*

a. *The sequence* $\{\mathbf{x}^k\}$ *converges to* $\mathbf{x}$ *in some norm.*

b. *The sequence* $\{\mathbf{x}^k\}$ *converges to* $\mathbf{x}$ *in every norm.*

c. The components of the sequence $\{\mathbf{x}^k\}$ all converge to the corresponding components of $\mathbf{x}$; that is, $x_i^k \to x_i$ as $k \to \infty$ for $i = 1, \ldots, n$.

As a consequence of this result – sometimes known as the *norm equivalence theorem* – when we speak of the convergence of a sequence of vectors, it is not necessary that we specify the norm.

Matrix Norms

Any vector norm gives rise to a corresponding matrix norm by means of the definition

$$||A|| = \max_{\mathbf{x} \neq 0} \frac{||A\mathbf{x}||}{||\mathbf{x}||} = \max_{||\mathbf{x}||=1} ||A\mathbf{x}||. \qquad (2.3.7)$$

The properties (2.3.2) also hold for a matrix norm; in addition, there is the multiplicative property $||AB|| \leq ||A||\,||B||$. The geometric interpretation of a matrix norm is that $||A||$ is the maximum length of a unit vector after multiplication by A; this is depicted in Figure 2.3.2 for the l_2 norm.

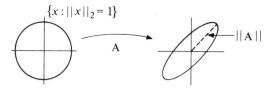

Figure 2.3.2: *The l_2 Matrix Norm*

As with vectors, the convergence of a sequence of matrices may be defined component-wise or, equivalently, in terms of any matrix norm. Thus, we write $A_k \to A$ as $k \to \infty$ if in some norm $||A_k - A|| \to 0$ as $k \to \infty$. Again, convergence in some norm implies convergence in any norm.

The matrix norms corresponding to the l_1 and l_∞ vector norms are easily computed by

$$||A||_1 = \max_{1 \leq j \leq n} \sum_{i=1}^{n} |a_{ij}|, \quad ||A||_\infty = \max_{1 \leq i \leq n} \sum_{j=1}^{n} |a_{ij}|. \qquad (2.3.8)$$

That is, $||A||_1$ is the maximum absolute value column sum of the elements of A, and $||A||_\infty$ is the maximum row sum. The Euclidean matrix norm is much more difficult to compute. For any $n \times n$ matrix B with eigenvalues $\mu_1, \ldots, \mu_n$, the *spectral radius* of B is $\rho(B) = \max_{1 \leq i \leq n} |\mu_i|$. Then

$$||A||_2 = [\rho(A^T A)]^{1/2}. \qquad (2.3.9)$$

2.3 Norms

If A is symmetric, (2.3.9) reduces to

$$||A||_2 = \rho(A), \qquad (2.3.10)$$

which is still difficult to compute but is more directly related to the matrix A.

It follows immediately from (2.3.7) that if λ is any eigenvalue of A and $\mathbf{x}$ a corresponding eigenvector, then

$$|\lambda|\,||\mathbf{x}|| = ||\lambda\mathbf{x}|| = ||A\mathbf{x}|| \le ||A||\,||\mathbf{x}||,$$

so that

$$|\lambda| \le ||A||. \qquad (2.3.11)$$

Thus, any norm of the matrix A gives a bound on all eigenvalues of A. In general, however, $\rho(A)$ will not equal $||A||$.

Gerschgorin's Theorem

By (2.3.11) and (2.3.8), any eigenvalue λ of A satisfies

$$|\lambda| \le \max_{1 \le i \le n} \sum_{j=1}^{n} |a_{ij}|. \qquad (2.3.12)$$

Much better estimates of the eigenvalues may usually be found by the following technique. Let

$$r_i = \sum_{\substack{j=1 \\ j \ne i}}^{n} |a_{ij}|, \qquad i = 1, \ldots, n,$$

be the sum of the absolute values of the off-diagonal elements in the ith row of A, and define disks in the complex plane centered at a_{ii} and with radius r_i:

$$\Lambda_i = \{z : |z - a_{ii}| \le r_i\}, \qquad i = 1, \ldots, n.$$

We then have the following:

> GERSCHGORIN'S THEOREM *All the eigenvalues of A lie in the union of the disks $\Lambda_1, \ldots, \Lambda_n$. Moreover, if S is a union of m disks such that S is disjoint from all the other disks, then S contains exactly m eigenvalues of A (counting multiplicities).*

As a simple example of the use of Gerschgorin's theorem, consider the matrix

$$A = \frac{1}{16} \begin{bmatrix} -8 & -2 & 4 \\ -1 & -4 & 2 \\ 2 & 2 & -10 \end{bmatrix}. \qquad (2.3.13)$$

By (2.3.11), all eigenvalues of A are less than $\frac{7}{8}$ in absolute value; thus, they lie in a disk of radius $\frac{7}{8}$ centered at the origin of the complex plane. By use of Gerschgorin's Theorem, however, we can conclude that the eigenvalues lie in the union of the disks shown in Figure 2.3.3, which gives much more information than (2.3.11). In particular, no eigenvalue can be zero and therefore by Theorem 2.2.1 the matrix is nonsingular.

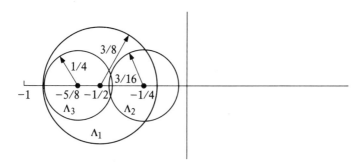

Figure 2.3.3: *Gerschgorin's Disks in the Complex Plane*

To illustrate the second part of Gerschgorin's Theorem, suppose that the second row of the matrix of (2.3.13) is changed to $\frac{1}{16}(-1,6,2)$. Then the disk Λ_2 is centered at $+\frac{3}{8}$, again with radius $\frac{3}{16}$. Since Λ_2 is now disjoint from the other two disks, it contains exactly one eigenvalue of A. Moreover, since any complex eigenvalues of A must occur in conjugate pairs, this eigenvalue must be real and therefore lies in the interval $[\frac{3}{16}, \frac{9}{16}]$.

The proof of the first part of Gerschgorin's Theorem is very easy. Let λ be any eigenvalue of A, and $\mathbf{x}$ a corresponding eigenvector. Then, by definition,

$$(\lambda - a_{ii})x_i = \sum_{\substack{j=1 \\ j \neq i}}^{n} a_{ij}x_j, \qquad i = 1, \ldots, n.$$

If we let x_k be the component of $\mathbf{x}$ of largest absolute value, then

$$|\lambda - a_{kk}| \leq \sum_{\substack{j=1 \\ j \neq k}}^{n} |a_{kj}| \frac{|x_j|}{|x_k|} \leq \sum_{\substack{j=1 \\ j \neq k}}^{n} |a_{kj}|.$$

Thus, λ is in the disk centered at a_{kk} and therefore in the union of all the disks. The proof of the second part of the theorem is more complicated and relies on the fact that the eigenvalues of a matrix are continuous functions of the elements of the matrix.

By a simple similarity transformation, it is sometimes possible to use Gerschgorin's Theorem to extract additional information about the eigenvalues.

2.3 Norms

For example, consider the matrix

$$A = \begin{bmatrix} 8 & 1 & 0 \\ 1 & 12 & 1 \\ 0 & 1 & 10 \end{bmatrix}.$$

Since A is symmetric its eigenvalues are real, and by Gerschgorin's Theorem we conclude that they lie in the union of the intervals $[7, 9]$, $[10, 14]$, $[9, 11]$. Since these intervals are not disjoint we cannot yet conclude that any of them contains an eigenvalue. However, if we do a similarity transformation with the matrix $D = \text{diag}(d, 1, 1)$, we obtain

$$DAD^{-1} = \begin{bmatrix} 8 & d & \\ d^{-1} & 12 & 1 \\ & 1 & 10 \end{bmatrix}.$$

By Gerschgorin's Theorem, the eigenvalues of this matrix (which are the same as those of A) lie in the union of the intervals $[8-d, 8+d]$, $[11-d^{-1}, 13+d^{-1}]$, $[9, 11]$. As long as $1 > d > \frac{1}{2}[3 - \sqrt{5}]$, the first interval is disjoint from the others and thus contains exactly one eigenvalue. In particular, the interval $[7.6, 8.4]$ contains one eigenvalue.

Another important use of Gerschgorin's Theorem is in ascertaining the change in the eigenvalues of a matrix due to changes in the coefficients. Let A be a given $n \times n$ matrix with eigenvalues $\lambda_1, \ldots, \lambda_n$ and suppose that E is a matrix whose elements are small compared to those of A; for example, E may be the rounding errors committed in entering the matrix A into a computer. Suppose that $\mu_1, \ldots, \mu_n$ are the eigenvalues of $A + E$. Then, what can one say about the changes $|\lambda_i - \mu_i|$? We next give a relatively simple result in the case that A has n linearly independent eigenvectors.

THEOREM 2.3.2 *Assume that $A = PDP^{-1}$, where D is the diagonal matrix of eigenvalues of A, and let $d = \|P^{-1}EP\|_\infty$. Then every eigenvalue of $A + E$ is within d of an eigenvalue of A.*

The proof of this theorem is a simple consequence of Gerschgorin's Theorem. Set $C = P^{-1}(A + E)P$. Then C has the same eigenvalues $\mu_1, \ldots, \mu_n$ as $A + E$. Let $B = P^{-1}EP$. Then $C = D + B$, and the diagonal elements of C are $\lambda_i + b_{ii}$, $i = 1, \ldots, n$. Hence, by Gerschgorin's Theorem, the eigenvalues $\mu_1, \ldots, \mu_n$ are in the union of the disks

$$\{z : |z - \lambda_i - b_{ii}| \le \sum_{\substack{j=1 \\ j \ne i}}^{n} |b_{ij}|\}$$

Therefore, given any μ_k, there is an i such that

$$|\mu_k - \lambda_i - b_{ii}| \leq \sum_{\substack{j=1 \\ j \neq i}}^{n} |b_{ij}|,$$

or

$$|\mu_k - \lambda_i| \leq \sum_{j=1}^{n} |b_{ij}| \leq d,$$

which was to be shown.

We end this section by applying Gerschgorin's Theorem to show that an important type of matrix is non-singular. An $n \times n$ matrix A is *row diagonally dominant* if

$$|a_{ii}| \geq \sum_{\substack{j=1 \\ j \neq i}}^{n} |a_{ij}|, \qquad i = 1, \ldots, n, \tag{2.3.14}$$

and *strictly row diagonally dominant* if strict inequality holds in (2.3.14) for all i. A is *(strictly) column diagonally dominant* if A^T is (strictly) row diagonally dominant. A is (row or column) *irreducibly diagonally dominant* if it is irreducible (see Section 2.1) and strict inequality holds in (2.3.14) for at least one i.

Diagonal dominance is not enough to preclude singularity of the matrix; for example, a 2×2 matrix with all elements equal to 1 is both row and column diagonally dominant, but singular. Strict or irreducible diagonal dominance is sufficient, however.

> THEOREM 2.3.3 *If the $n \times n$ matrix A is strictly row or column diagonally dominant, or irreducibly diagonally dominant, then it is nonsingular.*

We will prove this theorem for the case that A is strictly row diagonally dominant. By the strict diagonal dominance, no Gerschgorin disk can contain the origin. Hence, no eigenvalue can be zero and A is nonsingular. If A is strictly column diagonally dominant, A^T is strictly row diagonally dominant and thus nonsingular. But since $\det A = \det A^T$, A is then nonsingular also.

Supplementary Discussion and References: Chapter 2

Much of the material of this chapter can be found in most introductory books on linear algebra. For more advanced treatments, see Horn and Johnson [1985], Lancaster and Tismenetsky [1985] and Ortega [1987]. Many books on numerical analysis also have good background material on linear algebra in a form most suitable for scientific computing. See, for example, Golub and Van

2.3 Norms

Loan [1989], Ortega [1990], Parlett [1980], Stewart [1973], Varga [1962] and Wilkinson [1965].

EXERCISES 2.3

2.3.1. Compute the ℓ_1, ℓ_2 and ℓ_∞ norms of the vectors

(a) $\begin{bmatrix} 2 \\ 1 \end{bmatrix}$ (b) $\begin{bmatrix} 1 \\ 0 \end{bmatrix}$ (c) $\begin{bmatrix} 1 \\ -1 \end{bmatrix}$

2.3.2. Compute the ℓ_1, ℓ_2 and ℓ_∞ norms of the matrices

(a) $\begin{bmatrix} 0 & 1 \\ -1 & 0 \end{bmatrix}$ (b) $\begin{bmatrix} 1 & 2 \\ 2 & 2 \end{bmatrix}$ (c) $\begin{bmatrix} 1 & 4 \\ 1 & 2 \end{bmatrix}$

2.3.3. Let B be an $n \times n$ real symmetric and indefinite matrix. Show that $(\mathbf{x}^T B \mathbf{x})^{1/2}$ is not a norm.

2.3.4. If $\alpha_1, \ldots, \alpha_n$ are positive numbers, show that norms can be defined by

$$\sum_{i=1}^{n} \alpha_i |x_i| \quad \text{and} \quad \max_{1 \le i \le n} \alpha_i |x_i|$$

2.3.5. Verify that the unit spheres of Figure 2.3.1 are correct.

2.3.6. Show that $|\|\mathbf{x}\| - \|\mathbf{y}\|| \le \|\mathbf{x} - \mathbf{y}\|$ for all $\mathbf{x}, \mathbf{y}$.

2.3.7. Let A be an $n \times n$ matrix and define the *Frobenius norm*

$$\|A\| = \left(\sum_{i,j=1}^{n} |\mathbf{a}_{ij}|^2 \right)^{1/2}$$

Show that the properties (2.3.2) hold but that this is not a norm in the sense of (2.3.7) since (2.3.7) implies that $\|I\| = 1$ in any norm.

2.3.8. Ascertain if $f(\mathbf{x}) \equiv (\mathbf{x}_1^2 + 2x_1 x_2 + 4x_2^2)^{1/2}$ defines a norm.

2.3.9. If $\mathbf{u}$ and $\mathbf{v}$ are column vectors, show that $\|\mathbf{u}\mathbf{v}^T\|_2 = \|\mathbf{u}\|_2 \|\mathbf{v}\|_2$.

2.3.10. Verify the *polarization formula*

$$(\mathbf{x}, \mathbf{y}) = \frac{1}{4}(\|\mathbf{x} + \mathbf{y}\|^2 - \|\mathbf{x} - \mathbf{y}\|^2)$$

if the norm is defined in terms of the inner product.

2.3.11. For the l_1 and l_∞ norms, show that $\|\mathbf{x}\|_\infty \le \|\mathbf{x}\|_1 \le n\|\mathbf{x}\|_\infty$.

2.3.12. Find the Gerschgorin disks for the matrix

$$A = \begin{bmatrix} 4 & 2 & 2 \\ 1 & 8 & 1 \\ 1 & 1 & 12 \end{bmatrix}.$$

Use the fact that A and A^T have the same eigenvalues to show that A has an eigenvalue that satisfies $|\lambda - 4| \leq 2$ by applying Gerschgorin's Theorem to A^T.

2.3.13. Assume that $A = A^T$. In Theorem 2.3.2, give an upper bound for $d = \|P^{-1}EP\|_\infty$.

2.3.14. Prove Theorem 2.3.3 by assuming that A is singular and thus $A\mathbf{x} = 0$ for some $\mathbf{x} \neq 0$. Let $|x_k| = \max\{|x_i|\}$ and show that you can obtain a contradiction.

2.3.15. Show that $\|A^k\| \leq \|A\|^k$ for any norm and any positive integer k. Hence, if $\|A\| < 1$, conclude that $A^k \to 0$ as $k \to \infty$.

2.3.16. Show that $\|A\|_2 \leq \|A\|_1^{1/2} \|A\|_\infty^{1/2}$.

2.3.17. Show that any vector norm is a continuous function of its elements.

2.3.18. If B is an $n \times n$ matrix such that $\rho(B) < 1$, then the geometric series (also called the *Neumann expansion*)

$$(I - B)^{-1} = \sum_{i=0}^{\infty} B^i$$

is valid. Use this to show that if $\|B\| < 1$, then

$$\|(I - B)^{-1}\| \leq \frac{1}{1 - \|B\|}.$$

Chapter 3

Parallel and Vector Computing

3.1 Parallel and Vector Computers

In the early 1970s, computers began to appear that consisted of a number of separate processors operating in parallel or that had hardware instructions for operating on vectors. The latter type of computer we will call a *vector computer* while the former we will call a *parallel computer*.

Vector Computers

Vector computers utilize *pipelining*, which is the explicit segmentation of an arithmetic unit into different parts, each of which performs a subfunction on a pair of operands. This is illustrated in Figure 3.1.1 for floating point addition.

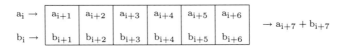

Figure 3.1.1: *A Floating Point Pipeline*

In the example of Figure 3.1.1, a floating point adder is segmented into six sections, each of which does one part of the overall floating point addition. Each segment works on one pair of operands, so that six pairs of operands can be in the pipeline at a given time. The advantage of this segmentation is that results are being computed at a rate that is 6 times faster (or, in general,

K times, where K is the number of segments) than an arithmetic unit that accepts a pair of operands and computes the result before accepting the next pair of operands. However, in order to utilize this capability, the data must reach the arithmetic units rapidly enough to keep the pipeline full. Generally, each suboperation in Figure 3.1.1 will be done in one *machine cycle*, typically a few nanoseconds ($1ns = 10^{-9}$ seconds). Thus, new data must be ready to enter the pipeline every cycle. To help achieve this, some machines have hardware instructions for, say, vector addition that eliminate the need for separate load and store instructions for the data; a single hardware instruction will control the loading of the operands and storing of the results.

Data Access

One of the earliest vector computers was the CDC STAR-100 built by the Control Data Corporation in 1973. This machine evolved into the CYBER 203 in the late 1970s, the CYBER 205 in the early 1980s and the ETA-10 in the late 1980s. All of these machines, which are no longer being built, took their operands directly from main memory and stored the result back into main memory. Thus, they were known as *memory-to-memory* machines.

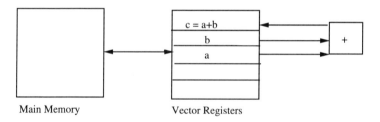

Figure 3.1.2: *Register-to-Register Addition*

In the mid 1970s, Cray Research, Inc. began producing what was to become by far the most commercially successful line of vector supercomputers. All machines built by Cray have used *vector registers*, which have now become standard for many other manufacturers. These registers are very fast memory that hold the operands and results for vector operations. Hardware instructions for vector operations operate only on operands from the vector registers, and these machines are sometimes called *register-to-register* machines. This is illustrated in Figure 3.1.2 for vector addition. In Figure 3.1.2 each vector register is assumed to hold a certain number of words. For example, on the Cray machines, there are eight vector registers, each of which holds 64 floating point numbers. Operands for a vector addition are obtained from two vector registers and the result is put into another vector register. Prior to the vector addition, the vector registers must be loaded from main memory and, at some point, results stored into main memory from vector registers. It is usually

3.1 Parallel and Vector Computers

desirable to use data in the vector registers as much as possible while it is available; several examples of this will be given in later sections.

Vector registers play somewhat the same role as cache memory on conventional computers. More recent vector computers may have a more complex memory hierarchy. Some machines have a cache memory as well as vector registers, and the Cray-2, in addition to vector registers, has a 16,000-word fast local memory in each processor. Other machines such as the Cray Y-MP series have back-up storage (the Solid-State Storage Device) that is slower than main memory but faster than disk. And, of course, all the machines will have disk units. The challenge is to use these various types of storage in such a way as to have the data ready for the arithmetic units when it is needed.

Vectors

On register-to-register machines, a vector for arithmetic instructions will be a sequence of contiguous elements in a vector register, usually starting in the first element of a register. An important consideration for these machines is what constitutes a vector in main memory for the purpose of loading a vector register.

Elements that are sequentially addressable always constitute a suitable vector. Henceforth, we will use *contiguous* as a synonym for sequentially addressable, although sequentially addressable elements are usually not physically contiguous in memory; rather, they are stored in different memory banks. Elements with a constant stride also form a vector. By *stride* we mean the address separation between elements. Thus, elements with addresses $a, a+s, a+2s, \ldots$ have a constant stride equal to s. In the special case that $s = 1$, the elements are sequentially addressable.

For elements that are not stored with a constant stride it is necessary to use auxiliary hardware instructions or software to load the vector registers. A *gather* operation will load into a register elements specified by a list of their addresses. A *scatter* operation, the inverse of gather, stores elements of a register into positions prescribed by an associated address list. These operations add to the overhead of the vector arithmetic operations. A primary consideration in developing algorithms for vector computers is to have the data arranged so as to minimize delays in accessing it from memory.

A major cause of delay is a *bank conflict*. The elements of a stride one vector are stored in different banks because the memory cannot be read every cycle. Typically, a memory bank can be read only every few cycles, the delay being called the *restoration time*. Suppose, for example, that there are four memory banks and we have a 4×4 matrix stored as shown in Figure 3.1.3. If we access columns of this matrix, the elements are stored in different banks and may be accessed without delay. If, however, we access elements by rows of the matrix, and each element can be read only every four cycles, then the access rate is four times slower than for columns. On current machines, there

Bank 1 a_{11} a_{12} a_{13} a_{14}
      2   a_{21} a_{22} a_{23} a_{24}
      3   a_{31} a_{32} a_{33} a_{34}
      4   a_{41} a_{42} a_{43} a_{44}

Figure 3.1.3: *Bank Storage for* 4×4 *Matrix*

are many more than four memory banks; this alleviates but does not eliminate the problem. For example, suppose there are 64 banks and we have a 64×64 matrix. Then again all the elements of a row are in the same bank and the same delay occurs in reading the matrix by rows.

Arithmetic Units

Cray machines, as well as most others, have separate pipeline units for addition and multiplication, as well as other functions. (Some other machines have multiple units, for example, four for addition, and four for multiplication.) Vector hardware operations are always provided for the addition of two vectors, the element-wise product of two vectors, and either the element-wise quotient of two vectors or the reciprocals of the elements of a vector. There may also be vector instructions for other operations.

Most (but not all) machines allow the *chaining* of arithmetic units together so that results from one unit are routed directly to another without first returning to a register. This is illustrated in Figure 3.1.4 for an *axpy* operation of the form vector plus scalar times vector. Loading and storing of vector registers can also be chained with the arithmetic units.

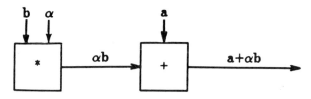

Figure 3.1.4: *Chaining*

Most vector computers provide separate units for scalar arithmetic. These units may also be pipelined but do not accept vector operands as the vector pipelines do. They can run concurrently with the vector pipelines, and usually produce scalar results 5-10 times slower than the maximum rates of the vector pipelines.

3.1 Parallel and Vector Computers

Vector Start-up Times

The use of vector operations incurs an overhead penalty as shown in the following approximate formula for the time T for a vector operation:

$$T = S + KN. \tag{3.1.1}$$

In (3.1.1), N is the length of the vectors involved, K is the time interval at which results are leaving the pipeline, and S is the *start-up time*. S is the time for the pipeline to become full before the first result is produced. After that, results leave the pipeline every K units where K is usually one cycle time. If the vector operands are already in registers the start-up time is relatively small, usually a half-dozen or so cycles. If the operands are in main memory, the start-up time will include the time to begin loading the registers and will increase to 50-100 cycles, depending on the machine.

The formula (3.1.1) is only approximate for vectors whose length exceeds the register length. However, we will assume that (3.1.1) is correct for the following discussion. The *result rate* is the number of results per unit time and is given by

$$R = \frac{N}{S + KN}. \tag{3.1.2}$$

If $S = 0$, or as $N \to \infty$, (3.1.2) yields

$$R_\infty = \frac{1}{K}, \tag{3.1.3}$$

which is called the *asymptotic result rate*. This is the (not achievable) maximum result rate when the start-up overhead is ignored. For example, if K is 10 ns (nanoseconds), then the asymptotic rate is $R_\infty = 10^8$ results per second, or 100 mflops, where "mflops" denotes *megaflops* or one million floating point operations per second.

A number of some interest is $N_{1/2}$, which is defined to be the vector length for which half the asymptotic rate is achieved. For example, if $K = 10$ ns, it follows from (3.1.2) that $N_{1/2} = 100$ for $S = 1000$, while if $S = 100$, then $N_{1/2} = 10$. Another important number is the *cross-over point*, N_c, at which vector arithmetic becomes faster than scalar arithmetic.

Parallel Computers

The basic idea of a parallel computer is that a number of processors work in cooperation on a single task. The motivation is that if it takes one processor an amount of time t to do a task, then p processors can do the task in time t/p. Only for very special situations can this perfect "speedup" be achieved, however, and it is our goal to devise algorithms that can take as much advantage as possible, for a given problem, of multiple processors.

The processors of a parallel computer can range from very simple ones to very powerful vector processors. Most of our discussions will be directed towards the case in which the processors are powerful sequential processors, or vector processors.

MIMD and SIMD Machines

A first important dichotomy in parallel systems is how the processors are controlled. In a Single-Instruction-Multiple-Data (SIMD) system, all processors are under the control of a master processor, called the *controller*, and the individual processors all do the same instruction (or nothing) at a given time. Thus, there is a single instruction stream operating on multiple data streams, one for each processor. The Illiac IV, the first large parallel system (which was completed in the early 1970s), was a SIMD machine, as were the first Connection Machines built by Thinking Machines, Inc. For example, the CM-2 is a SIMD machine with 64,936 simple 1-bit processors. Vector computers may also be conceptually included in the class of SIMD machines by considering the elements of a vector as being processed individually under the control of a vector hardware instruction.

Most parallel computers built since the Illiac IV have been Multiple-Instruction-Multiple-Data (MIMD) systems. Here, the individual processors run under the control of their own program, which allows great flexibility in the tasks the processors can do at any given time. It also introduces the problem of synchronization. In an SIMD system, synchronization of the individual processors is carrried out by the controller, but in an MIMD system other mechanisms must be used to ensure that the processors are doing their tasks in the correct order with the correct data. Synchronization will be discussed more later.

For many problems, the programs in the individual processors of an MIMD system may be identical (or nearly so). Thus, all programs are carrying out the same operations on different sets of data, just as SIMD machines would do. This gives rise to the Single-Program Multiple-Data (SPMD) model of computation, also known as the *data parallel* model.

Shared Versus Distributed Memory

Another important dichotomy in parallel computers is *shared* versus *distributed* memory. A shared memory system is illustrated in Figure 3.1.5. Here, all the processors have access to a common memory. (Hereafter, we will use terms "shared memory" and "common memory" interchangeably.) Each processor can also have its own local memory for program code and intermediate results. The common memory would then be used for data and results that are needed by more than one processor. All communication between individual processors is through the common memory. A major advantage of a shared

3.1 Parallel and Vector Computers

memory system is potentially very rapid communication of data between processors. A serious disadvantage is that different processors may wish to use the common memory simultaneously, in which case there will be a delay until the memory is free. This delay, called *contention time*, can increase as the number of processors increases. Typically, shared memory has been used for systems with a small number of processors such as Cray machines with up to 16 processors.

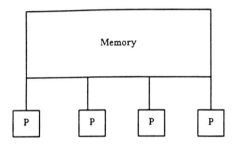

Figure 3.1.5: *A Shared Memory System*

An alternative to shared memory systems are *distributed memory* systems, in which each processor can address only its own local memory. Communication between processors takes place by *message passing*, in which data or other information are transferred between processors.

Interconnection Schemes

An important and interesting aspect of parallel computers is how the individual processors communicate with one another. This is particularly important for distributed memory systems but it is also important for shared memory systems since the connection to the shared memory can be implemented by different communication schemes. We shall next discuss briefly a number of the more common connections.

Completely Connected

In a completely connected system, each processor has a direct connection to every other processor. This is theoretically the ideal connection scheme but is impractical for a large number p of processors since it requires $p - 1$ lines emanating from each processor.

Switches

Another approach to a completely connected system is through a *crossbar switch* in which each processor can be connected to each memory through the use of switches. This has the advantage of allowing any processor access to any memory with only a small number of connection lines. This is illustrated in Figure 3.1.6.

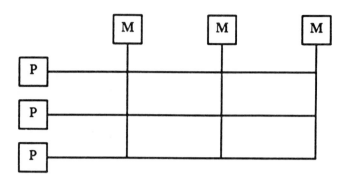

Figure 3.1.6: *A Crossbar Switch*

A disadvantage of a crossbar switch is that p^2 switches are required to connect p processors to p memories. This is impractical for large p but can be mitigated to some extent by a *switching network*. A simple switching network is depicted in Figure 3.1.7. On the left are shown eight processors and on the right are eight memories. Each box represents a two-way switch and the lines are the transmission links. By means of this switching network each of the processors can address any of the memories. For example, suppose P_1 wishes to address M_8. Switch 1,1 is then set to accept the link from P_1 and set the path to Switch 2,2, which sets the path to Switch 3,4, which sets the path to M_8.

Figure 3.1.7 represents a switching network implementation of a shared memory system, in the sense that each processor can access all memory. Alternatively, the memories on the right could be the processors themselves, as indicated by the processor numbers in parentheses. In this case, the figure would depict a message-passing distributed memory system.

Using the two-way switches shown in Figure 3.1.7, we would need $p/2$ switches at each switching stage and $\log_2 p$ stages for a total of $\frac{1}{2} p \log_2 p$ switches. This compares very favorably with the p^2 switches needed for the crossbar switch of Figure 3.1.6; for example, with $p = 2^{10}$, only 5×2^{10} switches would be needed, as compared with 2^{20}.

Mesh Connection

One of the most popular interconnection schemes has been to have each processor connected to only a few neighboring processors. The simplest example of this is a *linear array* illustrated in Figure 3.1.8. Here, each processor is

3.1 Parallel and Vector Computers

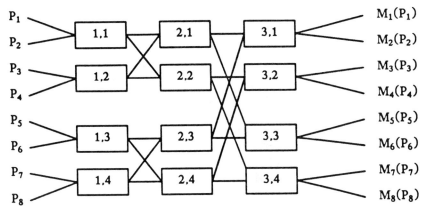

Figure 3.1.7: *A Switching Network*

connected to two nearest neighbors. The end processors may be connected to only one other processor, or there may be a "wrap-around" connection between P_1 and P_p; in this latter case, one speaks of a *ring array*.

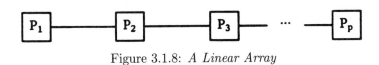

Figure 3.1.8: *A Linear Array*

There are certain similarities between the linear array of Figure 3.1.8 and a *bus network*; in which the processors obtain information from the bus. If the ends of the bus are connected, we obtain a *ring network*. One problem with bus networks is *bus contention* in which several processors are attempting to send information on the bus at the same time.

A linear array is different than a bus in that data being passed between two distant processors needs to go through the intermediate processors. For example, in Figure 3.1.8, if processor P_1 wishes to send data to processor P_p, the data must first be sent to P_2, and then transmitted to P_3, and so on. Thus, $p - 1$ transmissions of the data must be made. The maximum number of transmissions that must be made to communicate between any two processors of the system is called the *communication length* or *diameter* of the system. The problem of communication between distant processors has been mitigated to a large extent recently by *direct-connect* or *wormhole* routing, in which transmission can be made with minimal interruption of the intermediate processors.

Most mesh-connected arrays that have actually been built have used a two-dimensional connection pattern. One of the simplest such connection schemes is illustrated in Figure 3.1.9, in which the processors are laid out in a regular two-dimensional grid and each processor is connected to its north, south, east, and west neighbors. In addition, the edge processors can be connected in a wrap-around fashion. This connection pattern was used by the Illiac IV for 64 processors arranged in an 8 × 8 array. Other connection patterns, such as eight nearest neighbors or six nearest neighbors in three dimensions, could be used.

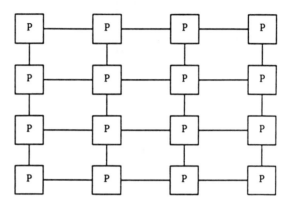

Figure 3.1.9: *A Mesh-Connected Array*

The connection pattern illustrated in Figure 3.1.9 again has the disadvantage in that transmission of data between distant processors must pass through a succession of intermediate processors. The communication length for p processors arranged in a square array increases as $\sqrt{p}$ but, again, direct-connect routing can substantially alleviate the problem. Since there are several possible paths between two distant processors, the first step in direct-connect routing is to establish a path. Once this is done, the message is sent with minimal interruption to the intermediate processors. There is, however, a potential problem of *edge contention*, which occurs when two messages simultaneously attempt to use the same communications link between two processors.

Hypercube Connection

An interesting variation of the mesh connection principle is to use, conceptually, local connections in higher dimensions. Consider first a cube in three dimensions and imagine the processors located at the vertices of the cube. The edges of the cube are the communication paths between processors so that each processor is connected to its three nearest neighbors in the sense of nearest vertices on the cube.

Now imagine the analogous connection scheme using a cube in k dimensions. Again, the processors would be located at the 2^k vertices of the k-dimensional cube. Each processor is connected to its k adjacent vertices, along

3.1 Parallel and Vector Computers

the edges of the cube; this is called a *hypercube* connection. Of course, a k-dimensional cube cannot actually be constructed for $k > 3$ and the connection pattern must be mapped back into no more than three dimensions for fabrication. For a 4-cube, a cube in four dimensions, there are 16 processors, with each processor connected to four others. It turns out in this case that Figure 3.1.9 of a mesh-connected array also shows the 4-cube interconnection pattern. Figure 3.1.10 illustrates a three-dimensional way of visualizing the connection pattern for a 4-cube and also shows that the 4-cube can be constructed by connecting corresponding vertices of two 3-cubes. In general, we can construct a k-cube by connecting all corresponding vertices of two $(k-1)$-cubes.

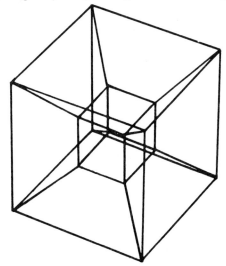

Figure 3.1.10: *Connection Pattern for a 4-Cube*

In the hypercube connection scheme, the number of connections from each processor grows as the number of processors increases and the communication length is $\log_2 p$. For example, for 64 processors - a 6-cube - each processor is connected to six other processors and the communication length is 6, while for 1024 processors – a 10-cube – each processor is connected to ten other processors and the communication length is 10. Moreover, the hypercube contains some of the other connection patterns that have been discussed. For example, ring or mesh-connected arrays can be included by ignoring some of the local links. On the other hand, as the size of the cube is increased, the complexity and the number of lines emanating from each processor also increase and a practical limit to the cube size will eventually be reached. Several commercial hypercube systems have been built but the trend now seems towards mesh-connected arrays for systems with more than 500 processors. We note that direct-connect routing was first used on hypercubes.

Clusters

A *cluster* scheme is illustrated in Figure 3.1.11. Here there are n clusters, each consisting of m processors. Within each cluster, the processors are connected in some fashion (any of the previous schemes could be used), and then the clusters are connected by, for example, a bus. The communication within each cluster is *local* while communication between clusters is called *global*. The hope in such a scheme is that there will be a suitable balancing of these two types of communication so that most communication for a given processor will be local within its own cluster and less frequently a cluster will need to communicate with another cluster.

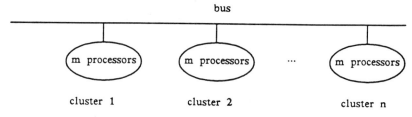

Figure 3.1.11: *A Cluster Connection*

Clearly, there is a large number of ways a cluster scheme could be implemented. Within a cluster, the connections could be by bus, ring, mesh, hypercube, and so on. The connections between clusters could also use any of these schemes rather than the bus of Figure 3.1.11. Moreover, the clustering concept can be applied repeatedly to obtain clusters of clusters of clusters, and so on.

More on Communication

We next discuss communication in distributed memory systems in a little more detail. Consider a typical problem of sending n floating point numbers from the memory of one processor, P_1, to the memory of another processor, P_2. Generally, this communication will be accomplished by a combination of hardware and software, and a typical scenario might be the following. First, the data are loaded into a buffer memory. Then a *send* command will be executed to transfer the data to a buffer memory of P_2. P_2 will execute a *receive* command and the data will be routed to their final location in the memory of P_2. Co-processors may be available to relieve the main processors of much of this work.

Many details (largely dependent on the system) have been omitted in the above simplified description, but the main points are that to send the data, they must first be collected from the memory of the sending processor, information must be provided as to where they are to be sent, the data must be physically transmitted between the processors, and, finally, they must be put in the

3.1 Parallel and Vector Computers

correct memory locations of the receiving processor. On many systems, the time for this communication is approximated by

$$t = s + \alpha n, \qquad (3.1.4)$$

where s is a start-up time (latency) and α is the incremental time necessary for each of the n words to be sent. Note that this has the same form as the time for vector instructions in (3.1.1). On some machines, s and α in (3.1.4) may depend on the size of the message being sent. Moreover, (3.1.4) may not be a good approximation for "multi-hop" messages, that is, for messages between distant processors.

Effects of Memory Delays

Many parallel systems are being constructed with high performance reduced instruction set (RISC) processors, whose memory system will usually include a data cache as well as main memory. Although the details vary from processor to processor, typical times in machine cycles are:

a) Floating point operations: 1 - 3 cycles;

b) Cache read or write: 1 cycle;

c) Main memory read or write: 2 cycles.

Floating point operations will obtain their operands from registers and the times in b) and c) are for loading or storing registers.

The read/write times in c) do not include "page access" delays. Memory will usually be partitioned into *pages*, which are blocks of contiguous memory locations. The address of a memory word will then consist of an upper part that determines the page and a lower part that determines location within the page. If a word from memory has just been read, a subsequent read from the same page is called a *near read* and incurs no additional page access delay. On the other hand, a subsequent read from a different page is called a *far read* and the total memory delay might be as high as 10 or more cycles. The same dichotomy holds for writes, with roughly the same penalty for a far write as for a far read.

The consequences of these page access delays can be seen in the example of a vector add:

$$z_i = x_i + y_i, \qquad i = 1, \ldots, n. \qquad (3.1.5)$$

Suppose that $n = 5000$ and the vectors **x** and **y** are each stored in contiguous memory locations. For a nominal page size of, say, 4096, each of the two reads as well as the write in (3.1.5) will then be to different pages. Thus, the time for each add in (3.1.5) will increase from the few cycles that would be necessary with no memory delays to approximately 35-40 cycles. Obviously, the best performance would be obtained if the vectors were stored in the interleaved fashion $x_1, y_1, z_1, x_2, y_2, z_2, \ldots$ so that only occasional reads/writes to a new page would be necessary. However, this interleaved storage pattern may well be at odds with the overall code of which (3.1.5) is presumably a small part.

Supplementary Discussion and References: 3.1

Many of the basic principles of parallel and vector machines are given in Stone [1990]. Details on particular machines may be obtained from manufacturers such as Cray Research, Inc., Intel Corp., and Thinking Machines Inc. Hockney and Jesshope [1988] discuss machine architectures as well as such concepts as $N_{1/2}$. See also Dongarra and Duff et al. [1990]. Saad and Schultz [1988, 1989a,b] gives further information on properties of hypercubes, as well as communication on hypercubes and other parallel architectures.

3.2 Basic Concepts of Parallel Computing

In this section we will collect a number of basic concepts and techniques in parallel computing.

Parallelism and Load Balancing

Consider the problem of adding two n-vectors **a** and **b**. The additions

$$a_i + b_i, \qquad i = 1, \ldots, n, \qquad (3.2.1)$$

are all independent and can be done in parallel. Thus, this problem has perfect mathematical parallelism. On the other hand, it may not have perfect parallelism on a parallel computer because of poor load balancing. By *load balancing* we mean the assignment of tasks to the processors of the system so as to keep each processor doing useful work as much as possible. Suppose, for example, that there are $p = 32$ processors and $n = 100$ in (3.2.1). Then, the processors can work in perfect parallelism on 96 additions but only 4 processors will be active during the remaining 4 additions. Thus, there is not perfect load balancing to match the perfect mathematical parallelism.

In general, load balancing may be done either statically or dynamically. In *static load balancing*, tasks (and, perhaps, data for distributed memory systems) are assigned to processors at the beginning of a computation. In *dynamic load balancing*, tasks (and data) are assigned to processors as the computation proceeds. A useful concept for dynamic load balancing is that of a *pool of tasks*, from which a processor obtains its next task when it is ready to do so. In general, dynamic load balancing is more efficiently implemented on shared memory systems than on distributed memory systems since, on the

3.2 Basic Concepts of Parallel Computing

latter, data transfers between local memories may also be required as part of a task assignment.

Related to load balancing is the idea of *granularity*. *Large-scale granularity* means large tasks that can be performed independently in parallel. An example is the solution of six different large systems of linear equations, whose solutions will be combined at a later stage of the computation. *Small-scale granularity* means small tasks that can be performed in parallel; an example is the addition of two vectors where each task is the addition of two scalars.

Summation and Fan-in

Next, consider the problem of summing n numbers $a_1, \ldots, a_n$. The usual serial algorithm

$$s = a_1, \quad s \leftarrow s + a_i, \quad i = 2, \ldots, n, \tag{3.2.2}$$

is unsuitable for parallel computation. However, there is a good deal of parallelism in the problem itself, as shown in Figure 3.2.1 for the addition of eight numbers in three stages. In the first stage, four additions are done in parallel, then two are done in the next stage, and finally one in the last stage. This illustrates the general principal of *divide and conquer*. We have divided the summation problem into smaller problems, which can be done independently. It is interesting that this approach also has better rounding error properties than (3.2.2) and can be used beneficially on serial machines.

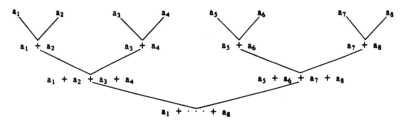

Figure 3.2.1: *Fan-in for Addition*

The graph of Figure 3.2.1 is called a *fan-in graph* and is widely applicable, as we shall see. In particular, the same idea can be used to obtain the product $a_1 a_2 \ldots a_n$ of n numbers; in Figure 3.2.1 simply replace $+$ by $\times$. Similarly, to find the maximum of n numbers we can replace the $+$ operation by max; for example, in Figure 3.2.1, $a_1 + a_2$ would be replaced by $\max(a_1, a_2)$. We note that the fan-in graph of Figure 3.2.1 is a *binary tree* and the fan-in operation is sometimes called a *tree operation*.

Communication and Synchronization

In general, in a distributed memory system, exchange of data between processors will be necessary at various times during an overall computation, and to the extent that processors are not doing useful computation during the communication, this constitutes an overhead. In shared memory systems, there may be contention delays (see Section 3.1) that give the same effect as communication time: processors are idle or underutilized while waiting for the data necessary to continue.

Synchronization is necessary when certain parts of a computation must be completed before the overall computation can proceed. There are two aspects of synchronization that contribute to overhead. The first is the time to do the synchronization; usually this requires that all processors perform certain checks. The second aspect is that some, or even almost all, processors may become idle, waiting for clearance to proceed with the computation.

There are several ways to implement synchronization. A common situation is that there is a *critical section* of code, which is a sequential part of an overall code. A *fork* usually follows a critical section and initiates code segments that operate in parallel. At a *join* these parallel segments return to a critical section. This is illustrated in Figure 3.2.2. Although synchronization, communication, and memory contention delays are quite different, their effect on the overall computation is the same: a delay while data are made ready to continue the computation.

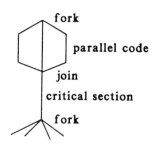

Figure 3.2.2: *A Critical Section with Forks and Joins*

We will illustrate communication and synchronization delays by means of the example of addition of n numbers by the fan-in algorithm. Suppose that we have a distributed memory system with $\frac{n}{2}$ processors, and a_1 and a_2 are stored in processor 1, a_3 and a_4 in processor 2, and so on. At the first stage, the additions $a_i + a_{i+1}$ are all done in parallel in the corresponding processors. Before the additions of the second stage can be done, we will need to send a_3+a_4 to processor 1, a_5+a_6 to processor 2, and so on. The addition of (a_1+a_2)

3.2 Basic Concepts of Parallel Computing

and $(a_3 + a_4)$ in processor 1 cannot be done until $a_3 + a_4$ has been received by processor 1. This receive thus serves as a synchronization mechanization in this problem.

Speed-up

Ideally, we could solve a problem p times as fast on p processors as on a single processor. This ideal is rarely achieved; what is achieved is called the *speed-up of the parallel algorithm* defined by

$$S_p = \frac{\text{execution time for a single processor}}{\text{execution time using } p \text{ processors}}. \tag{3.2.3}$$

Closely related to speed-up is *efficiency* defined by

$$E_p = \frac{S_p}{p}. \tag{3.2.4}$$

Since $S_p \leq p$, we have $E_p \leq 1$ and an efficiency of $E_p = 1$ corresponds to a perfect speedup of $S_p = p$.

Consider again the problem of addition of two vectors of length n in the case that $n = 100$ and $p = 32$. Assume t is the time for one addition so that $100t$ is the single processor time. In parallel, 96 of the additions can be done with perfect parallelism in a time of $3t$ but the remaining four additions also require time t. Thus,

$$S_p = \frac{100t}{4t} = 25, \quad E_p = \frac{25}{32} < 1,$$

and the loss of perfect speed-up is due to the less than perfect load balancing.

Consider next the addition of n numbers by the fan-in algorithm, where $p = \frac{n}{2}$. For simplicity, assume that $n = 2^q$. Then the fan-in algorithm has $q = \log_2 n$ stages with $\frac{n}{2}$ additions at the first stage, $\frac{n}{4}$ at the second and so on to a single addition at the last stage. If t is again the time for an addition, the addition time at each stage is t for a total of qt. However, we need to account for the communication time, as discussed previously; we assume that this is αt for each communication, where α will usually be greater than 1. We assume that all the communications at each stage are done in parallel so that the time for each stage is now $(1 + \alpha)t$ for a total of $q(1 + \alpha)t$. Since the time for a single processor is $(n - 1)t$, and using $p = \frac{n}{2}$ and $n = 2^q$, the speed-up is

$$S_p = \frac{(n-1)t}{q(1+\alpha)t} = \frac{1}{(1+\alpha)} \frac{(n-1)}{\log_2 n} = \frac{1}{(1+\alpha)} \frac{(2p-1)}{(1+\log_2 p)}. \tag{3.2.5}$$

If we ignore communication costs, so that $\alpha = 0$, there is a logarithmic degradation of speed-up due to the lack of perfect parallelism. For $\alpha > 0$, there is further degradation of the speed-up by the factor $1+\alpha$. If α is approximately equal to 1, so that communication takes about as much time as arithmetic, then the speedup is roughly halved from that with $\alpha = 0$. On the other hand, if α is large, say $\alpha = 10$, then communication time dominates the computation, and the speed-up is reduced accordingly.

More realistic values of n can affect the speed-up in different ways. If n is not a power of 2 some parallelism is lost. For example, if $n = 7$ and $p = 4$ only three processors do additions at the first stage, which further degrades the speed-up. On the other hand, if $n \gg p$, a realistic situation, the speed-up is improved. For example, if $p = 4$ and $n = 128$, each processor holds 32 numbers, and the 4 processors can add these with perfect parallelism and no communication; thus, only the final fan-in of 4 additions degrades from the perfect parallelism. (See Exercise 3.2.1 for an elaboration of this example.)

The speed-up S_p is a measure of how a given algorithm compares with itself on 1 and p processors. However, the parallel algorithm may not be the best algorithm on a single processor, as we will see in several cases in later chapters. Hence, a better measure of what is gained by parallel computation is given by the alternative definition

$$S'_p = \frac{\text{execution time on a single processor of fastest serial algorithm}}{\text{execution time of the parallel algorithm on } p \text{ processors}}.$$
(3.2.6)

Both of the measures S_p and S'_p are useful and the context of discussion will determine which to use.

Amdahl's Law

We have seen, with the two simple examples of vector addition and summation of numbers, how speed-up may be degraded because of various factors. We can try to encapsulate these factors into a formal model of speed-up in which

$$S_p = \frac{T_1}{(\alpha_1 + \alpha_2/k + \alpha_3/p)T_1 + t_d},$$
(3.2.7)

where T_1 is the time for a single processor, α_1 is the fraction of operations done with one processor, α_2 is the fraction of operations done with average parallelism $k < p$, α_3 is the fraction of operations done with perfect parallelism, and t_d is the total time delay due to communication, synchronization or contention.

Although (3.2.7) is already a considerable simplification of most realistic situations, a further simplification is very instructive. Let $\alpha_1 = \alpha, \alpha_2 = 0, \alpha_3 = 1 - \alpha$ and $t_d = 0$. Then (3.2.7) reduces to

$$S_p = \frac{1}{\alpha + (1-\alpha)/p},$$
(3.2.8)

which is known as *Amdahl's Law* or *Ware's Law*. The assumption is that all operations are done either with perfect parallelism or none, and there are no

3.2 Basic Concepts of Parallel Computing

delays. Suppose that in a given problem, half of the operations can be done in parallel and the other half not. Then $\alpha = 1/2$ and (3.2.8) becomes

$$S_p = \frac{2}{(1+p^{-1})} < 2.$$

Thus, no matter how many processors there are, and ignoring all communication, synchronization, and contention delays, the speed-up is always less than 2. More generally, Figure 3.2.3 gives a plot of (3.2.8) as a function of α for $p = 100$. Note the very rapid decrease of S_p for small values of α. If even 1% of the operations can be done only on a single processor, the speedup is halved from 100 to 50.

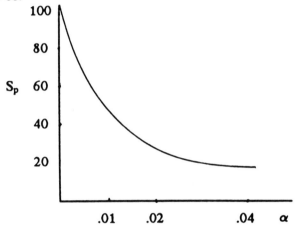

Figure 3.2.3: *Degradation of Speed-Up by Amdahl's Law*

Scaled Speed-up

There are potential difficulties in applying the definitions (3.2.3) and (3.2.6) of S_p and S'_p in practice. On parallel machines with distributed memory, the more processors that are used the larger the problem that can be run. In particular, problems that can be run on a large number of processors may not fit into the memory of a single processor so that the time on a single processor would have to include I/O time to secondary memory. In one sense, this is an entirely legitimate measure of speed-up, but it may not be an appropriate one for some considerations.

On the other hand, if we restrict ourselves to problems whose size is suitable for a single processor, we will possibly obtain very poor speedups when using several processors. A first indication of this may be obtained from the model (3.2.7); no matter what the values of α_1, α_2, and α_3, for sufficiently large t_d we can have $S_p < 1$. Thus, in a problem with a lot of communication, contention, or synchronization it is possible that using more than one processor is actually worse than just a single processor. This is a rather extreme case, but for many problems one does reach a point of diminishing returns in terms of employing additional processors. Indeed, a speed-up curve such as shown in Figure 3.2.4 is not at all unusual.

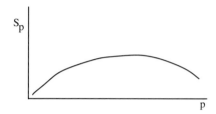

Figure 3.2.4: *Speed-up for Fixed Problem Size*

The reason that speed-up curves such as Figure 3.2.4 frequently occur is because the problem size is fixed. However, the primary purpose of parallel computing is to be able to handle problems much larger than can be done on a single processor. Thus, the more processors used, the larger the problem that can be solved. This leads to another approach to speed-up, called *scaled speed-up*, in which we measure the size of the problem that can be done in a fixed amount of time as the problem size increases. This is in contrast to holding the problem size constant and measuring the time required, as in the definition (3.2.3) of S_p.

As an example of this approach, consider the problem of matrix addition $A + B$, where A and B are $n \times n$. Then n^2 operations are required. Suppose, for a nominal problem size, the computation runs in 30 seconds on a single processor. Then if we double n, the number of operations increases by a factor of 4, and if this problem runs on four processors in 30 seconds we have achieved perfect speed-up.

Data Flow Analysis

As we shall see in subsequent sections, it is often fairly easy to detect the inherent parallelism in an algorithm. Other times it is not so obvious. A systematic tool for uncovering the parallelism is a *precedence graph* for the computation. This is illustrated in Figure 3.2.5 for the computation

$$f(x,y) = x^2 + y^2 + x^3 + y^4 + x^2y^2. \tag{3.2.9}$$

The serial computation for (3.2.9) might proceed as

$$x, y, x^2, y^2, x^2 + y^2, x^3, (x^2 + y^2) + x^3, y^4, x^2y^2, y^4 + x^2y^2,$$
$$(x^2 + y^2 + x^3) + (y^4 + x^2y^2)$$

3.2 Basic Concepts of Parallel Computing

which requires ten arithmetic operations. These ten operations can be done in four parallel steps as indicated in Figure 3.2.5.

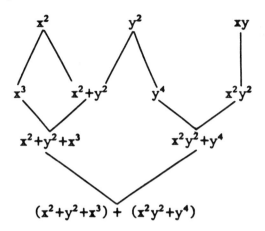

Figure 3.2.5: *Precedence Graph*

Vector Computing

Although most of the discussion of this section applies only to parallel computing, some of it has applicability to vector computing also. For example, the summation of n numbers on a vector computer can also use the fan-in algorithm. In this case, again assume for simplicity that $n = 2^q$ and suppose that $a_1, \ldots, a_n$ are the elements of the vector $\mathbf{a}$. Let $\mathbf{a}_1$ be the subvector starting at position $\frac{n}{2} + 1$ of $\mathbf{a}$. Then the $\frac{n}{2}$-long vector add of $\mathbf{a}_1$ and the first $\frac{n}{2}$ components of $\mathbf{a}$ is the first stage of a fan-in process. If $\mathbf{a}_2$ is the result vector of this addition, we can then add the first $\frac{n}{4}$ components of $\mathbf{a}_2$ to the last $\frac{n}{4}$ and this is the second stage of the fan-in, and so on. An extension of this process is given in Exercise 3.2.2.

The idea of speed-up in vector computing would be the ratio of the time required using only scalar operations to that using vector operations when possible. This concept is not nearly as useful as in parallel computing although Amdahl's law (3.2.8) is of some interest. Here, the interpretation of (3.2.8) would be that a fraction α of the computation is done in scalar arithmetic and the rest in "optimum" vector arithmetic. Graphs similar to Figure 3.2.2 may be plotted and the important point is that even a small proportion of scalar arithmetic seriously degrades the overall performance. Again, the special case $\alpha = \frac{1}{2}$ is instructive. If 50% of the computation must be done in scalar arithmetic, a speed-up of at most a factor of 2 over a scalar code is possible, no matter how fast the vector operations.

Supplementary Discussion and References: 3.2

1. The reorganization of computations, such as the fan-in algorithm of Figure 3.2.1 versus the serial algorithm (3.2.1), raises the question of numerical stability. That is, if a computation is numerically stable, will it still be stable if the order of operations is changed? Rönsch [1984] addresses this problem and shows, in particular, that the fan-in algorithm is more stable than (3.2.2).

2. Synchronization can be, and has been, implemented by a variety of techniques. We note that many of the concepts and terminolgy of synchronization have arisen in operating systems theory; a good overall review may be found in Andrews and Schneider [1983]. An idea related to a critical section is a *barrier* (Axelrod [1986], Jordan [1986]), which is a logical point in the control flow at which all processes must arrive before any can proceed further. On shared memory machines, synchronization is usually accomplished through the use of shared variables, which can be referenced by more than one process or processor. For example, in *busy-waiting*, a process tests a shared variable for a certain value that signals that it may proceed. While the process is waiting for the correct value, it is said to be *spinning* and the shared variable or variables are called *spin locks*.

3. The idea of using megaflop rates to compute speed-ups in situations where the problem is too large to fit in the memory of a single processor is given in Moler [1986]. The concept of scaled speed-up is due to Gustafson et al. [1988]. The speed-up formula (3.2.8) is given in Ware [1973] and is implied in Amdahl [1967].

EXERCISES 3.2

3.2.1 Let $p = 2^r$ and $n = 2^q$ where $q >> r$. Find S_p for the addition of n numbers assuming as in (3.2.5) that αt is the communication time. Show that $S_p \to p$ for fixed r as $q \to \infty$.

3.2.2. Formulate the fan-in algorithm of Figure 3.2.1 for the sum of $s = 2^r$ vectors $\mathbf{a}_1, \ldots, \mathbf{a}_s$ of length n. Write a pseudocode to do this on a vector machine.

3.2.3. Formulate the fan-in algorithm of Figure 3.2.1 so as to compute $a_1 \ldots a_n$ and $\max\{a_1, \ldots, a_n\}$. Also, find both the max and min of $a_1, \ldots, a_n$.

3.2.4. Let $q(x) = a_0 + a_1 x + \cdots + a_n x^n$ be a polynomial of degree n, where $n = r 2^r$. Write a parallel code for evaluating $q(x)$ by first writing q as

$$q(x) = a_0 + q_1(x) + x^r q_2(x) + x^{2r} q_3(x) + \cdots + x^{(s-1)r} q_s(x)$$

where $s = 2r$ and

$$q_i(x) = a_k x + \cdots + a_{k+r-1} x^r, \qquad k = (i-1)2^r + 1.$$

3.3 Matrix Multiplication

We consider in this section the problem of computation of $A\mathbf{x}$ and AB, where A and B are matrices and $\mathbf{x}$ is a vector. These basic operations will be important in the following chapters. Moreover, the study of matrix multiplication allows us to explore many of the basic issues of parallel and vector computation in the context of a very simple mathematical problem. We assume first that A and B are full matrices and then treat sparse matrices of various types later in the section. We will first work in terms of vector operations and then consider parallel computers.

Matrix-Vector Multiplication

Assume that A is an $m \times n$ matrix and $\mathbf{x}$ an n-vector. Then

$$A\mathbf{x} = \begin{bmatrix} (\mathbf{a}_1, \mathbf{x}) \\ \vdots \\ (\mathbf{a}_m, \mathbf{x}) \end{bmatrix}, \qquad (3.3.1)$$

where $\mathbf{a}_i$ is the ith row of A and $(\mathbf{x}, \mathbf{y}) = \sum_{i=1}^{n} x_i y_i$ is the usual inner product. Thus, the evaluation of $A\mathbf{x}$ requires the computation of m inner products. Before examining this approach in more detail, we consider the other standard way to view matrix-vector multiplication, as a linear combination of the columns of A. Thus,

$$A\mathbf{x} = \sum_{i=1}^{n} x_i \mathbf{a}_i, \qquad (3.3.2)$$

where now $\mathbf{a}_i$ denotes the ith column of A.

The difference in (3.3.1) and (3.3.2) can be viewed as two different ways of accessing the data, as shown in the codes given in Figure 3.3.1. In Figure 3.3.1(a), for each i the j loop computes the inner product of the ith row of A and $\mathbf{x}$, so that this is (3.3.1). Figure 3.3.1(b) corresponds to (3.3.2). Note that the final arithmetic statement is the same in both codes of Figure 3.3.1, so that the difference is in the loop order of the indices. We will see in the sequel that the same type of reordering of indices gives various possible algorithms for matrix-matrix multiplication and the solution of linear equations.

We now consider an inner product calculation in more detail. The inner product $(\mathbf{x}, \mathbf{y})$ of two vectors may be computed as

$$t_i = x_i y_i, \quad i = 1, \ldots, n, \quad (\mathbf{x}, \mathbf{y}) = \sum_{i=1}^{n} t_i. \qquad (3.3.3)$$

The first step of (3.3.3) is a vector-vector multiplication, while the second involves a summation which, as we saw in the previous section, may be carried

$$\boxed{\begin{array}{l} \mathbf{y} = 0 \\ \text{For } i = 1 \text{ to } m \\ \quad \text{For } j = 1 \text{ to } n \\ \quad\quad y_i = y_i + a_{ij} x_j \end{array}} \qquad \boxed{\begin{array}{l} \mathbf{y} = 0 \\ \text{For } j = 1 \text{ to } n \\ \quad \text{For } i = 1 \text{ to } m \\ \quad\quad y_i = y_i + a_{ij} x_j \end{array}}$$

(a) inner product (b) linear combination

Figure 3.3.1: *Two Forms of Matrix-Vector Multiplication*

out by a fan-in operation with some loss of parallelism or, equivalently, increasingly shorter vectors. On the other hand, for the linear combination approach, (3.3.2), the algorithm will be

$$\mathbf{y} = x_1 \mathbf{a}_1, \qquad \text{for } i = 2 \text{ to } n \text{ do } \mathbf{y} = \mathbf{y} + x_i \mathbf{a}_i. \tag{3.3.4}$$

As noted previously, an operation of the form vector plus scalar times vector is called an axpy operation.

The linear combination algorithm avoids the fan-in of the inner product computation and utilizes vectors of maximum length. Therefore, it will usually be more efficient although this will depend on the sizes of m and n as well as the storage. For example, suppose that m is small and n is large. Then the vector lengths in (3.3.4) will be small while those in (3.3.1) are large and probably (3.3.1) would be the more efficient algorithm. Storage of A will also affect the efficiency. Suppose that A is stored by columns, which is the usual Fortran convention for storing two-dimensional arrays. Then the vectors required for the linear combination algorithm (3.3.4) are sequentially addressable locations in storage, while for the inner product algorithm (3.3.1) the rows of A will be vectors with a stride of m. As noted previously, with stride greater than one the speed of the vector operation may be degraded. Thus, if the matrix A is stored by columns, then this is an advantage for the linear combination algorithm. On the other hand, if the matrix A is already stored by rows, then the storage is better for the inner product algorithm and this may dictate it to be the algorithm of choice. Only a detailed analysis for the particular machine to be used can lead to a clear choice.

Loop Unrolling

We next consider how the linear combination algorithm (3.3.4) utilizes vector registers. A straightforward implementation of (3.3.4) would have the following sequence of operations:

3.3 Matrix Multiplication

Load $\mathbf{a}_1$ into vector register.

$\mathbf{y} = x_1 \mathbf{a}_1$. Result goes to register.

Store $\mathbf{y}$ in main memory.

Load $\mathbf{a}_2$ into vector register.

Multiply $x_2 \mathbf{a}_2$. Result goes to register.

Load $\mathbf{y}$ into vector register.

Add $\mathbf{y} = \mathbf{y} + x_2 \mathbf{a}_2$. Result goes to register.

Store $\mathbf{y}$ in main memory.

Load $\mathbf{a}_3$ into vector register.

$\vdots$

In the above, we have assumed that the vectors fit in the registers; if not, we would process them in segments, which is known as *strip mining*.

The above code is very inefficient. Each stage requires two vector loads, a store, and two vector arithmetic operations. For the arithmetic operations, we should incorporate chaining, provided the hardware allows this. Chaining was illustrated in Figure 3.1.4 and allows the result $x_1 \mathbf{a}_1$ to go directly to the addition unit. Addition of $\mathbf{y}$ can then begin while the multiplications are still being carried out, so that both vector operations are being done almost concurrently. We next note that the store and load of $\mathbf{y}$ between the vector operations is unnecessary; we can simply access the current $\mathbf{y}$ from its register. Finally, most machines will allow vector register loads to occur concurrently with arithmetic operations. Thus, we can be loading $\mathbf{a}_2$ while the operation $\mathbf{y} + x_1 \mathbf{a}_1$ is being done. (Note, however, that we would use a different vector register for $\mathbf{a}_2$ than for $\mathbf{a}_1$, so that the code must "flip-flop" on its access of the $\mathbf{a}_i$.)

With these changes, the code would be as follows:

Load $\mathbf{a}_1$ into vector register.

Form $\mathbf{y} = x_1 \mathbf{a}_1$. Load $\mathbf{a}_2$.

Form $\mathbf{y} = \mathbf{y} + x_2 \mathbf{a}_2$ by chaining. Load $\mathbf{a}_3$.

$\vdots$

Now, each stage will require only slightly longer than the time for a single vector arithmetic operation and this code will run almost five times faster than the original.

74   Chapter 3 Parallel and Vector Computing

Compilers will typically recognize when a chaining operation can be done. Unfortunately, compilers have not always recognized when intermediate results could be left in vector registers nor when loads and arithmetic operations could be overlapped. Of course, this problem can be corrected in assembly language. In Fortran, it can be mitigated to a large extent by the idea of *loop unrolling*. For example, if we replace the loop

$$\text{For } i = 1 \text{ to } n$$
$$\mathbf{y} = \mathbf{y} + x_i \mathbf{a}_i$$

by

$$\text{For } i = 2 \text{ to } n, \text{ steps of } 2$$
$$\mathbf{y} = \mathbf{y} + x_{i-1}\mathbf{a}_{i-1} + x_i\mathbf{a}_i$$

the vector $\mathbf{y}$ will be stored and reloaded only half as many times as in the original; moreover, the loading of $\mathbf{a}_i$ will be done concurrently with the arithmetic of $\mathbf{y} + x_{i-1}\mathbf{a}_{i-1}$. (Note that an extra statement is required to process $\mathbf{a}_n$ if n is odd.) In this example, the loop has been unrolled to a *depth* of two. Unrolling to a depth of three is done by

$$\text{For } i = 3 \text{ to } n, \text{ steps of } 3$$
$$\mathbf{y} = \mathbf{y} + x_{i-2}\mathbf{a}_{i-2} + x_{i-1}\mathbf{a}_{i-1} + x_i\mathbf{a}_i$$

and similarly for unrolling to greater depth. (Again, additional statements are required if n is not a multiple of the depth.) Loop unrolling has been employed in practice up to depths of eight or even higher.

Multiplication of Matrices

The previous discussion of matrix-vector multiplication extends in a natural way to the multiplication of matrices, although we will see that other alternatives are now available. Let A and B be $m \times n$ and $n \times q$ matrices, respectively, so that AB is $m \times q$.

The inner product form of matrix-vector multiplication extends to the *inner product* algorithm

$$C = AB = \begin{bmatrix} \mathbf{a}_1 \\ \vdots \\ \mathbf{a}_m \end{bmatrix} (\mathbf{b}_1, \ldots, \mathbf{b}_q) = (\mathbf{a}_i \mathbf{b}_j). \qquad (3.3.5)$$

Here, A is partitioned into rows and B into columns and the product AB requires the formation of the mq inner products $\mathbf{a}_i \mathbf{b}_j$ of the rows of A and columns of B. Note that if $q = 1$, so that B is a vector, (3.3.5) reduces to the inner product algorithm for matrix-vector multiplication. The inner product algorithm has the same advantages and disadvantages as the corresponding matrix-vector algorithm, and we shall not discuss it further.

3.3 Matrix Multiplication

The next algorithm is based on repeated matrix-vector multiplications. Let $\mathbf{a}_i$ and $\mathbf{b}_i$ denote the ith columns of A and B. Then

$$C = AB = (A\mathbf{b}_1, \ldots, A\mathbf{b}_q) = (\sum_{k=1}^{n} b_{k1}\mathbf{a}_k, \ldots, \sum_{k=1}^{n} b_{kq}\mathbf{a}_k). \qquad (3.3.6)$$

The jth column of C is just A times the jth column of B, and these matrix-vector operations are done as linear combinations of the columns of A. A pseudocode for (3.3.6) is given in Figure 3.3.2. The algorithm of Figure 3.3.2 is sometimes called the middle product algorithm; we prefer to call it the *linear combination* algorithm, since this is the basic operation. Ideally, A is stored by columns. If B is stored by rows, and $\mathbf{a}_i$ and $\mathbf{b}_i$ now denote the ith rows of A and B, then an alternative, which we call the *dual linear combination algorithm*, is

$$C = AB = \begin{bmatrix} \mathbf{a}_1 B \\ \vdots \\ \mathbf{a}_m B \end{bmatrix} = \begin{bmatrix} \sum a_{1k}\mathbf{b}_k \\ \vdots \\ \sum a_{mk}\mathbf{b}_k \end{bmatrix}. \qquad (3.3.7)$$

Here, the jth row of C is a linear combination of the rows of B. If $m = 1$ in (3.3.7), so that A is a row vector, then (3.3.7) is the linear combination algorithm for a row vector times a matrix.

> Set $C = 0$
> For $j = 1$ to q
>   For $k = 1$ to n
> $\mathbf{c}_j = \mathbf{c}_j + b_{kj}\mathbf{a}_k$

Figure 3.3.2: *Linear Combination Matrix Mutliply Algorithm*

Another algorithm is based on outer products. Recall from Section 2.1 that the outer product of an m-long column vector $\mathbf{u}$ and a q-long row vector $\mathbf{v}$ is the $m \times q$ matrix

$$\mathbf{uv} = (v_1\mathbf{u}, \ldots, v_q\mathbf{u}) = (u_i v_j). \qquad (3.3.8)$$

The *outer product algorithm* for matrix multiplication based on (3.3.8) is

$$C = AB = (\mathbf{a}_1, \ldots, \mathbf{a}_n) \begin{bmatrix} \mathbf{b}_1 \\ \vdots \\ \mathbf{b}_n \end{bmatrix} = \sum_{i=1}^{n} \mathbf{a}_i \mathbf{b}_i = \sum_{i=1}^{n} (b_{i1}\mathbf{a}_i, \ldots, b_{iq}\mathbf{a}_i), \qquad (3.3.9)$$

so that C is the sum of outer products of columns of A and rows of B. A pseudocode is given in Figure 3.3.3; ideally, A is stored by columns. If B is stored by rows, an alternative is

$$C = AB = \sum_{i=1}^{n} \mathbf{a}_i \mathbf{b}_i = \sum_{i=1}^{n} \begin{bmatrix} a_{1i} \mathbf{b}_i \\ \vdots \\ a_{mi} \mathbf{b}_i \end{bmatrix}, \quad (3.3.10)$$

which we call the *dual outer product algorithm*. In this case, C is also the sum of n outer products and (3.3.9) and (3.3.10) differ only in how the outer products are computed.

Set $C = 0$
For $k = 1$ to n
  For $j = 1$ to q
    $\mathbf{c}_j = \mathbf{c}_j + b_{kj} \mathbf{a}_k$

Figure 3.3.3: *Outer Product Matrix Multiplication*

We note that, corresponding to Figure 3.3.1, all of the above matrix multiplication algorithms can be viewed as different orderings of the loop variables i, j, and k in the basic code

For _____

For _____

For _____

$$c_{ij} = c_{ij} + a_{ik} b_{kj}$$

where the arithmetic statement remains fixed. The six possible codes, sometimes called the *ijk forms* of matrix multiplication, are given in Exercise 3.3.2.

Comparison of Algorithms

We next wish to compare the linear combination and outer product algorithms under the assumption that A is stored by columns. Analogous considerations apply to the dual algorithms if B is stored by rows. For the outer-product algorithm, Figure 3.3.3, the inner loop forms the outer product $\mathbf{a}_i \mathbf{b}_i$, adding it to the accumulating product in C. The basic step is an axpy for vectors of length m. For the linear combination algorithm of Figure 3.3.2, again the basic operation is an axpy of length m. For small m neither algorithm

3.3 Matrix Multiplication

will be attractive; if B is stored by rows and $q > m$, the dual forms of the algorithms may be preferred.

Although the basic vector operation in both algorithms is an axpy of length m, there are important differences in memory access, which can affect proper use of vector registers or cache memory. We first discuss the use of vector registers and then show how the same considerations apply to cache. For simplicity, we assume that the registers are of sufficient size to hold a column of A; if this is not the case, obvious modifications by use of stripmining can be made.

Note first that the inner loop of the linear combination algorithm completely forms a column of C, while in the outer-product algorithm the columns of C are accumulated over the whole calculation. In the outer-product algorithm $\mathbf{a}_i$ can be used q times since it is in a register. However, the columns of C must be continuously loaded and stored as they are updated. For example, $\mathbf{c}_j$ is updated by $\mathbf{c}_j + b_{ij}\mathbf{a}_i$, then stored in main memory, $\mathbf{c}_{j+1}$ is loaded from main memory, updated by $\mathbf{c}_{j+1} + b_{ij+1}\mathbf{a}_i$, and so on. On machines for which loads and stores may be performed concurrently (for example, the Cray Y-MP) this is no problem but on machines for which only a load or a store may be performed at a given time (Cray-1, Cray-2), there will be a delay.

On the other hand, in the linear combination algorithm, Figure 3.3.2, the successive columns of $\mathbf{a}_j$ of A must be loaded for each vector operation but a store is not required until a column of C has been completely computed. Thus, on machines that can only load or store there is a storage delay only once for every column of C. The linear combination algorithm has a decided advantage on such machines and is at least as good on machines that can load and store simultaneously.

The same general considerations apply to cache memory. The cache will usually be considerably larger than vector register memory but not so large that entire matrices can be stored if they are of a reasonable size. Hence, again, in the outer product algorithm columns of C will have to be loaded and stored continuously so that this algorithm will require much more memory traffic than the linear combination algorithm.

Block Algorithms

The above discussion leads to the general principle that once data are in vector registers or cache they should be used as much as possible. A second general principle is that this amount of reusable data should be as large as possible. In the previous algorithms, it is only a single vector (column of A or column of C) that is being reused. On the other hand, *block algorithms* that utilize submatrices rather than just columns have the potential for larger amounts of reusable data.

Consider, for example, the linear combination algorithm of Figure 3.3.2. For the formation of the ith column of the product, the column $\mathbf{c}_i$ can remain in cache as it is being accumulated by

$$\mathbf{c}_j = b_{1j}\mathbf{a}_1, \qquad \mathbf{c}_j = \mathbf{c}_j + b_{kj}\mathbf{a}_k, \qquad k = 2,\ldots,n.$$

Each b_{kj} is used for just one axpy operation and needs to be retained in cache only for that operation. But if the matrices are of large size, A will not be able to remain in cache and will need to be brought from main memory for the computation of each column of C. Suppose, however, that the cache is large enough to hold r columns of C and a column of A. Then, as each column of A is brought from main memory it can be used to update r columns of C. Thus, only q/r loads of A are needed (q is the number of columns of C), which can have a significant effect on the overall computation time. Mathematically, the above algorithm can be summarized by the partitioning

$$C = AB = (AB_1, AB_2, \ldots, AB_\ell), \tag{3.3.11}$$

where B_1 is the submatrix of the first r columns of B, B_2 is the second r columns, and so on.

An alternative approach is to reverse the roles of A and C in the cache. First, we form part of the first column, $\mathbf{c}_1$, of C by

$$\mathbf{c}_1 = \sum_{k=1}^{p} b_{k1}\mathbf{a}_k.$$

With $\mathbf{a}_1,\ldots,\mathbf{a}_p$ now in cache, we partially form all the remaining columns of C by

$$\mathbf{c}_j = \sum_{k=1}^{p} b_{kj}\mathbf{a}_k, \qquad j = 2,\ldots,q. \tag{3.3.12}$$

It is assumed here that each $\mathbf{c}_j$ is being replaced in cache after it is formed and that the $\mathbf{a}_1,\ldots,\mathbf{a}_p$ remain in cache throughout the operations (3.3.12). At the end of the operations (3.3.12), $\mathbf{a}_1,\ldots,\mathbf{a}_p$ do not need to be used again. The next stage commences with the analogous computations using $\mathbf{a}_{p+1},\ldots,\mathbf{a}_{2p}$, and so on. Thus, A needs to be brought from main memory only once but now C needs to be brought q/r times. (B needs to be brought only once, as before.) Mathematically, this alternative corresponds to the partitioning

$$C = (A_1,\ldots,A_\ell) \begin{bmatrix} B_1 \\ \vdots \\ B_\ell \end{bmatrix} = \sum_{i=1}^{\ell} A_i B_i$$

of A by columns and B by rows.

3.3 Matrix Multiplication

Other algorithms may be based on more general partitionings of both A and B. Thus, we may write $C = AB$ in the partitioned form

$$C = \begin{bmatrix} A_{11} & \cdots & A_{1r} \\ \vdots & & \vdots \\ A_{s1} & \cdots & A_{sr} \end{bmatrix} \begin{bmatrix} B_{11} & \cdots & B_{1t} \\ \vdots & & \vdots \\ B_{r1} & \cdots & B_{rt} \end{bmatrix}. \quad (3.3.13)$$

Here, of course, it is assumed that the submatrices are of commensurate sizes so that the products $A_{ik}B_{kj}$ are well defined. The partitioning (3.3.13) gives rise to several possible algorithms; in particular, there is a block counterpart of each of the ijk forms previously discussed.

We will illustrate the use of (3.3.13) by modifying the algorithm based on (3.3.11). Suppose, for example, that A and B are 1000×1000 and the cache can hold 8000 floating point numbers. Then, in the algorithm based on (3.3.11) we could keep 6 columns of C and one column of A in cache (with some room to spare). Thus, $1000/6 = 167$ loads of A and one load of B would be required. Suppose, however, that we take $s = 4$ in (3.3.13) so that C is partitioned as

$$C = \begin{bmatrix} C_{11} & \cdots & C_{1t} \\ \vdots & & \vdots \\ C_{41} & \cdots & C_{4t} \end{bmatrix},$$

where each C_{ij} has 250 rows. Thus, if C_{ij} is 250×30 we can hold it in cache. This corresponds to taking $t = 1000/30 = 34$ in (3.3.13). To compute C_{ij} we use the columns of $A_{i1}, A_{i2}, \ldots, A_{ir}$ and elements of $B_{1j}, \ldots, B_{rj}$. Only one column of an A_{ij} needs to be in cache at a time as well as single elements of the B_{ij}. Thus, for each column of an A_{ij} that is brought to cache, 30 columns of a C_{ij} can be updated. It follows that only $1000/30 = 34$ loads of A are now required as compared with 167. On the other hand, 4 loads of B are required, compared with only one for (3.3.11), but there is still a considerable reduction in the overall memory traffic.

The above examples are meant to be only illustrative of the use of data in cache or vector registers. Actual algorithms will depend on characteristics of the machine that is to be used.

The BLAS

The BLAS (Basic Linear Algebra Subroutines) are a collection of subroutines for performing certain vector and matrix operations. The original BLAS, now called BLAS1, consisted of only vector operations such as the inner product of two vectors, vector plus scalar times vector (axpy), scale a vector by a constant, and so on. Provision is allowed for single and double precision real and complex arithmetic. For example, the axpy operation in the four types of arithmetic has the names saxpy, daxpy, caxpy, and zaxpy. (It is for this reason that we have used the generic term axpy rather than the more common saxpy or daxpy.) The second level BLAS, BLAS2, are matrix-vector operations such as matrix-vector multiply. BLAS3, the third level BLAS, contains matrix-matrix operations such as matrix multiplication. It is assumed that the BLAS are implemented very efficiently in assembly language and can form the building blocks for more complicated procedures.

Fast Matrix Multiply

All of the algorithms discussed so far require $O(2n^3)$ arithmetic operations to multiply two $n \times n$ matrices, but there are algorithms that require only $O(cn^p)$ operations where $p < 3$. (Here, and henceforth, we include the leading constant in the big O term since it is this constant that is sometimes of critical importance.) Most of these are rather complicated and their practical utility is not yet clear. We will discuss only the original such algorithm, known as *Strassen's method*, in which the operation count is $O(4.7n^{2.81})$. This algorithm provides a useful decrease in computation for large n. For example, for $n = 1000$, $2n^3 = 2 \times 10^9$ while $4.7n^{2.81} \doteq 1.27 \times 10^9$.

Strassen's method is based on the observation that multiplication of two 2×2 matrices can be achieved with only 7 multiplications rather than 8. Consider

$$\begin{bmatrix} C_{11} & C_{12} \\ C_{21} & C_{22} \end{bmatrix} = \begin{bmatrix} A_{11} & A_{12} \\ A_{21} & A_{22} \end{bmatrix} \begin{bmatrix} B_{11} & B_{12} \\ B_{21} & B_{22} \end{bmatrix} \quad (3.3.14)$$

and compute

$$P_1 = (A_{11} + A_{22})(B_{11} + B_{22}), \quad P_2 = (A_{21} + A_{22})B_{11} \quad (3.3.15a)$$

$$P_3 = A_{11}(B_{12} - B_{22}), \quad P_4 = A_{22}(B_{21} - B_{11}), \quad P_5 = (A_{11} + A_{12})B_{22} \quad (3.3.15b)$$

$$P_6 = (A_{21} - A_{11})(B_{11} + B_{12}), P_7 = (A_{12} - A_{22})(B_{21} + B_{22}). \quad (3.3.15c)$$

Then

$$C_{11} = P_1 + P_4 - P_5 + P_7, \quad C_{12} = P_3 + P_5 \quad (3.3.16a)$$

$$C_{21} = P_2 + P_4, \quad C_{22} = P_1 + P_3 - P_2 + P_6. \quad (3.3.16b)$$

The seven multiplications required are those in (3.3.15). The number of additions, however, is 18, as compared with 4 for the usual matrix multiplication. Thus, unless multiplication is many times slower than addition, this is not a good way to multiply 2×2 matrices.

We now apply the same idea to $n \times n$ matrices A, B and C. Assuming that n is even, let (3.3.14) be a partitioning into $\frac{n}{2} \times \frac{n}{2}$ matrices. The formulas (3.3.15) and (3.3.16) are still valid for these submatrices. Assuming that the matrix

3.3 Matrix Multiplication

multiplications in (3.3.15) are done in the conventional way, they will each require approximately $2(\frac{n}{2})^3$ operations while the additions will each require $(\frac{n}{2})^2$ operations. Hence the total is approximately

$$7 \times 2(\frac{n}{2})^3 + 18(\frac{n}{2})^2 = \frac{7}{4}n^3 + \frac{9}{2}n^2. \qquad (3.3.17)$$

This count will be less than $2n^3$ provided that $n > 18$, and the larger n is, the more the savings. Provided that n is sufficiently large, we can use the same idea for each of the matrix multiplications in (3.3.15), and continue to apply the Strassen algorithm until the size of the submatrices is less than 18.

Parallel Computers

We next discuss algorithms for parallel computers. Consider matrix-vector multiplication by the linear combination algorithm, (3.3.2), where, again, A is $m \times n$. For simplicity, assume first that $n = p$, the number of processors, and let x_i and $\mathbf{a}_i$ be assigned to processor i as illustrated in Figure 3.3.4. All of the products $x_i \mathbf{a}_i$ are formed with perfect parallelism, and then the additions are done by the fan-in algorithm of Section 3.2, applied now to vectors (see Exercise 3.2.2). For a distributed memory system we interpret Figure 3.3.4 as meaning that the indicated data are in the local memories of the processors. In this case, the fan-in algorithm will require transfer of data between processors as the algorithm proceeds. On a shared memory system, we interpret Figure 3.3.4 as meaning a task assignment; that is, P_i will do the multiplication $x_i \mathbf{a}_i$. In this case, no data are transferred since all are in the global memory.

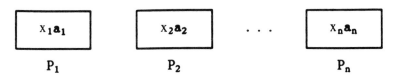

Figure 3.3.4: *Parallel Linear Combination Algorithm*

Synchronization is needed to ensure that the necessary multiplications are complete before addition begins. On a distributed memory machine, this will be accomplished by the receive of the data. For example, suppose that P_1 is to add $x_1 \mathbf{a}_1$ and $x_2 \mathbf{a}_2$. When P_2 has completed its multiplications it will send $x_2 \mathbf{a}_2$ to P_1. P_1 will begin the addition after it has completed its multiplications, and if the data have not yet been received from P_2, it will wait. It is this wait that effects the necessary synchronization. Note that without the wait, if P_1 started adding from the memory locations reserved for $x_2 \mathbf{a}_2$ before the correct data arrived, then erroneous results would be produced. In a shared memory system, synchronization can be achieved in a variety of ways (Section 3.2), and the best way will depend on both the hardware and software of the system. As an example of how the synchronization might be done, assume that when P_2 has finished its multiplication it sets a "flag" (for example, a Boolean variable can be set to True), and P_1 will then test this flag before it begins the addition.

The inner product algorithm is more attractive. Assume that $p = m$ and that **x** and $\mathbf{a}_i$, which is now the ith row of A, are assigned to processor i as illustrated in Figure 3.3.5. Each processor does one inner product and there is perfect parallelism. No fan-in nor data transfers are required, and synchronization is needed only at the end of the computation. We note the seeming paradox, relative to the vector codes previously discussed, that the linear combination algorithm requires a fan-in while the inner product algorithm does not.

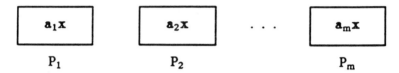

Figure 3.3.5: *Parallel Inner Product Algorithm*

Although the inner product algorithm exhibits perfect parallelism, which of the two algorithms one uses will usually depend on other considerations. A matrix-vector multiplication will inevitably be done as part of a larger calculation, and on a distributed memory system the storage of A and **x** at the time the multiplication is required will play a major role in determining the algorithm. For example, if x_i and the column $\mathbf{a}_i$ are already stored in the ith processor, as in Figure 3.3.4, the linear combination algorithm will probably be used, although its parallelism is less than the other algorithm. Another consideration is the desired location of the result at the end of the multiply; the first algorithm will leave the result vector in a single processor, while the second leaves the result distributed over the processors.

The above discussion assumes that the number of processors is equal to the number of columns or rows of A. Usually n and/or m will be considerably larger than the number of processors and each processor will be assigned many rows or columns. Ideally, for the inner product algorithm, p would divide m and m/p rows would be assigned to each processor. Mathematically, we would perform the multiplication in the partitioned form

$$A\mathbf{x} = \begin{bmatrix} A_1 \\ \vdots \\ A_p \end{bmatrix} \mathbf{x} = \begin{bmatrix} A_1\mathbf{x} \\ \vdots \\ A_p\mathbf{x} \end{bmatrix}, \quad (3.3.18)$$

where A_i contains m/p rows of A. If A_i and **x** are assigned to processor i,

3.3 Matrix Multiplication

then the products $A_1\mathbf{x}, \ldots, A_p\mathbf{x}$ are done with perfect parallelism. Note that for the multiplication $A_i\mathbf{x}$, it may not make any difference how this is carried out; it could be done by inner products or by a linear combination of columns. However, if the processors are vector processors, then our previous discussion would be relevant. Similar considerations apply to partitioning A into groups of columns (Exercise 3.3.3). If p does not divide m (or n), then we would try to distribute the rows (or columns) across the processors as evenly as possible.

Matrix Multiplication

Similar considerations apply to the multiplication of two matrices A and B, and it is left to Exercise 3.3.4 to discuss the inner product, outer product, and linear combination algorithms for parallel computers. The partitioning (3.3.13) also leads to a variety of possible algorithms. For example, if $p = st$, the number of blocks in C, then these st blocks can be computed in parallel provided A and B are assigned suitably to the processors (see Exercise 3.3.5). Special cases are $s = 1$, so that A is partitioned into groups of columns, and $t = 1$, where B is partitioned into groups of rows. Other special cases of interest are $s = t = 1$, which gives a "block inner product" type algorithm

$$AB = \sum_{j=1}^{r} A_{1j} B_{j1} \qquad (3.3.19)$$

and $r = 1$, which gives a "block outer product" algorithm

$$AB = (A_{i1} B_{1j}). \qquad (3.3.20)$$

For example, (3.3.20) would be useful if there were $p = st$ processors with A_{i1} and B_{1j} assigned to the processors as indicated in Figure 3.3.6. Here, the processor configuration mirrors the block structure of the product C and the individual products $A_{i1} B_{1j}$ are all formed in parallel.

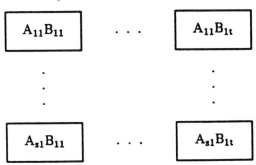

Figure 3.3.6: *Parallel Block Multiplication*

Banded Matrices

So far in this section we have assumed that the matrices were full so that all, or most, elements are nonzero. However, many matrices that arise in practice are *sparse*, with the majority of the elements being zero. We will now examine the multiplication algorithms for various types of sparse matrices and see that additional algorithms are needed in certain cases. We begin with banded matrices.

The $n \times n$ matrix A is *banded* if

$$a_{ij} = 0, \qquad i - j > \beta_1, \qquad j - i > \beta_2 \qquad (3.3.21)$$

as illustrated in Figure 3.3.7. For simplicity, we will consider only the case in which A is *symmetrically banded*: $\beta_1 = \beta_2 = \beta$; in this case, β is called the *semibandwidth*. Thus, A has nonzero elements only on the main diagonal and on the 2β adjacent diagonals above and below the main diagonal. Note that a symmetrically banded matrix is not necessarily symmetric.

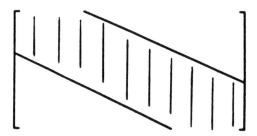

Figure 3.3.7: *Banded Matrix*

Consider first the matrix-vector multiplication $A\mathbf{x}$ by the inner product and linear combination algorithms (3.3.1) and (3.3.2). Assuming that A is stored by columns, the vector lengths in (3.3.2) range from $\beta + 1$ to $2\beta + 1$. Typical values of β and n in some problems might be $n = 10^4$ and $\beta = 10^2$, in which case the vector lengths for (3.3.2) are reasonably large. On the other hand, for small β the vector lengths are small. For example, if $\beta = 1$, the matrix is *tridiagonal* and the vector lengths are at most 3.

Similar considerations hold for the inner products in (3.3.1), for which the vector lengths are the same as in (3.3.2). However, (3.3.1) has almost perfect parallelism in the formation of the n inner products as illustrated in Figure 3.3.5 and discussed previously, independent of the bandwidth. Hence, the inner product algorithm is attractive on parallel systems, even for very small β.

Multiplication by Diagonals

As discussed above, the algorithms given so far are not efficient on vector computers for matrices with small bandwidth. This is also the case for matrices

3.3 Matrix Multiplication

that have only a few nonzero diagonals that are not clustered near the main diagonal. An example is given in Figure 3.3.8, in which an $n \times n$ matrix is shown that has only four diagonals that contain nonzero elements. We will call a matrix that has relatively few nonzero diagonals a *diagonally sparse matrix*. Such matrices arise frequently in practice, especially in the solution of partial differential equations by finite difference or finite element methods, as we will see in more detail later.

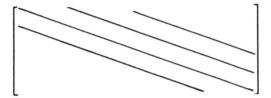

Figure 3.3.8: *Diagonally Sparse Matrix*

A diagonally sparse matrix may have a relatively large bandwidth, but the algorithms we have discussed so far are unsatisfactory if there are large numbers of zeros in each row or column. Indeed, the storage of such matrices by row or column is unsatisfactory for that reason and a natural storage scheme is by diagonal. Thus, the nonzero diagonals of the matrix become the vectors to be used in multiplication algorithms and we next show how this multiplication can be done.

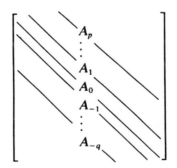

Figure 3.3.9: *Matrix by Diagonals*

Consider first the matrix-vector multiplication $A\mathbf{x}$, where the $n \times n$ matrix A is written in terms of its diagonals as shown in Figure 3.3.9. We do not assume at this point that A is diagonally sparse or symmetrically banded; indeed, A could be a full matrix. We denote in Figure 3.3.9 the main diagonal by A_0, the diagonals below A_0 by $A_{-1}, \ldots, A_{-q}$ and the diagonals above A_0 by $A_1, \ldots, A_p$. Then it is easy to see (Exercise 3.3.9) that $A\mathbf{x}$ may be represented as

$$Ax = A_0 \mathbf{x} \stackrel{\wedge}{+} A_1 \mathbf{x}^2 \stackrel{\wedge}{+} \cdots \stackrel{\wedge}{+} A_p \mathbf{x}^{p+1} + A_{-1} \mathbf{x}_{n-1} \stackrel{\vee}{+} \cdots \stackrel{\vee}{+} A_{-q} \mathbf{x}_{n-q} \quad (3.3.22)$$

where

$$\mathbf{x}^j = (x_j, \ldots, x_n), \qquad \mathbf{x}_{n-j} = (x_1, \ldots, x_{n-j}). \quad (3.3.23)$$

The multiplications in (3.3.22) are to be interpreted as elementwise vector multiplication of the vectors A_i, which are the diagonals of A, and the vectors of (3.3.23). The vectors appearing in (3.3.22) are not all the same length (for example, $A_1 \mathbf{x}^2$ is $n-1$ long) and $\stackrel{\wedge}{+}$ means add the shorter vector to the first components of the longer one; for example, $A_1 \mathbf{x}^2$ would add to the first $n-1$ components of $A_0 \mathbf{x}$. Likewise $\stackrel{\vee}{+}$ denotes adding the shorter vector to the last components of the longer one.

We next discuss various special cases of (3.3.22) on vector computers. If A is a full matrix ($p = q = n - 1$), then there are $2n - 1$ vector multiplications in (3.3.22). Moreover, the vectors range in length from 1 to n. Thus, (3.3.22) is unattractive for full matrices compared with the linear combination algorithm. Only if A were already stored by diagonal would we consider (3.3.22), and this storage scheme for a full matrix is unlikely.

At the other extreme, suppose that $p = q = 1$ so that A is tridiagonal. Then (3.3.22) reduces to

$$Ax = A_0 \mathbf{x} \stackrel{\wedge}{+} A_1 \mathbf{x}^2 \stackrel{\vee}{+} A_{-1} \mathbf{x}_{n-1}, \quad (3.3.24)$$

which consists of vector lengths of $n, n-1$ and $n-1$ so that we have almost perfect vectorization for large n. More generally, (3.3.22) is very attractive for small bandwidth systems but becomes increasingly less so vis-à-vis the linear combination algorithm as the bandwidth increases. There will be some bandwidth β_0, depending on the computer and n, for which (3.3.22) will be the fastest algorithm for $\beta < \beta_0$ and the linear combination algorithm will be fastest for $\beta > \beta_0$. (See Exercise 3.3.10.)

The algorithm (3.3.22) is also very attractive for diagonally sparse matrices. For example, suppose that A_0, A_1, A_{30}, A_{-1}, and A_{-30} are the only nonzero diagonals. Then (3.3.22) becomes

$$Ax = A_0 \mathbf{x} \stackrel{\wedge}{+} A_1 \mathbf{x}^2 \stackrel{\wedge}{+} A_{30} \mathbf{x}^{31} \stackrel{\vee}{+} A_{-1} \mathbf{x}_{n-1} \stackrel{\vee}{+} A_{-30} \mathbf{x}_{n-30}. \quad (3.3.25)$$

As with the tridiagonal case, $A_1 \mathbf{x}^2$ and $A_{-1} \mathbf{x}_{n-1}$ have vector lengths of $n-1$, while $A_{30} \mathbf{x}^{31}$ and $A_{-30} \mathbf{x}_{n-30}$ have vector lengths of $n - 30$. If n is large, these vector lengths are still quite satisfactory. On the other hand, if A_{30} and A_{-30} were replaced by A_{n-p} and $A_{-(n-p)}$ then the vector lengths for these diagonals are small if p is small, and the corresponding vector operations are inefficient.

3.3 Matrix Multiplication

However, for such a matrix the linear combination algorithm does not provide a viable alternative.

On parallel computers, the computation (3.2.22) could be carried out as follows. The nonzero diagonals of A are distributed among the processors so that the total data assigned to the processors is balanced as much as possible. The vector **x**, or at least that portion of it needed by each processor, will also have to be distributed. Then the multiplications of the diagonals and **x** can be done in a highly parallel fashion, but, at some point, additions must be done with a fan-in across the processors, with a loss of parallelism. Moreover, there is less motivation to consider multiplication by diagonals on parallel computers since, as we have already discussed, the implementation of the inner product algorithm can be quite satisfactory for small bandwidth systems. The same is true of diagonally sparse matrices. Here, again, the rows of A can be distributed across the processors and the inner products done in parallel. Only the nonzero elements in each row, together with an index list giving their position in the row, need be stored provided the processors are sequential computers. Thus, there would seem to be little use for the multiplication by diagonal algorithm on parallel systems unless the individual processors are themselves vector computers.

It is part of the folklore of parallel and vector computing that matrix multiplication is "easy." As we have seen in this section, this is true only if proper attention is paid to the structure of the matrices and if we adapt our algorithms to that structure.

Supplementary Discussion and References: 3.3

1. The different loop formulations of Figure 3.3.1 and Exercise 3.3.2 are given in Dongarra, Gustavson, and Karp [1984].

2. The terms "symmetrically banded" and "diagonally sparse" are not standard, but no other suitable terms are in common use for matrices of these kinds.

3. The multiplication by diagonals algorithm was developed by Madsen et al. [1976]. Another advantage of storing a matrix by diagonals is that its transpose is much more readily available than if the matrix is stored by rows or columns. Indeed, the transpose may be obtained simply by changing the pointers to the first elements of the diagonals and no change in storage of the matrix elements need be made.

4. Melhem [1987] considers a generalization of diagonals called *stripes*. Roughly speaking, stripes may be considered to be "bent" diagonals. He then gives matrix multiplication algorithms based on this type of storage. A similar "jagged diagonal" storage scheme is given in Anderson and Saad [1989].

5. Loop unrolling is discussed in more detail in Dongarra and Hinds [1979]. As pointed out in this paper, it is also a useful technique for serial computers since it can reduce loop overhead as well as give other benefits.

6. See Dongarra and Duff et al. [1990] for further information and references for the BLAS.

7. Strassen's algorithm (Strassen [1969]) is discussed for vector computers in Bailey et al. [1990]. See also Higham [1990].

EXERCISES 3.3

3.3.1. Write pseudocodes, corresponding to Figures 3.3.2 and 3.3.3, for the dual algorithms (3.3.7) and (3.3.10).

3.3.2. The matrix multiplication $C = AB$, where A is $m \times n$ and B is $n \times q$, can be described by any one of the six loops

for $i = 1$ to m	for $j = 1$ to q	for $i = 1$ to m
for $j = 1$ to q	for $i = 1$ to m	for $k = 1$ to n
for $k = 1$ to n	for $k = 1$ to n	for $j = 1$ to q
ijk form	jik form	ikj form

for $j = 1$ to q	for $k = 1$ to n	for $k = 1$ to n
for $k = 1$ to n	for $i = 1$ to m	for $j = 1$ to q
for $i = 1$ to m	for $j = 1$ to q	for $i = 1$ to m
jki form	kij form	kji form

The arithmetic statement in all cases is $c_{ij} = c_{ij} + a_{ik}b_{kj}$. Show that the jik, jki, and kji forms correspond to the inner product, linear combination, and outer product algorithms, respectively, while the ikj and kij forms correspond to the dual linear combination and outer product algorithms. Show also that the ijk form gives a "dual inner product" algorithm that was not discussed in the text.

3.3.3. Formulate the algorithm illustrated by Figure 3.3.4 in the case that $p \neq n$.

3.3.4. Discuss in detail the inner product, linear combination, and outer product algorithms for matrix multiplication on p processors.

3.3.5. Discuss the storage requirements to carry out the algorithm of Figure 3.3.6 on $p = st$ processors.

3.3.6. Let A be an $m \times n$ matrix and B an $n \times q$ matrix. Assume that there are $p = nmq$ processors. Give a storage distribution of A and B across the processors so that all nmq multiplications needed to form AB can be done in parallel. (Note that this will require the elements of A and/or B to be stored in more than one processor.) Then show that $O(\log n)$ fan-in addition steps are required to form the necessary sums. Hence, conclude that the matrix multiplication can be done in $1 + O(\log n)$ parallel steps.

3.3 Matrix Multiplication

3.3.7. Assume that a parallel system has p processors. Discuss the distribution of work for the inner product algorithm (3.3.1) for a matrix A of semibandwidth β so as to achieve maximum utilization of the processors. Consider, in particular, the case that p does not divide n.

3.3.8. Let A and B be $n \times n$ matrices with semibandwidths α and β, respectively. Show that AB will, in general, have semibandwidth $\alpha + \beta$.

3.3.9. Verify the formula (3.3.22).

3.3.10. Determine the semibandwidth β for which the algorithm (3.3.22) for matrix-vector multiplication ceases to be better than the linear combination algorithm. Assume that A is symmetrically banded, the storage is suitable for both algorithms, and the vector instructions obey the timing formula $T = (100 + 5N)$ns.

3.3.11. Let $A = P + iQ, B = U + iV$ be two complex matrices. Give an algorithm for computing AB. Let $L = P + Q, M = U - V$. Can you use L and M to give an algorithm that requires fewer matrix multiplies?

Chapter 4

Polynomial Approximation

4.1 Taylor Series, Interpolation and Splines

For many purposes, some of which will be discussed in subsequent chapters, we may wish to approximate a given function by a simpler function such as a polynomial. One advantage of a polynomial is that it can be easily evaluated in terms of the usual computer operations of addition (or subtraction) and multiplication. Also, a polynomial may be easily integrated or differentiated to obtain approximations to the integral or derivative of the original function; integration will be discussed further in Section 5.1. A disadvantage of approximation by a polynomial p is that $|p(x)| \to \infty$ as $x \to \infty$ so that it cannot approximate for large $|x|$ a function that is bounded as $|x| \to \infty$. (However, such functions may sometimes be approximated by a rational function, which is the ratio of two polynomials.) Also, in many cases, the function to be approximated may have properties, such as being positive or being monotone, that the approximating functions should also satisfy. This imposes constraints that may make the approximation problem much more difficult and, again, polynomials may not be suitable.

One of the basic approaches to finding an approximating polynomial is by the use of Taylor series. Recall from the calculus that if f is sufficiently differentiable, the Taylor series expansion of f about a point x_0 is

$$f(x) = f(x_0) + f'(x_0)(x - x_0) + \frac{1}{2}f''(x_0)(x - x_0)^2 + \cdots$$
$$+ \frac{1}{n!}f^{(n)}(x_0)(x - x_0)^n + \frac{1}{(n+1)!}f^{(n+1)}(z)(x - x_0)^{n+1}, \quad (4.1.1)$$

where in the last term - the *remainder* term - z is some point between x and

x_0. If we drop the remainder term in (4.1.1), we have the approximation

$$f(x) \doteq p(x) \equiv f(x_0) + f'(x_0)(x - x_0) + \cdots + \frac{1}{n!} f^{(n)}(x_0)(x - x_0)^n, \quad (4.1.2)$$

where p is a polynomial of degree n in x. Note that f and its derivatives are all evaluated at a single point x_0. Later, we shall consider approximations based on values of f at several points.

As an example of Taylor series approximation, let $f(x) = e^x$ and $x_0 = 0$. Then $f^{(i)}(x) = e^x$ for all $i \geq 0$ so that $f^{(i)}(x_0) = 1$ for all $i \geq 0$. Thus, (4.1.2) is

$$p(x) = 1 + x + \frac{x^2}{2} + \cdots + \frac{x^n}{n!}, \quad (4.1.3)$$

which is just the polynomial of degree n consisting of the first $n + 1$ terms of the series expansion of e^x. Note that the approximation p does not have all the same properties as f. For example, $e^x > 0$ for any real x but $p(x)$ may be negative for negative x.

As a second example, let $f(x) = x^3$ and $x_0 = 1$. Then $f(x_0) = 1, f'(x_0) = 3$ and $f''(x_0) = 6$ so that

$$p(x) = 1 + 3(x - 1) + 3(x - 1)^2 \quad (4.1.4)$$

is a quadratic approximation to x^3. Here p is not in standard form but can easily be put in that form by collecting coefficients of powers of x:

$$p(x) = 1 + 3x - 3 + 3x^2 - 6x + 3 = 1 - 3x + 3x^2.$$

This is simply another representation of the polynomial (4.1.4). As we shall see, there are many ways to represent the same polynomial and one representation might have computational or other advantages over another.

Approximation Error

In any approximation, there will be an error and it is important to understand the nature of this error and obtain error estimates if possible. The error in the approximation (4.1.2) arises from dropping the remainder term in (4.1.1):

$$f(x) - p(x) = \frac{1}{(n+1)!} f^{(n+1)}(z)(x - x_0)^{n+1}. \quad (4.1.5)$$

In general it is difficult to obtain a precise estimate of this error since even if $f^{(n+1)}$ is known, the point z is not. Hence, it is customary to obtain error bounds in terms of the maximum of $|f^{(n+1)}|$ over some interval of interest. For example, suppose that we wish to use the approximation in the interval $[a, b]$, where x_0 is some point in this interval. If we assume that

$$|f^{(n+1)}(z)| \leq M, \quad \text{for all } z \in [a, b], \quad (4.1.6)$$

4.1 Taylor Series, Interpolation and Splines

then from (4.1.5) we have

$$|f(x) - p(x)| \leq \frac{M}{(n+1)!}|x - x_0|^{n+1}, \qquad \text{for all } x \in [a, b]. \tag{4.1.7}$$

Clearly, the error will be small for x close to x_0 but may be large otherwise. For example, for the approximation (4.1.3) to e^x, assume that the interval of interest is $[0, 1]$ so that $a = 0$ and $b = 1$. Then $M = e$ and (4.1.7) becomes

$$|f(x) - p(x)| \leq \frac{e}{(n+1)!}|x|^{n+1} \leq \frac{e}{(n+1)!} \tag{4.1.8}$$

since $|x| \leq 1$. Thus, the error may be guaranteed to be as small as we like by choosing n sufficiently large (this just reflects the convergence of the exponential series). On the other hand, for fixed n and an unbounded interval, the error will increase rapidly as $x \to \infty$. This is illustrated in Figure 4.1.1 for $n = 2$ (a quadratic approximation). In other problems, the error (4.1.5) may oscillate between positive and negative and an estimate of the form (4.1.7) would be very pessimistic for at least some x.

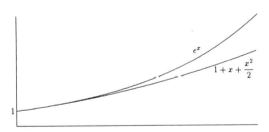

Figure 4.1.1: *A Quadratic Approximation to the Exponential*

Interpolation

A Taylor series approximation is found by utilizing the values of f and its derivatives at some point x_0. Another approach to finding a polynomial approximation uses only values of f, and not derivatives of f. Assume that we have $n + 1$ points $x_0, \ldots, x_n$ and the corresponding function values $f_i = f(x_i), i = 0, \ldots, n$. Then we wish to find a polynomial p, called the *interpolating polynomial*, so that

$$p(x_i) = f_i, \qquad i = 0, \ldots, n. \tag{4.1.9}$$

For example, if $n = 1$, we have two points and it is clear that the linear polynomial

$$p(x) = f_1 + \frac{(f_1 - f_0)}{(x_1 - x_0)}(x - x_1) \qquad (4.1.10)$$

satisfies (4.1.9).

The Vandermonde Matrix

In general, if the polynomial is

$$p(x) = a_0 + a_1 x + \ldots + a_n x^n, \qquad (4.1.11)$$

then the condition (4.1.9) is

$$a_0 + a_1 x_i + a_2 x_i^2 + \ldots + a_n x_i^n = f_i, \qquad i = 0, \ldots, n. \qquad (4.1.12)$$

Since the x_i and f_i are known, this is a linear system of equations for $a_0, \ldots, a_n$, which we may write in matrix-vector form as

$$\begin{bmatrix} 1 & x_0 & x_0^2 & \ldots & x_0^n \\ 1 & x_1 & x_1^2 & \ldots & x_1^n \\ \vdots & \vdots & \vdots & & \vdots \\ 1 & x_n & x_n^2 & \ldots & x_n^n \end{bmatrix} \begin{bmatrix} a_0 \\ a_1 \\ \vdots \\ a_n \end{bmatrix} = \begin{bmatrix} f_0 \\ f_1 \\ \vdots \\ f_n \end{bmatrix}. \qquad (4.1.13)$$

The coefficient matrix of (4.1.13) is called the *Vandermonde matrix*. Provided that it is nonsingular, we can obtain the interpolating polynomial by solving (4.1.13). In practice, there are better ways to obtain this polynomial and we next examine two of these.

Lagrange Polynomials

If $x_0, x_1, \ldots, x_n$ are distinct points, we can define the *Lagrange polynomials*

$$\begin{aligned} l_j(x) &= \frac{(x - x_0)(x - x_1) \ldots (x - x_{j-1})(x - x_{j+1}) \ldots (x - x_n)}{(x_j - x_0)(x_j - x_1) \ldots (x_j - x_{j-1})(x_j - x_{j+1}) \ldots (x_j - x_n)} \\ &= \prod_{\substack{k=0 \\ k \neq j}}^{n} \left(\frac{x - x_k}{x_j - x_k} \right), \qquad j = 0, 1, \ldots, n. \end{aligned} \qquad (4.1.14)$$

It is easy to verify that these polynomials, which are all of degree n, satisfy

$$l_j(x_i) = \begin{cases} 1 & \text{if } i = j \\ 0 & \text{if } i \neq j. \end{cases} \qquad (4.1.15)$$

4.1 Taylor Series, Interpolation and Splines

Therefore, $l_j(x_i)f_j$ has the value 0 at all nodes x_i, $i = 0, 1, \ldots, n$, except for x_j, where $l_j(x_j)f_j = f_j$. Thus, by defining

$$p(x) = \sum_{j=0}^{n} l_j(x)f_j, \qquad (4.1.16)$$

we have a polynomial of degree n or less that satisfies (4.1.9).

As an example of the use of Lagrange polynomials, we will determine the polynomial $p(x)$ of degree 2 or less that satisfies $p(-1) = 4$, $p(0) = 1$, and $p(1) = 0$. The interpolating polynomial (4.1.16) is then

$$\begin{aligned}
p(x) &= \frac{(x-0)(x-1)}{(-1-0)(-1-1)}4 + \frac{(x-(-1))(x-1)}{(0-(-1))(0-1)}1 + \frac{(x-(-1))(x-0)}{(1-(-1))(1-0)}0 \\
&= 2x^2 - 2x + 1 - x^2 + 0 = x^2 - 2x + 1. \qquad (4.1.17)
\end{aligned}$$

The Newton Form

The Lagrange polynomials are not convenient if a node is added or dropped from the data. For example, if (x_{n+1}, f_{n+1}) were added to the set of data (x_i, f_i), $i = 0, 1, \ldots, n$, and we wished to compute the polynomial of degree $n + 1$ that interpolates this data, then the Lagrange polynomials would all have to be recomputed. There is another representation of the interpolating polynomial that is very useful in this context; this is the *Newton form*, which we now describe.

We assume now that the points x_i are equally spaced with spacing h. We define forward differences of the f_i by means of $\Delta f_i = f_{i+1} - f_i$, and higher differences by repeated application of this:

$$\Delta^2 f_0 = \Delta f_1 - \Delta f_0 = f_2 - 2f_1 + f_0$$
$$\Delta^3 f_0 = \Delta^2 f_1 - \Delta^2 f_0 = f_3 - 3f_2 + 3f_1 - f_0$$
$$\vdots \qquad (4.1.18)$$
$$\Delta^n f_0 = f_n - \binom{n}{1}f_{n-1} + \binom{n}{2}f_{n-2} - \ldots + (-1)^n f_0,$$

where the binomial coefficients are given by

$$\binom{n}{i} = \frac{n(n-1)\ldots(n-i+1)}{i!}.$$

In terms of the differences (4.1.18), we define a polynomial of degree n by

$$\begin{aligned}
p_n(x) &= f_0 + \frac{(x-x_0)}{h}\Delta f_0 + \frac{(x-x_0)(x-x_1)}{2h^2}\Delta^2 f_0 \qquad (4.1.19) \\
&\quad + \ldots + \frac{(x-x_0)(x-x_1)\ldots(x-x_{n-1})}{n!\, h^n}\Delta^n f_0.
\end{aligned}$$

Note that (4.1.10) is the linear polynomial p_1.
To show that (4.1.19) satisfies (4.1.9) we have first that $p_n(x_0) = f_0$ since all remaining terms in (4.1.19) vanish. Similarly, since $x_1 - x_0 = h$,

$$p_n(x_1) = f_0 + \frac{(x_1 - x_0)}{h}(f_1 - f_0) = f_1,$$

and

$$\begin{aligned} p_n(x_2) &= f_0 + \frac{(x_2 - x_0)}{h}(f_1 - f_0) + \frac{(x_2 - x_0)(x_2 - x_1)}{2h^2}(f_2 - 2f_1 + f_0) \\ &= f_0 + 2(f_1 - f_0) + (f_2 - 2f_1 + f_0) = f_2. \end{aligned}$$

It is easy to verify in an analogous way that $p_n(x_i) = f_i$, $i = 3, \ldots, n$, although the computations become increasingly tedious.

The polynomial p_n of (4.1.19) is analogous to the first $n+1$ terms of a Taylor expansion about x_0. Now, suppose that we add (x_{n+1}, f_{n+1}) to the data set. Then, the polynomial p_{n+1} that satisfies $p_{n+1}(x_i) = f_i$, $i = 0, 1, \ldots, n+1$, is given by

$$p_{n+1}(x) = p_n(x) + \frac{(x - x_0)(x - x_1) \ldots (x - x_n)}{(n+1)! \, h^{n+1}} \Delta^{n+1} f_0,$$

and it is this feature of the Newton form of the interpolating polynomial that is sometimes useful in practice. This is similar to taking one more term in a Taylor expansion.

Uniqueness and Representations

We have given three different ways of computing the interpolating polynomial, but it is important to realize that in all cases we obtain the same polynomial. This is a consequence of the following basic theorem.

> THEOREM 4.1.1 *If $x_0, \ldots, x_n$ are distinct points, then for any $f_0, \ldots, f_n$ there exists a unique polynomial p of degree n or less such that (4.1.9) holds.*

Proof: We have already shown the existence of the interpolating polynomial by means of the Lagrange polynomials.

To prove uniqueness, suppose, on the contrary, that there is another interpolating polynomial of degree n or less, say $q(x)$. By defining

$$r(x) = p(x) - q(x)$$

we obtain a polynomial, r, of degree n or less that is equal to zero at the $n+1$ distinct points $x_0, x_1, \ldots, x_n$. By the fundamental theorem of algebra, such a

4.1 Taylor Series, Interpolation and Splines

polynomial must be identically equal to zero, and it follows that $p(x) = q(x)$. Thus, uniqueness is proved.

The conclusion of Theorem 4.1.1 is intuitively clear for small degree polynomials. For example, if there are three points x_0, x_1, x_2, no polynomial of degree 1 can satisfy (4.1.9) unless the f_i are such that the points (x_i, f_i) are on a straight line. On the other hand, for any f_i there is a unique quadratic that fits the data and infinitely many polynomials of degree 3.

We note that Theorem 4.1.1 gives an indirect proof that the Vandermonde matrix of (4.1.13) is nonsingular provided that the x_i are distinct. For, by Theorem 2.1.2, if it were singular, either (4.1.13) would have no solution, and thus there would be no interpolating polynomial, or it would have infinitely many solutions, in which case there would be infinitely many interpolating polynomials.

Another consequence of Theorem 4.1.1 is that the polynomials obtained by Lagrange polynomials, the Vandermonde matrix equation or the Newton form (in the case of equally spaced points) are all the same; only the representations of the polynomials are different. In fact, all of these representations are of the form

$$p(x) = \alpha_0 \phi_0(x) + \cdots + \alpha_n \phi_n(x) \qquad (4.1.20)$$

for different *basis functions* ϕ_i. For the polynomial in standard form (4.1.11), $\phi_i(x) = x^i, i = 0, \ldots, n$. For (4.1.16), ϕ_i is the Lagrange polynomial l_i, and for the Newton form the basis functions are $\phi_0(x) = 1$ and

$$\phi_{i+1}(x) = (x - x_0) \ldots (x - x_i), \qquad i = 0, \ldots, n-1.$$

Other basis polynomials could be used in (4.1.20). Indeed, other basis functions different than polynomials can be used. For example, the ϕ_i could be exponential functions like $\phi_i(x) = e^{\beta_i x}$ for given β_i, or trigonometric functions like $\phi_i(x) = \sin \beta_i \pi x$. Such functions might mirror the behavior of f better than polynomials. In any case, the condition (4.1.9) leads to the system of equations

$$\alpha_0 \phi_0(x_i) + \cdots \alpha_n \phi_n(x_i) = f_i, \qquad i = 0, \ldots, n,$$

analogous to (4.1.12). In matrix-vector form, this system is

$$\begin{bmatrix} \phi_0(x_0) & \phi_1(x_0) & \cdots & \phi_n(x_0) \\ \phi_0(x_1) & \phi_1(x_1) & \cdots & \phi_n(x_1) \\ \vdots & & & \\ \phi_0(x_n) & \phi_1(x_n) & \cdots & \phi_n(x_n) \end{bmatrix} \begin{bmatrix} \alpha_0 \\ \alpha_1 \\ \vdots \\ \alpha_n \end{bmatrix} = \begin{bmatrix} f_0 \\ f_1 \\ \vdots \\ f_n \end{bmatrix}, \qquad (4.1.21)$$

and a necessary and sufficient condition for a unique interpolating function of the form (4.1.20) to exist is that the coefficient matrix of (4.1.21) be nonsingular. A necessary condition for this is, of course, that the points x_i be distinct, or else two or more rows of the coefficient matrix are the same.

Error in Interpolation

We next consider the error in polynomial interpolation. We earlier saw that the error in the approximation of a function by a nth degree polynomial obtained from the Taylor series was given in terms of the $(n+1)$st derivative of f. A similar result holds for interpolation. We state the following basic error theorem without proof.

THEOREM 4.1.2 (Polynomial Interpolation Error) *Let $f(x)$ be a function with $n+1$ continuous derivatives on an interval containing the distinct nodes $x_0 < x_1 < \ldots < x_n$. If $p(x)$ is the unique polynomial of degree n or less satisfying (4.1.9), then for any x in the interval $[x_0, x_n]$,*

$$f(x) - p(x) = \frac{(x - x_0)(x - x_1) \cdots (x - x_n)}{(n+1)!} f^{n+1}(z) \qquad (4.1.22)$$

for some z, depending on x, in the interval (x_0, x_n).

As in the remainder term of a Taylor series, $f^{(n+1)}$ is evaluated at some unknown point z. Thus, we might attempt to obtain a bound on $f^{(n+1)}$ over the whole interval. However, if n is at all large (even 4 or 5), it will probably be difficult, if not impossible, to compute the $(n+1)$th derivative of f. Even if n is only 1 (linear interpolation) so that only the second derivative of f is needed, this also may be impossible if f is an unknown function for which only its values at some discrete points are known; at best, we might be able to estimate some bound for the second derivative on the basis of our assumed knowledge of f. In any case, it will almost never be the case that (4.1.22) can be used to give a very precise bound on the error. It can, however, be useful in providing various insights into the errors that are produced.

Theorem 4.1.2 gives an error estimate under the assumption that the interpolating polynomial p is known exactly. In most cases, however, we will have rounding (and perhaps other) errors in determining p. For example, suppose that p is obtained by solving the Vandermonde system (4.1.13). The numerical solution of this system will cause the coefficients a_i to be in error because of rounding; this will be discussed in more detail in Chapter 6. Similarly, rounding errors will prevent the exact determination of p in the Lagrange and Newton approaches.

Piecewise Polynomials

The error (4.1.22) shows that the difference between the given function f and the approximating polynomial p may be large if $f^{n+1}(z)$ is large. Indeed, it is the case that we cannot always approximate f arbitrarily closely by taking higher and higher degree interpolating polynomials (see the Supplementary Discussion). An alternative strategy is to approximate f by using functions

4.1 Taylor Series, Interpolation and Splines

made up of low degree polynomials. If we wish to approximate f over an interval $[a, b]$, we first partition the interval into m subintervals $[x_i, x_{i+1}]$, where $a = x_0 < x_1 < \cdots < x_{m-1} < x_m = b$. We then define a function g such that the g is a polynomial on each of the intervals $[x_1, x_{i+1}]$. Such a function g is a *piecewise polynomial*.

The simplest continuous piecewise polynomial is piecewise linear as illustrated in Figure 4.1.2. (Piecewise constant functions, also called step functions, are not continuous.) In this case, on each interval f is approximated by a linear interpolating function, which is a first degree polynomial.

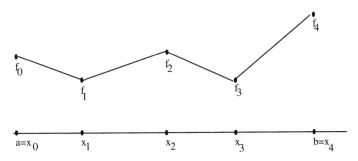

Figure 4.1.2: *A Piecewise Linear Function*

The next simplest piecewise polynomial is made up of quadratic functions. As an example, suppose that values of f at selected points in the interval $[0, 1]$ are given by

x	0	1/6	1/3	1/2	2/3	5/6	1
f	1	3	2	1	0	2	1

Then, the function g defined by

$$\begin{aligned} g(x) &= -54x^2 + 21x + 1, & 0 \leq x \leq \tfrac{1}{3}, \\ &= -6x + 4, & \tfrac{1}{3} \leq x \leq \tfrac{2}{3}, \\ &= -54x^2 + 93x - 38, & \tfrac{2}{3} \leq x \leq 1, \end{aligned} \quad (4.1.23)$$

is the piecewise quadratic function on $[0, 1]$ that equals f at the given nodes, is continuous on the whole interval, and is a quadratic on each of the subintervals $[0, \tfrac{1}{3}], [\tfrac{1}{3}, \tfrac{2}{3}], [\tfrac{2}{3}, 1]$. This function is shown in Figure 4.1.3.

Consider now the error in approximating the function f by the function g of (4.1.23). Suppose that M is a bound for the third derivative of f on the entire interval $[0, 1]$. Then, on each of the intervals $[0, \tfrac{1}{3}], [\tfrac{1}{3}, \tfrac{2}{3}]$, and $[\tfrac{2}{3}, 1]$, the error bound (4.1.22) can be applied; here, $h = \tfrac{1}{6}$, and $n = 2$. Therefore

$$|f(x) - g(x)| \leq \frac{h^3 M}{3} = \frac{M}{3 \cdot 6^3}, \quad 0 \leq x \leq 1. \quad (4.1.24)$$

100   Chapter 4 Polynomial Approximation

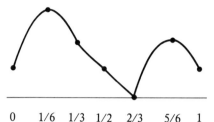

0   1/6   1/3 1/2 2/3   5/6   1
Figure 4.1.3: *A Piecewise Quadratic Function*

Without further information on M, this estimate does not furnish a quantitative bound. It does, however, show how the spacing h enters the error estimate. In particular, approximation by piecewise quadratics has an error estimate that is $O(h^3)$, where we use the standard notation that $O(h^p) \doteq \text{constant} \times h^p$ as $h \to 0$. Thus, if we use another piecewise quadratic $\bar{g}$ consisting of six quadratics and with $h = \frac{1}{12}$, the error estimate in place of (4.1.24) will be

$$|f(x) - \bar{g}(x)| \leq \frac{M}{8 \cdot 3 \cdot 6^3}, \qquad 0 \leq x \leq 1, \qquad (4.1.25)$$

so that the error bound is now one-eighth that of (4.1.24). It is easy to see (Exercise 4.1.8) that a piecewise linear function gives an error estimate that is $O(h^2)$, while a piecewise cubic function gives $O(h^4)$.

Splines

The piecewise quadratic function defined by (4.1.23) and shown in Figure 4.1.3 has the deficiency that it is not necessarily differentiable at the points where the different quadratics join. (The same lack of differentiabilty is true for the piecewise linear function of Figure 4.1.2.) If we wish to have a piecewise quadratic function that is everywhere differentiable, then this property needs to be built into the definition.

To illustrate the approach, let $n = 4$, $I_i = [x_i, x_{i+1}]$, $i = 1, 2, 3$, and

$$q_i(x) = a_{i2}x^2 + a_{i1}x + a_{i0}, \qquad i = 1, 2, 3. \qquad (4.1.26)$$

We will define a piecewise quadratic function q such that $q(x) = q_i(x)$ if $x \in I_i$, $i = 1, 2, 3$, as illustrated in Figure 4.1.4. In order that q be continuous and take on the prescribed values y_i at the nodes requires that

$$\begin{aligned} q_1(x_1) &= f_1, & q_1(x_2) &= f_2, & q_2(x_2) &= f_2, \\ q_2(x_3) &= f_3, & q_3(x_3) &= f_3, & q_3(x_4) &= f_4. \end{aligned} \qquad (4.1.27)$$

If we also wish that q be differentiable at the nodes, then q'_1 must equal q'_2 at x_2, and q'_2 must equal q'_3 at x_3:

$$q'_1(x_2) = q'_2(x_2), \qquad q'_2(x_3) = q'_3(x_3). \qquad (4.1.28)$$

4.1 Taylor Series, Interpolation and Splines

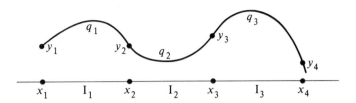

Figure 4.1.4: *A Differentiable Piecewise Quadratic Function*

The function q is determined by the nine coefficients in (4.1.26) that define q_1, q_2, and q_3. The relations (4.1.27) and (4.1.28) give only eight conditions that these nine coefficients must satisfy, and hence another condition must be specified to determine q uniquely. Usually, a value of q' at some node is specified, for example,

$$q_1'(x_1) = d_1, \qquad (4.1.29)$$

where d_1 is an approximation to $f'(x_1)$. The nine relations (4.1.27), (4.1.28), and (4.1.29) are then a system of nine linear equations for the coefficients of the q_i. This approach is easily extended to an arbitrary number, n, of nodes, in which case there will be $n-1$ intervals I_i and $n-1$ quadratics q_i defined on these intervals. The conditions (4.1.27) and (4.1.28) become

$$q_i(x_i) = f_i \qquad q_i(x_{i+1}) = f_{i+1}, \qquad i = 1, \ldots, n-1, \qquad (4.1.30)$$

and

$$q_i'(x_{i+1}) = q_{i+1}'(x_{i+1}), \qquad i = 1, \ldots, n-2. \qquad (4.1.31)$$

These relations give $3n-4$ linear equations for the $3n-3$ unknown coefficients of the polynomials $q_1, \ldots, q_{n-1}$. Again, one additional condition is needed, and we can use, for example, (4.1.29). Thus, to determine the piecewise quadratic, we need to solve a system of $3n-3$ linear equations. These equations will have a very special form with only a few non-zero elements in each row of the coefficient matrix (see Exercise 4.1.10).

Cubic Splines

The piecewise quadratic function discussed above will fail in general to have a second derivative at the points where the subintervals join. If we wish a piecewise polynomial function that is twice continuously differentiable on the whole interval, it is necessary to use third degree polynomials. We thus seek a function that is a cubic polynomial on each subinterval $[x_i, x_{i+1}]$ and is twice continuously differentiable on the whole interval. Such a function is called a *cubic spline*, and is particularly important.

We could proceed, as in the case of quadratic splines, to develop a system of equations for the coefficients of each cubic polynomial. Instead we will illustrate a different approach in which the desired function is obtained from a set of relatively simple cubic splines. These are called *cubic basis splines*, or cubic *B-splines* for short, since other cubic splines may be formed as a linear combination of them. Let $x_1, \ldots, x_n$ be now equally spaced points with spacing h. The cubic B-spline B_i centered at x_i is defined in Figure 4.1.5 and illustrated in Figure 4.1.6. It is straightforward (Exercise 4.1.9) to verify that this function is a cubic spline with the function values

$$B_i(x_i) = 1, \qquad B_i(x_{i\pm 1}) = \frac{1}{4}, \qquad (4.1.32)$$

and zero at the other nodes.

$$\frac{1}{4h^3}(x - x_{i-2})^3, \qquad x_{i-2} \leq x \leq x_{i-1},$$

$$\frac{1}{4} + \frac{3}{4h}(x - x_{i-1}) + \frac{3}{4h^2}(x - x_{i-1})^2 - \frac{3}{4h^3}(x - x_{i-1})^3, \qquad x_{i-1} \leq x \leq x_i,$$

$$\frac{1}{4} + \frac{3}{4h}(x_{i+1} - x) + \frac{3}{4h^2}(x_{i+1} - x)^2 - \frac{3}{4h^3}(x_{i+1} - x)^3, \qquad x_i \leq x \leq x_{i+1},$$

$$\frac{1}{4h^3}(x_{i+2} - x)^3, \qquad x_{i+1} \leq x \leq x_{i+2},$$

$$0, \quad \text{otherwise.}$$

Figure 4.1.5: *Definition of Cubic B-spline*

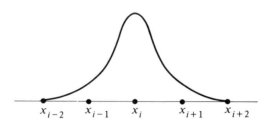

Figure 4.1.6: *A Cubic B-spline*

4.1 Taylor Series, Interpolation and Splines

We now show how one can use cubic B-splines to define another spline, c, that satisfies

$$c(x_i) = f_i, \qquad i = 1, \ldots, n, \qquad (4.1.33)$$

where $f_1, \ldots, f_n$ are given. Define

$$c(x) = \sum_{i=1}^{n} \alpha_i B_i(x). \qquad (4.1.34)$$

Then, since $B_i(x_j) = 0$ whenever $|i - j| \geq 2$, the conditions (4.1.33) become

$$\alpha_1 B_1(x_1) + \alpha_2 B_2(x_1) = f_1$$
$$\alpha_1 B_1(x_2) + \alpha_2 B_2(x_2) + \alpha_3 B_3(x_2) = f_2$$
$$\vdots \qquad \vdots \qquad (4.1.35)$$
$$\alpha_{n-2} B_{n-2}(x_{n-1}) + \alpha_{n-1} B_{n-1}(x_{n-1}) + \alpha_n B_n(x_{n-1}) = f_{n-1}$$
$$\alpha_{n-1} B_{n-1}(x_n) + \alpha_n B_n(x_n) = f_n.$$

This is a linear system of equations to determine the α_i. Using (4.1.32), we may write it in matrix-vector form as

$$\frac{1}{4} \begin{bmatrix} 4 & 1 & & & \\ 1 & 4 & \ddots & & \\ & \ddots & \ddots & 1 & \\ & & 1 & 4 \end{bmatrix} \begin{bmatrix} \alpha_1 \\ \alpha_2 \\ \vdots \\ \alpha_n \end{bmatrix} = \begin{bmatrix} f_1 \\ f_2 \\ \vdots \\ f_n \end{bmatrix}. \qquad (4.1.36)$$

The tridiagonal coefficient matrix of (4.1.36) is strictly diagonally dominant; thus, by Theorem 2.3.3 it is non-singular, so that (4.1.36) has a unique solution. (We will discuss in Chapter 6 methods to solve tridiagonal systems numerically). With the α_i determined as the solution of (4.1.36), the function c of (4.1.34) satisfies (4.1.33). Clearly, c is a cubic polynomial on each interval $[x_i, x_{i+1}]$ since it is a linear combination of cubics on each interval.

Supplementary Discussion and References: 4.1

1. The theory and practice of Taylor series approximation and interpolation is covered in more detail in most elementary books on numerical analysis; see, for example, Young and Gregory [1990].

2. For further reading on spline functions, see Prenter [1975] and de Boor [1978]. Note that the system of equations (4.1.36) was derived under the assumption that the x_i are equally spaced. In the general case, one is

also led to the solution of tridiagonal systems of equations but with the unknowns the second derivatives of the spline function at the nodes. For details of this approach, see Golub and Ortega [1991].

3. As mentioned in the text, it is not always the case that higher degree interpolating polynomials give better approximations. Almost 100 years ago, C. Runge gave the following example. Let $f(x) = (1+x^2)^{-1}$ and with $h = 10/n$, let $-5, -5+h, -5+2h, \ldots, -5+10$ be equally spaced points in the interval $[-5, 5]$. If p_n is the polynomial that interpolates f at these points, then p_n does not converge to f as $n \to \infty$, even though f is infinitely differentiable.

EXERCISES 4.1

4.1.1. Let $p(x)$ be the approximation (4.1.2). Show that $p^{(i)}(x_0) = f^{(i)}(x_0), i = 0, \ldots, n$.

4.1.2. Compute the polynomial p of degree 2 that satisfies $p(0) = 0$, $p(1) = 1$, $p(2) = 0$ by all three methods: Lagrange polynomials, the Vandermonde matrix, and the Newton representation. Conclude that the polynomial is the same in all three cases.

4.1.3. Let $f(x) = \sin \pi x/2$, and let p be the polynomial of Exercise 4.1.2 that agrees with f at the points $x = 0, 1, 2$. Use (4.1.22) to compute a bound for $|f(x) - p(x)|$ on the interval $[0, 2]$. Compare this bound with the actual error at selected points in the interval, and in particular at $x = \frac{1}{4}$ and $\frac{3}{4}$.

4.1.4. Assume that f is a given function for which the following values are known: $f(1) = 2$, $f(2) = 3$, $f(3) = 5$, $f(4) = 3$. For these data find the interpolating polynomial of degree 3 and write it in the form $a_0 + a_1 x + a_2 x^2 + a_3 x^3$.

4.1.5. Find the piecewise linear and piecewise quadratic functions that agree with the following data:

x	0	1/6	1/3	1/2	2/3	5/6	1
f	1	4	1	-1	2	4	0

Compute error bounds for these functions on the interval $[0, 1]$, assuming that the function f satisfies $|f''(z)| \leq 4$, $|f'''(z)| \leq 10, 0 \leq z \leq 1$.

4.1.6. For the function f of Exercise 4.1.4 find the quadratic spline function that satisfies the condition $q'(1) = 0$. (*Hint:* Start from the left.)

4.1.7. Find a polynomial of degree 3 that agrees with $\sqrt{x}$ at $0, 1, 3, 4$. Compare the approximation $p(2)$ with $\sqrt{2} = 1.414216$.

4.1 Taylor Series, Interpolation and Splines

4.1.8. Let f be sufficiently differentiable. Show that the error in the linear piecewise function is $O(h^2)$ and the error in the cubic piecewise function is $O(h^4)$.

4.1.9. Verify that the function B_i defined in Figure 4.1.5 is a cubic spline and satisfies (4.1.32).

4.1.10. Write out in detail the system of linear equations that determine the coefficients a_{ij} of the quadratics (4.1.26) from the conditions (4.1.29) to (4.1.31).

4.1.11. Let $f(x) = \sin x$ and let p and q be two polynomials of degree 3 that satisfy $p(k/3) = q(k/3) = f(k/3)$, $k = 0, 1, 2, 3$. Compute a bound for $|p(x) - q(x)|$ that holds on the whole interval $[0, 1]$.

4.1.12. For given $y_0, \ldots, y_n$ and distinct $x_0, \ldots, x_n$, let p be the polynomial of degree n that satisfies $p(x_i) = y_i$, $i = 0, \ldots, n$. Suppose that we wish to write p in the form $p(x) = c_0 q_0(x) + \cdots + c_n q_n(x)$, where $q_0(x) \equiv 1$ and $q_i(x) = (x - x_0) \cdots (x - x_{i-1})$. Give an algorithm for finding $c_0, \ldots, c_n$.

4.1.13. Find a polynomial p of degree 3 and a number a with $0 < a < 1$ so that $|\frac{d\sqrt{x}}{dx} - p'(x)| \leq 10^{-3}$ on the interval $(a, 1)$.

4.1.14. Let p be a polynomial that satisfies $p(-2) = -5$, $p(-1) = 1$, $p(0) = 1$, $p(1) = 1$, $p(2) = 7$, $p(3) = 25$. What can you say about the degree of p?

4.1.15. Suppose that we are approximating a function on $[0, 1]$ by using only two points x_0, x_1. Show how to choose x_0 and x_1 so that the term $(x - x_0)(x - x_1)$ in the error estimate (4.1.22) is minimized.

4.1.16. Approximate the function $f(x) = (1 + 25x^2)^{-1}$ on the interval $[-1, 1]$ by:

a. Write a program for computing the interpolating polynomial p_n for the points $x_{in} = -1 + (2i)/n$, $i = 0, 1, \ldots, n$, for $n = 5, 10, 25$.

b. Do the same but use the points $x_{in} = \cos \frac{2i\pi}{n}$.

Plot the error function $f(x) - p_n(x)$ in each case and comment on the difference in the two approaches.

4.1.17. Let V_n be the Vandermonde matrix of (4.1.13) for the points $x_0, \ldots, x_n$. Show that

$$\det V_1 = x_1 - x_0, \qquad \det V_2 = (x_1 - x_0)(x_2 - x_1)(x_2 - x_0).$$

Can you prove the general result that

$$\det V_n = \ell_0 \ell_1 \ldots \ell_{n-1}, \qquad \ell_i = (x_{i+1} - x_i)(x_{i+1} - x_{i-1}) \ldots (x_{i+1} - x_0)?$$

4.1.18. (Hermite Interpolation) Let $x_0, \ldots, x_n$ be distinct points and $\ell_0, \ldots, \ell_n$ the Lagrange polynomials (4.1.14). For $i = 0, \ldots, n$ define

$$h_i(x) = (x - x_i)\ell_i^2(x), \qquad H_i(x) = [1 - 2\ell_i'(x_i)(x - x_i)]\ell_i^2(x).$$

For a given differentiable function f, show that

$$p(x) = \sum_{i=0}^{n}[H_i(x)f(x_i) + h_i(x)f'(x_i)]$$

is a polynomial of degree $2n+1$ that satisfies

$$p(x_i) = f(x_i), \qquad p'(x_i) = f'(x_i), \qquad i = 0, \ldots, n.$$

4.2 Least Squares Approximation

In the previous section we saw that if $x_0, x_1, \ldots, x_n$ are $n+1$ distinct points and f is a given function, then there is a unique polynomial p of degree n such that

$$p(x_i) = f(x_i), \qquad i = 0, \ldots, n.$$

Now, suppose that f itself is a polynomial of degree n, and that we wish to determine its coefficients. Then, by the preceding interpolation result, it suffices to know f at $n+1$ distinct points provided that the determination of the values $f(x_i)$ can be made exactly. In many situations, however, the values of f can be found only by measurements and may be in error. In this case, it is common to take many more than $n+1$ measurements in the hope that these measurement errors will "average out." The way in which these errors average out is determined by the method used to combine the measurements to obtain the coefficients of f. For both computational and statistical reasons, the method of choice is often that of *least squares*, and this method also enjoys an elegant mathematical simplicity.

We now assume that we have m points $x_1, \ldots, x_m$, where $m \geq n+1$ and at least $n+1$ of the points are distinct. Let $f_1, \ldots, f_m$ be approximate values of a function f, not necessarily a polynomial, at the points $x_1, \ldots, x_m$. Then, we wish to find a polynomial $p(x) = a_0 + a_1 x + \cdots + a_n x^n$ such that

$$\sum_{i=1}^{m} w_i[f_i - p(x_i)]^2 \qquad (4.2.1)$$

is a minimum over all polynomials of degree n. That is, we wish to find $a_0, a_1, \ldots, a_n$ so that the weighted sum of the squares of the "errors" $f_i - p(x_i)$ is minimized. In (4.2.1), the w_i are given positive constants, called *weights*, that may be used to assign greater or lesser emphasis to the terms of (4.2.1). For example, if the f_i are measurements and we have great confidence in, say, $f_1, \ldots, f_{10}$, but rather little confidence in the rest, we might set $w_1 = w_2 = \cdots = w_{10} = 5$ and $w_{11} = \cdots = w_m = 1$. On the other hand, for some problems arising in signal processing it is common to take $w_j = \alpha^j$ for some α between 0 and 1.

4.2 Least Squares Approximation

The simplest case of a least squares problem is when $n = 0$, so that p is just a constant. Suppose, for example, that we have m measurements $l_1, \ldots, l_m$ of the length of some object. Here, the points $x_1, \ldots, x_m$ are all identical and do not enter explicitly. By the least-squares principle, we then wish to minimize

$$g(l) = \sum_{i=1}^{m} w_i(l_i - l)^2.$$

From the calculus, we know that g takes on a (relative) minimum at a point $\hat{l}$ that satisfies $g'(\hat{l}) = 0$ and $g''(\hat{l}) \geq 0$. Since

$$g'(l) = -2\sum_{i=1}^{m} w_i(l_i - l), \qquad g''(l) = 2\sum_{i=1}^{m} w_i,$$

it follows that

$$\hat{l} = \frac{1}{s}\sum_{i=1}^{m} w_i l_i, \qquad s = \sum_{i=1}^{m} w_i,$$

and because this is the only solution of $g'(l) = 0$, it must be the unique point that minimizes g. Thus, if the weights w_i are all 1, so that $s = m$, the least-squares approximation to l is just the average of the measurements $l_1, \ldots, l_m$.

The next-simplest situation is if we use a linear polynomial $p(x) = a_0 + a_1 x$. Problems of this type arise very frequently under the assumption that the data obey some linear relationship. In this case, the function (4.2.1) is

$$g(a_0, a_1) = \sum_{i=1}^{m} w_i(f_i - a_0 - a_1 x_i)^2, \qquad (4.2.2)$$

which we wish to minimize over the coefficients a_0 and a_1. Again from the calculus, we know that a necessary condition for g to be minimized is that the partial derivatives of g at the minimizer must vanish:

$$\frac{\partial g}{\partial a_0} = -2\sum_{i=1}^{m} w_i(f_i - a_0 - a_1 x_i) = 0,$$

$$\frac{\partial g}{\partial a_1} = -2\sum_{i=1}^{m} w_i x_i(f_i - a_0 - a_1 x_i) = 0.$$

Collecting coefficients of a_0 and a_1 gives the system of two linear equations

$$\left(\sum_{i=1}^{m} w_i\right) a_0 + \left(\sum_{i=1}^{m} w_i x_i\right) a_1 = \sum_{i=1}^{m} w_i f_i \qquad (4.2.3)$$

$$\left(\sum_{i=1}^{m} w_i x_i\right) a_0 + \left(\sum_{i=1}^{m} w_i x_i^2\right) a_1 = \sum_{i=1}^{m} w_i x_i f_i$$

for the unknowns a_0 and a_1.

The Normal Equations

For polynomials of degree n, the function (4.2.1) that we wish to minimize is

$$g(a_0, a_1, \ldots, a_n) = \sum_{i=1}^{m} w_i (a_0 + a_1 x_i + \cdots + a_n x_i^n - f_i)^2. \quad (4.2.4)$$

Proceeding as in the $n = 2$ case, a necessary condition for g to be minimized is that

$$\frac{\partial g}{\partial a_j}(a_0, a_1, \ldots, a_n) = 0, \quad j = 0, 1, \ldots, n.$$

Writing these partial derivatives out explicitly gives the conditions

$$\sum_{i=1}^{m} w_i x_i^j (a_0 + a_1 x_i + \cdots + a_n x_i^n - f_i) = 0, \quad j = 0, 1, \ldots, n,$$

which is a system of $n+1$ linear equations in the $n+1$ unknowns $a_0, a_1, \ldots, a_n$; these equations are known as the *normal equations*. Collecting the coefficients of the a_j and rewriting the equations in matrix-vector form gives the system

$$\begin{bmatrix} s_0 & s_1 & s_2 & \cdots & s_n \\ s_1 & s_2 & & & \\ s_2 & & \ddots & & \vdots \\ \vdots & & & & \\ s_n & & \cdots & & s_{2n} \end{bmatrix} \begin{bmatrix} a_0 \\ a_1 \\ \vdots \\ a_n \end{bmatrix} = \begin{bmatrix} c_0 \\ c_1 \\ \vdots \\ c_n \end{bmatrix}, \quad (4.2.5)$$

where

$$s_j = \sum_{i=1}^{m} w_i x_i^j, \qquad c_j = \sum_{i=1}^{m} w_i x_i^j f_i. \quad (4.2.6)$$

Equation (4.2.3) is the special case of (4.2.5) for $n = 1$. Note that the matrix in (4.2.5) is determined by only $2n + 1$ quantities $s_0, \ldots, s_{2n}$, and the "cross-diagonals" of the matrix are constant. Such a matrix is called a *Hankel matrix* and has many interesting properties. The quantities s_j are known as *moments*.

The system (4.2.5) can also be written in the form

$$E^T W E \mathbf{a} = E^T W \mathbf{f}, \quad (4.2.7)$$

where W is a diagonal matrix containing the weights and

$$E = \begin{bmatrix} 1 & x_1 & \cdots & x_1^n \\ 1 & x_2 & \cdots & x_2^n \\ \vdots & & & \vdots \\ 1 & x_m & \cdots & x_m^n \end{bmatrix}, \quad \mathbf{a} = \begin{bmatrix} a_0 \\ a_1 \\ \vdots \\ a_n \end{bmatrix}, \quad \mathbf{f} = \begin{bmatrix} f_1 \\ f_2 \\ \vdots \\ f_m \end{bmatrix}. \quad (4.2.8)$$

4.2 Least Squares Approximation

The matrix E is $m \times (n+1)$ and is of Vandermonde type; in particular, if $m = n+1$, it is precisely the Vandermonde matrix of (4.1.13), which we showed was nonsingular. We now extend that argument to show that the matrix of (4.2.7) is nonsingular provided that at least $n+1$ of the points x_i are distinct. Recall from Chapter 2 that a symmetric matrix A is positive definite if

$$\mathbf{x}^T A \mathbf{x} > 0 \quad \text{for all} \quad \mathbf{x} \neq 0. \tag{4.2.9}$$

A positive definite matrix is nonsingular since its eigenvalues are all positive (Theorem 2.2.5). We now show that the matrix $E^T W E$ of (4.2.7) is symmetric positive definite. Clearly, it is symmetric since it is the matrix of (4.2.5). Let $\mathbf{y}$ be any m-vector and consider

$$\mathbf{y}^T W \mathbf{y} = \sum_{i=1}^m w_i y_i^2. \tag{4.2.10}$$

Since the w_i are assumed positive, $\mathbf{y}^T W \mathbf{y} \geq 0$ and is equal to zero if and only if $\mathbf{y} = 0$. Thus, if we set $\mathbf{y} = E\mathbf{a}$, we conclude that

$$\mathbf{a}^T E^T W E \mathbf{a} > 0 \quad \text{for all} \quad \mathbf{a} \neq 0$$

provided that $E\mathbf{a} \neq 0$. But if $E\mathbf{a} = 0$, this implies that

$$a_0 + a_1 x_i + \cdots + a_n x_i^n = 0, \quad i = 1, \ldots, m.$$

Recalling that we have assumed that at least $n+1$ of the x_i are distinct, the nth degree polynomial $a_0 + a_1 x + \cdots + a_n x^n$ would have at least $n+1$ distinct roots. This contradiction proves that $E^T W E$ is positive definite. Thus, the system (4.2.7) has a unique solution, and the resulting polynomial with coefficients a_i is the unique polynomial of degree n which minimizes (4.2.1).

General Linear Least Squares Problems

We next consider more general least-squares problems in which the approximating function is not necessarily a polynomial but is a linear combination

$$\phi(x) = \sum_{i=0}^n a_i \phi_i(x) \tag{4.2.11}$$

of given functions $\phi_0, \phi_1, \ldots, \phi_n$. If $\phi_j(x) = x^j$, $j = 0, \ldots, n$, then ϕ is the polynomial considered previously. Other common choices for the "basis functions" ϕ_j are

$$\phi_j(x) = \sin j\pi x, \quad j - 0, 1, \ldots, n,$$

and

$$\phi_j(x) = e^{\alpha_j x}, \quad j = 0, 1, \ldots, n,$$

where the α_j are given real numbers. We could also use piecewise polynomials or spline functions.

The *general linear least-squares problem* is to find $a_0, \ldots, a_n$ such that

$$g(a_0, a_1, \cdots, a_n) = \sum_{i=1}^{m} w_i[\phi(x_i) - f_i]^2 \qquad (4.2.12)$$

is minimized, where ϕ is given by (4.2.11). We can proceed exactly as before in obtaining the normal equations for (4.2.12). The first partial derivatives of g are

$$\frac{\partial g}{\partial a_j} = 2\sum_{i=1}^{m} w_i\phi_j(x_i)[a_0\phi_0(x_i) + a_1\phi_1(x_1) + \cdots + a_n\phi_n(x_i) - f_i].$$

Setting these equal to zero, collecting coefficients of the a_i, and writing the resulting linear system in matrix-vector form gives the system

$$\begin{bmatrix} s_{00} & s_{10} & \cdots & s_{n0} \\ s_{01} & s_{11} & & \\ \vdots & & \ddots & \vdots \\ s_{0n} & \cdots & & s_{nn} \end{bmatrix} \begin{bmatrix} a_0 \\ a_1 \\ \vdots \\ a_n \end{bmatrix} = \begin{bmatrix} c_0 \\ c_1 \\ \vdots \\ c_n \end{bmatrix}, \qquad (4.2.13)$$

where

$$s_{ij} = \sum_{k=1}^{m} w_k \phi_i(x_k) \phi_j(x_k), \qquad c_j = \sum_{k=1}^{m} w_k \phi_j(x_k) f_k.$$

The matrix of (4.2.13) is a *Grammian* matrix but it is not necessarily a Hankel matrix. As before, we can write (4.2.13) in the form (4.2.7), where now

$$E = \begin{bmatrix} \phi_0(x_1) & \phi_1(x_1) & \cdots & \phi_n(x_1) \\ \phi_0(x_2) & \phi_1(x_2) & \cdots & \phi_n(x_2) \\ \vdots & & & \vdots \\ \phi_0(x_m) & \cdots & & \phi_n(x_m) \end{bmatrix}, \qquad \mathbf{f} = \begin{bmatrix} f_1 \\ f_2 \\ \vdots \\ f_m \end{bmatrix}.$$

Clearly, E^TWE is again symmetric but in order to conclude that it is positive definite, suitable conditions must be imposed on the functions $\phi_0, \ldots, \phi_n$ as well as $x_1, \ldots, x_m$.

Orthogonal Polynomials

The normal equations are very useful for theoretical purposes, or for computation when n is small. But they have a tendency to become very ill-conditioned (see Chapter 6) as n increases. We now describe an alternative approach to computing the least-squares polynomial by means of orthogonal

4.2 Least Squares Approximation

polynomials. In what follows we assume that $w_i = 1$ although the inclusion of weights presents no problem (Exercise 4.2.5).

Let $q_0, q_1, \ldots, q_n$ be polynomials of degree $0, 1, \ldots, n$, respectively. Then, we will say that the q_i are *orthogonal* with respect to the points $x_1, \ldots, x_m$ if

$$\sum_{i=1}^{m} q_k(x_i) q_j(x_i) = 0, \quad k, j = 0, 1, \ldots, n, \quad k \neq j. \tag{4.2.14}$$

We shall return shortly to the question of how one obtains such a set of orthogonal polynomials. For the moment, assume that we have them and take $\phi_i = q_i$, $i = 0, 1, \ldots, n$, in the normal equations (4.2.13), with the weights w_i all equal to 1. Then, because of (4.2.14), all elements of the coefficient matrix of (4.2.13) off the main diagonal are zero, and the system of equations reduces to

$$\sum_{i=1}^{m} [q_k(x_i)]^2 a_k = \sum_{i=1}^{m} q_k(x_i) f_i, \quad k = 0, 1, \ldots, n.$$

Thus,

$$a_k = \frac{1}{\gamma_k} \sum_{i=1}^{m} q_k(x_i) f_i, \quad k = 0, 1, \ldots, n, \tag{4.2.15}$$

where

$$\gamma_k = \sum_{i=1}^{m} [q_k(x_i)]^2. \tag{4.2.16}$$

Therefore, the least-squares polynomial is

$$q(x) = \sum_{k=0}^{n} a_k q_k(x). \tag{4.2.17}$$

An obvious question is whether the polynomial q of (4.2.17) is the same as the polynomial obtained from the normal equations (4.2.5) (with the $w_i = 1$). The answer is yes, under our standard assumption that at least $n+1$ of the points x_i are distinct. This follows from the fact that – as shown earlier – there is a unique polynomial of degree n or less that minimizes (4.1.1). Therefore, the polynomial q of (4.2.17) is this same minimizing polynomial, but just represented differently.

The use of orthogonal polynomials reduces the normal equations to a diagonal system of equations, which is trivial to solve. However, the burden is shifted to the computation of the q_i and we now describe how to construct these. Let

$$q_0(x) \equiv 1, \quad q_1(x) \equiv x - \alpha_1, \tag{4.2.18}$$

where α_1 is to be determined so that q_0 and q_1 are orthogonal with respect to the x_i. Thus, we must have

$$0 = \sum_{i=1}^{m} q_0(x_i)q_1(x_i) = \sum_{i=1}^{m}(x_i - \alpha_1) = \sum_{i=1}^{m} x_i - \sum_{i=1}^{m} \alpha_1 = \sum_{i=1}^{m} x_i - m\alpha_1$$

so that

$$\alpha_1 = \frac{1}{m} \sum_{i=1}^{m} x_i. \tag{4.2.19}$$

Now, let

$$q_2(x) = xq_1(x) - \alpha_2 q_1(x) - \beta_1,$$

where α_2 and β_1 are to be determined so that q_2 is orthogonal to both q_0 and q_1; that is,

$$\sum_{i=1}^{m}[x_i q_1(x_i) - \alpha_2 q_1(x_i) - \beta_1] = 0$$

$$\sum_{i=1}^{m}[x_i q_1(x_i) - \alpha_2 q_1(x_i) - \beta_1]q_1(x_i) = 0.$$

Noting that $\sum q_1(x_i) = 0$, these relations reduce to

$$\sum_{i=1}^{m} x_i q_1(x_i) - m\beta_1 = 0, \qquad \sum_{i=1}^{m} x_i [q_1(x_i)]^2 - \alpha_2 \gamma_1 = 0,$$

where γ_1 is given by (4.2.16). Thus,

$$\beta_1 = \frac{1}{m} \sum_{i=1}^{m} x_i q_1(x_i), \qquad \alpha_2 = \frac{1}{\gamma_1} \sum_{i=1}^{m} x_i [q_1(x_i)]^2.$$

The computation for the remaining q_i proceeds in an analogous fashion. Assume that we have determined $q_0, q_1, \ldots, q_j$, and define q_{j+1} by the *three-term recurrence relation*

$$q_{j+1}(x) = xq_j(x) - \alpha_{j+1} q_j(x) - \beta_j q_{j-1}(x), \tag{4.2.20}$$

where α_{j+1} and β_j are to be determined from the orthogonality requirements

$$\sum_{i=1}^{m} q_{j+1}(x_i) q_j(x_i) = 0, \qquad \sum_{i=1}^{m} q_{j+1}(x_i) q_{j-1}(x_i) = 0. \tag{4.2.21}$$

4.2 Least Squares Approximation

If these two relations are satisfied, then q_{j+1} must also be orthogonal to all the previous q_k, $k < j-1$, since by (4.2.20)

$$\sum_{i=1}^{m} q_{j+1}(x_i)q_k(x_i) = \sum_{i=1}^{m} x_i q_j(x_i)q_k(x_i) - \alpha_{j+1}\sum_{i=1}^{m} q_j(x_i)q_k(x_i) \quad (4.2.22)$$

$$-\beta_j \sum_{j=1}^{m} q_{j-1}(x_i)q_k(x_i).$$

The last two terms in (4.2.22) are zero by assumption, whereas $xq_k(x)$ is a polynomial of degree $k+1$ and can be expressed as a linear combination of $q_0, q_1, \ldots, q_{k+1}$. Hence, the first term on the right-hand side of (4.2.22) is zero.

Returning to (4.2.21) and putting in q_{j+1} from (4.2.20) leads to

$$\alpha_{j+1} = \frac{1}{\gamma_j} \sum_{i=1}^{m} x_i [q_j(x_i)]^2, \quad (4.2.23)$$

$$\beta_j = \frac{1}{\gamma_{j-1}} \sum_{i=1}^{m} x_i q_j(x_i) q_{j-1}(x_i),$$

for α_{j+1} and β_j, where the γ's are as in (4.2.16). The β_i can be computed in a better way if we substitute for $x_i q_{j-1}(x_i)$ using (4.2.20), and then note that

$$\sum_{i=1}^{m} q_j(x_i)[q_j(x_i) + \alpha_j q_{j-1}(x_i) + \beta_{j-1} q_{j-2}(x_i)] = \sum_{i=1}^{m} [q_j(x_i)]^2 = \gamma_j$$

by the orthogonality of the q's. Thus,

$$\beta_j = \frac{\gamma_j}{\gamma_{j-1}}. \quad (4.2.24)$$

From (4.2.16) the denominator in (4.2.24) can vanish only if $q_{j-1}(x_i) = 0$, $i = 1, \ldots, m$. But since at least $n+1$ of the x_i are assumed to be distinct, this would imply that q_{j-1} is identically zero, which contradicts the definition of q_{j-1}. Hence, $\gamma_{j-1} \neq 0$.

We can summarize the orthogonal polynomial algorithm as follows:

1. Set $q_0(x) \equiv 1$, $q_1(x) = x - \frac{1}{m}\sum_{i=1}^{m} x_i$; $\gamma_0 = m$.

2. Compute for $j = 1$ to $n-1$

   $\gamma_j = \sum_{i=1}^{m} [q_j(x_i)]^2$,

   $\alpha_{j+1} = \frac{1}{\gamma_j} \sum_{i=1}^{m} x_i [q_j(x_i)]^2$, $\quad \beta_j = \frac{\gamma_j}{\gamma_{j-1}}$

   $q_{j+1}(x) = xq_j(x) - \alpha_{j+1} q_j(x) - \beta_j q_{j-1}(x)$

3. Compute the coefficients $a_0, a_1, \ldots, a_n$ of the least-squares polynomial $a_0 q_0(x) + a_1 q_1(x) + \cdots + a_n q_n(x)$ by (4.2.15).

As mentioned previously, this approach is to be preferred numerically because it avoids the necessity of solving the (possibly ill-conditioned) system (4.2.5). Another advantage is that we are able to build up the least-squares polynomial degree by degree. For example, if we do not know what degree polynomial we wish to use, we might start with a first-degree polynomial, then a second-degree, and so on, until we obtain a fit that we believe is suitable. With the orthogonal polynomial algorithm, the coefficients a_i are independent of n, and as soon as we compute q_j, we can compute a_j, and hence the least-squares polynomial of degree j.

A Numerical Example

We now give a simple example, using the data

$$x_1 = 0 \quad x_2 = \tfrac{1}{4} \quad x_3 = \tfrac{1}{2} \quad x_4 = \tfrac{3}{4} \quad x_5 = 1$$
$$f_1 = 1 \quad f_2 = 2 \quad f_3 = 1 \quad f_4 = 0 \quad f_5 = 1.$$

Since there are five points x_i, these data will uniquely determine a fourth-degree interpolating polynomial. We will compute the linear and quadratic least-squares polynomials by both the normal equations and the orthogonal polynomial approaches.

For the normal equations for the linear polynomial, we will need the following quantities:

$$\sum_{i=1}^{5} x_i = \tfrac{5}{2}, \quad \sum_{i=1}^{5} x_i^2 = \tfrac{15}{8}, \quad \sum_{i=1}^{5} f_i = 5, \quad \sum_{i=1}^{5} x_i f_i = 2. \qquad (4.2.25)$$

Then, the coefficients a_0 and a_1 are the solutions of the system (4.2.3) (with the $w_i = 1$):

$$a_0 = \frac{7}{5}, \quad a_1 = \frac{-4}{5}.$$

Thus, the linear least-squares polynomial is

$$p_1(x) = \tfrac{7}{5} - \tfrac{4}{5}x. \qquad (4.2.26)$$

If we compute the same polynomial by orthogonal polynomials, the polynomial is given in the form

$$a_0 q_0(x) + a_1 q_1(x) = a_0 + a_1(x - \alpha_1), \qquad (4.2.27)$$

where a_0 and a_1 are given by (4.2.15), and α_1 by (4.2.19):

$$a_0 = 1, \quad a_1 = \frac{-4}{5}, \quad \alpha_1 = \tfrac{1}{2}.$$

4.2 Least Squares Approximation

Therefore, the polynomial (4.2.27) is $1 - \frac{4}{5}(x - \frac{1}{2})$, which is, as expected, the same as (4.2.26).

To compute the least-squares quadratic polynomial by the normal equations, we need to solve the system (4.2.5) for $n = 2$, which for our data is

$$\begin{bmatrix} 640 & 320 & 240 \\ 320 & 240 & 200 \\ 240 & 200 & 177 \end{bmatrix} \begin{bmatrix} a_0 \\ a_1 \\ a_2 \end{bmatrix} = \begin{bmatrix} 640 \\ 256 \\ 176 \end{bmatrix}.$$

The solution of this system is

$$a_0 = \tfrac{7}{5}, \qquad a_1 = \tfrac{-4}{5}, \qquad a_2 = 0. \tag{4.2.28}$$

Thus, the best least-squares polynomial approximation turns out to be the linear least-squares polynomial; that is, no improvement to the linear approximation can be made by adding a quadratic term. That (4.2.28) is correct is verified by computing the least-squares quadratic by orthogonal polynomials. The orthogonal polynomial representation will be

$$a_0 q_0(x) + a_1 q_1(x) + a_2 q_2(x) = \tfrac{7}{5} - \tfrac{4}{5}x + a_2[x(x - \tfrac{1}{2}) - \alpha_2(x - \tfrac{1}{2}) - \beta_1],$$

where α_2 and β_1 are computed from (4.2.23) and (4.2.24) as

$$\alpha_2 = \frac{\sum x_i(x_i - \tfrac{1}{2})^2}{\sum(x_i - \tfrac{1}{2})^2} = \tfrac{1}{2}, \qquad \beta_1 = \tfrac{1}{5}\sum x_i(x_i - \tfrac{1}{2}) = \tfrac{1}{8}.$$

Thus, $q_2(x) = x^2 - x - \tfrac{1}{8}$, and from (4.3.15), we find that $a_2 = 0$.

Supplementary Discussion and References: 4.2

1. For further discussions of least squares problems see Golub and Van Loan [1989] and Lawson and Hanson [1974].

2. An alternative approach to solving the least-squares polynomial problem is to deal directly with the system of linear equations $E\mathbf{a} = \mathbf{f}$, where E and $\mathbf{f}$ are given in (4.2.8). This is an $m \times (n+1)$ system, where m is usually greater that $n+1$; hence, the matrix E is not square. Techniques for dealing with this type of system are given in Chapter 6.

EXERCISES 4.2

4.2.1. Assume that f is a given function for which the following values are known: $f(1) = 2$, $f(2) = 3$, $f(3) = 5$, $f(4) = 3$. Find the constant, linear, and quadratic least-squares polynomials by both the normal-equation and orthogonal-polynomial approaches.

4.2.2. Write a computer program to obtain the least-squares polynomial of degree n using $m \geq n+1$ data points by the orthogonal-polynomial approach. Test your program on the polynomials of Exercise 4.2.1.

4.2.3. Let $y_1, \ldots, y_m$ be a set of data that we wish to approximate by a constant c. Determine c so that

    a. $\sum_{i=1}^{m} |c - y_i| = \text{minimum}$.

    b. $\max_{1 \leq i \leq m} |c - y_i| = \text{minimum}$.

4.2.4. If $y_1, \ldots, y_m$ are m observations, define the mean and variance by

$$\bar{y} = \frac{1}{m} \sum_{i=1}^{m} y_i, \qquad v = \sum_{i=1}^{m} (y_i - \bar{y})^2.$$

Let $y_{m+1}, \ldots, y_{m+n}$ be n additional observations with mean and variance y_a and v_a. Show how to combine v and v_a by $\alpha_1 v + \alpha_2 v_a$ to obtain the variance of the combined set of observations. Specialize this to $n = 1$ so as to obtain an updating formula for the variance each time a new observation is added.

4.2.5. Modify the orthogonal polynomial approach to handle the problem (4.2.1) with given weights w_i. Replace the relation (4.2.14) by $\sum w_i q_k(x_i) q_j(x_i)$ and then modify (4.2.15), (4.2.23) and (4.2.24) accordingly.

4.2.6. Let f be a twice differentiable function on the interval $[0, 8]$, and let $x_i = (i - 9)$, $i = 1, \ldots, 17$, and $f_i = f(x_i)$.
a. Using orthogonal polynomials, find the polynomial of degree 5 which minimizes (4.2.1) with the weights given by $w_i = \gamma |i - 9|$, where $0 < \gamma \leq 1$.
b. Show how $f'(0)$ and $f''(0)$ can be approximated by using the recursion relation (4.2.20) for the orthogonal polynomials. In particular, show how to obtain coefficients a_i so that $f'(0)$ is approximated by $\sum_{i=1}^{17} a_i f_i$.

4.2.7. If $\phi_j(x) = x^j$, then $\phi_{i+j}(x) = \phi_i(x) \phi_j(x)$. Give another set of basis functions that satisfy this property.

4.2.8. Let $S_j = \sum_{i=1}^{m} (f_i - a_0 q_0(x_i) - \cdots - a_j q_j(x_i))^2$, where the q_j are orthogonal in the sense of (4.2.14). Show how to compute S_{j+1} from S_j.

4.2.9. Suppose that we wish to approximate a function on the interval $[a, b]$ by least squares polynomials and we wish that the polynomial p_n satisfy $p_n(a) = 0$. Show how this can be accomplished by suitable choice of the weights.

4.3 Application to Root-Finding

We now consider the problem of finding a solution of a nonlinear equation

$$f(x) = 0, \tag{4.3.1}$$

where f is a given nonlinear function of x. For example, f may be the polynomial

$$f(x) = a_n x^n + a_{n-1} x^{n-1} + \cdots + a_1 x + a_0. \tag{4.3.2}$$

Other examples are

$$f(x) = x - \sin x + 2, \tag{4.3.3}$$

$$f(x) = e^x - x - 4. \tag{4.3.4}$$

In the case of the polynomial (4.3.2) we know from the fundamental theorem of algebra that f has exactly n real or complex roots if we count multiplicities of the roots. For a general function f it is usually difficult to ascertain how many solutions equation (4.3.1) has: there may be none, one, finitely many, or infinitely many. A simple condition that ensures that there is at most one solution in a given interval (a, b) is that

$$f'(x) > 0 \quad \text{for all} \quad x \in (a, b) \tag{4.3.5}$$

(or $f'(x) < 0$ in the interval), although this does not guarantee that a root exists in the interval. (The proof of these statements is left to Exercises 4.3.1 and 4.3.2.) If, however, f is continuous and

$$f(a) < 0, \quad f(b) > 0, \tag{4.3.6}$$

then it is intuitively clear (and rigorously proved by a theorem of the calculus) that f must have at least one root in the interval (a, b).

Most of the methods for approximating a solution of (4.3.1) replace f by an approximating low degree polynomial in the neighborhood of a solution, and then obtain a next approximation to a solution as a root of this polynomial. The simplest such procedure is when the approximating polynomial is a first degree polynomial and one way to obtain such polynomial is by a Taylor expansion.

Linearization and Newton's Method

From (4.1.1) for $n = 1$ we have

$$f(x) \doteq p(x) = f(x_0) + f'(r_0)(x - x_0). \tag{4.3.7}$$

This approximation of f by a linear polynomial is sometimes referred to as *linearization* of f. Geometrically, we are approximating the curve of f by a

Chapter 4 Polynomial Approximation

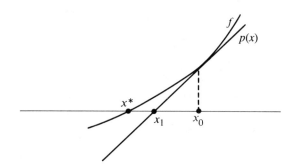

Figure 4.3.1: *Linear Approximation and Newton Iterate*

line tangent to the curve of f at x_0, as shown in Figure 4.3.1. The line is tangent at x_0 because $p(x_0) = f(x_0)$ and $p'(x_0) = f'(x_0)$.

The linear approximation (4.3.7) forms the basis of *Newton's method* for approximating roots of f. Given x_0 as an approximation to a root $x^\star$, the next approximation, x_1, is found by solving the linear equation $p(x) = 0$; thus,

$$x_1 = x_0 - \frac{f(x_0)}{f'(x_0)}. \quad (4.3.8)$$

This is illustrated in Figure 4.3.1.

We can use the error estimate (4.1.5) to obtain information on how well x_1 approximates the root $x^\star$. We evaluate (4.1.5) for $x = x^\star$ and $n = 1$; since $f(x^\star) = 0$, this gives

$$f(x^\star) - p(x^\star) = -f(x_0) - f'(x_0)(x^\star - x_0) = \frac{1}{2}f''(z)(x^\star - x_0)^2. \quad (4.3.9)$$

From (4.3.8), we have $f'(x_0)x_0 - f(x_0) = f'(x_0)x_1$ and making this substitution in (4.3.9) shows that

$$-f'(x_0)(x^\star - x_1) = \frac{1}{2}f''(z)(x^\star - x_0)^2. \quad (4.3.10)$$

Assuming that $f'(x_0) \neq 0$, we then have

$$x^\star - x_1 = c_0(x^\star - x_0)^2, \quad c_0 = \frac{f''(z)}{2f'(x_0)}. \quad (4.3.11)$$

Thus, the error in x_1 is proportional to the square of the error in x_0. This will be discussed more shortly.

We can now repeat the above process by forming a linear approximation to f at x_1 and solving for the root of this linear function to obtain a new

4.3 Application to Root-Finding

approximation, x_2, to $x^\star$. Continuing in this way gives the *Newton iteration*

$$x_{k+1} = x_k - \frac{f(x_k)}{f'(x_k)}, \qquad k = 0, 1, \ldots. \tag{4.3.12}$$

Note that we are assuming here that $f'(x_k) \neq 0$; if $f'(x_k) = 0$ the tangent line to f at x_k is horizontal and the corresponding linear function has no root. Thus, Newton's method stops (more dramatically, the next Newton iterate is at infinity and the method blows up).

To ascertain the convergence of the sequence (4.3.12) to the root $x^\star$, we can first proceed exactly as in the derivation of (4.3.11) to obtain the relation

$$x^\star - x_k = c_k(x^\star - x_k)^2, \qquad c_k = \frac{f''(z_k)}{2f'(x_k)}, \tag{4.3.13}$$

where z_k is now between x_k and $x^\star$. We assume that on some interval $I = [x^\star - \alpha, x^\star + \alpha]$

$$|f''(x)| \leq M, \qquad |f'(x)| \geq m > 0, \quad \text{for all } x \in I. \tag{4.3.14}$$

Let $c = M/(2m)$. Then if

$$c|x^\star - x_0| \leq \gamma < 1, \qquad |x^\star - x_0| \leq \alpha, \tag{4.3.15}$$

we have from (4.3.13) that

$$|x^\star - x_1| \leq c|x^\star - x_0|^2 \leq \gamma|x^\star - x_0|.$$

This shows that $x_1 \in I$ and also satisfies the first inequality of (4.3.15). Hence, we can proceed by induction to conclude that all $x_i \in I$ and

$$|x^\star - x_k| \leq \gamma|x^\star - x_{k-1}| \leq \gamma^k|x^\star - x_0| \tag{4.3.16}$$

which shows, since $\gamma < 1$, that $x_k \to x^\star$ as $k \to \infty$. Moreover, from (4.3.13)

$$|x^\star - x_{k+1}| \leq c|x^\star - x_k|^2 \tag{4.3.17}$$

so that when $|x^\star - x_k| < 1$, the rate of convergence begins to be very rapid. For example, if $c \doteq 1$ and $|x^\star - x_k| \doteq .1$, then $|x^\star - x_{k+1}| \doteq 10^{-2}$ and $|x^\star - x_{k+2}| \doteq 10^{-4}$. Thus, the number of correct decimal digits in the x_i are approximately doubling at each iteration. This is the property of *quadratic convergence*.

We may summarize the preceding discussion in the following basic result:

THEOREM 4.3.1 (Newton Convergence) *If f is twice-continuously differentiable in a neighborhood of a root $x^\star$ and if $f'(x^\star) \neq 0$, then provided that x_0 is sufficiently close to $x^\star$, the Newton iterates (4.3.12) will converge and the rate of convergence will be quadratic.*

As an illustration of quadratic convergence in Newton's method, consider the problem of finding the zeros of $f(x) = 1/x + lnx - 2$. This function is defined for all positive values of x and has two zeros: one between $x = 0$ and $x = 1$ and the other between $x = 6$ and $x = 7$. Table 4.3.1 contains a summary of the first six iterations of Newton's method using the starting value of $x = 0.1$. Note that once an approximation is "close enough" (in this case, after three iterations), the number of correct digits doubles in each iteration, which shows the quadratic convergence.

Table 4.3.1: *Convergence of Newton's Method for* $f(x) = 1/x + lnx - 2$

Iteration	x_{i-1}	$f(x_{i-1})$	x_i	Number of Correct Digits
1	0.1	5.6974149	0.16330461	0
2	0.16330461	2.3113878	0.23697659	0
3	0.23697659	0.7800322	0.29438633	1
4	0.29438633	0.1740346	0.31576121	2
5	0.31576121	0.0141811	0.31782764	4
6	0.31782764	0.0001134	0.31784443	8

Theorem 4.3.1 shows that provided x_0, or some iterate x_k, is sufficiently close to x^* the Newton iterates will always converge; this is the property of *local convergence* and is an intrinsic property of Newton's method. The condition that $f(x^*) \neq 0$ is necessary for quadratic convergence but if $f'(x^*) = 0$, convergence will still occur (see Exercises 4.3.3,4). On the other hand, when an iterate is not sufficiently close to a solution, various types of "bad" behavior can occur with Newton's method, as shown in Figure 4.3.2. Figure 4.3.2(a) indicates the possibility of "cycling" in which $x_{i+2} = x_i$ and this cycle then repeats (see Exercise 4.3.5); thus, there is no convergence but no divergence either. Cycles of order higher than 2 are also possible. Figure 4.3.2(b) shows divergence to infinity as would be the case if x_i is outside the domain of convergence to the solution of interest and the function behaves like, for example, e^{-x} as $x \to \infty$.

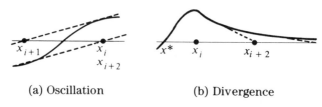

(a) Oscillation   (b) Divergence

Figure 4.3.2: *Possible "Bad" Behavior of Newton's Method*

4.3 Application to Root-Finding

Another problem is when to stop the iteration. The usual simplest tests are $|f(x_i)| < \varepsilon$ or $|x_{i+1} - x_i| < \varepsilon$, where ε is some given tolerance. The first can be misleading when the function f is very "flat" near the root, and the second can fail in a variety of situations depending on the iterative method. For example, for Newton's method, it can fail when the derivative is very large at the current iterate. These two possibilities are depicted in Figure 4.3.3.

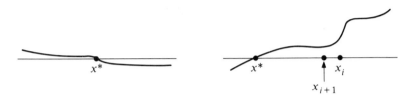

Figure 4.3.3: *Defeating Convergence Tests*

Convexity

In contrast to the above instances of bad behavior, there are situations in which Newton's method will converge for any starting approximation, no matter how far from the solution. In this case, there is *global convergence*. Perhaps the simplest functions for which global convergence is obtained are those that are convex. A function is *convex* if it satisfies any one of the following equivalent properties, depending upon the differentiability of the function:

$$f''(x) \geq 0, \quad \text{for all } x, \tag{4.3.18a}$$

$$f'(y) \geq f'(x), \quad \text{if } y \geq x, \tag{4.3.18b}$$

$$f(\alpha x + (1-\alpha)y) \leq \alpha f(x) + (1-\alpha)f(y), \tag{4.3.18c}$$

where (4.3.18c) holds for any $\alpha \in (0,1)$ and all x, y.

A linear function $f(x) = ax + b$ is always convex, as is easily checked by any of the definitions of (4.3.18). However, we are interested in functions that actually "bend upwards" as illustrated in Figure 4.3.4. Such functions are *strictly convex* and satisfy (4.3.18b,c) with strict inequality whenever $x \neq y$. Strict inequality in (4.3.18a) is also sufficient for strict convexity, but not necessary; the function $f(x) = x^4$ is strictly convex although $f''(0) = 0$.

A convex function may have infinitely many roots ($f(x) \equiv 0$) and even a strictly convex function may have no roots (for example, $f(x) = e^{-x}$). In the sequel, we will assume that f is strictly convex, $f'(x) > 0$ for all x and $f(x) = 0$ has a solution. These assumptions are illustrated by the function in Figure 4.3.4. In this case, if x_0 is to the right of the solution, the Newton iterates converge monotonically to the solution, as is intuitively clear by drawing the tangent lines to the curve (see also Exercise 4.3.10). If x_0 is to the left of

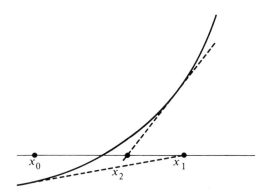

Figure 4.3.4: *Convergence of Newton's Method for a Convex Function*

the solution, as shown in Figure 4.3.4, then the next Newton iterate is to the right of the solution, and thereafter the Newton iterates again converge monotonically to the solution. Figure 4.3.4 shows a function for which $f'(x) > 0$. If $f'(x) < 0$, the corresponding situation holds but monotone convergence is now from left to right (Exercise 4.3.11). Similar convergence statements can be made if f is *concave*, that is, if $-f$ is convex.

The above discussion assumes that the properties of f hold for all x, in which case we obtain global convergence. They may, however, hold only in a neighborhood of a solution and this will again ensure monotone convergence of the Newton iterates for suitable starting approximations x_0. For example, in Figure 4.3.2(b), there will be an interval $[x^\star, b]$, for which the Newton iterates will converge monotonically to $x^\star$ if $x_0 \in [x^\star, b]$. See also Exercises 4.3.5 and 4.3.6.

The Secant Method

Instead of obtaining an approximation to f by a first-order Taylor expansion, we can use interpolation. Consider the linear interpolating polynomial $p(x)$ of (4.1.10) as an approximation to f, as illustrated in Figure 4.3.5. The solution of $p(x) = 0$, is

$$x_2 = x_1 - \frac{f_1}{d_1}, \qquad d_1 = \frac{f_1 - f_0}{x_1 - x_0}, \qquad (4.3.19)$$

and x_2 may be considered as a new approximation to the root $x^\star$.

There are two approaches to continuing this process to obtain further approximations to the root. The first simply repeats the formula (4.3.19):

$$x_{i+1} = x_i - \frac{f_i}{d_i}, \qquad d_i = \frac{f_i - f_{i-1}}{x_i - x_{i-1}}, \quad i = 1, 2, \ldots . \qquad (4.3.20)$$

4.3 Application to Root-Finding

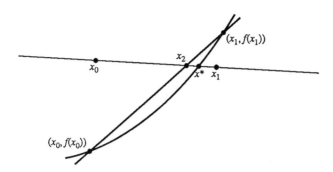

Figure 4.3.5: *Root Approximation by Linear Interpolation*

This is the *secant method*. Note that d_i is an approximation to $f'(x_i)$, and thus (4.3.20) is an approximate Newton iteration. In general, the rate of convergence of the secant iterates is rapid but not as rapid as in Newton's method. If $f'(x^*) \neq 0$, then assuming that f is sufficiently differentiable it is possible to show that the errors satisfy

$$|x^\star - x_{i+1}| \leq (c_1 + c_2)|x^\star - x_i|^2 + c_2|x^\star - x_i\|x^\star - x_{i-1}| \qquad (4.3.21)$$

when the iterates are sufficiently close to the solution $x^\star$. The last term in (4.3.21) prevents quadratic convergence. It can be shown that the order of convergence is $\frac{1}{2}(1+\sqrt{5}) \doteq 1.62$, as opposed to 2 for Newton's method, so that in the limit $|x^\star - x_{i+1}| \doteq c|x^\star - x_i|^{1.62}$. But the secant method requires only one evaluation of f each iteration, as opposed to both f and f' for Newton's method, and so may be more efficient.

Bisection and Regula Falsi

The second way to continue the process (4.3.19) may be viewed as a variation of the following *bisection method*. Assume we have two points a and b such that $f(a)$ and $f(b)$ have different signs. Then, as mentioned previously, if f is continuous, there is at least one root in the interval (a,b). For simplicity in the following discussion, assume that there is just one root, $x^\star$. Let $x_1 = (a+b)/2$ be the midpoint of the interval (a,b). If $f(a)$ and $f(x_1)$ have different signs, then the root $x^\star$ is in the interval (a,x_1); otherwise, it is in (x_1,b). We then repeat the process, always keeping the interval in which $x^\star$ is known to lie and evaluating f at its midpoint in order to obtain the next interval. For example, a typical sequence of steps might be:

$f(x_1) < 0$. Hence, $x^\star \in (x_1, b)$. Set $x_2 = \frac{1}{2}(x_1 + b)$;
$f(x_2) > 0$. Hence, $x^\star \in (x_1, x_2)$. Set $x_3 = \frac{1}{2}(x_1 + x_2)$;
$f(x_3) < 0$. Hence, $x^\star \in (x_3, x_2)$. Set $x_4 = \frac{1}{2}(x_2 + x_3)$;
$f(x_4) < 0$. Hence, $x^\star \in (x_4, x_2)$. Set $x_5 = \frac{1}{2}(x_2 + x_4)$.

Clearly, each step of the bisection procedure reduces the length of the interval known to contain $x^\star$ by a factor of 2. Therefore, after m steps, the length of the interval will be $(b-a)2^{-m}$, and this provides a bound on the error in the current approximation to the root:

$$|x_m - x^\star| \leq \frac{|b-a|}{2^m}. \qquad (4.3.22)$$

One drawback of the bisection method is that it may be rather slow. To reduce the initial interval by a large factor, say 10^6, which may correspond to about six-decimal-digit accuracy, we would expect to require, from the error bound (4.3.22), approximately

$$m = \frac{6}{\log_{10} 2} \doteq 20$$

evaluations of f. When each evaluation is expensive, we would like to keep the number of evaluations as small as possible. It is here that linear interpolation may be useful. As shown in the somewhat favorable situation of Figure 4.3.5, a new approximation obtained from (4.3.19) may be considerably better than the midpoint of the interval. This is the *Regula Falsi* method. In contrast with the secant method, the Regula Falsi method retains x_{i+1} and either x_i or x_{i-1} so that the function values at the two retained points have different signs.

Errors

The effect of errors in the evaluation of f is most easily seen in conjunction with the bisection method. The bisection method does not use the value of $f(x_i)$ but only the sign of $f(x_i)$; therefore, the method is impervious to errors in evaluating f as long as the sign of $f(x_i)$ is evaluated correctly. One might think that round-off error could not be so severe as to change the sign of the function, but this is not the case when the function values become sufficiently small. If the sign of $f(x_i)$ is incorrect, a wrong decision will be made in choosing the next subinterval, and the error bound (4.3.22) does not necessarily hold.

It is clear that if one makes a maximum error of e in evaluating f at any point in the interval (a, b), then the sign of f will be correctly evaluated as long as

$$|f(x)| > |e|.$$

4.3 Application to Root-Finding

Since the function f will be close to zero near the root $x^\star$, we can also argue the converse: there will be an *interval of uncertainty*, $(x^\star - \varepsilon, x^\star + \varepsilon)$, about the root in which the sign of f may not be correctly evaluated. When the approximations reach this interval, their further progress toward the root is at best problematical. Unfortunately, it is extremely difficult to determine this interval in advance. It depends on the unknown root $x^\star$, the "flatness" of f in the neighborhood of the root, and the magnitude of the errors made in evaluating f. On the other hand, the interval is usually detectable during the course of the computation by an erratic behavior of the iterates; when this occurs, there is no longer any point in continuing the computation.

The fact that the sign of the function f may not be evaluated correctly near the root affects not only the bisection method but also Newton's method and the secant method. Consider Newton's method (4.3.12). If the sign of $f(x_k)$ is evaluated incorrectly, but that of $f'(x_k)$ correctly (a reasonable assumption if $f'(x^\star)$ is not particularly small), then the computed value of $f(x_k)/f'(x_k)$ has the wrong sign, and the computed next iterate moves in the wrong direction. The same thing can happen with the secant and Regula Falsi methods. Thus, the notion of an interval of uncertainty about the root $x^\star$ applies to all of these methods.

Ill-conditioning

The effect of errors in f can be greatly magnified in some cases. In general, a problem is said to be *ill-conditioned* if small changes in the data of the problem can lead to large changes in the solution. Consider, for example, determining the linear interpolating polynomial (4.1.10) when the points x_0 and x_1 are very close together. The slope of the interpolating line is given by

$$\frac{f_1 - f_0}{x_1 - x_0}.$$

Suppose, for example, that $x_1 - x_0 = 10^{-6}$ and the exact values of f_1 and f_0 are $f_1 = 1$ and $f_0 = 1 + 10^{-6}$ so that the exact slope is 1. If, however, f_0 is evaluated incorrectly to be $1 + 2 \times 10^{-6}$, the computed slope will be 2. Note that the percentage error in the computed value of f_0 is only

$$100[(1 + 2 \times 10^{-6}) - (1 + 10^{-6})]\% = 10^{-4}\%,$$

but this has been magnified into a factor of two change in the computed slope.

The problem of ill-conditioning also can affect the determination of roots of a function. The simplest example of this is given by the trivial polynomial equation

$$x^n = 0,$$

which has an n-fold root equal to zero, and the polynomial equation

$$x^n = \varepsilon, \qquad \varepsilon > 0,$$

whose n roots are $\varepsilon^{1/n}$ times the nth roots of unity, and therefore all have absolute value of $\varepsilon^{1/n}$. If, for example, $n = 10$ and $\varepsilon = 10^{-10}$, the roots of the second polynomial have absolute value 10^{-1}; thus, a change of 10^{-10} in one coefficient (the constant term) of the original polynomial has caused changes 10^9 times as much in the roots.

This simple example is a special case of the general observation that if a root $x^\star$ of a polynomial f is of multiplicity m, then small changes of order ε in the coefficients of f may cause a change of order $\varepsilon^{1/m}$ in $x^\star$; an analogous result holds for functions other than polynomials by expanding in a Taylor series about $x^\star$. A necessary condition for a multiple root at $x^\star$ is that $f'(x^\star) = 0$. If $f'(x^\star) \neq 0$ but $f'(x)$ is small in the neighborhood of $x^\star$, then small changes in f can still cause large changes in $x^\star$, as Figure 4.3.6 illustrates.

Figure 4.3.6: *A Large Change in $x^\star$ Due to a Small Change in f*

Perhaps the most famous example of very ill-conditioned simple roots is given by the following. Let $f = (x-1)(x-2)\ldots(x-20)$ be the polynomial of degree 20 with roots $1, \ldots, 20$, and let $\hat{f}$ be the same polynomial but with the coefficient of x^{19} changed by $2^{-23} \doteq 10^{-7}$. Then, the roots of $\hat{f}$ to one decimal place are given by

$$1.0 \quad 2.0 \quad 3.0 \quad 4.0 \quad 5.0 \quad 6.0 \quad 7.0 \quad 8.0 \quad 8.9$$

$$10.1 \pm 0.6\mathrm{i} \quad 11.8 \pm 1.7\mathrm{i} \quad 14.0 \pm 2.5\mathrm{i} \quad 16.7 \pm 2.8\mathrm{i} \quad 19.5 \pm 1.9\mathrm{i} \quad 20.8.$$

Since the coefficient of x^{19} in $f(x)$ is 210, we see that a change in this one coefficient of about $10^{-7}\%$ has caused such large changes in the roots that some have even become complex!

Linearization in Several Variables

Suppose now that f is a function of several variables $x_1, \ldots, x_n$. It is possible to write a Taylor series expansion for such functions but we will discuss in detail only the first term of the expansion. This expansion about a point $\bar{x}_1, \ldots, \bar{x}_n$ is

$$f(x_1, x_2, \ldots, x_n) \doteq f(\bar{x}_1, \ldots, \bar{x}_n) + \sum_{j=1}^{n} \frac{\partial f}{\partial x_j}(\bar{x}_1, \ldots, \bar{x}_n)(x_j - \bar{x}_j), \quad (4.3.23)$$

4.3 Application to Root-Finding

where $\frac{\partial f}{\partial x_j}$ denotes the partial derivative of f with respect to x_j. The function on the right side of (4.3.23) is linear in the variables $x_1, \ldots, x_n$; thus, (4.3.23) is a linear approximation to f or a *linearization* of f about the point $\bar{x}_1, \ldots, \bar{x}_n$. For example, let

$$f(x_1, x_2) = 2x_1^2 + x_1 x_2 + 2x_2^2.$$

Then

$$\frac{\partial f}{\partial x_1}(x_1, x_2) = 4x_1 + x_2, \quad \frac{\partial f}{\partial x_2}(x_1, x_2) = x_1 + 4x_2.$$

If we take $\bar{x}_1 = 1, \bar{x}_2 = 2$, then the approximation (4.3.23) becomes

$$f(x_1, x_2) \doteq 12 + 6(x_1 - 1) + 9(x_2 - 2). \tag{4.3.24}$$

The right-hand side of (4.3.24) defines a plane in 3-space that is tangent at the point $(1, 2)$ to the surface defined by f. This is true, more generally, for (4.3.23): The right-hand side defines a "hyperplane" that is tangent at $\bar{x}_1, \ldots, \bar{x}_n$ to the surface defined by f.

Newton's Method for Systems of Equations

The linearization (4.3.23) provides the mechanism for extending Newton's method to systems of nonlinear equations. Let $f_1, \ldots, f_n$ be n functions of n variables, and consider the system of equations

$$f_i(x_1, \ldots, x_n) = 0, \quad i = 1, \ldots, n. \tag{4.3.25}$$

It will be convenient to write this system, using vector notation, as

$$\mathbf{F}(\mathbf{x}) = 0 \tag{4.3.26}$$

where $\mathbf{x} = (x_1, \ldots, x_n)^T$ and $\mathbf{F}(\mathbf{x})$ is the n-vector with components $f_1(\mathbf{x}), \ldots, f_n(\mathbf{x})$. For example, if

$$f_1(x_1, x_2) = x_1^2 + x_2^2 - 1, \quad f_2(x_1, x_2) = x_1^2 - x_2,$$

then

$$\mathbf{F}(\mathbf{x}) = \begin{bmatrix} x_1^2 + x_2^2 - 1 \\ x_1^2 - x_2 \end{bmatrix}. \tag{4.3.27}$$

Let $\mathbf{x}^\star$ be a solution of the system (4.3.26) and $\mathbf{x}^0$ an approximation to this solution. We expand each function f_i about $\mathbf{x}^0$ by (4.3.23):

$$f_i(\mathbf{x}) \doteq f_i(\mathbf{x}^0) + \sum_{j=1}^n \frac{\partial f_i}{\partial x_j}(\mathbf{x}^0)(x_j - x_j^0) \equiv \ell_i(\mathbf{x}). \tag{4.3.28}$$

We then solve the linear system of equations

$$\ell_i(\mathbf{x}) = 0, \quad i = 1, \ldots, n, \tag{4.3.29}$$

to obtain a new approximation $\mathbf{x}^1$ to $\mathbf{x}^\star$. This is *Newton's method* for systems of equations. Each $\ell_i(\mathbf{x})$ defines a tangent hyperplane to f_i at $\mathbf{x}^0$, and the common intersection of these n hyperplanes with the $\mathbf{x} = 0$ hyperplane is the new approximation $\mathbf{x}^1$.

We can write Newton's method in a slightly different form. The matrix

$$\mathbf{F}'(\mathbf{x}) = \left(\frac{\partial f_i(\mathbf{x})}{\partial x_j} \right) \qquad (4.3.30)$$

is the *Jacobian matrix* of $\mathbf{F}$. The ith row of this matrix consists of the partial derivatives of f_i and, hence, the summation term in (4.3.28) is the ith row of $\mathbf{F}'(\mathbf{x})$ times $\mathbf{x} - \mathbf{x}^0$. Therefore, the system (4.3.29) may be written as

$$\mathbf{F}'(\mathbf{x}^0)(\mathbf{x} - \mathbf{x}^0) = -\mathbf{F}(\mathbf{x}^0). \qquad (4.3.31)$$

After this system is solved to obtain a new approximation $\mathbf{x}^1$, the process can be repeated. Thus, the Newton iteration may be written in the form

Solve $\mathbf{F}'(\mathbf{x}^k)\boldsymbol{\delta}^k = -\mathbf{F}(\mathbf{x}^k)$ for $\boldsymbol{\delta}^k$, $\qquad$ (4.3.32a)

Set $\mathbf{x}^{k+1} = \mathbf{x}^k + \boldsymbol{\delta}^k$, $\qquad$ (4.3.32b)

for $k = 0, 1, \ldots$.

As an example, consider the function $\mathbf{F}$ of (4.3.27). The two curves $x_1^2 + x_2^2 - 1 = 0$ and $x_1^2 - x_2 = 0$ intersect at two points that are the two real solutions of the system $\mathbf{F}(\mathbf{x}) = 0$. These equations can be solved "exactly" by substituting $x_2 = x_1^2$ into the first equation to give the quadratic equation $x_2^2 + x_2 - 1 = 0$ with positive root

$$x_2 = -\tfrac{1}{2} + \tfrac{1}{2}\sqrt{5} = 0.61803399. \qquad (4.3.33a)$$

Then

$$x_1 = \sqrt{x_2} = \pm 0.78615138. \qquad (4.3.33b)$$

We now use Newton's method to obtain one of the solutions (4.3.33). The Jacobian matrix for (4.3.27) is

$$\mathbf{F}'(\mathbf{x}) = \begin{bmatrix} 2x_1 & 2x_2 \\ 2x_1 & -1 \end{bmatrix}.$$

Hence, the Newton system (4.3.32a) is

$$\begin{bmatrix} 2x_1^{(k)} & 2x_2^{(k)} \\ 2x_1^{(k)} & -1 \end{bmatrix} \begin{bmatrix} \delta_1^{(k)} \\ \delta_2^{(k)} \end{bmatrix} = - \begin{bmatrix} (x_1^{(k)})^2 + (x_2^{(k)})^2 - 1 \\ (x_1^{(k)})^2 - x_2^{(k)} \end{bmatrix}. \qquad (4.3.34)$$

4.3 Application to Root-Finding

Table 4.3.2: *Convergence of Newton's Method for (4.3.27)*

Iteration	x_1	x_2	Number of Correct Digits
0	0.5	0.5	0,0
1	0.87499999	0.62499999	0,1
2	0.79067460	0.61805555	1,4
3	0.78616432	0.61803399	4,8
4	0.78615138	0.61803399	8,8

Given $\mathbf{x}^k$, we solve this system for the correction $\boldsymbol{\delta}^{k+1}$, and then update $\mathbf{x}^k$ by (4.3.32b) to obtain the next iterate. Table 4.3.2 shows the first four iterations starting from the initial approximation $x_1^{(0)} = .5, x_2^{(0)} = .5$.

The last column of Table 4.3.2 shows the number of correct digits in the current approximate solution. Once the approximate solutions are close to the solution, the number of correct digits begins to double at each iteration. This is the property of quadratic convergence that was discussed for a single equation and also holds for systems of equations. We state the following basic theorem without proof.

THEOREM 4.3.2 *If* $\mathbf{F}$ *is two times continuously differentiable in a neighborhood of a solution* $\mathbf{x}^\star$ *of* $\mathbf{F}(\mathbf{x}) = 0$ *and* $\mathbf{F}'(\mathbf{x}^\star)$ *is nonsingular, then the Newton iterates (4.3.32) will converge to* $\mathbf{x}^\star$ *provided that* $\mathbf{x}^0$ *is sufficiently close to* $\mathbf{x}^\star$, *and they will have the property of quadratic convergence:*

$$||\mathbf{x}^{i+1} - \mathbf{x}^\star||_2 \leq c||\mathbf{x}^i - \mathbf{x}^\star||_2^2. \tag{4.3.35}$$

for some constant c.

The critical condition in this theorem is that $\mathbf{F}'(\mathbf{x}^\star)$ is nonsingular. If this is not true, quadratic convergence is lost although the iterates may still converge to a solution. Nonsingularity of the Jacobian matrix is the correct extension to systems of equations of the condition that $f'(x^\star) \neq 0$ for a single equation.

The computationally intensive parts of Newton's method are the evaluation of $\mathbf{F}(\mathbf{x}^k)$ and $\mathbf{F}'(\mathbf{x}^k)$, and the solution of the linear system (4.3.32a). The computation of $\mathbf{F}'(\mathbf{x}^k)$ requires, in general, the evaluation of the n^2 partial derivatives $\partial f_i / \partial x_j$. If n is large and/or the functions f_i are complicated, it can be drudgery to work out by hand – and then convert to computer code – the expressions for these derivatives; this can sometimes be mitigated by the use of symbolic differentiation techniques as discussed in Chapter 1. Another

130   Chapter 4 Polynomial Approximation

commonly used approach is to approximate the partial derivatives by

$$\frac{\partial f_i}{\partial x_j}(\mathbf{x}) \doteq \frac{1}{h}[f_i(x_1,\ldots,x_{j-1},x_j+h,x_{j+1},\ldots,x_n) - f_i(\mathbf{x})]. \qquad (4.3.36)$$

This has the advantage of requiring only the expressions for the f_i, which are needed in any case. But the actual numerical evaluation of the Jacobian matrix, either by expressions for the partial derivatives or by approximations such as (4.3.36), can be costly in computer time. On the other hand, these evaluations lend themselves naturally to parallel computation: each of the functions $f_i(\mathbf{x}^k)$ as well as all of the partial derivatives can be evaluated independently. However, there may be several common expressions in these functions and derivatives (for example, the two functions of (4.3.27) have the common expression x_1^2), in which case redundant work may be performed in a straight forward parallel implementation. If the Jacobian matrix is sparse with most of the partial derivatives identically zero, then different approaches may be necessary.

Another common modification of Newton's method when the Jacobian matrix is costly to evaluate is to do so only every k iterations. The extreme case of this, known as the *simplified Newton* method, evaluates $F'(\mathbf{x})$ only once so that in (4.3.32) $F'(\mathbf{x}^k)$ is replaced by $F'(\mathbf{x}^0)$. This iteration, however, may have a slow rate of convergence.

The second computationally intensive part of Newton's method is the solution of the linear system (4.3.32a). Solution of linear systems will be discussed in Chapters 6 - 9.

Supplementary Discussion and References: 4.3

1. A thorough treatment of the theory of iterative methods for roots of a single equation is given in Traub [1964]. For systems of equations, see Ortega and Rheinboldt [1970] and Dennis and Schnabel [1983].

2. An attractive alternative to symbolic differentiation or approximation of the partial derivatives in the Jacobian matrix by finite differences is *automatic differentiation*; see Griewank and Corliss [1991].

3. Many systems of equations arise in the attempt to minimize (or maximize) a function g of n variables. From the calculus, we know that if g is continuously differentiable, then a necessary condition for a local minimum is that the gradient vector vanishes:

$$\left(\frac{\partial g}{\partial x_1},\ldots,\frac{\partial g}{\partial x_n}\right) = 0.$$

4.3 Application to Root-Finding

Thus, by solving this system of equations, one obtains a possible local minimizer of g, and, in many situations, it will be known that this solution must indeed minimize g. Alternatively, if we are given an arbitrary system of equations $f_i(\mathbf{x}) = 0$, $i = 1, \ldots, n$, we can convert the solution of this system to a minimization problem by defining the function

$$g(\mathbf{x}) = \sum_{i=1}^{n} [f_i(\mathbf{x})]^2.$$

Clearly, g takes on a minimum value of zero only when all $f_i(\mathbf{x})$ are zero. This conversion, however, is usually not recommended for obtaining a numerical solution of the system since the ill-conditioning of the problem will be increased. (See Chapter 6.)

4. In many problems, the equations are to be solved for various values of one or more parameters. Suppose there is a single parameter α and we write the system of equations as

$$\mathbf{F}(\mathbf{x}; \alpha) = \mathbf{0}. \tag{4.3.37}$$

Assume that we wish solutions $\mathbf{x}_0^\star, \cdots, \mathbf{x}_N^\star$ for values $\alpha_0 < \alpha_1 < \cdots < \alpha_N$, where α_0 corresponds to an "easy" problem; for example, the equations for α_0 may be linear. If $\mathbf{x}_0^\star$ can be computed and if $|\alpha_1 - \alpha_0|$ is small, then we hope that $\mathbf{x}_0^\star$ is sufficiently close to $\mathbf{x}_1^\star$ so that $\mathbf{x}_0^\star$ is a suitable starting approximation for Newton's method for the equation $\mathbf{F}(\mathbf{x}; \alpha_1) = \mathbf{0}$. Continuing in this way, we use each previous solution as a starting approximation for the next problem. This is called the *continuation method*.

If the equations to be solved do not contain a parameter, we can always introduce one. For example, let $\mathbf{F}(\mathbf{x}) = \mathbf{0}$ be the system and let $\mathbf{x}^0$ be our best approximation to the solution (but not good enough that the Newton iteration will converge). Define a new set of equations depending on a parameter α by

$$\hat{\mathbf{F}}(\mathbf{x}; \alpha) = \mathbf{F}(\mathbf{x}) + (\alpha - 1)\mathbf{F}(\mathbf{x}^0) = \mathbf{0}, \qquad 0 \leq \alpha \leq 1. \tag{4.3.38}$$

Then, $\hat{\mathbf{F}}(\mathbf{x}; 0) = \mathbf{F}(\mathbf{x}) - \mathbf{F}(\mathbf{x}^0) = \mathbf{0}$, for which $\mathbf{x}^0$ is a solution, and $\hat{\mathbf{F}}(\mathbf{x}; 1) = \mathbf{F}(\mathbf{x}) = \mathbf{0}$, which is the system to be solved. Hence, we proceed as in the previous paragraph for parameters $0 = \alpha_0 < \alpha_1 < \cdots < \alpha_N = 1$.

The continuation method is closely related to *Davidenko's method*. Consider (4.3.38) and assume that for each $\alpha \in [0, 1]$, the equation defines a solution $\mathbf{x}(\alpha)$ that is continuously differentiable in α. Then, if we differentiate

$$\mathbf{F}(\mathbf{x}(\alpha)) + (\alpha - 1)\mathbf{F}(\mathbf{x}^0) = \mathbf{0}$$

with respect to α, we obtain by the chain rule

$$\mathbf{F}'(\mathbf{x}(\alpha))\mathbf{x}'(\alpha) + \mathbf{F}(\mathbf{x}^0) = 0,$$

or, assuming that the Jacobian matrix $\mathbf{F}'(\mathbf{x}(\alpha))$ is nonsingular,

$$\mathbf{x}'(\alpha) = -[\mathbf{F}'(\mathbf{x}(\alpha))]^{-1}\mathbf{F}(\mathbf{x}^0),$$

with the initial condition $\mathbf{x}(0) = \mathbf{x}^0$. The solution $\mathbf{x}(\alpha)$ of this initial-value problem at $\alpha = 1$ will, we hope, be the desired solution of the original system of equations $\mathbf{F}(\mathbf{x}) = 0$. In practice, we will have to solve the differential equations numerically, and we can, in principle, use any of the methods of Section 5.2. Although Davidenko's method and the continuation method are attractive possibilities, their reliability in practice has been less than desired. In particular, it is possible that the Jacobian matrix will become singular for some $\mathbf{x}(\alpha)$ with $\alpha < 1$, or even that the solution curve itself will blow up prematurely. For a review of ways of potentially overcoming some of these difficulties, see Allgower and Georg [1990].

EXERCISES 4.3

4.3.1. Let f be a continuously differentiable function. Use the mean-value theorem to show that if $f'(x) > 0$ for all x in the interval (a, b), then f has at most one root in the interval (a, b).

4.3.2. Let $f(x) = e^x$. Show that $f'(x) > 0$ for all x but f does not have any finite roots.

4.3.3. Let $f(x) = (x - 1)^2$. Show that $f'(1) = 0$ but the Newton iterates still converge to 1. What can you say about the rate of convergence?

4.3.4. Let f be twice continuously differentiable and suppose that $f'(x^*) = 0$ at a root x^* of f, but $f'(x) \neq 0$ in a neighborhood of x^*. Show that the Newton iterates converge to x^* if x^0 is sufficiently close to x^*.

4.3.5. Consider the function $f(x) \equiv x - x^3$ with roots 0 and ± 1.

  a. Show that Newton's method is locally convergent to each of the three roots.

  b. Carry out several steps of Newton's method starting with the initial approximation $x_0 = 2$. Discuss the rate of convergence that you observe in your computed iterates.

  c. Carry out several steps of both the bisection and secant methods starting with the interval $(\frac{3}{4}, 2)$. Compare the rate of convergence of the iterates from these methods with that of the Newton iterates.

  d. Determine the set of points S for which the Newton iterates will converge (in the absence of rounding errors) to the root 1 for any starting approximation x_0 in S. Do the same for the roots 0 and -1.

4.3 Application to Root-Finding

4.3.6. Consider the equation $x - 2\sin x = 0$.

   a. Show graphically that this equation has precisely three roots: 0, and one in each of the intervals $(\pi/2, 2)$ and $(-2, -\pi/2)$.

   b. Show that the iterates $x_{i+1} = 2\sin x_i$, $i = 0, 1, \ldots$, converge to the root in $(\pi/2, 2)$ for any x_0 in this interval.

   c. Apply the Newton iteration to this equation and ascertain for what starting values the iterates will converge to the root in $(\pi/2, 2)$. Compare the rate of convergence of the Newton iterates with those of part **b**.

4.3.7. Let n be a positive integer and α a positive number. Show that Newton's method for the equation $x^n - \alpha = 0$ is

$$x_{k+1} = \frac{1}{n}\left[(n-1)x_k + \frac{\alpha}{x_k^{n-1}}\right], \qquad k = 0, 1, \cdots,$$

and that this Newton sequence converges for any $x_0 > 0$. For the case $n = 2$, show the quadratic convergence relation

$$x_{k+1} - \sqrt{\alpha} = \frac{x_{k+1} + \sqrt{\alpha}}{(x_k + \sqrt{\alpha})^2}(x_k - \sqrt{\alpha})^2.$$

4.3.8. Ascertain whether the following statements are true or false in general and prove your assertions:

   a. Let $\{x_k\}$ be a sequence of Newton iterates for a continuously differentiable function f. If for some i, $|f(x_i)| \leq 0.01$ and $|x_{i+1} - x_i| \leq 0.01$, then x_{i+1} is within 0.01 of a root of $f(x) = 0$.

   b. The Newton iterates converge to the unique solution of $x^2 - 2x + 1 = 0$ for any $x_0 \neq 1$. (Ignore rounding error.)

4.3.9. Consider the equation $x^2 - 2x + 2 = 0$. What is the behavior of the Newton iterates for various real starting values?

4.3.10. Show that the Newton iterates converge to the unique solution of $e^{2x} + 3x + 2 = 0$ for any starting value x_0.

4.3.11. Assume that f is differentiable, convex, and $f'(x) < 0$ for all x. If $f(x) = 0$ has a solution x^*, show that x^* is unique and that the Newton iterates converge monotonically upward to x^* if $x_0 < x^*$. What happens if $x_0 > x^*$?

4.3.12. If $p \geq 2$ is an integer, show that the Newton iterates for the equation $x^p = 0$ converge to the solution $x^* = 0$ only linearly with an asymptotic convergence factor of $(p-1)/p$.

4.3.13. Newton's method can be used for determining the reciprocal of numbers when division is not available.

a. Show how Newton's method can be applied to the equation

$$f(x) = \frac{1}{x} - a,$$

without using division.

b. Give an equation for the error term, $e_k = x_k - a^{-1}$, and show that the convergence is quadratic.

c. Give conditions on the initial approximation so that $x_k \to a^{-1}$ as $k \to \infty$. If $0 < a < 1$, give a numerical value of x_0 that will guarantee convergence.

4.3.14. Show graphically that the system of equations $x_1^2 + x_2^2 = 1$, $x_1^2 - x_2 = 0$ has precisely two real solutions.

4.3.15. Compute the Jacobian matrix $\mathbf{F}'(\mathbf{x})$ for

$$\mathbf{F}(\mathbf{x}) = \begin{bmatrix} x_1^2 + x_1 x_2 x_3 + x_3^3 \\ x_1^3 x_2 + x_2 x_3^2 \\ x_1/x_2^3 \end{bmatrix}.$$

4.3.16. For the functions of Exercise 4.3.14, compute the tangent planes at $x_1 = 2$, $x_2 = 2$.

4.3.17. For the equations of Exercise 4.3.14, for what points $\mathbf{x}$ is the Jacobian matrix nonsingular?

4.3.18. Let $f(x) = (x-a)^2 p(x)$, where a is real and p is a polynomial. Show that the rate of convergence of Newton's method to the root a is only linear but the rate of convergence of the method

$$x_{k+1} = x_k - 2\frac{f(x_k)}{f'(x_k)}$$

is quadratic. Extend this to the case that $f(x) = (x-a)^m p(x)$.

4.3.19. (Bairstow's Method) Suppose that p is a polynomial of even degree n and that all roots are complex. Show how to modify Newton's method so as to obtain a quadratic factor $x^2 + \alpha x + \beta$ with real α and β.

4.3.20. Let $q_0(x), \ldots, q_n(x)$ be a set of orthogonal polynomials that satisfy the three-term recurrence relation (4.2.20). Suppose that

$$p(x) = \sum_{j=0}^{n} \gamma_j q_j(x).$$

Show how to apply Newton's method to obtain solutions of $p(x) = 0$, taking advantage of (4.2.20).

4.3 Application to Root-Finding

4.3.21. Let
$$\mathbf{F}(\mathbf{x}) = \frac{1}{3} \begin{bmatrix} x_1^3 + 3x_1 x_2^2 \\ 3x_1^2 x_2 + x_2^3 \end{bmatrix}.$$
Show that for any $\mathbf{x}^0$ with $|x_1^{(0)}| \neq |x_2^{(0)}|$, the Newton iterates are well-defined and converge to zero.

4.3.22. Denote the roots of the quadratic $x^2 + bx + c$ by $x_\pm = x_\pm(b,c)$, considered as functions of b and c. Show that
$$\frac{\partial x_\pm}{\partial b} = -\frac{1}{2} \pm \frac{1}{2} \frac{b}{\sqrt{b^2 - 4c}}, \qquad \frac{\partial x_\pm}{\partial c} = \mp \frac{1}{\sqrt{b^2 - 4c}}$$
and, therefore, these derivatives are infinite at a multiple root. Discuss the relationship of this result to ill conditioning of the roots.

Chapter 5

Continuous Problems Solved Discretely

5.1 Numerical Integration

By a "continuous problem," we mean one that uses data on a whole interval or, more generally, some region of higher dimensional space. Perhaps the simplest example is the problem of finding the integral

$$I(f) = \int_a^b f(x)dx, \qquad (5.1.1)$$

which depends on values of f on the whole interval $[a, b]$, an "uncountably infinite" number of points. Although it is possible to integrate certain simple functions in "closed form," for example

$$\int_a^b x^2 dx = \frac{1}{3}(b^3 - a^3),$$

many functions that arise in applications cannot be integrated in this way and their integral must be approximated. This might also be the case even if an explicit expression such as an infinite series were given for the integral, but this expression was too complicated to evaluate. Moreover, in many cases the function f may not even be given explicitly, although it can be computed for any value in $[a, b]$ by a computer program. Or, sometimes, only a table of values $\{x_i, f(x_i)\}$ is given for a fixed, finite number of points x_i in the interval.

In this section, we will develop approximations of the form

$$\int_a^b f(x)dx \doteq \sum_{i=0}^n \alpha_i f(x_i) \qquad (5.1.2)$$

to the integral (5.1.1). Thus, the integral is approximated by a linear combination of values of the function f at a finite number of points x_i in the interval $[a, b]$. One way to obtain such approximations, often called *quadrature rules*, is to replace the function $f(x)$ by some other function, whose integral is relatively easy to evaluate. Any class of simple functions may be used to approximate $f(x)$, such as polynomials, piecewise polynomials, and trigonometric, exponential, or logarithmic functions. The choice of the class of functions used may depend on some particular properties of the integrand but the most common choices, which we will use here, are polynomials or piecewise polynomials.

The Newton-Cotes Formulas

The simplest polynomial is a constant. In the *rectangle rule*, f is approximated by its value at the end point a (or, alternatively, at b) so that

$$I(f) \doteq R(f) = (b-a)f(a). \tag{5.1.3}$$

We could also approximate f by another constant obtained by evaluating f at a point interior to the interval; the most common choice is $(a+b)/2$, which gives the *midpoint rule*

$$I(f) \doteq M(f) = (b-a)f\left(\frac{a+b}{2}\right). \tag{5.1.4}$$

The rectangle and midpoint rules are illustrated in Figure 5.1.1, in which the areas of the shaded rectangles are the approximations to the integral.

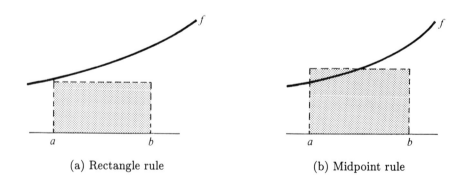

(a) Rectangle rule (b) Midpoint rule

Figure 5.1.1: *The Rectangle Rule (a) and Midpoint Rule (b)*

The next simplest polynomial is a linear function. If it is chosen so that it agrees with f at the end points a and b, then a trapezoid is formed, as illustrated in Figure 5.1.2(a). The area of this trapezoid – the integral of the

5.1 Numerical Integration

linear function — is the approximation to the integral of f and is given by

$$I(f) \doteq T(f) = \frac{(b-a)}{2}[f(a) + f(b)]. \tag{5.1.5}$$

This is known as the *trapezoidal rule*.

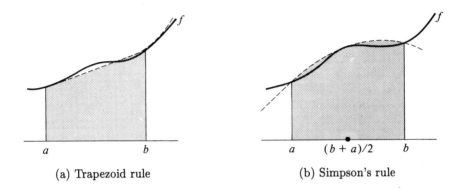

(a) Trapezoid rule   (b) Simpson's rule

Figure 5.1.2: *The Trapezoid Rule (a) and Simpson's Rule (b)*

To obtain one further formula, we next approximate f by an interpolating quadratic polynomial that agrees with f at the end points a and b and the midpoint $(a+b)/2$. The integral of this quadratic is given by (see Exercise 5.1.1)

$$I(f) \doteq S(f) = \frac{(b-a)}{6}\left[f(a) + 4f\left(\frac{a+b}{2}\right) + f(b)\right], \tag{5.1.6}$$

which is *Simpson's rule* and is illustrated in Figure 5.1.2(b). We note that Simpson's rule may also be written as a linear combination of the trapezoid rule and the midpoint rule since

$$\frac{1}{6}\left[f(a) + 4f\left(\frac{a+b}{2}\right) + f(b)\right] = \frac{1}{3}\left[\frac{f(a)+f(b)}{2}\right] + \frac{2}{3}f\left(\frac{a+b}{2}\right).$$

We can continue the preceding method of generating quadrature formulas by using polynomials of still higher degree. The interval $[a,b]$ is divided by $n+1$ equally spaced points, including the two endpoints a and b, and an interpolating polynomial of degree n is constructed to agree with f at these $n+1$ points. This polynomial is then integrated from a to b to give an approximation to the integral. Such quadrature formulas are called the *Newton-Cotes formulas*, and may be written in the form (5.1.2).

140   Chapter 5 Continuous Problems Solved Discretely

Another approach to the Newton-Cotes formulas is to require that the formula (5.1.2) be exact for polynomials of degree n. Then, in particular, it must be exact for $1, x, x^2, \ldots, x^n$. Thus, we must have

$$\sum_{i=0}^{n} \alpha_i x_i^j = \frac{b^{j+1} - a^{j+1}}{j+1}, \qquad j = 0, 1, \ldots, n, \qquad (5.1.7)$$

where the right-hand sides are the exact integrals of the powers of x. The relations (5.1.7) constitute a system of linear equations for the unknown coefficients α_i. The coefficient matrix is the Vandermonde matrix of (4.1.13) and is nonsingular if the x_i are all distinct. If the x_i are equally spaced, then this approach again gives the Newton-Cotes formulas.

Composite Formulas

It is clear from Figures 5.1.1 and 5.1.2 that if f is approximated by a low degree polynomial, the error may be considerable. To obtain better accuracy one might be tempted to use higher degree polynomials but this is usually not satisfactory. Although for $n \leq 7$, the coefficients α_i in the Newton-Cotes formulas are all positive, beginning with $n = 8$, some coefficients will be negative. This has a deleterious effect on rounding error since cancellations will occur. Moreover, it is not guaranteed that the right-hand side of (5.1.2) will converge to the integral as $n \to \infty$, even for infinitely differentiable functions. (See the Supplementary Discussion of Section 4.1.)

An alternative, and usually much more satisfactory approach, is through the use of piecewise polynomial functions. Thus, we partition the interval $[a, b]$ into n subintervals $[x_{i-1}, x_i]$, $i = 1, \ldots, n$, where $x_0 = a$ and $x_n = b$. Then,

$$I(f) = \int_a^b f(x)dx = \sum_{i=1}^{n} \int_{x_{i-1}}^{x_i} f(x)dx.$$

If we apply the rectangle rule to each subinterval $[x_{i-1}, x_i]$, we obtain the *composite rectangle rule*

$$I(f) \doteq I_{CR}(f) = \sum_{i=1}^{n} h_i f(x_{i-1}), \qquad (5.1.8)$$

where $h_i = x_i - x_{i-1}$. Similarly, the *composite midpoint, trapezoid*, and *Simpson's rules* are obtained in the same way by applying the basic rule to each subinterval; they are given by

$$I_{CM}(f) = \sum_{i=1}^{n} h_i f\left(\frac{x_i + x_{i-1}}{2}\right), \qquad (5.1.9)$$

5.1 Numerical Integration

$$I_{CT}(f) = \sum_{i=1}^{n} \frac{h_i}{2}[f(x_{i-1}) + f(x_i)], \tag{5.1.10}$$

$$I_{CS}(f) = \frac{1}{6}\sum_{i=1}^{n} h_i \left[f(x_{i-1}) + 4f\left(\frac{x_{i-1}+x_i}{2}\right) + f(x_i)\right]. \tag{5.1.11}$$

These composite rules may all be obtained by approximating the integrand f on the interval $[a,b]$ by a piecewise polynomial function, and then integrating this function to obtain an approximation to the integral. For the midpoint and rectangle rules, the approximating function is piecewise constant, for the trapezoid rule, it is piecewise linear, and for Simpson's rule, it is piecewise quadratic.

Discretization Error

We next consider the errors that can occur in using the quadrature formulas that have been described. There are two sources of error. The first is rounding error in evaluating the formulas; we will return to this shortly. The second is the *discretization* (or *truncation*) *error* made in replacing the integral by a "discrete" approximation of the form (5.1.2), which involves only finitely many values of the integral f.

If f is approximated by an interpolating polynomial p of degree n over the interval $[a,b]$, the error in the approximation to the integral is

$$E = \int_a^b [f(x) - p(x)]dx. \tag{5.1.12}$$

By Theorem 4.1.2, this can be written as

$$E = \frac{1}{(n+1)!}\int_a^b (x-x_0)\cdots(x-x_n)f^{(n+1)}(z(x))dx, \tag{5.1.13}$$

where $x_0, x_1, \ldots, x_n$ are the interpolation points, and $z(x)$ is a point in the interval $[a,b]$ that depends on x.

For the rectangle rule (5.1.3), $n=0$ and $x_0 = a$; hence, (5.1.13) becomes

$$|E_R| = \left|\int_a^b (x-a)f'(z(x))dx\right| \leq M_1 \int_a^b (x-a)dx = \frac{M_1}{2}(b-a)^2, \tag{5.1.14}$$

where M_1 is a bound for $|f'(x)|$ over the interval $[a,b]$. For the trapezoid rule (5.1.5), $n=1$, $x_0 = a$, and $x_1 = b$. Hence, again applying (5.1.13), we have

$$|E_T| = \frac{1}{2}\left|\int_a^b (x-a)(x-b)f''(z(x))dx\right| \leq \frac{M_2}{12}(b-a)^3, \tag{5.1.15}$$

where M_2 is a bound on $|f''(x)|$ over $[a, b]$.

Consider next the midpoint rule (5.1.4), in which $n = 0$ and $x_0 = (a+b)/2$. If we apply (5.1.13) and proceed as in (5.1.14), we obtain

$$|E_M| = \left| \int_a^b \left[x - \frac{(a+b)}{2} \right] f'(z(x)) dx \right| \leq \frac{M_1}{4}(b-a)^2. \tag{5.1.16}$$

This, however, is not as good a bound as we can obtain. We shall instead expand the integrand of (5.1.12) in a Taylor series about $m = (a+b)/2$. Since the interpolating polynomial $p(x)$ is just the constant $f(m)$, this gives

$$f(x) - p(x) = f'(m)(x-m) + \tfrac{1}{2}f''(z(x))(x-m)^2,$$

where z is a point in the interval and depends on x. Thus, the error in the midpoint rule is

$$\begin{aligned}
|E_M| &= \left| \int_a^b [f'(m)(x-m)dx + \tfrac{1}{2}f''(z(x))(x-m)^2]dx \right| \quad (5.1.17) \\
&\leq \left| f'(m) \int_a^b (x-m)dx \right| + \frac{1}{2} \left| \int_a^b f''(z(x))(x-m)^2 dx \right| \\
&\leq \frac{M_2}{24}(b-a)^3,
\end{aligned}$$

since

$$\int_a^b (x-m)dx = 0, \qquad \int_a^b (x-m)^2 dx = \frac{(b-a)^3}{12}.$$

In a similar way, we can derive the following bound for the error in Simpson's rule (5.1.6), which we state without proof (M_4 is a bound for the fourth derivative):

$$|E_S| \leq \frac{M_4}{2880}(b-a)^5. \tag{5.1.18}$$

The above error bounds all involve powers of the length, $b - a$, of the interval, and unless this length is small, the bounds will not, in general, be small. However, we will only apply them to sufficiently small intervals that we obtain by subdividing the interval $[a, b]$. In particular, for the composite formulas (5.1.8) - (5.1.11) we can apply the error bounds on each subinterval $[x_{i-1}, x_i]$. For example, for the rectangle rule, we use (5.1.14) to obtain the following bound on the error in the composite rule:

$$E_{CR} \leq \frac{M_1}{2} \sum_{i=1}^n h_i^2. \tag{5.1.19}$$

5.1 Numerical Integration

Note that we have used the maximum M_1 of $|f'(x)|$ on the whole interval $[a, b]$, although a better bound in (5.1.19) could be obtained if we used the maximum of $|f'(x)|$ separately on each subinterval.

In the special case that the subintervals are all of the same length, $h_i = h = (b-a)/n$, (5.1.19) becomes

$$E_{CR} \leq \frac{M_1}{2}(b-a)h \quad \text{(Composite rectangle rule error)}, \quad (5.1.20)$$

which shows that the composite rectangle rule is a first-order method; that is, the error reduces only linearly in h. In a similar fashion, we can obtain bounds for the errors in the other composite rules by using (5.1.15), (5.1.17), and (5.1.18). The following bounds are given in the case that the intervals are all of the same length h:

$$E_{CM} \leq \frac{M_2}{24}(b-a)h^2 \quad \text{(Composite midpoint rule error)}, \quad (5.1.21)$$

$$E_{CT} \leq \frac{M_2}{12}(b-a)h^2 \quad \text{(Composite trapezoid rule error)}, \quad (5.1.22)$$

$$E_{CS} \leq \frac{M_4}{2880}(b-a)h^4 \quad \text{(Composite Simpson's rule error)}. \quad (5.1.23)$$

Thus, the composite midpoint and trapezoid rules are both second-order, whereas the composite Simpson rule is fourth-order. Because of its relatively high accuracy and simplicity, the composite Simpson's rule is an often-used method.

A practical difficulty with the above quadrature rules is that some choice of the step sizes, h_j, must be made. If the integrand f is varying rapidly in one part of the interval but slowly in another part, then it is intuitively clear that the lengths h_i should be much smaller where the function is rapidly varying; a constant h on the whole interval $[a, b]$ would not be efficient. However, it is sometimes difficult to know just how small to choose the h_i to achieve a suitable overall accuracy. In practice, high-quality quadrature software will employ an automatic adaptive scheme that will vary the step size depending on estimates of the error obtained during the computation. One simple technique is to divide the current h_i by 2 and revaluate the approximate integral. This process is then continued until there is suitable agreement between successive approximations.

Rounding Error

We next give a brief discussion of rounding error. Consider, for example, the composite trapezoid rule (5.1.10) with constant interval length h. In general, there will be rounding error in evaluating f at the points x_i, and there will be rounding error in forming the linear combination of these values of f. Thus, the computed approximation will be

$$\sum_{i=1}^{n} \frac{h}{2}[f(x_{i-1}) + \varepsilon_{i-1} + f(x_i) + \varepsilon_i] + \eta,$$

where ε_i is the error in evaluating $f(x_i)$ and η is the error in forming the linear combination. It is possible to bound the effects of these errors in terms of bounds on the ε_i and η. Rather than do this, however, we will content ourselves with the following intuitive discussion. As the error estimate (5.1.22) shows, provided that f has a bounded second derivative on $[a, b]$, the discretization error goes to zero as h goes to zero. Hence, we can make the discretization error as small as we like by making h sufficiently small. However, the smaller h is, the more evaluations of f and the more terms in the linear combination. Thus, in general, the smaller h is, the larger will be the rounding error. In practice, for a fixed word length in the computer arithmetic, there will be a size of h below which the rounding error will become the dominant contribution to the overall error. The situation is depicted schematically in Figure 5.1.3 in which h_0 is the practical minimum that can be used. This minimum h is very difficult to ascertain in advance, but for problems for which only a moderate accuracy is required, the h actually used will be far larger than this minimum, and the discretization error will be the dominant contributor to the error. The same general behavior occurs in all the integration methods, as well as the methods for differential equations to be discussed later, although the minimum h will vary from method to method and problem to problem.

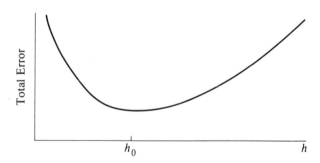

Figure 5.1.3: *Error in Trapezoid and Other Methods*

Parallel and Vector Computation

The general formula (5.1.2), or the more specific ones (5.1.8) - (5.1.11), are well-suited for vector and parallel computing. We consider first the vectorization. The same sequence of arithmetic operations to evaluate $f(x)$ can be

5.1 Numerical Integration

applied using vector instructions to evaluate $f(x_1), \ldots, f(x_n)$. For example, suppose that $f(x) = 3x^2 + x$. Then

$$\begin{bmatrix} f(x_1) \\ f(x_2) \\ \vdots \\ f(x_n) \end{bmatrix} = 3 \begin{bmatrix} x_1^2 \\ x_2^2 \\ \vdots \\ x_n^2 \end{bmatrix} + \begin{bmatrix} x_1 \\ x_2 \\ \vdots \\ x_n \end{bmatrix},$$

so that the vector operations would be

$$\mathbf{y} = \mathbf{x} \star \mathbf{x}, \qquad \mathbf{f} = 3\mathbf{y} + \mathbf{x}. \tag{5.1.24}$$

The $\star$ in (5.1.24) denotes the component-wise product of the vector $\mathbf{x}$ with itself, and the second vector operation of (5.1.24) is an axpy, as discussed in Chapter 3. Once the vector of function values has been obtained, the inner product $\sum_{i=1}^{n} \alpha_i f(x_i)$ is computed.

For parallel computation, assume that $x_1, \ldots, x_n$ have been assigned to the processors as indicated in Figure 5.1.4. Each processor will then compute the values $f(x_i)$ for those x_i assigned to it. This is essentially perfectly parallel although if the number of processors, p, does not divide n evenly, then there will be a small load imbalance. For example, if $n = 200$ and $p = 16$, then 12 of the x_i will be assigned to half the processors and 13 to the other half.

$$\boxed{x_o, \ldots, x_j} \quad \boxed{x_{j+1}, \ldots, x_k} \quad , \ldots \quad \boxed{x_\ell, \ldots, x_n}$$
$$P_1 \qquad\qquad P_2 \qquad\qquad\qquad P_p$$

Figure 5.1.4: *Assignment of Data to Processors*

Once the $f(x_i)$ have been computed, each processor will compute its portion of the inner product $\sum \alpha_i f(x_i)$. Then, these partial inner products can be added by a fan-in. On a distributed memory machine this addition will require communication between processors, as discussed in Chapter 3.

Gaussian Quadrature

We next consider an important class of quadrature formulas, which we discuss for integrals of the form

$$I(f) = \int_a^b f(x)\omega(x)dx. \tag{5.1.25}$$

Here, w is a non-negative function, and a and b need not be finite. We consider a quadrature formula of the type

$$I_G(f) = \sum_{i=1}^{n} \alpha_i f(x_i)\omega(x_i) \tag{5.1.26}$$

but, as opposed to the previous formulas we have discussed, we allow the nodes x_i to also be unknowns.

The criterion for determining both the α_i and the x_i in (5.1.26) is that the quadrature formula be exact for polynomials of as high a degree as possible. Since there are $2n$ unknowns in (5.1.26) we hope that the formula would be exact for polynomials of degree $2n - 1$. Thus, we wish that

$$\sum_{i=1}^{n} \alpha_i x_i^j w(x_i) = \mu_j = \int_a^b x^j w(x) dx, \quad i = 0, \ldots, 2n - 1, \qquad (5.1.27)$$

where the μ_j are called the *moments*. The equations (5.1.27), which correspond to (5.1.7) for the Newton-Cotes formulas, constitute a system of $2n$ equations in the $2n$ unknowns $\alpha_1, \ldots, \alpha_n$ and $x_1, \ldots, x_n$. These equations are nonlinear, but we will be able to solve them.

As an example, suppose that $n = 2$. Then (5.1.27) is the nonlinear system

$$\alpha_1 + \alpha_2 = \mu_0, \qquad \alpha_1 x_1 + \alpha_2 x_2 = \mu_1, \qquad (5.1.28a)$$

$$\alpha_2 x_1^2 + \alpha_2 x_2^2 = \mu_2, \qquad \alpha_1 x_1^3 + \alpha_2 x_2^3 = \mu_3. \qquad (5.1.28b)$$

Now let x_1 and x_2 be the roots of the *characteristic polynomial*

$$p(x) = (x - x_1)(x - x_2) \equiv x^2 + \beta_1 x + \beta_0, \qquad (5.1.29)$$

with $\beta_1 = -(x_1 + x_2)$ and $\beta_0 = x_1 x_2$. We can then reduce the system (5.1.28) to two equations for β_1 and β_0 (see Exercise 5.1.7):

$$\mu_0 \beta_0 + \mu_1 \beta_1 = -\mu_2, \qquad (5.1.30a)$$

$$\mu_1 \beta_0 + \mu_2 \beta_1 = -\mu_3. \qquad (5.1.30b)$$

Once the μ's are computed by (5.1.27), we can solve (5.1.30) for β_0 and β_1, then obtain x_1 and x_2 as the roots of (5.1.29), and finally determine α_1 and α_2 from the system (5.1.28a).

We next continue this example, for some specific data. Let $a = 0, b = 1$ and $w(x) = x(1-x)$. Then μ_j can be obtained from (5.1.27) as (see Exercise 5.1.8)

$$\mu_0 = \frac{1}{6}, \quad \mu_1 = \frac{1}{12}, \quad \mu_2 = \frac{1}{20}, \quad \mu_3 = \frac{1}{30}. \qquad (5.1.31)$$

The system (5.1.30) is then

$$5\beta_1 + 10\beta_0 = -3, \qquad 3\beta_1 + 5\beta_0 = -2,$$

which has the solution $\beta_1 = -1, \beta_0 = \frac{1}{5}$. Thus, $p(x) = x^2 - x + \frac{1}{5}$, which has the roots

$$x_1 = \frac{1}{2} + \frac{1}{10}\sqrt{5} \doteq 0.7236068, \qquad x_2 = \frac{1}{2} - \frac{1}{10}\sqrt{5} \doteq 0.2763932. \qquad (5.1.32)$$

5.1 Numerical Integration

Using these x_i, we obtain $\alpha_1 = \alpha_2 = \frac{1}{12}$ from (5.1.28a) and thus the quadrature formula (5.1.26) is

$$I_G(f) = \frac{1}{12}[f(x_1)x_1(1-x_1) + f(x_2)x_2(1-x_2)]$$

with x_1 and x_2 given by (5.1.32).

We next consider the general case. We assume that the μ_j of (5.1.27) can be computed. The system of equations corresponding to (5.1.30) is then

$$\sum_{i=0}^{n-1} \mu_{i+k}\beta_i = -\mu_{n+k}, \qquad k = 0, \ldots, n-1,$$

or
$$M\boldsymbol{\beta} = -\boldsymbol{\mu}, \tag{5.1.33}$$

where

$$M = \begin{bmatrix} \mu_o & \mu_1 & \cdots & \mu_{n-1} \\ \mu_1 & \mu_2 & \cdots & \mu_n \\ \vdots & & & \\ \mu_{n-1} & \cdots & & \mu_{2n-1} \end{bmatrix}$$

is the *moment matrix*, and is a Hankel matrix (see Section 4.2). The solution vector $\boldsymbol{\beta}$ of (5.1.33) gives the coefficients of the characteristic polynomial

$$p(x) = x^n + \beta_{n-1}x^{n-1} + \cdots + \beta_1 x + \beta_0, \tag{5.1.34}$$

whose roots are $x_1, \ldots, x_n$. Once the x_i are known, (5.1.27) then becomes a linear system for the α_i.

The basic properties of the x_i and α_i, which we state without proof, are

1. The x_i are real, distinct and lie in the interval (a, b).

2. The α_i are all non-negative.

Moreover, the error in the approximation $I_G(f)$ is given by

$$I(f) - I_G(f) = \frac{f^{(2n)}(z)}{2n!} \int_a^b [p(x)]^2 \omega(x) dx,$$

where z is some point in (a, b) and $p(x)$ is given by (5.1.34).

We note that for large n the above procedure is not the best way to carry out Gaussian quadrature computationally.

148 Chapter 5 Continuous Problems Solved Discretely

Supplementary Discussion and References: 5.1

1. We remarked in the text that Simpson's rule can be viewed as a linear combination of the trapezoid and midpoint rules. By taking suitable linear combinations of the composite trapezoid rule for different spacings h, we can also derive higher-order quadrature formulas. This is known as *Romberg integration* and is a special case of *Richardson extrapolation to the limit*, which will be discussed in the next section. The basis for the derivation of Romberg integration is that the composite trapezoid approximation can be shown to satisfy

$$T(h) = I(f) + c_2 h^2 + c_4 h^4 + \cdots + c_{2m} h^{2m} + O(h^{2m+2}), \qquad (5.1.35)$$

where the c_i depend on f and the interval, but are independent of h. The expansion (5.1.35) holds provided that f has $2m + 2$ derivatives. Now, define a new approximation to the integral by

$$T_1(h) = \tfrac{1}{3}\left[4T\left(\frac{h}{2}\right) - T(h)\right]. \qquad (5.1.36)$$

The coefficients of this linear combination have been chosen so that when the error in (5.1.36) is computed using (5.1.35), the coefficient of the h^2 term is zero. Thus,

$$T_1(h) = I(f) + c_4^{(1)} h^4 + \cdots + O(h^{2m+2}),$$

so that T_1 is a fourth-order approximation to the integral. One can continue the process by combining $T_1(h)$ and $T_1(h/2)$ in a similar fashion to eliminate the h^4 term in the error for T_1. More generally, we can construct the triangular array

$$T(h)$$
$$T(h/2) \quad T_1(h)$$
$$T(h/4) \quad T_1(h/2) \quad T_2(h)$$
$$\vdots \qquad \vdots \qquad \ddots$$

where

$$T_k\left(\frac{h}{2^{j-1}}\right) = \frac{4^j T_{k-1}\left(\frac{h}{2^j}\right) - T_{k-1}\left(\frac{h}{2^{j-1}}\right)}{4^j - 1}.$$

The elements in the ith column of this array converge to the integral at a rate depending on h^{2i}. But, provided that f is infinitely differentiable, the elements on the diagonal of the array converge at a rate faster than any power of h.

5.1 Numerical Integration

2. We have not touched at all upon several other important topics in numerical integration: use of splines or other functions to approximate the integrand, techniques for handling integrands with a singularity, multiple integrals, and adaptive procedures that attempt to fit the grid spacing automatically to the integrand. For a discussion of these matters, as well as further reading on the topics covered in this section, see Davis and Rabinowitz [1984] and Stroud [1971].

3. Although numerical integration has excellent properties for parallel computing, the power of parallel computers is not really needed unless the integrand is very complicated. For multiple integrals, however, much more computation may be required and parallel computers may then be very appropriate.

EXERCISES 5.1

5.1.1. Give the interpolating quadratic polynomial that agrees with f at the three points a, b, and $(a+b)/2$. Then integrate this quadratic from a to b to obtain Simpson's rule (5.1.6).

5.1.2. Show that the trapezoid rule integrates any linear function exactly and that Simpson's rule integrates any cubic polynomial exactly. (*Hint:* Expand the cubic about the midpoint.)

5.1.3. Apply the rectangle, midpoint, trapezoid, and Simpson's rules to the function $f(x) = x^4$ on the interval $[0, 1]$. Compare the actual error in the approximations to the bounds given by (5.1.14), (5.1.15), (5.1.17), and (5.1.18).

5.1.4. Based on the bound (5.1.22), how small would h need to be in order to guarantee an error no larger than 10^{-6} in the composite trapezoid rule approximation for $f(x) = x^4$ on $[0, 1]$. How small for the composite Simpson's rule?

5.1.5. Write a computer program to carry out the composite trapezoid and Simpson's rules for an "arbitrary" function on the interval $[a, b]$ and with an arbitrary subdivision of $[a, b]$. Test your program on $f(x) = x^4$ on $[0, 1]$ and find the actual h needed, in the case of an equal subdivision, to achieve an error of less than 10^{-6} with the composite trapezoid rule. Do the same for $f(x) = e^{-x^2}$.

5.1.6. Derive the four-point quadrature formula based on interpolation of the integrand by a cubic polynomial at equally spaced points. (Hint: The calculation can be simplified somewhat.)

5.1.7. Solve the two equations of (5.1.28a) for α_1 and α_2 in terms of μ_0, μ_1, x_1 and x_2. Then use these expressions for α_1 and α_2 in (5.1.28b) to obtain (5.1.30).

5.1.8. For $a = 0, b = 1$ and $w(x) = x(1-x)$, show that μ_j, as defined by (5.1.27), is $(j+1)^{-1}(j+2)^{-1}$, and then confirm (5.1.31).

5.1.9. Use the Hermite interpolation polynomial of Exercise 4.1.18 to obtain the Hermite Quadrature Formula

$$\int_0^b f(x)dx \doteq \int_a^b \sum_{i=0}^n [H_i(x)f(x_i) + h_i(x)f'(x_i)]dx.$$

Use this general formula to obtain a specific formula for $n = 1, x_0 = a, x_1 = b$ and compare with the other quadrature formulas of this section.

5.2 Initial Value Problems

The problem of obtaining the integral of f may be viewed as solving the differential equation

$$y'(x) = f(x), \qquad y(a) = 0, \qquad (5.2.1)$$

whose solution is

$$y(x) = \int_a^x f(s)ds. \qquad (5.2.2)$$

If the function f in (5.2.1) also depends on y, and we allow a general initial condition, then (5.2.1) becomes

$$y'(x) = f(x, y(x)), \qquad y(a) = \alpha, \qquad a \leq x. \qquad (5.2.3)$$

This is an *initial value problem* for an ordinary differential equation.

Initial value problems for ordinary differential equations arise in a large number of applications, including computation of rocket trajectories, chemical reactions and biological problems. In most applications, we will be concerned with systems of equations

$$y_i'(x) = f_i(x, y_1(x)), \ldots, y_n(x)), \qquad i = 1, \ldots, n, \qquad a \leq x, \qquad (5.2.4)$$

with initial conditions

$$y_i(a) = \alpha_i, \qquad i = 1, \ldots, n. \qquad (5.2.5)$$

A simple example is the system

$$y_1'(t) = c_{11}y_1(t) + c_{12}y_1(t)y_2(t) \qquad (5.2.6\text{a})$$

$$y_2'(t) = c_{21}y_2(t) + c_{22}y_1(t)y_2(t), \qquad (5.2.6\text{b})$$

5.2 Initial Value Problems

where the independent variable is time, and the constants c_{ij} satisfy $c_{ii} > 0, i = 1, 2, c_{ij} < 0, i \neq j$. These are the *Lotka-Volterra* equations, also called the *predator-prey* equations.

We will first develop methods for the single equation (5.2.3) and then show how they can be extended to systems of equations.

Euler's Method

The simplest method for (5.2.3) arises by approximating $y'(x)$ by the *finite-difference quotient*

$$y'(x) \doteq \frac{1}{h}[y(x+h) - y(x)]. \tag{5.2.7}$$

Using this approximation in (5.2.3) for $x = a$ gives

$$y(a + h) \doteq y(a) + hf(x, y(a)). \tag{5.2.8}$$

Since $y(a)$ is the known initial condition α, the right-hand side of (5.2.8) can be evaluated to give an approximation to the solution at $a + h$. The process can then be continued as follows. Let $x_k = a + kh, k = 0, 1, \ldots$, and define the approximations y_{k+1} to the solution at x_{k+1} by

$$y_{k+1} = y_k + hf(x_k, y_k), \qquad k = 0, 1, \ldots \,. \tag{5.2.9}$$

This is *Euler's method*.

Another derivation of Euler's method is the following. By the Taylor expansion of the solution of (5.2.3) about x_k, we have

$$y(x) = y(x_k) + (x - x_k)y'(x_k) + \frac{1}{2}y''(z_k)(x - x_k)^2, \tag{5.2.10}$$

where z_k is a point between x_k and x. By (5.2.3), the exact solution satisfies $y'(x_k) = f(x_k, y(x_k))$ so that if we drop the remainder term in (5.2.10), we have

$$y(x) \doteq y(x_k) + (x - x_k)f(x_k, y(x_k)). \tag{5.2.11}$$

If $x = x_{k+1}$, then this is the basis for (5.2.9). Note that the right-hand side of (5.2.11) is the equation of the tangent line to the solution curve at x_k.

Euler's method is very easy to carry out: at the k-th step, we evaluate $f(x_k, y_k)$ and use this in (5.2.9). Hence, essentially all of the computation required is in the evaluation of $f(x_k, y_k)$. We now consider the simple example

$$y'(x) = y^2(x) + 2x - x^4, \quad y(0) = 0. \tag{5.2.12}$$

It is easily verified that the exact solution of this equation is $y(x) = x^2$. Here, $f(x, y) = y^2 + 2x - x^4$, and therefore Euler's method for (5.2.12) becomes

$$y_{k+1} = y_k + h(y_k^2 + 2kh - k^4h^4), \quad k = 0, 1, \ldots, \qquad y_0 = 0, \tag{5.2.13}$$

152   Chapter 5 Continuous Problems Solved Discretely

Table 5.2.1: *Computed and Exact Solutions for* (5.2.12) *by Euler's Method*

x	Computed Solution	Exact Solution
0.1	0.00	0.01
0.2	0.02	0.04
0.3	0.06	0.09
0.4	0.12	0.16
0.5	0.20	0.25
0.6	0.30	0.36

since $x_k = kh$. In Table 5.2.1, we give some computed values for (5.2.13) for $h = 0.1$, as well as the corresponding values of the exact solution.

As Table 5.2.1 shows, the computed solution is in error, as is to be expected. In general, the error will come from two sources: the discretization error that results from the replacement of the differential equation (5.2.3) by the approximation (5.2.9); and the rounding error made in carrying out the arithmetic operations of (5.2.9).

The rounding error may be discussed in much the same way as for numerical integration in Section 5.1. Assuming that the discretization error goes to zero as h goes to zero, the number of arithmetic operations needed to obtain an approximate solution on a fixed interval $[a, b]$ will increase as $h \to 0$. Thus, we can expect a situation as depicted in Figure 5.1.3., in which there is a step size h_0 such that for $h < h_0$, the rounding error becomes dominant. As with numerical integration, for most problems we will not use an h that small and the discretization error will be the dominant error.

Discretization Error

If $y_1, \ldots, y_N$ is an approximate solution on an interval $[a, b]$, the *(global) discretization error* is defined as

$$e(h) = \max \{|y_i - y(x_i)| : \ i = 1, \ldots, N\}, \qquad (5.2.14)$$

where $b = a + Nh$ and $y(x_i)$ is the exact solution of the differential equation at x_i. We wish to know whether and how rapidly $e(h) \to 0$ as $h \to 0$. Note that $b = a + Nh$ is a fixed point so that if $h \to 0$, then $N \to \infty$. We only allow h to vary in such a way that N is an integer.

The first step in studying the discretization error is to define the *local discretization error*

$$L(h) = \max \{|L(x, h)| : \ a \le x \le b - h\}, \qquad (5.2.15a)$$

where if $y(x)$ is again the exact solution of (5.2.1)

5.2 Initial Value Problems

$$L(x, h) = \tfrac{1}{h}[y(x+h) - y(x)] - f(x, y(x)). \quad (5.2.15\text{b})$$

Since $y'(x) = f(x, y(x))$, if we assume that

$$|y''(x)| \leq M, \qquad a \leq x \leq b, \quad (5.2.16)$$

then the Taylor expansion (5.2.10) shows that

$$L(h) \leq \frac{M}{2} h = O(h). \quad (5.2.17)$$

The mathematical problem, which is beyond our scope, is now to show that $e(h)$ is also $O(h)$. We state the following basic theorem without proof.

THEOREM 5.2.1 (Euler Discretization Error) *If the function f has a bounded partial derivative with respect to its second variable and if the solution of (5.2.3) has a bounded second derivative, then the Euler approximations converge to the exact solution as $h \to 0$ and the discretization error satisfies $e(h) = O(h)$.*

The fact that the discretization error is $O(h)$ is expressed by saying that Euler's method is *first-order*. The practical consequence of this is that as we decrease h, we expect that the approximate solution will become more accurate and converge to the exact solution at a linear rate in h as h tends to zero. For example, if we halve the step size, h, we expect that the error will decrease by about a factor of 2. This error behavior is shown in the following example.

Consider the equation $y' = y$, $y(0) = 1$, for which the exact solution is $y(x) = e^x$. We compute the solution at $x = 1$ by Euler's method using various values of h (see Table 5.2.2). The exact solution at $x = 1$ is $e = 2.718\ldots$; the errors for the different step sizes are given in the middle column. The ratios of the errors for successive halvings of h are given in the right-hand column, and it is seen that these ratios are tending to $\tfrac{1}{2}$, as expected.

Table 5.2.2: *Error in Euler's Method*

h	Computed Value	Error	Error Ratio
1	2.000	0.718	
1/2	2.250	0.468	0.65
1/4	2.441	0.277	0.59
1/8	2.566	0.152	0.55
1/16	2.638	0.080	0.53

154   Chapter 5 Continuous Problems Solved Discretely

Higher-Order Methods

The very slow rate of convergence shown in Table 5.2.2 as h decreases is typical of first-order methods and militates against their use. Thus, we would like higher-order methods in which the discretization error is $O(h^p)$ for $p > 1$. The conceptually simplest approach to higher-order methods is to use more terms in the Taylor expansion (5.2.10). Suppose that one more term is taken so that

$$y(x_{k+1}) \doteq y(x_k) + hy'(x_k) + \frac{h^2}{2} y''(x_k), \qquad (5.2.18)$$

where the remainder term has been dropped. As before, $y'(x_k)$ can be replaced by $f(x_k, y(x_k))$ but we now need to replace $y''(x_k)$ also. If we differentiate the relation $y'(x) = f(x, y(x))$ we obtain

$$y''(x) = \frac{d}{dx} f(x, y(x)) = f_x(x, y(x)) + f_y(x, y(x)) y'(x), \qquad (5.2.19)$$

where the subscripts denote partial derivatives. Using this to replace $y''(x_k)$ in (5.2.18) leads to the method

$$y_{k+1} = y_k + hf(x_k, y_k) + \frac{h^2}{2} [f_x(x_k, y_k) + f_y(x_k, y_k) f(x_k, y_k)],$$

which can be shown to be second-order. To obtain still higher-order methods, one can use more terms in the Taylor expansion, but the methods become increasingly cumbersome.

Runge-Kutta Methods

There are two other commonly used approaches to obtaining higher-order methods. The first makes a number of additional evaluations of f at each step and then combines them. This gives a class of methods known as *Runge-Kutta methods*. The simplest such method is

$$y_{k+1} = y_k + \frac{h}{2} [f(x_k, y_k) + f(x_{k+1}, y_k + hf(x_k, y_k))], \qquad (5.2.20)$$

where we have replaced $f(x_k, y_k)$ in Euler's method by an average of f evaluated at two different places. It can be shown that the discretization error in this method satisfies $e(h) = O(h^2)$, and it is called the *second-order Runge-Kutta method*. It is also known as *Heun's method*.

The most famous of the Runge-Kutta methods is the fourth-order method given by

$$y_{k+1} = y_k + \frac{h}{6}(F_1 + 2F_2 + 2F_3 + F_4), \qquad (5.2.21)$$

5.2 Initial Value Problems

where

$$F_1 = f(x_k, y_k), \qquad F_2 = f\left(x_k + \frac{h}{2}, y_k + \frac{h}{2}F_1\right),$$

$$F_3 = f\left(x_k + \frac{h}{2}, y_k + \frac{h}{2}F_2,\right), \qquad F_4 = f(x_{k+1}, y_k + hF_3).$$

Here, the $f(x_k, y_k)$ in Euler's method has been replaced by a weighted average of f evaluated at four different points. Runge-Kutta methods of order higher than four may be obtained but at the cost of still additional evaluations of f. Indeed, methods of order p require $p+1$ evaluations of f for $5 \leq p \leq 7$ and $p+2$ evaluations for $p \geq 8$. This is in contrast to only p evaluations if $p \leq 4$.

It is interesting to note that if we apply Euler's method to the differential equation (5.2.1) we obtain the composite rectangle rule for numerical integration. Similarly, the second-order Runge-Kutta method is the composite trapezoid rule, and the fourth-order Runge-Kutta method is the composite Simpson's rule. (See Exercise 5.2.4)

Multistep Methods

Another approach to higher-order methods utilizes information already computed and does not require additional evaluations of f. One of the simplest such methods is

$$y_{k+1} = y_k + \frac{h}{2}(3f_k - f_{k-1}), \qquad (5.2.22)$$

for which the discretization error can be shown to satisfy $e(h) = O(h^2)$, and is known as the *second-order Adams-Bashforth* method. Note that (5.2.22) requires only the evaluation $f_k = f(x_k, y_k)$, the value of f_{k-1} being known from the previous step.

Methods of higher-order may be obtained by using additional prior values of f. Let $y_k, \ldots, y_{k-N}$ be the computed approximations to the solution at $x_k, \ldots, x_{k-N}$, let $f_i = f(x_i, y_i)$, and let p be the interpolating polynomial of degree N that satisfies

$$p(x_i) = f_i, \qquad i = k, k-1, \ldots, k-N. \qquad (5.2.23)$$

We may then consider $p(x)$ to be an approximation to $f(x, y(x))$. Since the solution of the differential equation satisfies $y'(x) = f(x, y(x))$, $p(x)$ is also an approximation to $y'(x)$ so that

$$y(x_{k+1}) - y(x_k) = \int_{x_k}^{x_{k+1}} y'(x)dx \doteq \int_{x_k}^{x_{k+1}} p(x)dx. \qquad (5.2.24)$$

This gives rise to the "method"

$$y_{k+1} = y_k + \int_{x_k}^{x_{k+1}} p(x)dx. \qquad (5.2.25)$$

We next exhibit some particular methods arising from (5.2.25) for various values of N. If $N = 0$, then p is just the constant f_k and (5.2.25) is Euler's method. If $N = 1$, p is the linear function that satisfies $p(x_k) = f_k$ and $p(x_{k-1}) = f_{k-1}$. In Section 4.1 we discussed three different ways to obtain interpolating polynomials. For our present purpose, it is convenient to use the Newton form (4.1.19) but applied backwards from (x_k, f_k); that is, with $\Delta f_k = f_{k-1} - f_k$

$$p(x) = p_1(x) = f_k - \frac{(x - x_k)}{h} \Delta f_k, \tag{5.2.26}$$

where the minus sign occurs since $h = |x_{k-1} - x_k|$. If we integrate (5.2.26) from x_k to x_{k+1}, (5.2.25) becomes

$$y_{k+1} = y_k + h f_k - \frac{h}{2} \Delta f_k = y_k + \frac{h}{2}(3 f_k - f_{k-1}), \tag{5.2.27}$$

which is the second-order Adams-Bashforth method (5.2.22). Note that the first form of (5.2.27) shows how Euler's method is modified to obtain the new method.

Similarly, if $N = 2$, then p is the interpolating quadratic polynomial for (x_{k-2}, f_{k-2}), (x_{k-1}, f_{k-1}), and (x_k, f_k). If we again use (4.1.19), this polynomial may be written as

$$p_2(x) = p_1(x) + \frac{(x - x_k)(x - x_{k-1})}{2 h^2} \Delta^2 f_k, \tag{5.2.28}$$

where $\Delta^2 f_k = f_k - 2 f_{k-1} + f_{k-2}$. Thus, by (5.2.25), the method is

$$y_{k+1} = y_k + h f_k - \frac{h}{2} \Delta f_k + \frac{5}{6} h \Delta^2 f_k. \tag{5.2.29}$$

This exhibits how the two-step formula (5.2.27) has been modified. We can also collect terms in (5.2.29) and write it as

$$y_{k+1} = y_k + \frac{h}{12}(23 f_k - 16 f_{k-1} + 5 f_{k-2}). \tag{5.2.30}$$

If $N = 3$, the interpolating polynomial is a cubic, and the method is

$$y_{k+1} = y_k + \frac{h}{24}(55 f_k - 59 f_{k-1} + 37 f_{k-2} - 9 f_{k-3}). \tag{5.2.31}$$

It can be shown that the discretization error of (5.2.30) is $O(h^3)$ and that of (5.2.31) is $O(h^4)$. Hence, these are called the *third-* and *fourth-order Adams-Bashforth methods*, respectively. We can, in principle, continue the preceding process to obtain Adams-Bashforth methods of arbitrarily high-order by increasing the number of prior points and the degree of the interpolating polynomial p. The formulas become increasingly complex as N increases, but the principle is still the same.

5.2 Initial Value Problems

The Adams-Bashforth methods are called *multistep methods* since prior values of the data are usual. The formula (5.2.27) is a two-step method, (5.2.30) is three-step and (5.2.31) is four-step. This is in contrast to the Runge-Kutta methods which use no prior data and are called *one-step methods*. It is also possible to devise methods that are combinations of multistep and one-step methods.

The Starting Problem

Multistep methods suffer from a problem not encountered with one-step methods. Consider the fourth-order Adams-Bashforth method of (5.2.31). The initial value y_0 is given, but for $k = 0$ in (5.2.31), information is needed at x_{-1}, x_{-2}, and x_{-3}, which doesn't exist. The problem is that multistep methods need "help" getting started. We cannot use (5.2.31) until $k \geq 3$, nor can we use (5.2.30) until $k \geq 2$. A common tactic is to use a one-step method, such as a Runge-Kutta method, of the same order of accuracy until enough values have been computed so that the multistep method is usable. The same problem arises if the step length is changed during the calculation. If, for example, at x_k the step length is changed from h to $\frac{h}{2}$, then we will need values of y and f at $x_k - \frac{h}{2}$, which we do not have. Again, a Runge-Kutta method starting at x_{k-1} with step length $\frac{h}{2}$ could be used. An attractive alternative is to use an interpolation formula, but care must be exercised to ensure that suitable accuracy is maintained. In general, it is much easier to change step-size with a one-step method than a multistep method.

Implicit and Predictor-Corrector Methods

The Adams-Bashforth methods have a major advantage over Runge-Kutta methods in that high-order methods still require only one evaluation of f at each step. However, the Adams-Bashforth methods can be unstable due to the fact they are obtained by integrating the interpolating polynomial outside the interval of the data that defines the polynomial. We can attempt to remedy this by also using data at x_{k+1} to determine the interpolating polynomial. This gives rise to the *Adams-Moulton methods*.

If p is the linear function that interpolates (x_k, f_k) and (x_{+1}, f_{k+1}), then (5.2.25) becomes

$$y_{k+1} = y_k + \frac{h}{2}(f_{k+1} + f_k), \qquad (5.2.32)$$

which is the *second-order Adams-Moulton method*. Similarly, if p is the cubic interpolating polynomial for the points (x_{k+1}, f_{k+1}), (x_k, f_k), (x_{k-1}, f_{k-1}), and (x_{k-2}, f_{k-2}), then (5.2.25) becomes

$$y_{k+1} = y_k + \frac{h}{24}(9f_{k+1} + 19f_k - 5f_{k-1} + f_{k-2}), \qquad (5.2.33)$$

which is the *fourth-order Adams-Moulton method*.

Now, note that in the formulas (5.2.32) and (5.2.33), f_{k+1} is not known since we need y_{k+1} to evaluate $f(x_{k+1}, y_{k+1}) = f_{k+1}$, but y_{k+1} is not yet known. Hence, the Adams-Moulton methods define y_{k+1} only implicitly. For example, (5.2.32) is really an equation

$$y_{k+1} = y_k + \frac{h}{2}[f(x_{k+1}, y_{k+1}) + f_k], \tag{5.2.34}$$

for the unknown value y_{k+1}, and similarly for (5.2.33). Thus, the Adams-Moulton methods are called *implicit* whereas the Adams-Bashforth methods are called *explicit* since no equation needs to be solved in order to obtain y_{k+1}.

In principle, we could attempt to solve (5.2.34) for y_{k+1} using Newton's method or the other methods discussed in Section 4.3 for the solution of equations. However, another use of implicit methods is to combine an explicit formula with an implicit one to form a *predictor-corrector method*. A commonly used predictor-corrector method is the combination of the fourth-order Adams methods (5.2.31) and (5.2.33):

$$y_{k+1}^{(p)} = y_k + \tfrac{h}{24}(55 f_k - 59 f_{k-1} + 37 f_{k-2} - 9 f_{k-3}), \tag{5.2.35a}$$

$$f_{k+1}^{(p)} = f(x_{k+1}, y_{k+1}^{(p)}), \tag{5.2.35b}$$

$$y_{k+1} = y_k + \tfrac{h}{24}(9 f_{k+1}^{(p)} + 19 f_k - 5 f_{k-1} + f_{k-2}). \tag{5.2.35c}$$

This method is entirely explicit. First, a "predicted" value $y_{k+1}^{(p)}$ of y_{k+1} is computed by the Adams-Bashforth formula, then $y_{k+1}^{(p)}$ is used to give an approximate value of f_{k+1}, which is used in the Adams-Moulton formula. The Adams-Moulton formula "corrects" the approximation given by the Adams-Bashforth formula. We could also take additional corrector steps in (5.2.35c). In fact, repeated use of the corrector formula gives an iterative method to solve the nonlinear equation (5.2.33).

Systems of Equations and A Numerical Example

We next indicate how the methods that have been discussed can be used for systems of equations. Consider the system (5.2.4), which we will write in the vector form

$$\mathbf{y}'(x) = \mathbf{f}(x, \mathbf{y}(x)). \tag{5.2.36}$$

Here, $\mathbf{y}(x)$ denotes the vector with components $y_1(x), \ldots, y_n(x)$, $\mathbf{y}'(x)$ is the vector with components $y_1'(x), \ldots, y_n'(x)$, and $\mathbf{f}$ is the vector with components $f_1(x, \mathbf{y}(x)), \ldots, f_n(x, \mathbf{y}(x))$. Then, Euler's method (5.2.9) can be written for the system (5.2.36) as

$$\mathbf{y}_{k+1} = \mathbf{y}_k + h\mathbf{f}(x_k, \mathbf{y}_k), \quad k = 0, 1, \ldots, \tag{5.2.37}$$

5.2 Initial Value Problems

where $\mathbf{y}_1, \mathbf{y}_2, \ldots$ are vector approximations to the solution $\mathbf{y}$ and $\mathbf{y}_0$ is the vector of initial conditions. We could, of course, write out (5.2.37) in component form; for $n = 2$, this would be

$$\left. \begin{array}{rcl} y_{1,k+1} & = & y_{1,k} + hf_1(x_k, y_{1,k}, y_{2,k}) \\ y_{2,k+1} & = & y_{2,k} + hf_2(x_k, y_{1,k}, y_{2,k}) \end{array} \right\}, \quad k = 0, 1 \ldots.$$

Clearly, the succinct vector notation (5.2.37) is advantageous, especially for a large number of equations.

Similarly, the second-order Runge-Kutta method (5.2.20) can be written in vector form for (5.2.36) as

$$\mathbf{y}_{k+1} = \mathbf{y}_k + \frac{h}{2}[\mathbf{f}(x_k, \mathbf{y}_k) + \mathbf{f}(x_{k+1}, \mathbf{y}_k + h\mathbf{f}(x_k, \mathbf{y}_k))], \quad (5.2.38)$$

and the second-order Adams-Bashforth method (5.2.22) can be written as

$$\mathbf{y}_{k+1} = \mathbf{y}_k + \frac{h}{2}(3\mathbf{f}_k - \mathbf{f}_{k-1}). \quad (5.2.39)$$

We next give the results of some numerical calculations for the predator-prey equations (5.2.6), a system of two equations in the unknown functions y_1 and y_2. We have used the following values for the parameters in (5.2.6): $c_{11} = 0.25, c_{12} = -0.01, c_{21} = -1.0, c_{22} = 0.01$. The initial values $y_1(0) = 80, y_2(0) = 30$ were used in all cases.

Figures 5.2.1 and 5.2.2 give approximations to the solution of (5.2.6), generated by several of the methods in this section. In all cases, we have plotted y_1 (the prey) versus y_2 (the predator), both as functions of time, t. The motion is in a clockwise direction as t increases.

Figure 5.2.1 demonstrates the dependency of the approximate solution on the value of the step size, h, for Euler's method. The values of h are $1, 0.5$, and 0.25. One sees that as the step size is halved the error is roughly halved, suggesting $O(h)$ convergence. Clearly, the errors are rather large even for $h = 0.25$. The "exact" solution used for comparison was obtained by a high-order Runge-Kutta method, and the solution so obtained may be considered to be exact for the purpose of comparing with the lower-order methods.

Figure 5.2.2(a) shows the effect of using a second-order method rather than the first-order Euler method. Here, the error for a step size of $h = 1$ is less than that for Euler's method with a step size of $h = 0.25$. Note that, as with Euler's method, the approximate solution is spiraling out away from the exact solution.

Figure 5.2.2(b) shows the second-order Adams-Bashforth method and a predictor corrector method based on the second-order Adams-Bashforth and Adams-Moulton methods. This predictor-corrector method follows the principle of (5.2.35) and is given explicitly in Exercise 5.2.12. Again, the step-size for

160   Chapter 5 Continuous Problems Solved Discretely

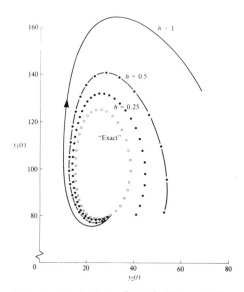

Figure 5.2.1: *Euler's Method for* (5.2.6) *Using Different Step Sizes*

both methods was $h = 1$. Note the strong effect that the correction step has: the accuracy is improved somewhat but, more noticeably, the approximate solution now spirals in rather than out.

The errors in the Runge-Kutta and Adams methods are comparable although Runge-Kutta is more accurate for this problem. These second-order methods both require two evaluations of f per step but for the corresponding fourth-order methods, Runge-Kutta would require four evaluations of f whereas the Adams predictor-corrector method still requires only two.

Unstable Solutions

One potential difficulty in solving initial value problems is that the solution of the differential equation may be unstable. Consider the second-order differential equation
$$y'' - 10y' - 11y = 0, \qquad (5.2.40)$$
with the initial conditions
$$y(0) = 1, \qquad y'(0) = -1. \qquad (5.2.41)$$
The solution of (5.2.40), (5.2.41) is $y(x) = e^{-x}$, as is easily verified. Now, suppose we change the first initial condition by a small quantity ε so that the initial conditions are
$$y(0) = 1 + \varepsilon, \qquad y'(0) = -1. \qquad (5.2.42)$$

5.2 Initial Value Problems

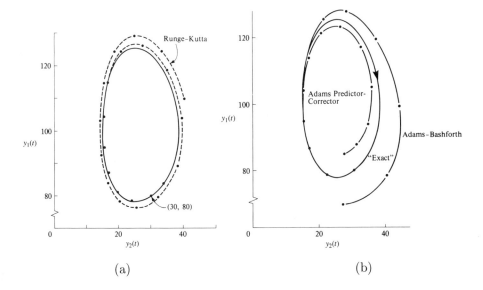

Figure 5.2.2: *(a) Second-Order Runge-Kutta (Spiraling Outward) with $h = 1$ (b) Second-Order Adams-Bashforth and Adams Predictor-Corrector*

Then, as is again easily verified, the solution of (5.2.40) with the initial conditions (5.2.42) is

$$y(x) = (1 + \tfrac{11}{12}\varepsilon)e^{-x} + \frac{\varepsilon}{12}e^{11x}. \tag{5.2.43}$$

Therefore, for any $\varepsilon > 0$, no matter how small, the second term in (5.2.43) causes the solution to tend to infinity as $x \to \infty$. We say that the solution $y(x) = e^{-x}$ of the problem (5.2.40), (5.2.41) is *unstable*, since arbitrarily small changes in the initial conditions can produce arbitrarily large changes in the solution as $x \to \infty$. In scientific computing, one would also say that this problem is *ill-conditioned*; it is extremely difficult to obtain the solution numerically because rounding and discretization error will cause the same effect as changing the initial conditions and the approximate solution will tend to diverge to infinity (see Exercise 5.2.18).

Unstable Methods

The previous example illustrated instability of solutions of the differential equation itself. We now turn to possible instabilities in the numerical method. Consider the method

$$y_{n+1} = y_{n-1} + 2hf_n, \tag{5.2.44}$$

which is similar to Euler's method but is a multistep method and is second-order accurate, as is easy to verify (Exercise 5.2.19).

162   Chapter 5 Continuous Problems Solved Discretely

We now apply (5.2.44) to the problem

$$y' = -2y + 1, \qquad y(0) = 1, \qquad (5.2.45)$$

whose exact solution is

$$y(x) = \tfrac{1}{2}e^{-2x} + \tfrac{1}{2}. \qquad (5.2.46)$$

This solution is stable since if the initial condition is changed to $y(0) = 1 + \varepsilon$, the solution becomes

$$y(x) = (\tfrac{1}{2} + \varepsilon)e^{-2x} + \tfrac{1}{2},$$

and the change in (5.2.46) is only εe^{-2x}. The method (5.2.44) applied to (5.2.45) is

$$y_{n+1} = y_{n-1} + 2h(-2y_n + 1) = 4hy_n + y_{n-1} + 2h, \quad y_0 = 1, \qquad (5.2.47)$$

with y_0 taken as the initial condition. However, since (5.2.44) is a two-step method, we need to supply y_1 also in order to start the process, and we will take y_1 to be the exact solution (5.2.46) at $x = h$:

$$y_1 = \tfrac{1}{2}e^{-2h} + \tfrac{1}{2}. \qquad (5.2.48)$$

If, for any fixed $h > 0$, we now generate the sequence $\{y_n\}$ by (5.2.47) and (5.2.48), we will see that $|y_n| \to \infty$ as $n \to \infty$ rather than mirroring the behavior as $x \to \infty$ of the solution (5.2.46) of the differential equation. Thus, the method (5.2.44), although second-order accurate, exhibits unstable behavior.

Stability of Methods

All of the multistep methods described so far are special cases of what are called *linear multistep* methods of the form

$$y_{k+1} = \sum_{i=1}^{m} \alpha_i y_{k+1-i} + h \sum_{i=0}^{m} \beta_i f_{k+1-i}, \qquad (5.2.49)$$

where, as usual, $f_j = f(x_j, y_j)$, and m is some fixed integer. The method (5.2.49) is called linear since y_{k+1} is a linear combination of the y_i and f_i. If $\beta_0 = 0$, the method is explicit, and if $\beta_0 \neq 0$, then the method is implicit. In all of the Adams methods, $\alpha_1 = 1$ and $\alpha_i = 0$, $i > 1$; in the Adams-Bashforth methods, $\beta_0 = 0$; for Adams-Moulton, $\beta_0 \neq 0$.

Associated with the method (5.2.49) is the polynomial

$$\rho(\lambda) \equiv \lambda^m - \alpha_1 \lambda^{m-1} - \cdots \alpha_{m-1}\lambda - \alpha_m. \qquad (5.2.50)$$

Then, the method (5.2.49) is *stable* provided that all roots λ_i of p satisfy $|\lambda_i| \leq 1$, and any root for which $|\lambda_i| = 1$ is simple. The method is *strongly stable* if, in addition, $m - 1$ roots of p satisfy $|\lambda_i| < 1$. (We note that some

5.2 Initial Value Problems

authors use the terms *weakly stable* and *stable* in place of stable and strongly stable.)

Any method that is at least first-order accurate must satisfy the condition $\sum_{i=1}^{m} \alpha_i = 1$ so that 1 is a root of (5.2.50). In this case, a strongly stable method will then have one root of (5.2.50) equal to 1 and all the rest strictly less than 1 in absolute value. For an m-step Adams method, $\rho(\lambda) = \lambda^m - \lambda^{m-1}$, so the other $m - 1$ roots are zero, and these methods are strongly stable. It is possible to extend the above framework so as to include Runge-Kutta methods, and these methods are also always strongly stable.

For the method (5.2.44), the polynomial (5.2.50) is $\rho(\lambda) = \lambda^2 - 1$ with roots ± 1; hence, this method is stable but not strongly stable, and it is this lack of strong stability that gives rise to the unstable behavior of the sequence $\{y_k\}$ defined by (5.2.47) and (5.2.48). (See the Supplementary Discussion.)

The stability theory that we have just discussed is stability in the limit as $h \to 0$, and the example of instability that we gave shows what can happen for arbitrarily small h if the method is stable but not strongly stable and the interval is infinite. On a finite interval, a stable method will give accurate results for sufficiently small h. On the other hand, even strongly stable methods can exhibit unstable behavior if h is too large. Although, in principle, h can be taken sufficiently small to overcome this difficulty, it may be that the computing time then becomes prohibitive. This is the situation with differential equations that are known as *stiff*, and we next give a short discussion of such problems.

Stiff Equations

Consider the equation

$$y' = -100y + 100, \qquad y(0) = y_0 = 2. \tag{5.2.51}$$

The exact solution of this problem is

$$y(x) = e^{-100x} + 1. \tag{5.2.52}$$

It is clear that the solution is stable since if we change the initial condition to $2 + \varepsilon$, then the solution changes by εe^{-100x}. Euler's method applied to (5.2.51) is

$$y_n = y_{n-1} + h(-100y_{n-1} + 100) = (1 - 100h)y_{n-1} + 100h. \tag{5.2.53}$$

Applying the corresponding formulas for y_{n-1}, y_{n-2} and so on, it is possible to represent y_n by

$$y_n = (1 - 100h)^n + 1. \tag{5.2.54}$$

Note that $y(x)$ decreases very rapidly from $y_0 = 2$ to its limiting value of 1; for example, $y(0.1) \doteq 1 + 5 \times 10^{-5}$. Initially, therefore, we expect to require a small step size h to compute the solution accurately. Beyond, say, $x = 0.1$, the solution varies slowly and is essentially equal to 1, so intuitively, we would expect to obtain sufficient accuracy with Euler's method using a relatively large h. However, we see from (5.2.54) that $|1 - 100h| > 1$ when $h > 0.02$, and the approximation y_n grows at each step and shows an unstable behavior. The quantity $(1 - 100h)^n$ is an approximation to the exponential term e^{-100x} and is a good approximation for very small h, but becomes a poor approximation as h becomes as large as 0.02. Even though the exponential term contributes virtually nothing to the solution (5.2.52) after $x = 0.1$, Euler's method still requires that we approximate it sufficiently accurately to maintain stability. This is the typical problem with stiff equations: the solution contains a component that contributes very little to the solution, but the usual methods require that it be approximated accurately.

The general approach to the problem of stiffness is to use implicit methods. It is beyond the scope of this book to discuss this in detail, and we will only give an indication of the value of implicit methods in this context by applying one of the simplest such methods to the problem (5.2.51). For the general equation $y' = f(x, y)$, the method

$$y_{n+1} = y_n + hf(x_{n+1}, y_{n+1}) \qquad (5.2.55)$$

is known as the *backward Euler method*. It is of the same form as Euler's method except that f is evaluated at (x_{n+1}, y_{n+1}) rather than at (x_n, y_n); hence, the method is implicit. If we apply (5.2.55) to (5.2.51), we obtain

$$y_{n+1} = y_n + h(-100y_{n+1} + 100), \qquad (5.2.56)$$

which can be put in the form

$$y_{n+1} = (1 + 100h)^{-1}(y_n + 100h). \qquad (5.2.57)$$

Corresponding to (5.2.54), the approximations generated by (5.2.57) may be expressed as

$$y_n = \frac{1}{(1 + 100h)^n} + 1, \qquad (5.2.58)$$

and we see that there is no unstable behavior regardless of the size of h. Note that with Euler's method we are attempting to approximate the solution by a polynomial, and no polynomial (except 0) can approximate e^{-x} as $x \to \infty$. With the backward Euler method, we are approximating the solution by a rational function and such functions can, indeed, go to zero as $x \to \infty$.

The backward Euler method, like Euler's method itself, is only first-order accurate, and a better choice would be the second-order Adams-Moulton method (5.2.34), also known as the *trapezoid rule*:

$$y_{n+1} = y_n + \frac{h}{2}[f_n + f(x_{n+1}, y_{n+1})]. \qquad (5.2.59)$$

5.2 Initial Value Problems

The application of an implicit method to (5.2.51) was deceptively simple since the differential equation is linear in a single unknown, and hence we could easily solve for y_{n+1} in (5.2.56). If the differential equation had been nonlinear, however, the method would have required the solution of a nonlinear equation for y_{n+1} at each step. More generally, for a system of differential equations, the solution of a system of equations (linear or nonlinear, depending on the differential equations) would be needed at each step. This is costly in computer time, but the effective handling of stiff equations requires that some kind of implicitness be brought into the numerical method.

Parallel and Vector Computation

The solution of a single differential equation $y' = f(x, y)$ will not allow much parallelism unless f is composed of a number of terms that can be evaluated independently. For example, if

$$f(x,y) = x^2 + \cos x + y^2 + e^y + \sin xy, \qquad (5.2.60)$$

each of the terms could be computed in parallel and then the results added. However, if the terms in (5.2.60) are allocated to five separate processors, there will be a load imbalance since the trigonometric and exponential functions will require more time for evaluation than the other terms.

There is usually a greater possibility for parallelism in systems of equations $\mathbf{y}' = \mathbf{f}(x, \mathbf{y})$. If there are n equations, it is tempting to allocate the n functions $f_1, \ldots, f_n$ to n processors for evaluation. However, here again, there may be bad load imbalances with this naive allocation since some of the f_i may be very time-consuming to evaluate but others may be very easy. Moreover, there may be redundancy in the functions that should be utilized. For example, consider the two functions

$$f_1(x, y_1, y_2) = x^2 + y_1^3 y_2^3 + y_2^2$$

$$f_2(x, y_1, y_2) = x^2 + y_1^3 y_2^3 + y_1^2$$

If these two functions were evaluated independently, much of the work would be repeated on the common terms.

In many systems of equations, the number of equations is small, say 3 to 5, and this does not allow much parallelism. In particular, if the system arises from a single higher-order equation, the number of equations would rarely be greater than 4 and most of these would be trivial (see the Supplementary Discussion). Moreover, even if there is a large number of equations, the f_i may be sufficiently different that vectorization is not possible. One situation in which there are both a large number of equations and good vectorization properties is in the "method of lines" for solving partial differential equations; this method will be discussed in Section 5.5.

Finally, another possible source of parallelism is in systems of stiff equations. If an implicit method is used, a system of nonlinear equations must be solved at each step. For this we might use Newton's method, which requires the solution of linear systems of equations. We will consider the parallel solution of linear systems in Chapters 7 – 9.

Supplementary Discussion and References: 5.2

1. There are a number of excellent books devoted to the numerical solution of initial value problems for ordinary differential equations. Two classic books are Henrici [1962] and Gear [1971]. More recent comprehensive treatments are given in Butcher [1987] and Hairer, Norsett and Wanner [1987]. For more information on the predator-prey equations and other topics in mathematical biology, see, for example, Rubinov [1975].

2. It is customary to convert high-order differential equations to a system of first-order equations. For example, the nth-order equation

$$y^{(n)}(x) = f(x, y(x), \ldots, y^{(n-1)}(x)) \qquad (5.2.61)$$

can be written as a first-order system by defining the variables

$$y_i(x) = y^{(i-1)}(x), \qquad i = 1, \ldots, n. \qquad (5.2.62)$$

In terms of these new variables, (5.2.61) becomes

$$y'_n(x) = f(x, y_1(x), \ldots, y_n(x)). \qquad (5.2.63)$$

Differentiating (5.2.62) gives

$$y'_i(x) = y_{i+1}(x), \qquad i = 1, \ldots, n-1, \qquad (5.2.64)$$

and (5.2.63) plus (5.2.64) give a system of n first-order equations.

3. For the general linear multistep method (5.2.49), the local discretization error $L(x, h)$ at x is defined by

$$\frac{1}{h}[y(x+h) - \sum_{i=1}^{m} \alpha_i y(x - (i-1)h)] - \sum_{i=0}^{m} \beta_i y'(x - (i-1)h), \qquad (5.2.65)$$

where $y(x)$ is the solution of the differential equation. For any choice of m and the constants α_i and β_i, one can compute the local discretization error by expansion of y and y' in Taylor series about x. In particular, under suitable assumptions on the differentiability of the solution, one can show that the Adams-Bashforth methods (5.2.30) and (5.2.31) are third- and fourth-order, respectively, whereas the Adams-Moulton methods (5.2.32) and (5.2.33) are second- and fourth-order. The verification

5.2 Initial Value Problems

of these statements is left to Exercise 5.2.14. Good Adams codes are implemented with the capability of changing not only the step-size but also the order of the method. For details on one particular collection of codes – ODEPACK – see Hindmarsh [1983]. For more discussion of the theory and practice of Adams methods see, for example, Gear [1971], and Butcher [1987].

4. Another approach to the derivation of multistep methods starts with the general linear method (5.2.49) and requires that it be exact when the solution y of the differential equation is a polynomial of degree q. This then implies that the method is order q. For example, if $q = 1$, then (5.2.65) must be zero whenever the solution is a constant; in this case, the y_i' all vanish, and we are left with the condition

$$1 = \sum_{i=1}^{m} \alpha_i. \qquad (5.2.66)$$

Similarly, the requirement that (5.2.49) be exact when the solution is $y(x) = x$ leads to the condition

$$m + 1 = \alpha_1 m + \alpha_2(m-1) + \cdots + \alpha_m + \sum_{i=0}^{m} \beta_i. \qquad (5.2.67)$$

The relations (5.2.66) and (5.2.67) for the coefficients α_i and β_i are known as the *consistency conditions* for the multistep method, and are necessary and sufficient conditions that the method be first-order. One can continue this process to obtain relations on the α_i and β_i that are necessary and sufficient that the method be of any given order. For further discussions of multistep methods, see, for example, Henrici [1962] and Butcher [1987].

5. In the previous section, we discussed Romberg integration. This is a special case of *Richardson extrapolation to the limit* which can be applied to differential equations, or a variety of other problems, to obtain higher-order methods. For the point x^*, denote the approximation to the solution by $y(x^*; h)$. Under certain assumptions, it can be shown that

$$y(x^*; h) = y(x^*) + c_1 h + c_2 h^2 + \cdots + c_p h^p + O(h^{p+1}),$$

where the c_i are functions of x^* but independent of h. Now suppose that we also compute the approximate solution with step length $h/2$. Then we can combine $y(x^*; h)$ and $y(x^*; h/2)$ to obtain a better approximation. In particular,

$$\bar{y}(x^*; h) \equiv 2y(x^*; \frac{h}{2}) - y(x^*; h) = y(x^*) + d_2 h^2 + \cdots + O(h^{p+1}),$$

so that $\bar{y}(x^*; h)$ is a second-order approximation. (Note that the more accurate solution is obtained only at the points with spacing h, and not at the intermediate points.) The process can then be repeated, if desired, to eliminate the coefficient d_2 and obtain a third-order approximation, and so on.

6. All good computer codes using Runge-Kutta methods employ some mechanism for automatically changing the step size h as the integration proceeds. The problem is to ascertain what the step size should be before the start of the next step. One approach to this is to estimate the local discretization error and, depending on its size, adjust the current step size either upward or downward. There are several ways to estimate the local error; two simple ones are to repeat the last step with a step size half as large, and then compare the two results, or to use two Runge-Kutta formulas of different order. Both of these ways are costly in evaluations of f, and an alternative approach is by means of the Runge-Kutta-Fehlberg formulas (see Butcher [1987] for further discussion). Here, one can use, for example, a Runge-Kutta method of order five to estimate the error in a fourth-order Runge-Kutta method in such a way that a total of only six evaluations of f are needed as opposed to ten if the usual Runge-Kutta formulas were used.

7. The fact that the y_n generated by (5.2.47) diverge as $n \to \infty$ can be ascertained by the theory of linear difference equations with constant coefficients. Such equations are of the form

$$y_{n+1} = a_m y_n + \cdots + a_1 y_{n-m+1} + a_0, \qquad n = m-1, m, m+1, \ldots, \quad (5.2.68)$$

with given constants $a_0, a_1, \ldots, a_m$, and their theory parallels that of differential equations. Associated with (5.2.68) is the *characteristic polynomial*

$$\lambda^m - a_m \lambda^{m-1} - \cdots - a_1 = 0. \qquad (5.2.69)$$

If we assume that the m roots $\lambda_1, \ldots, \lambda_m$ of (5.2.69) are distinct, then the *fundamental solutions* of (5.2.68) are $\lambda_1^k, \ldots, \lambda_m^k$, and the general solution of (5.2.68) is

$$y_k = \sum_{i=1}^m c_i \lambda_i^k + \frac{a_0}{1 - a_1 - \cdots - a_m}, \qquad k = 0, 1 \ldots . \qquad (5.2.70)$$

The arbitrary constants c_i in (5.2.70) can be determined by given initial conditions $y_0, y_1, \ldots, y_{m-1}$. Then, (5.2.70) requires that

$$\sum_{i=1}^m c_i \lambda_i^k + \frac{a_0}{1 - a_1 - \cdots - a_m} = y_k, \qquad k = 0, 1, \ldots, m-1, \qquad (5.2.71)$$

5.2 Initial Value Problems

which is a system of m linear equations in the m unknowns $c_1, \ldots, c_m$, and can be used to determine the c_i. If the characteristic equation has multiple roots then polynomial terms in n enter the solution in a manner analogous to that for differential equations. For more discussion of the theory of linear difference equations, see, for example, Ortega [1987].

We now apply the above theory to (5.2.47), which is an example given in Henrici [1962]. The characteristic equation (5.2.69) of (5.2.47) is $\lambda^2 + 4h\lambda - 1 = 0$ with roots

$$\lambda_1 = -2h + \sqrt{1 + 4h^2}, \quad \lambda_2 = -2h - \sqrt{1 + 4h^2}, \qquad (5.2.72)$$

and, thus, the solution of (5.2.47) is

$$y_n = c_1(-2h + \sqrt{1 + 4h^2})^n + c_2(-2h - \sqrt{1 + 4h^2})^n + \tfrac{1}{2}. \qquad (5.2.73)$$

For any fixed $h > 0$, it is evident that

$$0 < -2h + \sqrt{1 + 4h^2} < 1, \qquad 2h + \sqrt{1 + 4h^2} > 1.$$

Therefore, as n tends to infinity the first term in (5.2.73) tends to zero while the second tends to infinity, in an oscillatory way, unless $c_2 = 0$. (Which it is not for the initial condition (5.2.48).)

This example illustrates the concepts of stable and strongly stable in the following way. The stability polynomial (5.2.50) of (5.2.47) is $\lambda^2 - 1$ with roots ± 1. Hence, the method (5.2.47) is stable but not strongly stable. The sequence $\{y_k\}$ generated by (5.2.47) is meant to approximate the solution of the differential equation (5.2.45), which is a first-order equation with only one fundamental solution. This fundamental solution is approximated by λ_1^n, while λ_2^n is spurious and should rapidly go to zero. However, for any $h > 0$, it is the case that $|\lambda_2| > 1$, and hence λ_2^n tends to infinity and not zero; it is this that causes the instability. Now, note that λ_1 and λ_2 converge to the roots of the stability polynomial $\lambda^2 - 1$ as $h \to 0$; indeed, this polynomial is just the limit, as $h \to 0$, of the characteristic polynomial of (5.2.47). The idea of strong stability now becomes more evident. If all the roots except one of the stability polynomial are less than 1 in magnitude, then all but one of the roots of the characteristic equation of the method must be less than 1 for sufficiently small h, and powers of these roots – the spurious fundamental solutions of the difference equation – tend to zero and cause no instability.

8. The basic results of the theory of stability of multistep methods were developed by G. Dahlquist in the 1950s; for a detailed treatment of this theory, see Henrici [1962]. Since then, there have been a number of refined definitions of stability; in particular, the terms *stiffly-stable* and

A-stable deal with types of stability needed for methods to handle stiff equations. For more on different definitions of stability, see, for example, Gear [1971], and Butcher [1987]. There are a number of codes available for solving stiff equations. One of the best is the VODE package of Brown, Byrne and Hindmarsh [1989], which may also be used for problems which are not stiff.

9. An important modification to a system of differential equations is when there are side conditions. For example, a system of the form

$$y'_i = f_i(x, y, (x), \ldots, y_n(x)), \quad i = 1, \ldots, m,$$
$$0 = g_i(x, y, (x), \ldots, y_n(x)), \quad i = m+1, \ldots, n,$$

consists of m differential equations and $n - m$ non-differential equations. Such a system is called *differential-algebraic*. For further discussion, see, Hairer, Lubich and Roche [1989], and Brenan, Campbell and Petzold [1989].

EXERCISES 5.2

5.2.1. Rewrite the predator-prey equations (5.2.6) in the form (5.2.36); that is, give the functions f_1 and f_2.

5.2.2. Apply Euler's method (5.2.9) to the initial-value problem $y' = -y$, $0 \leq x \leq 1$, $y(0) = 1$, with $h = 0.25$. Compare your answers to the exact solution $y(x) = e^{-x}$. Repeat for $h/2$ and $h/4$. Next, apply the second-order Runge-Kutta method to this problem and compare with Euler's method.

5.2.3. Verify the calculations of Tables 5.2.1 and 5.2.2.

5.2.4. Show that Euler's method applied to (5.2.1) is just the composite rectangle rule, that the second-order Runge-Kutta method is the composite trapezoid rule, and the fourth-order Runge-Kutta method is the composite Simpson's rule.

5.2.5. Apply Euler's method and the second-order Runge-Kutta method to the problem $y'(x) = x^2 + [y(x)]^2$, $y(0) = 1$, $x \geq 0$. Compare your results.

5.2.6. Repeat the calculations of Figure 5.2.1 using Euler's method with step sizes 0.5, and 0.25. How small a step size do you have to use in order that the graph of the solution closes back on itself to visual accuracy?

5.2.7. Test the stability of the solution of the predator-prey equations (5.2.6) with respect to changes in the initial conditions by changing $x_0 = 80$, $y_0 = 30$ by a unit amount in each direction (four different cases) and repeating the calculation using the second-order Runge-Kutta method.

5.2 Initial Value Problems

5.2.8. Verify that p_1 of (5.2.26) is the linear interpolating polynomial for (x_k, f_k) and (x_{k-1}, f_{k-1}) and that (5.2.27) follows from (5.2.25) for this p_1. Similarly, verify that (5.2.28) is the quadratic interpolating polynomial for (x_k, f_k), (x_{k-1}, f_{k-1}) and (x_{k-2}, f_{k-2}). Then carry out the integration of p_2 to verify the formula (5.2.29). Finally, give the cubic interpolating polynomial p_3 by adding the appropriate cubic term (see (4.1.19)) to p_2. Then integrate p_3 in (5.2.25) to obtain the formula (5.2.30).

5.2.9. Write a computer program to carry out the second-order Adams-Bashforth method (5.2.22). Use the second-order Runge-Kutta method to supply the missing starting value y_1. Apply your program to the problems of Exercises 5.2.2 and 5.2.5 and compare your results with the Euler and second-order Runge-Kutta methods.

5.2.10. Do Exercise 5.2.9 using the fourth-order Adams-Bashforth method (5.2.31).

5.2.11. Carry out in detail the derivation of the Adams-Moulton method (5.2.32). Do the same for the method (5.2.33).

5.2.12. Use as much of your program of Exercise 5.2.9 as possible to write a program to carry out the predictor-corrector method

$$y_{k+1}^{(p)} = y_k + \frac{h}{2}(3f_k - f_{k-1}),$$

$$f_{k+1}^{(p)} = f(x_{k+1}, y_{k+1}^{(p)}),$$

$$y_{k+1} = y_k + \frac{h}{2}(f_{k+1}^{(p)} + f_k).$$

Apply this to the problem $y' = -y$, $y(0) = 1$, and compare your results with the methods of Exercise 5.2.9.

5.2.13. Give the coefficients α_i and β_i in the linear multistep formulation (5.2.49) for the Adams-Bashforth and Adams-Moulton methods of second, third, and fourth order.

5.2.14. Using the definition (5.2.65), compute the local discretization errors for the Adams-Bashforth methods (5.2.30) and (5.2.31) and show that they are third and fourth order, respectively. (Assume that the solution is sufficiently differentiable.) Do the same for the Adams-Moulton methods (5.2.32) and (5.2.33) and verify that they are second and fourth order, respectively.

5.2.15. Consider the method $y_{k+1} = y_{k-1} + (\frac{h}{2})(f_{k+1} + 2f_k + f_{k-1})$.

a. Find the order of the method.

b. Discuss how to apply this method to the system of equations $\mathbf{y}' = \mathbf{f}(x, \mathbf{y})$. What difficulties do you expect to encounter in carrying out the method?

5.2.16. Repeat the calculations of Figure 5.2.2 using the second-order Adams-Bashforth method and the predictor-corrector method of Exercise 5.2.12.

5.2.17. Consider the problem of evaluating the "normal density"

$$p(x) = \frac{1}{\sqrt{2\pi}} \int_0^x e^{-t^2/2} dt + \frac{1}{2}$$

by solving the differential equation

$$p'(x) = \frac{1}{\sqrt{2\pi}} e^{-x^2/2}, \qquad p(0) = \frac{1}{2}.$$

Use the second-order Runge-Kutta method and the second-order predictor-corrector method of Exercise 5.2.12 to solve this differential equation. Compare your results. Also, note that if $q(x) = (2\pi)^{-\frac{1}{2}} e^{-x^2/2}$, then $p'(x) = q(x)$ and $q'(x) = -xq$. Solve this system of equations.

5.2.18. By letting $z = y'$, show that the problem (5.2.40), (5.2.41) is equivalent to the first-order system $y' = z$, $z' = 10z + 11y$, with initial conditions $y(0) = 1$ and $z(0) = -1$. Attempt to solve this system numerically by any of the methods of this chapter and discuss your results.

5.2.19. Using the techniques of Exercise 5.2.14, verify that the method (5.2.44) is second-order accurate.

5.2.20. Carry out the algorithm (5.2.47), (5.2.48) numerically for various values of h. Discuss your results.

5.2.21. Consider the method $y_{n+1} = y_{n-3} + (4h/3)(2f_n - f_{n-1} + 2f_{n-2})$, which is known as *Milne's method*. Ascertain whether this method is stable and strongly stable.

5.2.22. Write a program to carry out Euler's method on (5.2.51) for different values of h both less than and greater than 0.02. Discuss your results.

5.2.23. The system $y' = z$, $z' = -100y - 101z$ is the first-order system equivalent to the second-order equation $y'' + 101y' + 100y = 0$. Using the initial conditions $y(0) = 2$ and $z(0) = -2$, apply Euler's method to this system and determine experimentally how small the step size h must be to maintain stability. Attempt to verify analytically your conclusion about the size of h.

5.2.24. The function $y(x) = e^{-x}$ is the solution of the problem $y'' = y$, $y(0) = 1$, $y'(0) = -1$. Is this solution stable?

5.2.25. Ascertain which of the following methods are stable and which are strongly stable, and also find their order.

    a. $y_{k+1} = y_k + \frac{h}{6}(6f_k - 3f_{k-1} + 3f_{k-2})$
    b. $y_{k+1} = y_{k-1} + \frac{h}{2}(f_{k+1} + 2f_k + f_{k-1})$
    c. $y_{k+1} = 3y_k - 2y_{k-1} + \frac{h}{2}(f_{k+1} + 2f_k + f_{k-1})$
    d. $y_{k+1} = \frac{1}{2}(y_k + y_{k-1}) + \frac{3h}{4}(3f_k - f_{k-1})$

5.2.26. Apply the trapezoid rule (5.2.59) to the equation (5.2.51).

5.3 Boundary Value Problems

In the previous section, we considered initial value problems for ordinary differential equations. But in many problems, there will be conditions on the solution at more than one point. For a single first-order equation $y' = f(x, y)$, data at one point completely determines the solution so that if conditions at more than one point are given, either higher-order equations or systems of equations must be treated. Consider the second-order equation

$$v''(x) = f(x, v(x), v'(x)), \qquad a \leq x \leq b. \tag{5.3.1}$$

Because it is second-order, such an equation requires two additional conditions, and the simplest possibility is to prescribe the solution values at the endpoints:

$$v(a) = \alpha, \qquad v(b) = \beta. \tag{5.3.2}$$

Equations (5.3.1) and (5.3.2) define a *two-point boundary value* problem. If the function f of (5.3.1) is nonlinear in either $v(x)$ or $v'(x)$, the boundary value problem is *nonlinear*. Linear problems may be written in the form

$$v''(x) = b(x)v'(x) + c(x)v(x) + d(x), \qquad a \leq x \leq b, \tag{5.3.3}$$

where b, c, and d are given functions of x. (There will be no confusion between the function $b(x)$ and the endpoint b of the interval.)

Finite Difference Approximations

We first consider the special case of (5.3.3) in which $b(x) \equiv 0$, so that the equation is

$$v''(x) = c(x)v(x) + d(x), \qquad a \leq x \leq b. \tag{5.3.4}$$

We will assume that $c(x) \geq 0$ for $a \leq x \leq b$; this is a sufficient condition for the problem (5.3.4), (5.3.2) to have a unique solution. To obtain a numerical solution we first divide the interval $[a, b]$ into a number of equal subintervals of length h as shown in Figure 5.3.1; x_0 and x_{n+1} are the *boundary* points, and $x_1, \ldots, x_n$ are the *interior* grid points.

Figure 5.3.1: *Grid Points*

We now need to approximate $v''(x)$ in (5.3.4). Let x_i be any interior grid point and approximate the first derivative at the intermediate points $x_i \pm \frac{h}{2}$ by

$$v'(x_i - \tfrac{h}{2}) \doteq \frac{[v(x_i) - v(x_{i-1})]}{h}, \qquad v'(x_i + \tfrac{h}{2}) \doteq \frac{[v(x_{i+1}) - v(x_i)]}{h}.$$

These *finite difference* approximations are then used to approximate the second derivative:

$$v''(x_i) \doteq \frac{v'(x_i + \tfrac{h}{2}) - v'(x_i - \tfrac{h}{2})}{h} \doteq \frac{v(x_{i+1}) - 2v(x_i) + v(x_{i-1})}{h^2}. \qquad (5.3.5)$$

If we now substitute this approximation into equation (5.3.4) and denote the functions c and d evaluated at x_i by c_i and d_i, we obtain

$$\frac{1}{h^2}[v(x_{i+1}) - 2v(x_i) + v(x_{i-1})] - c_i v(x_i) \doteq d_i, \quad i = 1, \ldots, n. \qquad (5.3.6)$$

What we have shown so far is that if we replace the second derivative of the solution v by finite difference approximations and put these approximations into the differential equation, we obtain the approximate relations (5.3.6) that the solution must satisfy. We now turn this procedure around. Suppose that we can find numbers $v_1, \ldots, v_n$ that satisfy the equations

$$\frac{1}{h^2}(v_{i+1} - 2v_i + v_{i-1}) - c_i v_i = d_i, \quad i = 1, \ldots, n, \qquad (5.3.7)$$

or

$$-v_{i+1} + 2v_i - v_{i-1} + c_i h^2 v_i = -h^2 d_i, \quad i = 1, \ldots, n, \qquad (5.3.8)$$

with $v_0 = \alpha$ and $v_{n+1} = \beta$. Then, we can consider $v_1, \ldots, v_n$ to be approximations at the grid points $x_1, \ldots, x_n$ to the solution v of the boundary-value problem (5.3.4), (5.3.2). We shall return shortly to the question of how accurate these approximations are.

The equations (5.3.8) form a system of n linear equations in the n unknowns $v_1, \ldots, v_n$, and can be written in matrix-vector form as

$$\begin{bmatrix} 2 + c_1 h^2 & -1 & & & \\ -1 & 2 + c_2 h^2 & \ddots & & \\ & \ddots & \ddots & & \\ & & & -1 & \\ & & & -1 & 2 + c_n h^2 \end{bmatrix} \begin{bmatrix} v_1 \\ v_2 \\ \vdots \\ \\ v_n \end{bmatrix} = \begin{bmatrix} -h^2 d_1 + \alpha \\ -h^2 d_2 \\ \vdots \\ -h^2 d_{n-1} \\ -h^2 d_n + \beta \end{bmatrix}. \qquad (5.3.9)$$

5.3 Boundary Value Problems

Thus, to obtain the approximate solution $v_1, \ldots, v_n$, we need to solve this system of linear equations, which has a tridiagonal coefficient matrix. In the case that the c_i are all zero, this matrix is

$$\begin{bmatrix} 2 & -1 & & & \\ -1 & 2 & \ddots & & \\ & \ddots & \ddots & -1 \\ & & -1 & 2 \end{bmatrix}, \quad (5.3.10)$$

which arises in many contexts.

The matrix (5.3.10) has many desirable properties. For example, it is obviously symmetric and it is also positive definite. The latter statement follows from the fact that the eigenvalues can be computed explicitly and shown to be positive (Exercise 5.3.5). Hence, by Theorem 2.2.5, positive definiteness follows. If $c_i \geq 0, i = 1, \ldots, n$, then the coefficient matrix in (5.3.9) is the sum of a positive definite matrix and a diagonal positive semidefinite matrix. Thus, by Exercise 2.2.17, it is also positive definite and, hence, nonsingular; therefore, the system (5.3.9) has a unique solution. We will see in Chapter 6 that positive definiteness is also a desirable property for the numerical solution of (5.3.9).

Discretization Error

We next consider the important question of the error in the approximations $v_1, \ldots, v_n$. Since the linear system (5.3.9) which determines these quantities will be solved numerically, the computed v_i will be in error because of rounding; this will be discussed in more detail in Chapter 6. For the present, we assume that the v_i are computed with no rounding error so that $v_1, \ldots, v_n$ is the exact solution of the system (5.3.9). Let $v(x_i)$ be again the exact solution of the boundary-value problem at x_i. Then, analogous to the definition for initial-value problems in Section 5.2,

$$\max_{1 \leq i \leq n} |v_i - v(x_i)| \quad (5.3.11)$$

is the *(global) discretization error*; this is the error caused by replacing the continuous boundary-value problem by the discrete analog (5.3.8). We now indicate how an analysis of the discretization error proceeds. We first define the *local discretization error*, in a manner analogous to that for initial-value problems, by

$$L(x, h) = \frac{1}{h^2}[v(x+h) - 2v(x) + v(x-h)] - c(x)v(x) - d(x), \quad (5.3.12)$$

where v is the exact solution of the differential equation (5.3.4). By means of (5.3.4) we can replace $cv + d$ in (5.3.12) by v'' so that

$$L(x, h) = \frac{1}{h^2}[v(x+h) - 2v(x) + v(x-h)] - v''(x). \tag{5.3.13}$$

Thus, the local discretization error is just the error in approximating v''. In order to estimate this error, we assume that v is four times continuously differentiable and expand $v(x+h)$ and $v(x-h)$ in Taylor series. After we collect terms (the details of which are left to Exercise 5.3.2), we obtain

$$L(x, h) = \frac{1}{12}v^{(4)}(x)h^2 + O(h^4) = O(h^2). \tag{5.3.14}$$

The problem now is to relate this local discretization error to the global error (5.3.11). To do this, we evaluate (5.3.12) at the grid points x_i, set $\sigma_i = L(x_i, h)$, and then subtract (5.3.7) from (5.3.12). Setting $e_i = v(x_i) - v_i$, this gives

$$\sigma_i = \frac{1}{h^2}[e_{i+1} - 2e_i + e_{i-1}] - c_i e_i, \quad i = 1, \cdots, n,$$

or

$$(2 + c_i h^2)e_i - e_{i+1} - e_{i-1} = -h^2 \sigma_i, \quad i = 1, \ldots, n, \tag{5.3.15}$$

where $e_0 = e_{n+1} = 0$. If A is the coefficient matrix of (5.3.9) and $\mathbf{e}$ and $\boldsymbol{\sigma}$ are vectors with components $e_1, \ldots, e_n$ and $\sigma_1, \ldots, \sigma_n$, we can write (5.3.15) as

$$A\mathbf{e} = -h^2 \boldsymbol{\sigma} \tag{5.3.16}$$

or, assuming that A^{-1} exists,

$$\mathbf{e} = -h^2 A^{-1} \boldsymbol{\sigma}. \tag{5.3.17}$$

This is the basic relationship between the global and local discretization errors. Note that the global discretization errors $e_1, \ldots, e_n$ and the approximate solutions $v_1, \ldots, v_n$ satisfy systems of equations with exactly the same coefficient matrix, but different right-hand sides.

The problem now is to study the behavior of A^{-1} as $h \to 0$. This is made more difficult by the fact that n, the order of A, tends to infinity as $h \to 0$. It is beyond the scope of this book to pursue this problem in any generality, but we can give a relatively simple analysis in the case that

$$c(x) \geq \gamma > 0, \quad a \leq x \leq b, \tag{5.3.18}$$

so that $c_i \geq \gamma$, $i = 1, \ldots, n$. If we set $e = \max|e_i|$ and $\sigma = \max|\sigma_i| = O(h^2)$, we obtain from (5.3.15) and (5.3.18) that

$$(2 + \gamma h^2)|e_i| \leq 2e + h^2 \sigma, \quad i = 1, \ldots, n. \tag{5.3.19}$$

5.3 Boundary Value Problems

Since (5.3.19) holds for all i we must have

$$(2 + \gamma h^2)e \leq 2e + h^2\sigma.$$

Therefore, since by (5.3.14) $\sigma = O(h^2)$, we conclude that

$$e \leq \frac{\sigma}{\gamma} = O(h^2), \tag{5.3.20}$$

which shows that the global discretization error is $O(h^2)$ provided that the local discretization is $O(h^2)$. It can be shown that the same result holds more generally, in particular for the important special case in which $c(x) \equiv 0$, but a more delicate analysis is required.

Equations with First Derivatives

We next consider Equation (5.3.3) in which v' is present. The standard centered difference approximation to $v'(x)$ is

$$v'(x) \doteq \frac{1}{2h}[v(x+h) - v(x-h)], \tag{5.3.21}$$

and it is easy to show (Exercise 5.3.3) that the error in this approximation is $O(h^2)$. If we replace $v'(x)$ in (5.3.3) at the grid points by (5.3.21) and proceed as before, the equations corresponding to (5.3.8) now become

$$\left(-1 - \frac{b_i h}{2}\right)v_{i-1} + (2 + c_i h^2)v_i + \left(-1 + \frac{b_i h}{2}\right)v_{i+1} = -h^2 d_i, \quad i = 1, \ldots, n,$$

or

$$r_i v_{i-1} + p_i v_i + q_i v_{i+1} = -h^2 d_i, \quad i = 1, \ldots, n,$$

where we have set

$$p_i = 2 + c_i h^2, \quad q_i = -1 + \frac{b_i h}{2}, \quad r_i = -1 - \frac{b_i h}{2}. \tag{5.3.22}$$

Then we can write these equations in matrix-vector form as

$$\begin{bmatrix} p_1 & q_1 & & & \\ r_2 & p_2 & q_2 & & \\ & \ddots & \ddots & \ddots & \\ & & & & q_{n-1} \\ & & & r_n & p_n \end{bmatrix} \begin{bmatrix} v_1 \\ v_2 \\ \vdots \\ v_n \end{bmatrix} = -h^2 \begin{bmatrix} d_1 + r_1 \alpha/h^2 \\ d_2 \\ \vdots \\ d_{n-1} \\ d_n + q_n \beta/h^2 \end{bmatrix}. \tag{5.3.23}$$

One desirable property of the coefficient matrix of (5.3.23) is diagonal dominance. Recall from Section 2.3 that a general $n \times n$ matrix $A = (a_{ij})$ is row diagonally dominant if

$$|a_{ii}| \geq \sum_{j \neq i} |a_{ij}|, \quad i = 1, \ldots, n; \tag{5.3.24}$$

and column diagonally dominant if A^T is row diagonally dominant. Diagonal dominance is important for a number of reasons. For example, from Theorem 2.3.3 a strictly or irreducibly diagonally dominant matrix is nonsingular. We will also see in Chapter 6 that diagonal dominance is a valuable property in the numerical solution of a linear system.

The matrix of (5.3.9) is diagonally dominant if the c_i are non-negative. However, in order for the matrix of (5.3.23) to be row diagonally dominant we need that

$$|p_i| \geq |r_i| + |q_i|, \quad i = 1, \ldots, n, \tag{5.3.25}$$

or, using (5.3.22),

$$|2 + c_i h^2| \geq |1 + \frac{b_i h}{2}| + |1 - \frac{b_i h}{2}|, \quad i = 1, \ldots, n. \tag{5.3.26}$$

If we assume that $c_i \geq 0$, then (5.3.26) holds if h is sufficiently small. In particular, if

$$|b_i h| \leq 2, \quad i = 1, \ldots, n, \tag{5.3.27}$$

then the absolute values of the quantities on the right side of (5.3.26) are the quantities themselves so that

$$|2 + c_i h^2| \geq 1 + \frac{b_i h}{2} + 1 - \frac{b_i h}{2} = 2.$$

(We note that the quantity $\frac{1}{2} \max |b_i h|$ is called the cell Reynolds number or cell Peclet number in fluid dynamics problems.)

The condition (5.3.27) on h, which also ensures column diagonal dominance, is a rather stringent one and can be avoided by using one-sided differences in place of the central difference (5.3.21) to approximate the first derivative. More precisely, we use the approximations

$$v'(x_i) \doteq \begin{cases} \frac{1}{h}(v_{i+1} - v_i) & \text{if } b_i < 0, \\ \frac{1}{h}(v_i - v_{i-1}) & \text{if } b_i \geq 0, \end{cases} \tag{5.3.28}$$

so that the direction of the one-sided difference is determined by the sign of b_i. Such differences are quite commonly used in fluid dynamics problems and in that context are called *upwind* (or *upstream*) differences. With (5.3.28), the ith row of the coefficient matrix (5.3.23) becomes

$$\begin{array}{cccc} -1 & 2 + c_i h^2 - b_i h & -1 + b_i h, & b_i \leq 0, \\ -(1 + b_i h) & 2 + c_i h^2 + b_i h & -1, & b_i > 0, \end{array} \tag{5.3.29}$$

and it is easy to verify that row diagonal dominance holds, independent of the size of h, assuming again that $c_i \geq 0$. We note, however, that although the

5.3 Boundary Value Problems

centered difference approximation (5.3.21) is second-order accurate, the one-sided approximations (5.3.28) are only first-order accurate, and this increase in the discretization error must be weighed against the better properties of the coefficient matrix. On the other hand, the second-order method is likely to give a "wiggly" approximate solution if diagonal dominance does not hold.

If we use (5.3.28) and the c_i are positive, then the matrix is strictly diagonally dominant and, hence, nonsingular. If the c_i are all zero, strict diagonal dominance does not hold but the matrix is irreducible since its directed graph is given in Figure 5.3.2 and is strongly connected (see Theorem 2.1.5 and Exercise 5.3.11). Moreover, the first row of the matrix has a single off-diagonal element so that (5.3.24) holds with strict inequality for $i = 1$. Hence, the matrix is irreducibly diagonally dominant and, by Theorem 2.3.3, nonsingular.

Figure 5.3.2: *Directed Graph of Tridiagonal Matrix*

Although we can achieve diagonal dominance by using (5.3.28), with either (5.3.28) or (5.3.21) symmetry in general is lost. For example, with (5.3.21) we would need

$$-1 + \frac{1}{2}b_i h = -1 - \frac{1}{2}b_{i+1}h, \qquad i = 1, \ldots, n-1.$$

Clearly, these relations will not usually hold. (See, however, the Supplementary Discussion.)

Other Boundary Conditions

The previous discussion has all been for the boundary conditions (5.3.2). In many problems, however, boundary conditions on the derivative rather than the function itself may be given, and we now consider the modifications that this requires.

Suppose that, for example, we have the boundary conditions

$$v'(a) = \alpha \qquad v'(b) = \beta \qquad (5.3.30)$$

in place of (5.3.2); that is, we specify derivative values rather than function values. Consider the difference equations (5.3.8). In the equation for $i = 1$, the value of v_0 is no longer known from the boundary condition at $x = a$. Instead, v_0 will be an additional unknown and we will need another equation that can be derived as follows. We approximate v'' at $x = a$ by

$$v''(a) \doteq \frac{1}{h^2}[v_{-1} - 2v_0 + v_1] \qquad (5.3.31)$$

using a grid point $-h$ outside the interval. Then, by the boundary condition

$$\alpha = v'(a) \doteq \frac{1}{2h}[v_1 - v_{-1}] \tag{5.3.32}$$

we have $v_{-1} \doteq v_1 - 2\alpha h$, and we can use this to eliminate v_{-1} in (5.3.31):

$$v''(a) \doteq \frac{1}{h^2}[2v_1 - 2v_0 - 2\alpha h]. \tag{5.3.33}$$

In general, this approximation is only first-order accurate, but in the important special case that $\alpha = 0$ it is second-order (Exercise 5.3.7). We can now use (5.3.33) to obtain an additional equation and we have $n + 1$ equations in the $n + 1$ unknowns $v_0, \ldots, v_n$. If $v(b) = \beta$ is specified as before, this is then the system of equations to be solved. If $v'(b)$ is specified, as in (5.3.30), then another equation would arise from the approximation

$$v''(b) \doteq \frac{1}{h^2}[2v_n - 2v_{n+1} + 2\beta h], \tag{5.3.34}$$

and v_{n+1} would be an additional unknown.

If only the function value is specified, as in (5.3.2), the boundary conditions are called *Dirichlet*, whereas if only the derivatives are specified, as in (5.3.30), the boundary conditions are called *Neumann*. More generally, mixed boundary conditions may be given as linear combinations of both function and derivative values at both endpoints:

$$\eta_1 v(0) + \eta_2 v'(0) = \alpha, \qquad \gamma_1 v(1) + \gamma_2 v'(1) = \beta.$$

In this case, approximations analogous to those discussed previously would be used.

We return to the boundary conditions (5.3.30) and the differential equation (5.3.4). The approximations (5.3.33) and (5.3.34) give the two equations

$$(2 + c_0 h^2) v_0 - 2v_1 = -2\alpha h - h^2 d_0,$$
$$(2 + c_{n+1} h^2) v_{n+1} - 2v_n = 2\beta h - h^2 d_{n+1},$$

which are added to the system (5.3.8) to give $n + 2$ equations in the $n + 2$ unknowns $v_0, \ldots, v_{n+1}$. The coefficient matrix of this system is

$$A = \begin{bmatrix} 2 + c_0 h^2 & -2 & & & & \\ -1 & 2 + c_1 h^2 & -1 & & & \\ & \ddots & \ddots & \ddots & & \\ & & & & & -1 \\ & & & & -2 & 2 + c_{n+1} h^2 \end{bmatrix}. \tag{5.3.35}$$

5.3 Boundary Value Problems

This matrix is no longer symmetric but it can be easily symmetrized if desired (Exercise 5.3.8). If $c_i > 0$, $i = 0, \ldots, n+1$, Theorem 2.3.3 shows that A is nonsingular. But, if the c_i are all zero, it is singular, since $A\mathbf{e} = 0$, where $\mathbf{e} = (1, 1, \ldots, 1)^T$. This singularity of A mirrors the non-uniqueness of solutions of the differential equation itself: if $c(x) \equiv 0$ in (5.3.4) and v is a solution satisfying the boundary conditions (5.3.30), then $v + \gamma$ is also a solution for any constant γ. Even though the solution is not unique, in many problems it may be desirable to be able to obtain some solution, but the solution of such singular problems is beyond the scope of this book.

Another type of boundary condition that leads to non-unique solutions is called *periodic*:
$$v(a) = v(b). \tag{5.3.36}$$

In this case, v_0 and v_{n+1} are both unknowns but since they are equal only one needs to be added to the system of equations. Again, we can use (5.3.31) to obtain an additional equation. By (5.3.36), we can assume that the solution v is extended periodically outside $[a, b]$; in particular, we can take $v_{-1} = v_n$ in (5.3.31) so that it becomes

$$v''(0) \doteq \frac{1}{h^2}[v_n - 2v_0 + v_1]. \tag{5.3.37}$$

Similarly, in the approximation at x_n, we can take $v_{n+1} = v_0$:

$$v''(x_n) \doteq \frac{1}{h^2}[v_{n-1} - 2v_n + v_{n+1}] = \frac{1}{h^2}[v_{n-1} - 2v_n + v_0]. \tag{5.3.38}$$

There will then be $n + 1$ unknowns $v_0, \ldots, v_n$, and the coefficient matrix for the problem (5.3.4) will be

$$A = \begin{bmatrix} 2 + c_0 h^2 & -1 & & -1 \\ -1 & 2 + c_1 h^2 & \ddots & \\ & \ddots & \ddots & -1 \\ -1 & & -1 & 2 + c_n h^2 \end{bmatrix}. \tag{5.3.39}$$

The tridiagonal structure has now been lost because of the outlying -1's. As with (5.3.35), it follows from Theorem 2.3.3 that the matrix A of (5.3.39) is nonsingular if all c_i are positive, but it is singular if all c_i are zero. (As before $A\mathbf{e} = 0$.) This singularity again reflects the differential equation since if $c(x) \equiv 0$ in (5.3.4) and v is a solution satisfying (5.3.36), then $v + \gamma$ is also a solution for any constant.

Nonlinear Problems

We now return to the original equation (5.3.1), for a function f that is nonlinear. For simplicity, we will consider a function that contains only v and not v'; thus, we consider an equation of the form

$$v''(x) = f(x, v(x)), \qquad a \le x \le b, \qquad (5.3.40)$$

where f is a given function of two variables. Again for simplicity, we will use only the Dirichlet boundary conditions (5.3.2).

As with the linear problem (5.3.4), we approximate $v''(x)$ in (5.3.40) by (5.3.5). This then leads to the system of equations, corresponding to (5.3.8),

$$-v_{i+1} + 2v_i - v_{i-1} + h^2 f(x_i, v_i) = 0, \qquad i = 1, \ldots, n, \qquad (5.3.41)$$

where $v_0 = \alpha$ and $v_{n+1} = \beta$ are known by the boundary conditions (5.3.2). This is a system of n equations in the n unknowns $v_1, \ldots, v_n$ and is a nonlinear system if the function f is nonlinear in v.

We can write the system (5.3.41) in matrix-vector form as

$$\mathbf{F}(\mathbf{v}) \equiv A\mathbf{v} + \mathbf{g}(\mathbf{v}) = 0, \qquad (5.3.42)$$

where $\mathbf{v}$ is the vector with components $v_1, \ldots, v_n$, A is the $(2, -1)$ tridiagonal matrix of (5.3.10) and $\mathbf{g}$ is the nonlinear function

$$\mathbf{g}(\mathbf{v}) = h^2 \begin{bmatrix} f(x_1, v_1) \\ \vdots \\ f(x_n, v_n) \end{bmatrix} - \begin{bmatrix} \alpha \\ 0 \\ \vdots \\ 0 \\ \beta \end{bmatrix}. \qquad (5.3.43)$$

As an example, consider the problem

$$v''(x) = 3v(x) + x^2 + 10[v(x)]^3, \qquad 0 \le x \le 1, \qquad v(0) = v(1) = 0. \quad (5.3.44)$$

Here,

$$f(x, v) = 3v + x^2 + 10v^3 \qquad (5.3.45)$$

and, with $h = 1/(n+1)$ and

$$x_i = ih, \qquad i = 0, 1, \ldots, n+1, \qquad (5.3.46)$$

the difference equations (5.3.41) are

$$-v_{i+1} + 2v_i - v_{i-1} + h^2(3v_i + i^2 h^2 + 10v_i^3) = 0, \qquad i = 1, \ldots, n, \quad (5.3.47)$$

5.3 Boundary Value Problems

where, from the boundary conditions, $v_0 = v_{n+1} = 0$. Hence, the ith component of the function $\mathbf{g}$ of (5.3.43) is $h^2(3v_i + i^2h^2 + 10v_i^3)$.

Now consider Newton's method (4.3.32) for (5.3.42). The Jacobian matrix will be (Exercise 5.3.12)

$$\mathbf{F}'(\mathbf{v}) = A + \mathbf{g}'(\mathbf{v}). \qquad (5.3.48)$$

Since the ith component, $g_i(\mathbf{v}) = h^2 g(x_i, v_i)$, of $\mathbf{g}$ depends on only v_i, we have that $\dfrac{\partial g_i}{\partial v_j} = 0$, $j \neq i$. Thus, the matrix $\mathbf{g}'(\mathbf{v})$ is diagonal and $\mathbf{F}'(\mathbf{v})$ is tridiagonal with a typical row given by

$$-1 \quad 2 + h^2 \frac{\partial f}{\partial v}(x_i, v_i) \quad -1.$$

The Newton iteration is then

$$\begin{aligned} &1.\ \text{Solve } [A + \mathbf{g}'(\mathbf{v}^k)]\boldsymbol{\delta}^k = -[A\mathbf{v}^k + \mathbf{g}(\mathbf{v}^k)], \\ &2.\ \text{Set } \mathbf{v}^{k+1} = \mathbf{v}^k + \boldsymbol{\delta}^k, \end{aligned} \qquad (5.3.49)$$

so that at each iteration a tridiagonal linear system is to be solved. If the function f is complicated, a major portion of the work of each Newton iteration will be the evaluations of $\mathbf{g}(\mathbf{v}^k)$ and $\mathbf{g}'(\mathbf{v}^k)$.

For the boundary-value problem (5.3.44), f is given by (5.3.45), so that

$$\frac{\partial f}{\partial v}(x, v) = 3 + 30v^2,$$

and the ith diagonal element of the Jacobian matrix (5.3.48) is $2 + h^2(3 + 30v_i^2)$. Since the $(2, -1)$ tridiagonal matrix A is diagonally dominant, it is clear that the addition of the positive terms $h^2(3 + 30v_i^2)$ to the diagonal only enhances the diagonal dominance. More generally, whenever

$$\frac{\partial f}{\partial v}(x, v) \geq 0, \qquad a \leq x \leq b, \qquad -\infty < v < \infty, \qquad (5.3.50)$$

and A is the $(2, -1)$ matrix, then

$$A + \mathbf{g}'(\mathbf{v}) \text{ is diagonally dominant}, \qquad (5.3.51)$$

$$A + \mathbf{g}'(\mathbf{v}) \text{ is symmetric positive-definite}. \qquad (5.3.52)$$

The validity of (5.3.52) follows from Exercise 2.2.17. Either of the conditions (5.3.51) or (5.3.52) is sufficient to ensure (but beyond the scope of this book to prove) that the system (5.3.42) has a unique solution. On the other hand, if (5.3.50) does not hold, the differential equation (5.3.40) need not have a unique solution, and this will be reflected in the discrete system (5.3.42). For example, if $f(x, v) = v^4$ there will be two solutions of both (5.3.40) and (5.3.42). This is explored further in Exercise 5.3.15.

The Shooting Method

We discuss briefly one other approach to the solution of nonlinear boundary value problems. Consider the equation (5.3.1) with boundary conditions (5.3.2). Let $v(x;s)$ denote the solution of the corresponding initial value problem

$$v''(x) = f(x, v(x), v'(x)), \qquad v(a) = \alpha, \quad v'(a) = s. \qquad (5.3.53)$$

If we can find a value of s, say $s^\star$, such that

$$v(b; s^\star) = \beta, \qquad (5.3.54)$$

then we will have solved the boundary value problem (5.3.1), (5.3.2). This suggests the following procedure. Choose a value s_0 and solve numerically the initial value problem (5.3.53) with $v'(a) = s_0$. (For this, we will probably convert (5.3.53) to a first-order system; see the Supplementary Discussion of Section 5.2). We then adjust the parameter s_0 and try again with $v'(a) = s_1$ which, hopefully, will give a more accurate solution. The process is illustrated in Figure 5.3.2, and is called the *shooting method* in analogy with an artillery gunner adjusting the angle of a gun to hit a desired target.

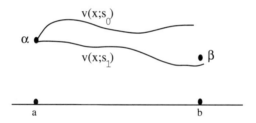

Figure 5.3.3: *Shooting*

The adjustment of the parameter s can be made systematic by recognizing that we are attempting to solve an equation in s. Define

$$g(s) = v(b; s) - \beta. \qquad (5.3.55)$$

Then, (5.3.54) is equivalent to finding a solution of the equation $g(s) = 0$. For this we can use any of the methods of Section 4.3, for example, the bisection method. Note, however, that each evaluation of g can be costly since it requires the solution of the initial value problem. Thus, the secant method, with its more rapid rate of convergence, would usually be better than bisection, provided that it converges. Newton's method is also possible; see the Supplementary Discussion. The shooting method is attractive but it may fail due to the initial value problem being unstable, as discussed in Section 5.2.

5.3 Boundary Value Problems

Systems of Equations

We consider finally first-order systems of equations

$$\mathbf{y}' = \mathbf{f}(x, \mathbf{y}), \qquad a \leq x \leq b, \qquad (5.3.56)$$

where $\mathbf{y}$ and $\mathbf{f}$ are n-vectors. Boundary conditions for this system may be given in different ways. The simplest situation is if $y_1, \ldots, y_m$ are prescribed at a and $y_{m+1}, \ldots, y_n$ at b; this gives the necessary n conditions. However, in many problems some of the variables y_i may be prescribed at both end points and others at neither point. More generally, linear combinations of values at the two end points may be prescribed. Thus, we will consider boundary conditions of the form

$$B_a \mathbf{y}(a) + B_b \mathbf{y}(b) = \mathbf{c}, \qquad (5.3.57)$$

where B_a and B_b are given $n \times n$ matrices.

A number of different methods may be considered for (5.3.56). One of the simplest is the *trapezoidal* scheme (see (5.2.59))

$$\mathbf{y}_{i+1} - \mathbf{y}_i = \frac{h_i}{2}[f(x_i, \mathbf{y}_i) + f(x_{i+1}, \mathbf{y}_{i+1})], \qquad i = 0, \ldots, N, \qquad (5.3.58)$$

where $x_0 = a, x_{N+1} = b$, and $h_i = x_{i+1} - x_i$. Then $\mathbf{y}_0 = \mathbf{y}(a), \mathbf{y}_{N+1} = \mathbf{y}(b)$ and (5.3.58) together with (5.3.57) gives a system of $N+2$ vector equations in the $N+2$ vector unknowns $\mathbf{y}_0, \ldots, \mathbf{y}_{N+1}$. In general, these equations will be nonlinear, and we can use Newton's method to solve them (see Exercise 5.3.16). The linear systems that need to be solved to carry out Newton's method will have the same form as for linear differential equations, which we now discuss.

Consider the general first order linear system of differential equations

$$\mathbf{y}' = A(x)\mathbf{y} + \mathbf{d}(x), \qquad a \leq x \leq b, \qquad (5.3.59)$$

where for each x, $A(x)$ is a given $n \times n$ matrix and $\mathbf{d}(x)$ a given vector. Then (5.3.58) becomes

$$\mathbf{y}_{i+1} - \mathbf{y}_i = \frac{h_i}{2}[A(x_i)\mathbf{y}_i + A(x_{i+1})\mathbf{y}_{i+1}] + \frac{h_i}{2}[\mathbf{d}(x_i) + \mathbf{d}(x_{i+1})] \qquad (5.3.60)$$

for $i = 0, \ldots, N$. The equations (5.3.60) together with (5.3.57) form a linear system of equations that may be written in matrix-vector form

$$\begin{bmatrix} G_0 & H_0 & & & \\ & G_1 & H_1 & & \\ & & \ddots & \ddots & \\ & & & G_N & H_N \\ B_a & & & & B_b \end{bmatrix} \begin{bmatrix} \mathbf{y}_0 \\ \mathbf{y}_1 \\ \vdots \\ \mathbf{y}_N \\ \mathbf{y}_{N+1} \end{bmatrix} = \begin{bmatrix} \mathbf{d}_0 \\ \mathbf{d}_1 \\ \vdots \\ \mathbf{d}_N \\ \mathbf{c} \end{bmatrix}, \qquad (5.3.61)$$

where

$$G_i = -h_i^{-1}I - \frac{1}{2}A(x_i), \qquad H_i = h_i^{-1}I - \frac{1}{2}A(x_{i+1}), \qquad (5.3.62)$$

and

$$\mathbf{d}_i = \frac{1}{2}[\mathbf{d}(x_i) + \mathbf{d}(x_{i+1})]. \qquad (5.3.63)$$

In the important special case that the boundary conditions (5.3.57) are *separated*, they may be written in the form

$$\hat{B}_a \mathbf{y}(a) = \mathbf{c}_1, \qquad \hat{B}_b \mathbf{y}(b) = \mathbf{c}_2, \qquad (5.3.64)$$

where $\hat{B}_a$ is $m_1 \times n$, $\hat{B}_b$ is $m_2 \times n$ and $m_1 + m_2 = n$. In this case, the system (5.3.60), (5.3.64) may be written as

$$\begin{bmatrix} \hat{B}_a & & & & \\ G_0 & H_0 & & & \\ & G_1 & & & \\ & & \ddots & \ddots & \\ & & & G_N & H_N \\ & & & & \hat{B}_b \end{bmatrix} \begin{bmatrix} \mathbf{y}_0 \\ \mathbf{y}_1 \\ \vdots \\ \mathbf{y}_N \\ \mathbf{y}_{N+1} \end{bmatrix} = \begin{bmatrix} \mathbf{c}_1 \\ \mathbf{d}_0 \\ \vdots \\ \mathbf{d}_N \\ \mathbf{c}_2 \end{bmatrix}. \qquad (5.3.65)$$

Linear systems of equations of the form (5.3.65) or (5.3.61) may be solved by the methods of Chapters 6 and 7 although their very special structure allows the use of more efficient techniques (see the Supplementary Discussion).

Supplementary Discussion and References: 5.3

1. For further reading on two-point boundary value problems, and in particular for special techniques for systems of the form (5.3.61) and (5.3.65), see Ascher et al. [1988].

2. We have discussed in the text only second-order discretizations of the differential equations but there are a number of approaches to higher-order methods. One is to use higher-order approximations to the derivatives. For example

$$v''(x_i) \doteq \frac{1}{12h^2}(-v_{i-2} + 16v_{i-1} - 30v_i + 16v_{i+1} - v_{i+2}) \qquad (5.3.66)$$

is a fourth-order approximation (the error is proportional to h^4) provided that v is sufficiently differentiable. One difficulty with applying approximations of this type to two-point boundary-value problems occurs near the boundary. For example, if we apply the approximation at

5.3 Boundary Value Problems

the first interior grid point x_1, it requires values of v not only at x_0 but also at x_{-1}, which is outside the interval. However, it is the case that second-order approximations can be applied at x_1 and x_{n-1} and still retain fourth-order accuracy of the solution.

3. Another approach to obtaining higher-order approximations, while using only second-order approximations to the derivatives, is Richardson extrapolation. This was discussed in Section 5.2 for initial value problems and applies also to boundary value problems. Still another approach is *deferred correction*. In this procedure, one computes an approximation to the local discretization error by means of the current approximate solution, and then uses this to obtain a new approximate solution with $O(h^4)$ accuracy. The process can then be repeated to obtain still higher-order approximations.

4. We noted in the text that symmetry of the coefficient matrix is lost when the differential equation contains first derivatives. However, for many important problems the differential equation is of the so-called "self-adjoint" form

$$[a(x)v'(x)]' = d(x). \tag{5.3.67}$$

In this case, symmetry of the coefficient matrix can be achieved by "symmetric differencing" as follows. As in the derivation of (5.3.5), we use the auxiliary grid points $x_i \pm \frac{h}{2}$ and approximate the outermost derivative of $(av')'$ at x_i by

$$[a(x)v'(x)]'_i \doteq \frac{1}{h}(a_{i+\frac{1}{2}} v'_{i+\frac{1}{2}} - a_{i-\frac{1}{2}} v'_{i-\frac{1}{2}}), \tag{5.3.68}$$

where the subscripts indicate the grid point at which the evaluation is done. Next, approximate the first derivatives by the centered differences

$$v'_{i+\frac{1}{2}} \doteq \frac{1}{h}(v_{i+1} - v_i) \qquad v'_{i-\frac{1}{2}} \doteq \frac{1}{h}(v_i - v_{i-1}),$$

and use these in (5.3.68) to obtain

$$\begin{aligned}[a(x)v'(x)]'_i &\doteq \frac{1}{h^2}[a_{i+\frac{1}{2}}(v_{i+1} - v_i) - a_{i-\frac{1}{2}}(v_i - v_{i-1})] \\ &= \frac{1}{h^2}[a_{i-\frac{1}{2}} v_{i-1} - (a_{i+\frac{1}{2}} + a_{i-\frac{1}{2}})v_i + a_{i+\frac{1}{2}} v_{i+1}].\end{aligned}$$

Then, the coefficient matrix of the system of difference equations is symmetric.

5. For proofs that the conditions (5.3.51) or (5.3.52) ensure the existence and uniqueness of a solution of the system (5.3.42), see Ortega and Rheinboldt [1970].

6. In order to apply Newton's method in the shooting method we need the derivative of the function g of (5.3.55); this is $g'(s) = v_s(b; s)$, where $v_s(b; s)$ is the partial derivative of $v(x; s)$ with respect to s and evaluated at $x = b$. We indicate how to compute this in the case that the function f of (5.3.53) depends only on v. Thus, $v''(x; s) = f(x, v(x; s))$ and we differentiate this with respect to s to obtain

$$\frac{\partial}{\partial s}(v''(x; s)) = f_v(x, v(x; s))v_s(x; s), \quad (5.3.69)$$

where $f_v(x, v)$ is the partial derivative of $f(x, v)$ with respect to v. Assuming that differentiation with respect to s and x can be interchanged on the left side of (5.3.69), we then have

$$v_s''(x; s) = f_v(x, v(x, s))v_s(x; s), \quad (5.3.70)$$

which is called the *adjoint equation* for v_s. In (5.3.70) we are assuming that s is held fixed so that this is a differential equation for $v_s(x; s)$ considered only as a function of x. The initial conditions for (5.3.70) are obtained by differentiating those of (5.3.53) with respect to s; thus,

$$v_s(a; s) = 0, \quad v_s'(a; s) = 1. \quad (5.3.71)$$

If we knew the exact solution $v(x; s)$ of (5.3.53), we could put this into f_v in (5.3.70) to obtain a known function of x, and then (5.3.70), (5.3.71) would be a linear initial-value problem for $v_s(x; s)$. Upon solving this initial value problem we obtain $g'(s) = v_s(b; s)$. Of course, we do not know the exact solution of (5.3.53), but we solve this initial-value problem approximately to obtain values $v_i \doteq v(x_i; s)$ at the grid points x_i. We can then use these approximate values to evaluate f_v in (5.3.70) and in this way we can obtain an approximate solution to (5.3.70) at the same time we obtain the approximate solution to (5.3.53). Thus, we will be able to carry out Newton's method, at least approximately, for $g(s) = 0$.

7. Parallel solution of boundary value problems requires the parallel solution of the linear systems of equations if the finite difference method is used for the discretization. This will be addressed in Chapter 7. For the shooting method, parallel solution of initial-value problems will be required, as discussed in Section 5.2. However, there is a modification of the shooting method that can bring in additional natural parallelism. This modification, known as *multiple shooting*, is also sometimes useful for serial computation to overcome stability problems with the shooting method. For further discussion of multiple and parallel shooting see Ascher et al. [1988].

5.3 Boundary Value Problems

EXERCISES 5.3

5.3.1. Consider the boundary value problem $u''(z) = g(z, u(z), u'(z))$ with $u(a) = \alpha$, $u(b) = \beta$. By the change of variable $t = (b-a)x + a$, show that this problem is equivalent to (5.3.1) and (5.3.2) with $a = 0, b = 1$ if $v(x) = u((b-a)x + a)$ and $f(x, v(x), v'(x)) = (b-a)^2 g((b-a)x + a, v(x), (b-a)^{-1}v'(x))$. Specialize this result to the linear problem (5.3.3). This shows that we can work on the interval $[0, 1]$ with no loss of generality.

5.3.2. Assume that the function $v(x)$ is suitably differentiable. By expanding $v(x+h)$ and $v(x-h)$ in Taylor series, verify that (5.3.14) holds.

5.3.3 Assume that the function $v(x)$ is twice differentiable. Show that the error in the centered difference approximation (5.3.21) is $O(h^2)$.

5.3.4. Consider the two-point boundary-value problem

$$v'' + 2xv' - x^2 v = x^2, v(0) = 1, v(1) = 0.$$

   a. Let $h = \frac{1}{4}$ and explicitly write out the difference equations (5.3.23).

   b. Repeat part **a** using the one-sided approximations (5.3.28) for v'.

   c. Repeat parts **a** and **b** for the boundary conditions $v'(0) = 1$, $v(1) = 0$, and then $v'(0) + v(0) = 1$, $v'(1) + \frac{1}{2}v(1) = 0$.

5.3.5. Use the trigonometric identity $\sin(\alpha \pm \beta) = \sin\alpha\cos\beta \pm \cos\alpha\sin\beta$ to verify that the eigenvalues of the matrix (5.3.10) are given by $\lambda_k = 2 - 2\cos kh$, $h = \pi/(n+1)$, with corresponding eigenvectors

$$\mathbf{x}_k = (\sin kh, \sin 2kh, \ldots, \sin nkh)^T.$$

That is, verify that $A\mathbf{x}_k = \lambda_k \mathbf{x}_k$, $k = 1, \ldots, n$.

5.3.6. Assume that the function a of (5.3.67) is twice differentiable. By the change of variable $w(x) = \sqrt{a(x)} v(x)$, show that (5.3.67) can be replaced by an equation of the form $w''(x) = c(x)w(x) + f(x)$. Give an expression for $c(x)$ and show how it can be approximated numerically.

5.3.7. Assuming that v is sufficiently differentiable, show that (5.3.33) is, in general, only a first-order approximation to $v''(a)$. However, if $\alpha = 0$, show that it is second-order by assuming that v has been extended outside the interval $[0, 1]$ so that $v(-h) = v(h)$.

5.3.8. Let A and B be the tridiagonal matrices

$$A = \begin{bmatrix} a_1 & b_1 & & & \\ c_1 & \ddots & \ddots & & \\ & \ddots & & & \\ & & & & b_{n-1} \\ & & & c_{n-1} & a_n \end{bmatrix}, \quad B = \begin{bmatrix} a_1 & \gamma_1 & & & \\ \gamma_1 & \ddots & \ddots & & \\ & \ddots & & & \\ & & & & \gamma_{n-1} \\ & & & \gamma_{n-1} & a_n \end{bmatrix},$$

where $b_i c_i > 0$ and $\gamma_i = \sqrt{b_i c_i}$, $i = 1, \ldots, n-1$. Show that if

$$D = \operatorname{diag}\left(1, \frac{b_1}{c_1}, \frac{b_1 b_2}{c_1 c_2}, \ldots, \frac{b_1 \cdots b_{n-1}}{c_1 \cdots c_{n-1}}\right),$$

then $B = D^{1/2} A D^{-1/2}$, where $D^{1/2} = \operatorname{diag}(d_1^{1/2}, d_2^{1/2}, \ldots, d_n^{1/2})$. Show also that DA is symmetric under only the conditions that $b_i c_i \neq 0, i = 1, \ldots, n-1$.

5.3.9. Derive the approximations (5.3.5) and (5.3.21) by means of interpolation polynomials: let l be the linear polynomial that satisfies $l(x \pm h) = v(x \pm h)$, and show that $l'(x)$ is the approximation (5.3.21). Then let q be the quadratic polynomial that satisfies $q(x) = v(x)$ and $q(x \pm h) = v(x \pm h)$ and show that $q''(x)$ gives the approximation (5.3.5).

5.3.10. Assume that the b_i in (5.3.22) are all equal to a constant b. Give a condition that ensures that the matrix of (5.3.23) can be symmetrized as in Exercise 5.3.8. Also, if the $c_i = 0$ in (5.3.22), derive an expansion for the quantities d_n of Exercise 5.3.8 as $n \to \infty$.

5.3.11. Let A be a tridiagonal matrix whose off-diagonal elements are all non-zero. Show that the directed graph of A is as shown in Figure 5.3.2. Conclude that the graph is strongly connected and, hence, by Theorem 2.1.5, that A is irreducible.

5.3.12. If $\mathbf{F}(\mathbf{v}) = A\mathbf{v} + \mathbf{g}(\mathbf{v})$ for some matrix A, verify that $\mathbf{F}'(\mathbf{v}) = A + \mathbf{g}'(\mathbf{v})$.

5.3.13. Write out the difference equations (5.3.41) and the corresponding Jacobian matrices for:

a. $f(x, v) = v + v^2$
b. $f(x, v) = xv^3$
c. $f(x, v) = e^v$

5.3.14. Write out the Newton iteration (5.3.49) explicitly for the difference equations (5.3.41) with the f's given by Exercise 5.3.13.

5.3.15. Consider the two-point boundary-value problem

$$v'' = v^4, \quad 0 \leq x \leq 1, \quad v(0) = 1, \quad v(1) = \frac{1}{2}.$$

a. Find an approximation to a solution by a third degree polynomial obtained by using the data $v(0), v(1), v''(0), v''(1)$.
b. Obtain an approximate solution by the shooting method using both bisection and the secant method for the resulting single nonlinear equation.

5.3.16. Apply Newton's method to the system (5.3.58), (5.3.57). Show that the Jacobian matrix has the same structure as the coefficient matrix of (5.3.61).

5.3.17. Show that the inverse of the $n \times n$ matrix (5.3.10) is

$$\frac{1}{n+1}\begin{bmatrix} n & n-1 & n-2 & \cdots & 1 \\ & 2(n-1) & 2(n-2) & & 2 \\ & & 3(n-2) & & \vdots \\ & \text{Symmetric} & & \ddots & \\ & & & & n \end{bmatrix}$$

More generally if A is a symmetric nonsingular tridiagonal matrix with non-zero diagonal elements $a_1, \ldots, a_n$ and non-zero off-diagonal elements $-b_1, \ldots, -b_{n-1}$, show that

$$A^{-1} = \begin{bmatrix} u_1 v_1 & u_1 v_2 & \cdots & u_1 v_n \\ u_1 v_2 & u_2 v_2 & \cdots & u_2 v_n \\ \vdots & & & \\ u_1 v_n & u_2 v_n & \cdots & u_n v_n \end{bmatrix}$$

where

$$u_1 = 1, \quad u_2 = \frac{a_1}{b_1}, \quad u_i = \frac{a_{i-1} u_{i-1} - b_{i-2} u_{i-2}}{b_{i-1}}, \quad i = 3, \ldots, n$$

$$v_n = \frac{1}{-b_{n-1} u_{n-1} + a_n u_n}, \quad v_i = \frac{1 + b_i u_i v_{i+1}}{a_i u_i - b_{i-1} u_{i-1}}, \quad i = n-1, \ldots, 2$$

and

$$v_1 = \frac{1 + b_1 v_2}{a_1}.$$

5.4 Space and Time

We now apply the techniques of the previous two sections to partial differential equations that satisfy both initial and boundary conditions. Perhaps the simplest such equation is the *heat* (or *diffusion*) equation

$$u_t = u_{xx}, \quad a \le x \le b, \quad 0 \le t, \tag{5.4.1}$$

where the subscripts denote partial derivatives. Initial and boundary conditions for (5.4.1) are specified by

$$u(0, x) = g(x), \quad a \le x \le b, \tag{5.4.2}$$

$$u(t, a) = \alpha, \quad u(t, b) = \beta, \quad t \ge 0, \tag{5.4.3}$$

where g is a given function, and α and β are given constants. As with two-point boundary value problems, boundary conditions involving the derivative u_x are possible but we will restrict our attention to (5.4.3).

The equation (5.4.1), which is the standard example of a *parabolic* partial differential equation (see the Supplementary Discussion), is the mathematical model of many important physical processes such as the diffusion of a gas. In particular, (5.4.1), together with (5.4.2) and (5.4.3), is the mathematical model of the temperature u in a thin wire with the temperature at the ends of the wire held at α and β. The solution $u(t,x)$ then gives the temperature at point x and time t starting from an initial temperature distribution given by $g(x)$.

Difference Equations for the Heat Equation

In order to solve (5.4.1) numerically, we set up a grid in the x,t plane with grid spacings Δx and Δt as illustrated in Figure 5.4.1. The basis of the simplest finite difference method for (5.4.1) is to replace the second derivative on the right-hand side of (5.4.1) by a central difference quotient in x, and replace u_t by a forward difference in time. If we let u_j^m denote the approximate solution at $x_j = j\Delta x$ and $t_m = m\Delta t$, then the finite difference analog of (5.4.1) is

$$\frac{u_j^{m+1} - u_j^m}{\Delta t} = \frac{1}{(\Delta x)^2}(u_{j+1}^m - 2u_j^m + u_{j-1}^m), \tag{5.4.4}$$

or

$$u_j^{m+1} = u_j^m + \mu(u_{j+1}^m - 2u_j^m + u_{j-1}^m), \qquad j = 1, \ldots, n, \tag{5.4.5}$$

where

$$\mu = \frac{\Delta t}{(\Delta x)^2}. \tag{5.4.6}$$

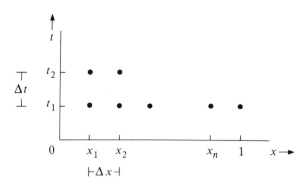

Figure 5.4.1: *Grid Spacings for Heat Equation*

The boundary conditions (5.4.3) give the values

$$u_0^m = \alpha, \qquad u_{n+1}^m = \beta, \qquad m = 0, 1, \ldots,$$

and the initial condition (5.4.2) furnishes

$$u_j^0 = g(x_j), \qquad j = 1, \ldots, n.$$

5.4 Space and Time

Therefore, (5.4.5) provides a prescription for marching the approximate solution forward one time step after another: first, the values u_j^1, $j = 1, \ldots, n$, are all obtained, and knowing these, we can compute u_j^2, $j = 1, \ldots, n$, and so on. Note that this process has excellent parallelism since the computations of u_j at the next time level are all independent.

Discretization Error

How accurate will be the approximate solution obtained by (5.4.5)? A rigorous answer to this question is a difficult problem that is beyond the scope of this book, but we will attempt to obtain some insight by considering two aspects of the error analysis.

Let $u(t, x)$ be the exact solution of (5.4.1) together with the initial and boundary conditions (5.4.2) and (5.4.3). If we put this exact solution into the difference formula (5.4.4), the amount by which the formula fails to be satisfied is called the *local discretization error*; at the point (t, x), the local discretization error, e, is

$$\frac{u(t + \Delta t, x) - u(t, x)}{\Delta t} - \frac{1}{(\Delta x)^2}[u(t, x + \Delta x) - 2u(t, x) + u(t, x - \Delta x)]. \quad (5.4.7)$$

This is entirely analogous to the previous definitions of local discretization error for ordinary differential equations.

It is easy to estimate the quantity e of (5.4.7) in terms of Δt and Δx. If we consider u as only a function of t for fixed x, we can apply the Taylor expansion

$$u(t + \Delta t, x) = u(t, x) + u_t(t, x)\Delta t + O[(\Delta t)^2]$$

to conclude that

$$\frac{u(t + \Delta t, x) - u(t, x)}{\Delta t} = u_t(t, x) + O(\Delta t).$$

Similarly, by Taylor expansions in x, we have

$$\frac{u(t, x + \Delta x) - 2u(t, x) + u(t, x - \Delta x)}{(\Delta x)^2} = u_{xx}(t, x) + O[(\Delta x)^2].$$

If we put these expressions into (5.4.7) and use $u_t = u_{xx}$ (since u is the exact solution of the differential equation), we obtain

$$e = O(\Delta t) + O[(\Delta x)^2]. \quad (5.4.8)$$

The fact that Δt appears to the first power and Δx to the second power in this expression for the local discretization error is usually described by the statement that the discretization (5.4.4) is *first-order accurate in time* and *second-order accurate in space*.

Stability

It is tempting to conclude from (5.4.8) that the discretization error in u_j^m converges to zero as Δt and Δx tend to zero. Unfortunately, this conclusion is not warranted. To show that the discretization error tends to zero on a whole time interval $[0, T]$ is difficult and, in general, requires additional conditions on how Δt and Δx tend to zero. A relationship of the type (5.4.8), or more generally a statement that the local discretization error tends to zero with Δt and Δx, is essentially a necessary condition for the global discretization error itself to tend to zero, and is called *consistency* of the difference scheme. The reason that consistency of the difference method does not necessarily imply convergence of the discretization error is connected with *stability* of the difference scheme, and we now discuss certain aspects of this for (5.4.5).

In order for (5.4.5) to be stable, Δt and Δx must satisfy the relation

$$\Delta t \leq \frac{1}{2}(\Delta x)^2, \qquad (5.4.9)$$

which is called the *stability condition* for (5.4.5). (See the Supplementary Discussion for a derivation of (5.4.9).) If this condition is not satisfied, unstable behavior in the u_j^m produced by (5.4.5) will develop. This is illustrated in Table 5.4.1, which gives the computed solution at selected t and x points for $a = 0, b = 1, \alpha = \beta = 0$ and $g(x) = \sin \pi x$. The exact solution of the differential equation tends to 0 as $t \to \infty$ and this behavior is evident in Table 5.4.1 up to about $t = .28$. However, starting at $t = .32$, noticeable instabilities have begun to develop which rapidly worsen. The calculation was done with $\Delta x = .1$ and $\Delta t = .04$ so that (5.4.9) is not satisfied. This instability has nothing to do with rounding error and will appear in exact arithmetic also.

Table 5.4.1: *Unstable Behavior*

t/x	0.2	0.4	0.6	0.8
0	0.59	0.95	0.95	0.59
0.16	0.08	0.13	0.13	0.08
0.24	0.03	0.05	0.05	0.03
0.28	0.02	0.03	0.02	0.004
0.32	0.01	0.02	0.09	0.20
0.36	0.005	−0.14	−1.22	−2.43
0.40	0.12	3.20	19.0	32.2
0.44	−3.97	−62.0	−286.9	−428.0

The condition (5.4.9) is not only needed to ensure that instabilities don't develop in the computation but it is also the correct condition for the global

5.4 Space and Time

discretization error to go to zero as Δt and Δx go to zero. This is a special case of a more general principle known as the *Lax Equivalence Theorem*, which states that for quite general differential equations and consistent difference schemes, the global discretization error will tend to zero if and only if the method is stable.

Table 5.4.2: *Maximum Time Steps for Given* Δx

Δx	Δt
0.1	$0.5 \cdot 10^{-2}$
0.01	$0.5 \cdot 10^{-4}$
0.001	$0.5 \cdot 10^{-6}$

The condition (5.4.9) imposes an increasingly stringent limitation on the time step Δt as the space increment Δx becomes small, as Table 5.4.2 shows. Thus, we may require a time step far smaller than otherwise necessary to resolve the time-dependent nature of the solution of the differential equation itself. Although the analysis that we have done has been restricted to the simplest differential equation and simplest difference scheme, the requirement of small time steps for explicit finite difference methods for parabolic and similar equations is a general problem and is a primary motivation for the implicit methods to be discussed next.

Implicit Methods

The finite difference method (5.4.5) is called *explicit* because the values of u_j^{m+1} at the next time level are obtained by an explicit formula in terms of the values at the previous time level. In contrast, consider again the heat equation and the difference approximation

$$\frac{u_j^{m+1} - u_j^m}{\Delta t} = \frac{1}{(\Delta x)^2}(u_{j+1}^{m+1} - 2u_j^{m+1} + u_{j-1}^{m+1}), \qquad j = 1, \ldots, n. \quad (5.4.10)$$

This is similar in form to (5.4.4) but has the important difference that the values of u_j on the right side are now evaluated at the $(m+1)$st time level rather than the mth. Consequently, even if we know u_j^m, $j = 1, \ldots, n$, the variables u_j on the right-hand side of (5.4.10) are all unknown, and (5.4.10) is a system of equations that implicitly defines the values u_j^{m+1}, $j = 1, \ldots, n$. This is one of the basic differences between implicit and explicit methods: in an explicit method, we have a formula for u_j^{m+1}, such as (5.4.5), in terms of known values of u_j at previous time levels, whereas with an implicit method we must solve equations to advance to the next time level. This is the same as the difference between explicit and implicit methods for ordinary differential equations, as discussed in Section 5.2.

If we again set $\mu = \Delta t/(\Delta x)^2$, then we can rewrite (5.4.10) as

$$(1 + 2\mu)u_j^{m+1} - \mu(u_{j+1}^{m+1} + u_{j-1}^{m+1}) = u_j^m, \qquad j = 1, \ldots, n, \qquad (5.4.11)$$

or in matrix-vector form,

$$(I + \mu A)\mathbf{u}^{m+1} = \mathbf{u}^m + \mathbf{b}, \qquad m = 0, 1, \ldots . \qquad (5.4.12)$$

Here, A is the $(2, -1)$ tridiagonal matrix of (5.3.10), and $\mathbf{u}^{m+1}$ and $\mathbf{u}^m$ are vectors with components u_i^{m+1} and u_i^m, $i = 1, \ldots, n$, respectively. By the boundary conditions (5.4.3), $u_0^k = \alpha$ and $u_{n+1}^k = \beta$ for $k = 0, 1, \ldots$; thus, the vector $\mathbf{b}$ in (5.4.12) is zero except for $\mu\alpha$ and $\mu\beta$ in the first and last components. We also assume the initial condition (5.4.2) so that, as before, $u_j^0 = g(x_j)$, $j = 1, \ldots, n$.

The implicit method is now carried out by solving the linear system of equations (5.4.12) at each time step to obtain $\mathbf{u}^{m+1}$ from $\mathbf{u}^m$. The matrix in (5.4.12) is tridiagonal and also diagonally dominant since $\mu > 0$. We will see in Chapter 6 that we can solve these tridiagonal systems very efficiently. Nevertheless, each time step of (5.4.12) is a little more costly than that of (5.4.5). However, in payment for this additional cost, we obtain a substantial benefit in the stability properties of the method, which, in many cases, will allow us to use a much larger time step than for the explicit method and thus greatly cut the overall computing costs.

Stability and Discretization Error

It can be shown that the stability condition for (5.4.10) is

$$0 < \frac{1}{1 + 2\mu(1 + \cos k\pi \Delta x)} < 1. \qquad (5.4.13)$$

Since $\mu > 0$ and $1 + \cos k\pi\Delta x > 0$, (5.4.13) is always satisfied. Most importantly, since $\mu = \Delta t/(\Delta x)^2$, (5.4.13) is true for *any* ratio of Δt and Δx. We say in this case that the method is *unconditionally stable*, meaning that it is stable without restrictions on the relative sizes of Δt and Δx.

The fact that the method (5.4.10) is unconditionally stable does *not* mean that we can expect to obtain a good approximate solution for any Δt and Δx; these must be chosen sufficiently small to control discretization error. It may be shown (Exercise 5.4.3) that (5.4.10), like the corresponding explicit method (5.4.4), is first-order accurate in time and second-order accurate in space; that is, the local discretization error will be

$$e = O(\Delta t) + O(\Delta x)^2. \qquad (5.4.14)$$

5.4 Space and Time

Suppose that
$$e = c_1 \Delta t + c_2 (\Delta x)^2.$$

Then, for the contributions to the total error from the discretization in time and the discretization in space to be commensurate, we require that

$$\Delta t \doteq c_3 (\Delta x)^2,$$

which is reminiscent of the stability condition (5.4.9) for the explicit method. Thus, we see that although the stability requirements for the implicit method do not impose any restrictions on the relative sizes of Δt and Δx, the accuracy requirements may.

The Crank-Nicolson Method

A potentially better implicit method in this regard is the *Crank-Nicolson method*, which is an average of the explicit method (5.4.4) and the implicit method (5.4.10):

$$u_j^{m+1} - u_j^m = \frac{\Delta t}{2(\Delta x)^2} (u_{j+1}^{m+1} - 2u_j^{m+1} + u_{j-1}^{m+1} + u_{j+1}^m - 2u_j^m + u_{j-1}^m). \quad (5.4.15)$$

This can be written in matrix-vector form as

$$(I + \frac{\mu}{2} A) \mathbf{u}^{m+1} = (I - \frac{\mu}{2} A) \mathbf{u}^m + \mathbf{b}, \qquad m = 0, 1, \ldots, \quad (5.4.16)$$

where A is again the $(2, -1)$ matrix. Hence, (5.4.15) is carried out by solving a tridiagonal system of equations at each time step. The advantage of (5.4.15) is that it is not only unconditionally stable, as is (5.4.10), but it is second-order accurate in time as well as in space (Exercise 5.4.3). These properties have made it one of the most often used methods for parabolic equations.

```
m + 1      •              •  •  •         •  •  •
m      •   •   •              •            •  •  •
     j-1  j  j+1         j-1  j  j+1      j-1  j  j+1

  (a) Explicit: (5.4.4)   (b) Fully implicit: (5.4.10)   (c) Crank-Nicolson: (5.4.15)
```

Figure 5.4.2: *Stencils for the Methods*

One easy way to keep in mind the three different methods (5.4.4), (5.4.10), and (5.4.15) is by their "stencils" of grid points as illustrated in Figure 5.4.2. These show which grid points enter into the difference method.

It has become common practice in the numerical solution of parabolic-type partial differential equations to use implicit methods since their good stability properties outweigh the additional work required per time step. Most of the methods in actual use are more complicated than the Crank-Nicolson method, but the principles are the same. However, for problems involving more than one space dimension, straightforward extensions of the implicit methods of this section are not satisfactory, and additional techniques are required. One such technique will be discussed in the next section.

The Method of Lines

We end this section by considering a conceptually different approach to the above methods. Suppose that we discretize (5.4.1) in the space variable only, leaving time continuous. Then the approximate relation

$$u_t(t, x_i) \doteq \frac{1}{(\Delta x)^2}[u(t, x_{i+1}) - 2u(t, x_i) + u(t, x_{i-1})] \tag{5.4.17}$$

holds at the grid points $x_1, \ldots, x_n$. We now seek n functions $v_1(t), \ldots, v_n(t)$ such that

$$v_i(t) \doteq u(t, x_i), \qquad i = 1, \ldots, n.$$

Thus, we attempt to approximate the solution of (5.4.1) by functions defined along the lines $x_i = a + ih$ in the (t, x) plane. The approximate relationship (5.4.17) suggests attempting to find these functions as the solution of the system of ordinary differential equations

$$v_i'(t) = \frac{1}{(\Delta x)^2}[v_{i+1}(t) - 2v_i(t) + v_{i-1}(t)], \qquad i = 1, \ldots, n, \tag{5.4.18}$$

in which, from the boundary conditions (5.4.3), the functions v_0 and v_{n+1} are taken to be identically equal to α and β, respectively. Also, from the initial condition (5.4.2), we will have

$$v_i(0) = g(x_i), \qquad i = 1, \ldots, n. \tag{5.4.19}$$

The system (5.4.18) can be written in matrix form as

$$\mathbf{v}'(t) = \frac{-1}{(\Delta x)^2} A\mathbf{v}(t), \tag{5.4.20}$$

where A is the $(2, -1)$ tridiagonal matrix (5.3.10). If we apply Euler's method to this system, we have

$$\mathbf{v}^{m+1} = \mathbf{v}^m - \frac{\Delta t}{(\Delta x)^2} A\mathbf{v}^m, \qquad m = 0, 1, \ldots.$$

Written out in component form, this is

$$v_i^{m+1} = v_i^m + \frac{\Delta t}{(\Delta x)^2}(v_{i+1}^m - 2v_i^m + v_{i-1}^m), \qquad i = 1, \ldots, n, \quad m = 0, 1, \ldots,$$

5.4 Space and Time

which is the explicit method (5.4.5). Similarly, the implicit method (5.4.10) is obtained by applying the backward Euler method (5.2.55) to (5.4.20), and the Crank-Nicolson method (5.4.15) arises by applying the trapezoid rule (5.2.59) to (5.4.20). More generally, we can apply other methods for initial value problems to (5.4.20). In particular, for more general parabolic equations the use of high quality packages for initial value problems for ordinary differential equations is a potentially effective way to obtain solutions of such partial differential equations with relatively little effort by the user. Note, however, that (5.4.20), and the corresponding equations for more general parabolic equations, tend to be rather stiff so that packages designed for stiff systems of ordinary differential equations should be used.

Supplementary Discussion and References: 5.4

1. Further discussion and analysis of the methods of this section, as well as higher-order methods, may be found in the books by Ames [1977], Hall and Porsching [1990], Isaacson and Keller [1966], Richtmyer and Morton [1967], and Strikwerda [1989]. See also Schiesser [1991] for the method of lines.

2. Consider a partial differential equation of the form

$$au_{xx} + bu_{xt} + cu_{tt} + du_x + eu_t + fu = g,$$

where the coefficients $a, b, \ldots$ are functions of x and t. Then the equation is *elliptic* if

$$[b(x,t)]^2 < a(x,t)c(x,t)$$

for all x, t in the region of interest, *hyperbolic* if $b^2 > ac$ and *parabolic* if $b^2 = ac$. For the heat equation (5.4.1), $b = c = d = f = g = 0$ and $e \neq 0$ so the equation is parabolic. The simplest example of a hyperbolic equation is the *wave* equation $u_{tt} = u_{xx}$, whereas the simplest example of an elliptic equation is *Laplace's* equation $u_{xx} + u_{tt} = 0$. In this latter equation, which will be discussed in Section 5.5, the variable t is usually a second space variable y. This classification scheme extends to more than two variables. However, many models of physical phenomena involve systems of equations rather than a single equation. The classification system of elliptic, hyperbolic, and parabolic can be extended to systems of equations, although relatively few systems that model realistic physical situations fit into this nice classification. For further discussion of the theory of partial differential equations, see, for example, Haberman [1983] and Keener [1988], and, for more advanced treatments, Courant and Hilbert [1953,1962], and Garabedian [1986].

200   Chapter 5 Continuous Problems Solved Discretely

3. Methods analogous to those discussed in the text for parabolic equations may be formulated for hyperbolic equations. For example, for the wave equation the simplest explicit finite difference scheme is

$$\frac{u_j^{m+1} - 2u_j^m + u_j^{m-1}}{(\Delta t)^2} = \frac{1}{(\Delta x)^2}(u_{j+1}^m - 2u_j^m + u_{j-1}^m). \tag{5.4.21}$$

It is also possible to formulate implicit methods but the stability condition for (5.4.21) is $\Delta t \leq \Delta x$ which does not impose a stringent condition on the time step. Consequently, implicit methods are rather little used for hyperbolic equations.

4. The stability condition (5.4.9) may be established by separation of variables for difference equations. We sketch the main ideas for the equation (5.4.5) with $\alpha = \beta = 0$ in (5.4.3). Assume that the u_j^m generated by (5.4.5) can be written as

$$u_j^m = v_m w_j, \quad j = 1, \ldots, n, \quad m = 0, 1, \ldots . \tag{5.4.22}$$

Substituting (5.4.22) into (5.4.5) and collecting terms yields

$$\frac{v_{m+1} - v_m}{\mu v_m} = \frac{w_{j+1} - 2w_j + w_{j-1}}{w_j}, \quad j = 1, \ldots, n, \quad m = 0, 1, \ldots .$$

Since the left side is independent of j and the right side is independent of m, both sides must be equal to some constant, say $-\lambda$; thus,

$$v_{m+1} - v_m = -\lambda \mu v_m, \quad m = 0, 1, \ldots, \tag{5.4.23a}$$

$$w_{j+1} - 2w_j + w_{j-1} = -\lambda w_j, \quad j = 1, \ldots, n, \tag{5.4.23b}$$

where $w_0 = w_{n+1} = 0$ from the boundary conditions. Equation (5.4.23b) represents the eigenvalue problem for the $(2, -1)$ tridiagonal matrix of (5.3.10). The eigenvalues of this matrix are (Exercise 5.3.5)

$$\lambda_k = 2 - 2\cos k\pi \Delta x, \quad k = 1, \ldots, n, \tag{5.4.24}$$

with corresponding eigenvectors

$$\mathbf{w} = [\sin(k\pi \Delta x), \sin(2k\pi \Delta x), \ldots, \sin(nk\pi \Delta x)]^T, \quad k = 1, \ldots, n,$$

where $\Delta x = 1/(n+1)$. Thus, for each $\lambda = \lambda_k$,

$$w_j = \sin(jk\pi \Delta x), \quad j = 0, 1, \ldots, n+1, \tag{5.4.25}$$

5.4 Space and Time

is a solution of (5.4.23b). Clearly,

$$v_m = (1 - \lambda\mu)^m v_0, \qquad m = 0, 1, \ldots,$$

is a solution of (5.4.23a) for any λ, so that

$$u_j^m = v_m w_j = (1-\lambda_k\mu)^m \sin(jk\pi\Delta x), \quad m = 0, 1, \ldots, \; j = 0, 1, \ldots, n+1,$$

satisfies (5.4.5) for each k. Thus, so does any linear combination

$$u_j^m = \sum_{k=1}^{n} a_k (1 - \lambda_k \mu)^m \sin(jk\pi\Delta x). \tag{5.4.26}$$

If the a_k are chosen so that

$$a_k = \sum_{l=1}^{n} g(x_l) \sin k\pi l \Delta x, \tag{5.4.27}$$

then u_j^m also satisfies the initial condition $u_j^0 = g(x_j), j = 1, \ldots, n$. We now use the representation (5.4.26) in the following way. From our previous discussion, the equation $u_t = u_{xx}$, together with the boundary conditions $u(t, 0) = u(t, 1) = 0$, is a model of the temperature distribution in a thin insulated rod whose ends are held at zero temperature. Since there is no source of heat, we expect that the temperature of the rod will decrease to zero so that $u(t, x) \to 0$ as $t \to \infty$. Therefore, it is reasonable to demand that the finite difference approximations u_j^m also tend to zero as m tends to infinity, for any initial conditions; by (5.4.26), this will be the case if and only if

$$|1 - \mu\lambda_k| < 1, \qquad k = 1, \ldots, n. \tag{5.4.28}$$

Since μ and all the λ_k are positive, (5.4.28) will hold if and only if

$$-(1 - \mu\lambda_k) < 1, \qquad k = 1, \ldots, n,$$

or

$$\mu < \min_k \frac{2}{\lambda_k} = \frac{1}{1 - \cos \pi n \Delta x} = \frac{1}{1 + \cos \pi \Delta x}, \tag{5.4.29}$$

since the largest λ_k is λ_n. Thus, with $\mu = \Delta t/(\Delta x)^2$, (5.4.29) becomes

$$\Delta t < \frac{(\Delta x)^2}{1 + \cos \pi \Delta x}. \tag{5.4.30}$$

and (5.4.9) guarantees that (5.4.30) will hold for any Δx.

EXERCISES 5.4

5.4.1. Use (5.4.5) to approximate a solution to $u_t = u_{xx}$ for the boundary and initial conditions $u(t,0) = 0$, $u(t,1) = 1$, $u(0,x) = \sin \pi x + x$. Use different values of Δt and Δx and discuss your approximate solutions. For what ratios of Δt and Δx do you conclude that your approximate solution is stable?

5.4.2. Repeat and verify the calculations of Table 5.4.1.

5.4.3. Expand in Taylor series to show that the local discretization error for (5.4.10) satisfies (5.4.14). Do the same for the method (5.4.15) to show that it is second-order in both time and space.

5.4.4. Consider the nonlinear parabolic equation

$$u_t = u_{xx} - u - x^2 - u^3$$

with boundary and initial conditions $u(t,0) = u(t,1) = 0$, $u(0,x) = \sin \pi x$.

a. Formulate an explicit method for this problem.

b. Formulate a completely implicit method and show how you would use Newton's method to advance from one time step to the next.

c. The corresponding *steady-state* equation is the two-point boundary-value problem $v'' = v + x^2 + v^3$, $v(0) = v(1) = 0$. If you are only interested in the steady-state solution, would it be better to attack this equation directly by the methods of Section 5.3 or integrate the partial differential equation to steady-state by the methods of parts **a** and **b**?

5.5 The Curse of Dimensionality

In the previous section, we considered partial differential equations in two independent variables: time and one space variable. Since physical phenomena occur in a three-dimensional world, mathematical models in only one space variable are usually considerable simplifications of the actual physical situation although, in many cases, they are sufficient for phenomena that exhibit various symmetries or in which events are happening in two of the three space dimensions at such a slow rate that those directions can be ignored. However, large-scale scientific computing is now increasingly concerned with more detailed analyses of problems in which all three space variables are of concern. This section, then, will be concerned with problems in more than one space dimension although, for simplicity of exposition, we will discuss mainly two-dimensional problems.

In the previous section, we considered the heat equation

$$u_t = u_{xx} \tag{5.5.1}$$

5.5 The Curse of Dimensionality

as a mathematical model of the temperature in a long, thin wire. If the body of interest is three-dimensional, (5.5.1) extends to three dimensions with partial derivatives in all three variables x, y, and z:

$$u_t = u_{xx} + u_{yy} + u_{zz}. \tag{5.5.2}$$

Equation (5.5.2) is a model of the temperature u as a function of time and at points within the interior of the body. As usual, to complete the model, we need to specify boundary conditions, and for this purpose, it is simplest for exposition to treat the corresponding problem in two space dimensions:

$$u_t = u_{xx} + u_{yy}. \tag{5.5.3}$$

We can consider (5.5.3) to be the mathematical model of the temperature in a flat, thin plate as shown in Figure 5.5.1, where we have taken the plate to be the unit square.

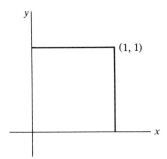

Figure 5.5.1: *Flat, Thin Plate*

The simplest boundary conditions are when the temperature is prescribed on the four sides of the plate:

$$u(t, x, y) = g(x, y), \quad (x, y) \text{ on boundary}, \tag{5.5.4}$$

where g is a given function. Another possibility is to assume that one of the sides, say $x = 0$, is perfectly insulated; thus, there is no heat loss across that side and no change in temperature, so the boundary condition is

$$u_x(t, 0, y) = 0, \quad 0 \le y \le 1, \tag{5.5.5}$$

combined with the specification (5.5.4) on the other sides. As with two-point boundary value problems, a specification of a derivative on a boundary is called a *Neumann condition*, and that of (5.5.4) is a *Dirichlet condition*. Clearly, various other combinations are possible, including periodic conditions. Boundary conditions for the three-dimensional problem can be given in a similar fashion.

We also have to specify a temperature distribution at some time, which we take to be $t = 0$; such an initial condition for (5.5.3) is of the form

$$u(0, x, y) = f(x, y). \tag{5.5.6}$$

Given the initial condition (5.5.6) and boundary conditions of the form (5.5.4) and/or (5.5.5), it is intuitively clear that the temperature distribution should evolve in time to a final steady state that is determined only by the boundary conditions. In many situations, it is this steady-state solution that is of primary interest and since it no longer depends on time, it should satisfy the equation (5.5.3) with $u_t = 0$:

$$u_{xx} + u_{yy} = 0. \tag{5.5.7}$$

This is Laplace's equation and, as mentioned in the previous section, is the prototype example of an elliptic equation. If we wish only the steady-state solution of the temperature distribution problem that we have been discussing, we can proceed, in principle, in two ways: solve equation (5.5.3) for u as a function of time until convergence to a steady state is reached, or solve (5.5.7) for only the steady-state solution.

Finite Differences for Poisson's Equation

We will return to the time-dependent problem shortly, after considering a finite difference method for (5.5.7) and, more generally, *Poisson's equation*

$$u_{xx} + u_{yy} = f, \tag{5.5.8}$$

where f is a given function of x and y. We assume that the domain of the problem is the unit square $0 \leq x, y \leq 1$, and that Dirichlet boundary conditions

$$u(x, y) = g(x, y), \quad (x, y) \text{ on boundary}, \tag{5.5.9}$$

are given, where g is a known function. We impose a mesh of grid points on the unit square with spacing h between the points in both the horizontal and vertical directions; this is illustrated in Figure 5.5.2.

The interior grid points are given by

$$(x_i, y_j) = (ih, jh), \quad i, j = 1, \ldots, N, \tag{5.5.10}$$

where $(N+1)h = 1$. At an interior grid point (x_i, y_j) we approximate u_{xx} and u_{yy} by the centered difference approximations

$$u_{xx}(x_i, y_j) \doteq \frac{1}{h^2}[u(x_{i-1}, y_j) - 2u(x_i, y_j) + u(x_{i+1}, y_j)], \tag{5.5.11a}$$

$$u_{yy}(x_i, y_j) \doteq \frac{1}{h^2}[u(x_i, y_{j-1}) - 2u(x_i, y_j) + u(x_i, y_{j+1})]. \tag{5.5.11b}$$

5.5 The Curse of Dimensionality

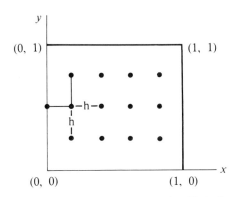

Figure 5.5.2: *Mesh Points on the Unit Square*

If we put these approximations into the differential equation (5.5.8), we obtain

$$u(x_{i-1}, y_j) + u(x_{i+1}, y_j) + u(x_i, y_{j-1}) + u(x_i, y_{j+1}) - 4u(x_i, y_j) \quad (5.5.12)$$
$$\doteq h^2 f(x_i, y_j),$$

which is an approximate relationship that the exact solution u of (5.5.8) satisfies at any grid point in the interior of the domain.

Analogously to what we did in Section 5.3, we now define approximations u_{ij} to the exact solution $u(x_i, y_j)$ at the N^2 interior grid points by requiring that they satisfy exactly the relationship (5.5.12); that is,

$$-u_{i-1,j} - u_{i+1,j} - u_{i,j+1} - u_{i,j-1} + 4u_{ij} = -h^2 f_{ij}, \quad i,j = 1,\ldots,N, \quad (5.5.13)$$

where we have multiplied (5.5.12) by -1. This is a linear system of equations in the $(N+2)^2$ variables u_{ij}. Note, however, that the variables $u_{0,j}$, $u_{N+1,j}$, $j = 0,\ldots,N+1$, and $u_{i,0}$, $u_{i,N+1}$, $i = 0,\ldots,N+1$, correspond to the grid points on the boundary and, hence, are given by the boundary condition (5.5.9):

$$\begin{aligned} u_{0,j} &= g(0, y_j), & u_{N+1,j} &= g(1, y_j), & j &= 0, 1, \ldots, N+1, \\ u_{i,0} &= g(x_i, 0), & u_{i,N+1} &= g(x_i, 1), & i &= 0, 1, \ldots, N+1. \end{aligned} \quad (5.5.14)$$

Therefore, (5.5.13) is a linear system of N^2 equations in the N^2 unknowns u_{ij}, $i, j = 1, \ldots, N$, corresponding to the interior grid points. It is easy to show (Exercise 5.5.1) that the local discretization error in the u_{ij} is $O(h^2)$. Note that (5.5.13) is the natural extension to two space variables of the discrete equations

$$-u_{i+1} + 2u_i - u_{i-1} = -h^2 f_i, \quad i = 1, \ldots, N,$$

obtained in Section 5.3 for the "one-dimensional Poisson equation" $u'' = f$.

We now wish to write the system (5.5.13) in matrix-vector form; for this purpose we will number the interior grid points in the manner shown in Figure

5.5.3, which is called the *natural* or *row-wise ordering*. Corresponding to this ordering of the grid points, we order the unknowns u_{ij} into the vector

$$(u_{11}, \ldots, u_{N1}, u_{12}, \ldots, u_{N2}, \ldots, u_{1N}, \ldots, u_{NN}), \tag{5.5.15}$$

and write the system of equations in the same order. For example, for $N = 2$

$$\begin{bmatrix} 4 & -1 & -1 & 0 \\ -1 & 4 & 0 & -1 \\ -1 & 0 & 4 & -1 \\ 0 & -1 & -1 & 4 \end{bmatrix} \begin{bmatrix} u_{11} \\ u_{21} \\ u_{12} \\ u_{22} \end{bmatrix} = -h^2 \begin{bmatrix} f_{11} \\ f_{21} \\ f_{12} \\ f_{22} \end{bmatrix} + \begin{bmatrix} u_{01} + u_{10} \\ u_{20} + u_{31} \\ u_{02} + u_{13} \\ u_{32} + u_{23} \end{bmatrix}, \tag{5.5.16}$$

in which we have put the u_{ij} known from the boundary values on the right-hand side of the equation (Exercise 5.5.2).

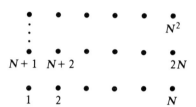

Figure 5.5.3: *Natural Ordering of the Interior Grid Points*

For general N, a typical row of the matrix will be

$$-1 \quad 0 \quad \cdots \quad 0 \quad -1 \quad 4 \quad -1 \quad 0 \quad \cdots \quad 0 \quad -1$$

where $N - 2$ zeros separate the -1's in both directions. The equations corresponding to an interior grid point adjacent to a boundary point will contain a known boundary value and this value will be moved to the right side of the equation, eliminating the corresponding -1 from the matrix. This happened in each equation in (5.5.16) because of the size of N. We show in Figure 5.5.4 the coefficient matrix for $N = 4$ (see Exercise 5.5.2).

Although Figure 5.5.4 illustrates the structure of the coefficient matrix, it is cumbersome for large N. Alternatively, we can write the matrix in block form. To do this, we define the $N \times N$ tridiagonal matrix

$$T = \begin{bmatrix} 4 & -1 & & & \\ -1 & \ddots & & & \\ & \ddots & \ddots & & \\ & & \ddots & & -1 \\ & & & -1 & 4 \end{bmatrix}, \tag{5.5.17}$$

and let I denote the $N \times N$ identity matrix. Then, the $N^2 \times N^2$ coefficient

5.5 The Curse of Dimensionality

$$\begin{bmatrix} 4 & -1 & & -1 & & & & & & & & & & & & \\ -1 & 4 & -1 & & -1 & & & & & & & & & & & \\ & -1 & 4 & -1 & & -1 & & & & & & & & & & \\ & & -1 & 4 & & & -1 & & & & & & & & & \\ -1 & & & & 4 & -1 & & & -1 & & & & & & & \\ & -1 & & & -1 & 4 & -1 & & & -1 & & & & & & \\ & & -1 & & & -1 & 4 & -1 & & & -1 & & & & & \\ & & & -1 & & & -1 & 4 & & & & -1 & & & & \\ & & & & -1 & & & & 4 & -1 & & & -1 & & & \\ & & & & & -1 & & & -1 & 4 & -1 & & & -1 & & \\ & & & & & & -1 & & & -1 & 4 & -1 & & & -1 & \\ & & & & & & & -1 & & & -1 & 4 & & & & -1 \\ & & & & & & & & -1 & & & & 4 & -1 & & \\ & & & & & & & & & -1 & & & -1 & 4 & -1 & \\ & & & & & & & & & & -1 & & & -1 & 4 & -1 \\ & & & & & & & & & & & -1 & & & -1 & 4 \end{bmatrix}$$

Figure 5.5.4: *Coefficient Matrix of* (5.5.13) *for* $N = 4$

matrix of (5.5.13) is the *block tridiagonal matrix*

$$A = \begin{bmatrix} T & -I & & & \\ -I & T & \ddots & & \\ & -I & \ddots & & \\ & & \ddots & & -I \\ & & & -I & T \end{bmatrix}. \tag{5.5.18}$$

The matrix of (5.5.16) is the special case of (5.5.18) for $N = 2$, and Figure 5.5.4 shows the matrix for $N = 4$. If we also define the vectors

$$\mathbf{u}_i = (u_{1i}, \ldots, u_{Ni})^T, \qquad \mathbf{f}_i = (f_{1i}, \ldots, f_{Ni})^T, \qquad i = 1, \ldots, N,$$
$$\mathbf{b}_1 = (u_{01} + u_{10}, u_{20}, \ldots, u_{N-1,0}, u_{N,0} + u_{N+1,1})^T,$$
$$\mathbf{b}_i = (u_{0i}, 0, \ldots, 0, u_{N+1,i})^T, \qquad i = 2, \ldots, N-1,$$
$$\mathbf{b}_N = (u_{0,N} + u_{1,N+1}, u_{2,N+1}, \ldots, u_{N-1,N}, u_{N,N+1} + u_{N+1,N})^T,$$

then we can write the system (5.5.13) in the form

$$\begin{bmatrix} T & -I & & \\ -I & \ddots & \ddots & \\ & \ddots & & -I \\ & & -I & T \end{bmatrix} \begin{bmatrix} \mathbf{u}_1 \\ \mathbf{u}_2 \\ \vdots \\ \mathbf{u}_N \end{bmatrix} = \begin{bmatrix} \mathbf{b}_1 - h^2 \mathbf{f}_1 \\ \mathbf{b}_2 - h^2 \mathbf{f}_2 \\ \vdots \\ \mathbf{b}_N - h^2 \mathbf{f}_N \end{bmatrix}. \tag{5.5.19}$$

We now make several comments about this system of equations. If N is of moderate size, say $N = 100$, then there are $N^2 = 10^4$ unknowns, and the matrix in (5.5.19) is $10,000 \times 10,000$. In each row of the matrix, there are at most five nonzero elements, regardless of the size of N, so the distribution of nonzero to zero elements is very "sparse" if N is at all large. For the corresponding three-dimensional problem there would be 10^6 equations with at most seven non-zero elements in each row of the matrix. Such matrices are called *large sparse matrices* and arise in a variety of ways besides the numerical solution of partial differential equations. It is the property of being sparse in a very structured way that allows such large systems of equations to be solved on today's computers with relative ease. Methods for the solution of such systems will be discussed primarily in Chapters 8 and 9.

Nonsingularity

We next show, by two different approaches, that the matrix (5.5.18) is nonsingular; thus, the system (5.5.19) has a unique solution. First, the directed graph of the matrix (5.5.18) is shown in Figure 5.5.5 for $N = 3$. This pattern persists for any N, and the graph is strongly connected (Exercise 5.5.3). Therefore, by Theorem 2.1.5, the matrix is irreducible. Clearly, it is also diagonally dominant, and since the first row has only two off-diagonal 1's it is irreducibly diagonally dominant. Thus, by Theorem 2.3.3 it is nonsingular.

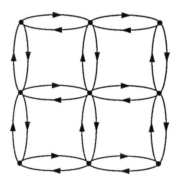

Figure 5.5.5: *Directed graph of* (5.5.18). $N = 3$

The matrix (5.5.18) is also positive definite; indeed, its eigenvalues may be computed explicitly as

$$4 - 2(\cos \frac{k\pi}{N+1} + \cos \frac{j\pi}{N+1}), \qquad j, k = 1, \ldots, N. \tag{5.5.20}$$

Clearly, these eigenvalues are all positive so that by Theorem 2.2.5 the matrix is positive definite. One way to prove (5.5.20) is by means of Kronecker products

5.5 The Curse of Dimensionality

and sums. The *Kronecker product* of two $m \times m$ matrices B and C is the $m^2 \times m^2$ matrix

$$B \otimes C = \begin{bmatrix} b_{11}C & \cdots & b_{1m}C \\ \vdots & & \\ b_{m1}C & \cdots & b_{mm}C \end{bmatrix}. \tag{5.5.21}$$

A *Kronecker sum* of two $m \times m$ matrices B and C is the $m^2 \times m^2$ matrix

$$(I \otimes B) + (C \otimes I), \tag{5.5.22}$$

where I is the $m \times m$ identity matrix. It is then easy to see (Exercise 5.5.4) that the matrix (5.5.18) may be written as

$$A = (I \otimes T) + (T \otimes I). \tag{5.5.23}$$

Eigenvalues of Kronecker products and sums are easily computed in terms of the eigenvalues of the constituent matrices. We state the following theorem without proof.

THEOREM 5.5.1. *Let B and C be $m \times m$ matrices with eigenvalues $\lambda_1, \ldots, \lambda_m$ and $\mu_1, \ldots, \mu_m$, respectively. Then the eigenvalues of the Kronecker product* (5.5.21) *are*

$$\lambda_i \mu_j, \qquad i, j = 1, \ldots, m, \tag{5.5.24}$$

and the eigenvalues of the Kronecker sum (5.5.22) *are*

$$\lambda_i + \mu_j, \qquad i, j = 1, \ldots, m. \tag{5.5.25}$$

As an application of this theorem to the matrix (5.5.23), the eigenvalues of the $(2, -1)$ matrix (5.3.10) of size N are (Exercise 5.3.5) $2 - 2\cos k\pi/(N+1)$; thus, the eigenvalues of T are $4 - 2\cos k\pi/(N+1)$. In Theorem 5.5.1, $B = C = T$ and therefore (5.5.25) gives the eigenvalues (5.5.20).

The Heat Equation

We end this section by applying the discretization of Poisson's equation to the heat equation (5.5.3) where, again for simplicity in exposition, we will assume that the x, y domain is the unit square of Figure 5.5.2 and that the Dirichlet boundary conditions (5.5.4) are given on the sides of the square. We also assume the initial condition (5.5.6).

Corresponding to the method (5.4.5) in the case of a single space variable, we can consider the following explicit method for (5.5.3):

$$u_{ij}^{m+1} = u_{ij}^m + \frac{\Delta t}{h^2}(u_{i,j+1}^m + u_{i,j-1}^m + u_{i+1,j}^m + u_{i-1,j}^m - 4u_{ij}^m), \tag{5.5.26}$$

210   Chapter 5 Continuous Problems Solved Discretely

for $m = 0, 1, \ldots,$ and $i, j = 1, \ldots, N$. Here, u_{ij}^m denotes the approximate solution at the i, j gridpoint and at the mth time level $m\Delta t$, and u_{ij}^{m+1} is the approximate solution at the next time level. The terms in parentheses on the right-hand side of (5.5.26) correspond exactly to the discretization (5.5.13) with $f_{ij} = 0$. The prescription (5.5.26) has the same properties as its one-dimensional counterpart (5.4.5): it is first-order accurate in time and second-order accurate in space, and it is easy to carry out. It is also subject to a similar stability condition

$$\Delta t \leq \frac{h^2}{4}, \tag{5.5.27}$$

and thus if h is small, very small time steps are required.

We can attempt to circumvent this restriction on the time step in the same way that we did in Section 5.4 by the use of implicit methods. For example, the implicit method (5.4.10) now becomes

$$u_{ij}^{m+1} = u_{i,j}^m + \frac{\Delta t}{h^2}(u_{i,j+1}^{m+1} + u_{i,j-1}^{m+1} + u_{i+1,j}^{m+1} + u_{i-1,j}^{m+1} - 4u_{ij}^{m+1}), \tag{5.5.28}$$

which is unconditionally stable. However, to carry out this method requires the solution at each time step of the system of linear equations

$$\left(4 + \frac{h^2}{\Delta t}\right) u_{ij}^{m+1} - u_{i,j+1}^{m+1} - u_{i,j-1}^{m+1} - u_{i+1,j}^{m+1} - u_{i-1,j}^{m+1} = \frac{h^2}{\Delta t} u_{ij}^m, \tag{5.5.29}$$

for $i, j = 1, \ldots, N$. This system has the same form as the system (5.5.13) for Poisson's equation, with the exception that the coefficient of u_{ij}^{m+1} is modified.

In the case of a single space variable, the use of an implicit method such as (5.4.10) does not cause much computational difficulty since the solution of tridiagonal systems of equations can be accomplished very rapidly, as we will see in Chapter 6. However, each time step of (5.5.29) requires the solution of a two-dimensional Poisson-type equation, which is a much harder computational problem. The Crank-Nicolson method (5.4.14) can also be easily extended (Exercise 5.5.5) to the equation (5.5.3) but suffers from the same difficulty that Poisson-type equations must be solved at each time step. We shall consider, instead, a different class of methods, in which the basic computational step is the solution of tridiagonal systems of equations. These are *time-splitting* methods, in which the time interval $(t, t + \Delta t)$ is further subdivided and, in essence, only one-dimensional problems solved at each timestep.

An Alternating Direction Method

One of the first time-splitting methods was the *Peaceman-Rachford alternating direction implicit (ADI)* method, in which

$$u_{ij}^{m+1/2} = u_{ij}^m \tag{5.5.30a}$$
$$+ \mu(u_{i+1,j}^{m+1/2} + u_{i-1,j}^{m+1/2} - 2u_{ij}^{m+1/2} + u_{i,j+1}^m + u_{i,j-1}^m - 2u_{ij}^m),$$

5.5 The Curse of Dimensionality

and then

$$u_{ij}^{m+1} = u_{ij}^{m+1/2} \qquad (5.5.30b)$$
$$+\mu(u_{i,j+1}^{m+1} + u_{i,j-1}^{m+1} - 2u_{ij}^{m+1} + u_{i+1,j}^{m+1/2} + u_{i-1,j}^{m+1/2} - 2u_{ij}^{m+1/2}),$$

where $\mu = \Delta t/(2h^2)$. This is a two-step method in which intermediate values $u_{ij}^{m+1/2}$, $i,j = 1,\ldots,N$, are computed at the first step (5.5.30a). These $u_{ij}^{m+1/2}$ are to be interpreted as approximate values of the solution at the intermediate time level $m + \frac{1}{2}$; thus, the factor $\frac{1}{2}$ appears in the definition of μ because the time step is $\frac{1}{2}\Delta t$.

With $\alpha = 2h^2/\Delta t$, for fixed j and $i = 1,\ldots,N$

$$(2+\alpha)u_{ij}^{m+1/2} - u_{i+1,j}^{m+1/2} - u_{i-1,j}^{m+1/2} = (\alpha-2)u_{ij}^m + u_{i,j+1}^m + u_{i,j-1}^m, \qquad (5.5.31)$$

is a tridiagonal system whose solution is $u_{ij}^{m+1/2}$, $i = 1,\ldots,N$. The computation (5.5.30a) requires the solution of N of these tridiagonal systems for $j = 1,\ldots,N$. The coefficient matrix of each of these systems is $\alpha I + A$, where A is the $(2,-1)$ tridiagonal matrix (5.3.10).

Once the intermediate values $u_{ij}^{m+1/2}$ have been computed, the final values u_{ij}^{m+1} are obtained from (5.5.30b) by solving the N tridiagonal systems

$$(2+\alpha)u_{ij}^{m+1} - u_{i,j+1}^{m+1} - u_{i,j-1}^{m+1} = (\alpha-2)u_{ij}^{m+1/2} + u_{i+1,j}^{m+1/2} + u_{i-1,j}^{m+1/2}, \qquad (5.5.32)$$

j=1,..., N, for each $i = 1,\ldots,N$. Again, the coefficient matrices are $\alpha I + A$. Thus, the computational process requires the solution of $2N$ tridiagonal systems of dimension N to move from the mth time level to the $(m+1)$st. It can be shown that this method is unconditionally stable.

The term *alternating direction* derives from the paradigm that we are, in some sense, approximating values of the solution in the x-direction by (5.5.30a), and then in the y-direction by (5.5.30b). There are many variants of ADI methods and, more generally, of time-splitting methods, and such methods are widely used for parabolic-type equations.

Supplementary Discussion and References: 5.5

1. The discussion of this section has been restricted to Poisson's equation in two variables on a square domain, and the corresponding heat equation. However, problems that arise in practice will generally deviate considerably from these ideal conditions: the domain may not be square; the equation may have nonconstant coefficients or even be nonlinear; the boundary conditions may be a mixture of Dirichlet, Neumann and periodic conditions; there may be more than a single equation – that is, there may be a coupled system of partial differential equations; the equations

may have derivatives of order higher than two; and there may be three or more independent variables. The general principles of finite difference discretization of this section still apply, but each of the preceding factors causes complications.

2. One of the classical references for the discretization of elliptic equations by finite difference methods is Forsythe and Wasow [1960]. See also Hall and Porsching [1990]. Discussions and analyses of alternating direction methods, and of related methods such as the method of fractional steps, are given in a number of books; see, for example, Varga [1962], Richtmeyer and Morton [1967], and Strikwerda [1989].

3. For further discussion of Kronecker products and sums, and a proof of Theorem 5.5.1, see Ortega [1987]. Extensive treatments of this topic may be found in Horn and Johnson [1985], [1991].

4. Finite element and other "projection-type" methods play an increasingly important role in the solution of elliptic- and parabolic-type equations. Although the mathematical basis of the finite element method goes back to the 1940s, its development into a viable procedure was carried out primarily by engineers in the 1950s and 1960s, especially for problems in structural analysis. Since then, the mathematical basis has been extended and broadened and its applicability to many partial differential equations well demonstrated. One of the method's main advantages is its ability to handle curved boundaries. For introductions to the finite element method, see Strang and Fix [1973], Becker, Carey and Oden [1981], Carey and Oden [1984], Axelsson and Barker [1984], Johnson [1987] and Hall and Porsching [1990].

EXERCISES 5.5

5.5.1. Assume that the function u is as many times continuously differentiable as needed. Expand u in Taylor series about (x_i, y_j) and show that the approximations (5.5.11) are second-order accurate:

$$u_{xx}(x_i, y_j) - \frac{1}{h^2}[u(x_{i-1}, y_j) - 2u(x_i, y_j) + u(u_{i+1}, y_j)] = O(h^2),$$

and similarly for the approximation to u_{yy}. Conclude that the local discretization error in the approximate solution of (5.5.13) is $O(h^2)$.

5.5.2. Verify that with the ordering (5.5.15), the system of equations (5.5.13) takes the form (5.5.16) for $N = 2$. Verify also that Figure 5.5.4 gives the coefficient matrix for $N = 4$. What is the right-hand side of the system in this case?

5.5 The Curse of Dimensionality

5.5.3. Extend the graph of Figure 5.5.5 for general N, and conclude that it is strongly connected.

5.5.4. Verify that the matrix (5.5.18) may be written as the Kronecker sum (5.5.23).

5.5.5. Formulate the Crank-Nicolson method for (5.5.3).

5.5.6. Consider the equation $au_{xx} + bu_{yy} = f$ with Dirichlet boundary conditions on the unit square, where a and b are constants. Discretize u_{xx} and u_{yy} by (5.5.11) to obtain a system of equations $A\mathbf{u} = \mathbf{b}$. Use Theorem 5.5.1 to show that the eigenvalues of A are

$$a(2 - 2\cos\frac{k\pi}{N+1}) + b(2 - 2\cos\frac{j\pi}{N+1}), \quad j, k = 1, \ldots, N$$

5.5.7. Show that $u(x, y) = x^2 + y^2$ is the exact solution of the Poisson equation (5.5.8) with $f \equiv 4$, as well as the discrete equations (5.5.13). (This provides a convenient test problem with known solution for both the differential and difference equations.)

Chapter 6

Direct Solution of Linear Equations

In previous chapters we have seen that a wide variety of problems lead ultimately to the need to solve a linear system of equations $A\mathbf{x} = \mathbf{b}$. A key consideration in attempting the solution of such systems is the structure of the matrix A. In some problems, for example least squares problems, A may be a full matrix; that is, it has few zero elements. On the other hand, we saw that for certain two-point boundary value problems, A was tridiagonal. And for boundary-value problems for partial differential equations, A was large and very sparse. In general, the zeros in the matrix A must be utilized as much as possible in order to have efficient algorithms.

There are two general approaches to the solution of linear systems. *Direct methods* obtain the exact solution (in real arithmetic) in finitely many operations whereas *iterative methods* generate a sequence of approximations that only converge in the limit to the solution. Generally, direct methods are best for full or banded matrices whereas iterative methods are best for very large and sparse matrices, especially those arising from partial differential equations in three space dimensions. Chapter 6 will consider direct methods and their basic properties and Chapter 7 their parallel and vector implementations. Chapters 8 and 9 will cover iterative methods.

6.1 Gaussian Elimination

We consider the linear system

$$A\mathbf{x} = \mathbf{b}, \tag{6.1.1}$$

216   Chapter 6   Direct Solution of Linear Equations

where A is a given $n \times n$ matrix assumed to be nonsingular, $\mathbf{b}$ is a given column vector, and $\mathbf{x}$ is the solution vector to be determined. The most basic method for solving (6.1.1) is *Gaussian elimination*. To describe this method, we write out (6.1.1) as

$$\begin{aligned} a_{11}x_1 + \cdots + a_{1n}x_n &= b_1 \\ a_{21}x_1 + \cdots + a_{2n}x_n &= b_2 \\ &\vdots \\ a_{n1}x_1 + \cdots + a_{nn}x_n &= b_n. \end{aligned} \qquad (6.1.2)$$

Assuming that $a_{11} \neq 0$, we first subtract a_{21}/a_{11} times the first equation from the second equation to eliminate the coefficient of x_1 in the second equation. Then, we subtract a_{31}/a_{11} times the first equation from the third equation, a_{41}/a_{11} times the first equation from the fourth equation, and so on until the coefficients of x_1 in the last $n-1$ equations have all been eliminated. This gives the modified system of equations

$$\begin{aligned} a_{11}x_1 + a_{12}x_2 + \cdots + a_{1n}x_n &= b_1 \\ a_{22}^{(1)}x_2 + \cdots + a_{2n}^{(1)}x_n &= b_2^{(1)} \\ &\vdots \\ a_{n2}^{(1)}x_2 + \cdots + a_{nn}^{(1)}x_n &= b_n^{(1)}, \end{aligned} \qquad (6.1.3)$$

where

$$a_{ij}^{(1)} = a_{ij} - a_{1j}\frac{a_{i1}}{a_{11}}, \qquad b_i^{(1)} = b_i - b_i\frac{a_{i1}}{a_{11}}, \qquad i,j = 2,\ldots,n. \qquad (6.1.4)$$

Precisely the same process is now applied to the last $n-1$ equations of the system (6.1.3) to eliminate the coefficients of x_2 in the last $n-2$ equations, and so on, until the entire system has been reduced to the *triangular form*

$$\begin{bmatrix} a_{11} & a_{12} & \cdots & a_{1n} \\ & a_{22}^{(1)} & \cdots & a_{2n}^{(1)} \\ & & \ddots & \vdots \\ & & & a_{nn}^{(n-1)} \end{bmatrix} \begin{bmatrix} x_1 \\ x_2 \\ \vdots \\ x_n \end{bmatrix} = \begin{bmatrix} b_1 \\ b_2^{(1)} \\ \vdots \\ b_n^{(n-1)} \end{bmatrix}. \qquad (6.1.5)$$

The superscripts indicate the number of times the elements have been changed. This completes the *forward reduction* (or *forward elimination* or *triangular reduction*) phase of the Gaussian elimination algorithm. Note that we have tacitly assumed that the $a_{ii}^{(i-1)}$ are all non-zero since we divide by these elements. We will consider in the next section the important question of how to handle zero or small divisors.

The Gaussian elimination method is based on the fact (usually established in an introductory linear algebra course) that replacing any equation of the

6.1 Gaussian Elimination

original system (6.1.2) by a linear combination of itself and another equation does not change the solution of (6.1.2). Thus the triangular system (6.1.5) has the same solution as the original system. The purpose of the forward reduction is to reduce the original problem to one that is easy to solve; this is a common theme in much of scientific computing. The second part of the Gaussian elimination method then consists of the solution of (6.1.5) by *back substitution*, in which the equations are solved in reverse order:

$$\left. \begin{aligned} x_n &= \frac{b_n^{(n-1)}}{a_{nn}^{(n-1)}} \\ x_{n-1} &= \frac{b_{n-1}^{(n-2)} - a_{n-1,n}^{(n-2)} x_n}{a_{n-1,n-1}^{(n-2)}} \\ &\vdots \\ x_1 &= \frac{b_1 - a_{12}x_2 - \cdots - a_{1n}x_n}{a_{11}}. \end{aligned} \right\} \quad (6.1.6)$$

The Gaussian elimination algorithm can be written in algorithmic form as shown in Figure 6.1.1. Note that in this algorithm the $a_{ij}^{(k)}$ are overwritten onto the same storage spaces occupied by the original elements a_{ij}. If this is done, the original matrix will, of course, be destroyed during the process. Similarly, the new $b_i^{(k)}$ may be overwritten on the original storage spaces of the b_i. The multiplier l_{ik} can be written into the corresponding storage space for a_{ik}, which is no longer needed after l_{ik} is computed.

For $k = 1, \ldots, n-1$
  For $i = k+1, \ldots, n$
    $l_{ik} = \frac{a_{ik}}{a_{kk}}$
    For $j = k+1, \ldots, n$
      $a_{ij} = a_{ij} - l_{ik} a_{kj}$
    $b_i = b_i - l_{ik} b_k$

(a) Forward Reduction

For $k = n, n-1, \ldots, 1$
  $x_k = b_k$
  For $i = k+1, \ldots, n$
    $x_k = x_k - a_{ki} x_i$
  $x_k = x_k / a_{kk}$

(b) Back Substitution

Figure 6.1.1: *Gaussian Elimination*

LU Factorization

Gaussian elimination is related to a factorization

$$A = LU \quad (6.1.7)$$

of the matrix A. Here U is the upper triangular matrix of (6.1.5) obtained in the forward reduction, and L is a *unit* lower triangular matrix (all main diagonal elements are 1) in which the subdiagonal element l_{ij} is the multiplier used in eliminating the jth variable from the ith equation. For example, if the original system is

$$\begin{bmatrix} 4 & -9 & 2 \\ 2 & -4 & 4 \\ -1 & 2 & 2 \end{bmatrix} \begin{bmatrix} x_1 \\ x_2 \\ x_3 \end{bmatrix} = \begin{bmatrix} 2 \\ 3 \\ 1 \end{bmatrix}, \tag{6.1.8}$$

then the reduced system (6.1.5) is

$$\begin{bmatrix} 4 & -9 & 2 \\ 0 & 0.5 & 3 \\ 0 & 0 & 4 \end{bmatrix} \begin{bmatrix} x_1 \\ x_2 \\ x_3 \end{bmatrix} = \begin{bmatrix} 2 \\ 2 \\ 2.5 \end{bmatrix}. \tag{6.1.9}$$

The multipliers used to obtain (6.1.9) from (6.1.8) are 0.5, -0.25, and -0.5. Thus

$$L = \begin{bmatrix} 1 & 0 & 0 \\ 0.5 & 1 & 0 \\ -0.25 & -0.5 & 1 \end{bmatrix}, \tag{6.1.10}$$

and A is the product of (6.1.10) and the matrix U of (6.1.9). The calculations needed for the above assertions are left to Exercise 6.1.1.

More generally, it is easy to verify (Exercise 6.1.2) that the elimination step that produces (6.1.3) from (6.1.2) is equivalent to multiplying (6.1.2) by the matrix

$$L_1 = \begin{bmatrix} 1 & & & \\ -l_{21} & 1 & & \\ \vdots & & \ddots & \\ -l_{n1} & & & 1 \end{bmatrix}, \tag{6.1.11}$$

where $l_{i1} = a_{i1}/a_{11}$. Continuing in this fashion, the reduced system (6.1.5) may be written as

$$\hat{L}A\mathbf{x} = \hat{L}\mathbf{b}, \qquad \hat{L} = L_{n-1} \cdots L_2 L_1, \tag{6.1.12}$$

where

$$L_i = \begin{bmatrix} 1 & & & & \\ & \ddots & & & \\ & & 1 & & \\ & & -l_{i+1,i} & & \\ & & \vdots & \ddots & \\ & & -l_{n,i} & & 1 \end{bmatrix}. \tag{6.1.13}$$

6.1 Gaussian Elimination

Each of the matrices L_i has determinant equal to 1 and so is nonsingular. Therefore, the product $\hat{L}$ is nonsingular. Moreover, $\hat{L}$ is unit lower-triangular and, therefore, so is $\hat{L}^{-1}$. By construction the coefficient matrix of (6.1.12) is U, and if we set $L = \hat{L}^{-1}$ we have

$$U = \hat{L}A = L^{-1}A, \qquad (6.1.14)$$

which is equivalent to (6.1.7). The verification of the above statements is left to Exercise 6.1.3.

The factorization (6.1.7) is known as the *LU factorization* (or *LU decomposition*) of A. The right-hand side of (6.1.12) is $L^{-1}\mathbf{b}$, which is the solution of $L\mathbf{y} = \mathbf{b}$. Thus, the Gaussian elimination algorithm for solving $A\mathbf{x} = \mathbf{b}$ is mathematically equivalent to the three-step process:

1. Factor $A = LU$. (6.1.15a)
2. Solve $L\mathbf{y} = \mathbf{b}$. (6.1.15b)
3. Solve $U\mathbf{x} = \mathbf{y}$. (6.1.15c)

This matrix formulation of Gaussian elimination is very useful for theoretical purposes and also is the basis for some computational variants of the elimination process. In particular, the classical Crout and Doolittle forms of LU decomposition are based on formulas for the elements of L and U obtained from equating LU and A. Another use of (6.1.15) is when there are many right-hand sides $\mathbf{b}$; in this case, step 1 is done once and the factors saved so that steps 2 and 3 may be performed repeatedly. This will be discussed in more detail later.

Operation Counts

An important question is the efficiency of the Gaussian elimination algorithm, and we next estimate the number of arithmetic operations needed to compute the solution vector $\mathbf{x}$. The major part of the work is in modifying the elements of A in the forward reduction by the arithmetic statement $a_{ij} = a_{ij} - l_{ik}a_{kj}$ in Figure 6.1.1(a). There is one addition and one multiplication in this statement and thus in the j loop of Figure 6.1.1(a) there are $n-k$ additions. The i loop repeats these $n-k$ times so that the total over the k loop is

$$\sum_{k=1}^{n-1}(n-k)^2 = \sum_{k=1}^{n-1}k^2.$$

Thus, using the summation formula in Exercise 6.1.4 we obtain

$$\text{Number of additions} = \sum_{k=1}^{n-1}k^2 = \frac{(n-1)(n)(2n-1)}{6} \doteq \frac{n^3}{3}, \qquad (6.1.16)$$

and the same count holds for the number of multiplications. The approximation in (6.1.16) holds when n is at all large since the other terms are order n^2 or lower. We also need the number of divisions to compute the l_{ik}, and the number of operations to modify the right-hand side **b** and to do the back substitution. All of these involve no more than order n^2 operations (see Exercise 6.1.5). Hence, for sufficiently large n the work involved in the triangular factorization makes the dominant contribution to the computing time and is proportional to n^3. Note that this implies that if n is doubled, the amount of work increases by approximately a factor of 8.

To obtain an understanding of the amount of time that Gaussian elimination might require on a moderate-size problem, suppose that $n = 100$ and that addition and multiplication times are $1\mu s$ and $2\mu s$, respectively (μs= microsecond $= 10^{-6}$ second). Then, the time for the additions and multiplications in the factorization is approximately

$$\frac{100^3}{3}(3\mu s) = 10^6 \mu s = 1 \text{ second}.$$

There will also be time required for the other arithmetical operations of the complete elimination process, but this will be much less than 1 second. More importantly, there will be various "overhead" costs such as moving data back and forth from storage, and indexing. This could easily double or triple the total computing time, but on a computer of this speed only a few seconds would be required to solve a 100×100 system.

Banded Matrices

The previous discussion has assumed that the matrix of the system is "full," that is, it has few zero elements. For many matrices that arise in practice, and particularly in the solution of differential equations, the elements of the matrix are primarily zero. The simplest examples of this are the tridiagonal matrices discussed in Section 5.3; here, there are no more than three non-zero elements in each row regardless of the size of n.

Tridiagonal matrices are special cases of banded matrices, which were discussed in Section 3.3. In Section 5.3 tridiagonal matrices arose in the finite difference solution of two-point boundary-value problems. When derivatives are approximated by higher-order difference approximations, the matrices will have larger bandwidths. For example, the fourth-order approximation given by (5.3.66) leads to a matrix with bandwidth 5. In most cases of interest, the matrix is symmetrically banded (Section 3.3) and we will assume that in the sequel.

The Gaussian elimination algorithm for banded matrices benefits by not having to deal with the zero elements outside of the band. If A has semi-bandwidth β, there will be β coefficients to be eliminated in the first column, and these eliminations will alter only the elements in the second through

6.1 Gaussian Elimination

($\beta + 1$)st rows and columns of A. The number of operations required will be β^2 multiplications and additions, and β divisions (not counting operations on the right-hand side of the system). The first reduced matrix will be a banded matrix with the same bandwidth, and hence the same count applies. After $n - \beta - 1$ of these reductions, there will remain a full $(\beta + 1) \times (\beta + 1)$ matrix to be reduced. Hence the number of additions (or multiplications) required in the triangular reduction is

$$(n - \beta - 1)\beta^2 + (\tfrac{1}{6}\beta)(\beta + 1)(2\beta + 1) \doteq n\beta^2 - \tfrac{2}{3}\beta^3. \quad (6.1.17)$$

If n is large with respect to β (for example, $n = 1000$, $\beta = 7$), the dominant term in this operation count is $n\beta^2$. As with full matrices, the number of operations to modify the right-hand side and perform the back substitution is of lower-order, namely $O(n\beta)$; see Exercise 6.1.7. However, as the bandwidth decreases, the number of operations for the right-hand side, the back substitution, and the divisions in the forward reduction constitute a larger fraction of the total operation count. In particular, for tridiagonal matrices ($\beta = 1$), the forward reduction requires only $n - 1$ addition/multiplication pairs and $n - 1$ divisions, whereas the right-hand side and back substitution require $(2n - 1)$ additions/multiplications and n divisions for a total operation count of

$$3(n - 1) \text{ additions } + 3(n - 1) \text{ multiplications } + (2n - 1) \text{ divisions.} \quad (6.1.18)$$

The storage of a banded matrix does not require reserving a full $n \times n$ two-dimensional array, which would be very inefficient. All that is needed is $2\beta + 1$ one-dimensional arrays, each holding one of the non-zero diagonals. In particular, a tridiagonal matrix may be stored in three one-dimensional arrays. However, if β is at all large it is probably better to store the diagonals of A as columns in a $(2\beta + 1) \times n$ two-dimensional array. The total storage required, including the right-hand side and the solution vector, is no more than $(2\beta+3)n$; in particular, for tridiagonal systems it is no more than $5n$.

For a banded matrix, the factors L and U of the LU decomposition retain the same bandwidth (Exercise 6.1.8). In particular, for tridiagonal matrices

$$L = \begin{bmatrix} 1 & & & \\ l_2 & 1 & & \\ & \ddots & \ddots & \\ & & l_n & 1 \end{bmatrix}, U = \begin{bmatrix} u_1 & a_{12} & & \\ & u_2 & a_{23} & \\ & & \ddots & \ddots \\ & & & & a_{n-1,n} \\ & & & & u_n \end{bmatrix}. \quad (6.1.19)$$

Note that the superdiagonal elements in U are the original elements of A. Matrices of the form (6.1.19) are called *bidiagonal*.

Determinants and Inverses

We return now to general matrices (not necessarily banded). We note first that the determinant of the coefficient matrix A is an easy by-product of the elimination process. By the LU decomposition of A, and using the facts (Theorem 2.1.1) that the determinant of a product of two matrices is the product of the determinants and that the determinant of a triangular matrix is the product of its diagonal elements, we have

$$\det A = \det LU = \det L \det U = u_{11} u_{22} \cdots u_{nn}, \qquad (6.1.20)$$

since L has 1's on its main diagonal and so $\det L = 1$. Thus, the determinant is just the product of the diagonal elements of the reduced triangular matrix and is computed by an additional $n - 1$ multiplications. Even if only the determinant of the matrix is desired – and not the solution of a linear system – the Gaussian elimination reduction to triangular form is still the best general method for its computation.

The Gaussian elimination process is also the best way, in general, to compute the inverse of A, if that is needed. Let $\mathbf{e}_i$ be the vector with 1 in the ith position and zeros elsewhere. Then $\mathbf{e}_i$ is the ith column of the identity matrix I, and from the basic relation $AA^{-1} = I$ it follows that the ith column of A^{-1} is the solution of the linear system of equations $A\mathbf{x} = \mathbf{e}_i$. Hence we can obtain A^{-1} by solving the n systems of equations

$$A\mathbf{x}_i = \mathbf{e}_i, \qquad i = 1, \ldots, n, \qquad (6.1.21)$$

where the solution vectors $\mathbf{x}_1, \ldots, \mathbf{x}_n$ will be the columns of A^{-1}.

We stress that one does *not* wish to solve a system $A\mathbf{x} = \mathbf{b}$ by computing A^{-1} and then forming $\mathbf{x} = A^{-1}\mathbf{b}$. This would generally require considerably more work than just solving the system.

Several Right-Hand Sides

The above procedure for computing A^{-1} extends to the more general problem of solving several systems with the same coefficient matrix:

$$A\mathbf{x}_i = \mathbf{b}_i, \qquad i = 1, \ldots, m. \qquad (6.1.22)$$

In terms of the LU decomposition of A, (6.1.22) can be carried out efficiently by the following modification of (6.1.15):

1. Factor $A = LU$. (6.1.23a)
2. Solve $L\mathbf{y}_i = \mathbf{b}_i$, $i = 1, \ldots, m$. (6.1.23b)
3. Solve $U\mathbf{x}_i = \mathbf{y}_i$, $i = 1, \ldots, m$. (6.1.23c)

Note that the matrix A is factored only once, regardless of the number of right-hand sides. Hence, the operation count is $O(n^3) + O(mn^2)$, the latter

6.1 Gaussian Elimination

term representing parts 2 and 3 in (6.1.23). Only when m becomes nearly as large as n does the amount of work in parts 2 and 3 approach that of the factorization, at least for full matrices. In the case of computing A^{-1}, $m = n$, but the total operation count is still $O(n^3)$. See also Exercise 6.1.12.

To carry out the intent of (6.1.23) in terms of the elimination process, we can either do parts 1 and 2 simultaneously, modifying the right-hand sides as the elimination proceeds, or we can first complete the factorization and save the multipliers l_{ij} to do 2.

Fill

In the first stage of Gaussian elimination, the element a_{ij} is modified to become $a_{ij}^{(1)}$ by

$$a_{ij}^{(1)} = a_{ij} - \frac{a_{i1}}{a_{11}} a_{1j}.$$

If $a_{ij} = 0$ but a_{i1} and a_{1j} are both non-zero, then $a_{ij}^{(1)} \neq 0$ and a non-zero element has been introduced into a position of the matrix that was originally zero. This is called *fill*, and can occur in any stage of Gaussian elimination. The more fill that occurs, the higher the operation count since elements that have become non-zero have to be eliminated later in the process. Ideally, no fill would occur, so that the operation count would be based on the original number of non-zero elements in A.

As a simple example of the effect of fill, consider the matrix (5.3.39) which arose for a two-point boundary problem with periodic boundary conditions. The non-zero structure of this matrix is represented by

$$\begin{bmatrix} * & * & & & & * \\ * & * & \ddots & & & \\ & \ddots & \ddots & & & \\ & & & & & * \\ & & & & * & * \\ * & & & & * & * \end{bmatrix}. \qquad (6.1.24)$$

If we apply Gaussian elimination to a system for which this is the coefficient matrix, then the elimination of the (2,1) element will introduce a non-zero element into the $(2,n)$ position when a multiple of the first row is subtracted from the second row. Similarly, the elimination of the $(n,1)$ element will introduce a non-zero element into the $(n,2)$ position. In the next stage of Gaussian elimination, the $(n,2)$ element must be eliminated and this causes fill in the $(n,3)$ position. This pattern persists at each stage so that all elements in the last row eventually fill and must be eliminated. Similarly, all elements in the last column fill. In terms of the LU decomposition, the last row of L

and last column of U are non-zero:

$$L = \begin{bmatrix} * & & & & \\ * & * & & & \\ & \ddots & \ddots & & \\ \bigcirc & & \ddots & \ddots & \\ * & * & \cdots & * & * \end{bmatrix}, U = \begin{bmatrix} * & * & \bigcirc & * \\ & * & \ddots & \vdots \\ & & \ddots & * \\ & & & * \end{bmatrix}, \qquad (6.1.25)$$

as opposed to the bidiagonal factors (6.1.19) for a tridiagonal matrix. Because of this fill, extra computation must be done to eliminate the elements that have filled in the last row, to compute the last column of U, and to solve the triangular system. Consequently (Exercise 6.1.21), the operation count for solving a system with the coefficient matrix (6.1.24) is

$$6(n-1) \text{ additions } + 8(n-1) \text{ multiplications } + (3n-2) \text{ divisions.} \qquad (6.1.26)$$

Thus, the addition of the two extra elements of (6.1.24) to a tridiagonal matrix has approximately doubled the operation count as compared to (6.1.18) for a tridiagonal matrix.

The Sherman-Morrision Formula

To solve a system with a coefficient matrix of the form (6.1.24) not only increases the operation count considerably over a tridiagonal matrix, but also requires a much more complicated code (Exercise 6.1.21). We next consider a technique that decreases the operation count slightly but, more importantly, utilizes a tridiagonal matrix code.

Recall that for any non-zero column vectors $\mathbf{u}$ and $\mathbf{v}$ of length n, the product $\mathbf{u}\mathbf{v}^T$ is an $n \times n$ rank-one matrix whose i, j entry is $u_i v_j$. Now let A be the matrix of (6.1.24) and B the tridiagonal part of the matrix. Assume that the $(1, n)$ element of A is c and the $(n, 1)$ element is d. Then, A may be written as

$$A = B + c\mathbf{e}_1\mathbf{e}_n^T + d\mathbf{e}_n\mathbf{e}_1^T, \qquad (6.1.27)$$

where $\mathbf{e}_i$ is the vector with 1 in the ith position and zero elsewhere. Thus, A may be written as the sum of its tridiagonal part plus two rank-one matrices that bring in the outlying elements. We may also write A as

$$A = T + \mathbf{u}\mathbf{v}^T \qquad (6.1.28)$$

where $T = B + \text{diag}(-c, 0, \ldots, 0, -d)$, $\mathbf{u} = c\mathbf{e}_1 + d\mathbf{e}_n$ and $\mathbf{v} = \mathbf{e}_1 + \mathbf{e}_n$. This changes the original tridiagonal matrix but has the advantage that A is now a tridiagonal matrix plus a single rank-one matrix.

Next, let C be any nonsingular matrix and $\mathbf{u}\mathbf{v}^T$ any rank-one matrix. Then, the *Sherman-Morrison* formula is

$$(C + \mathbf{u}\mathbf{v}^T)^{-1} = C^{-1} - \alpha^{-1}C^{-1}\mathbf{u}\mathbf{v}^T C^{-1}, \qquad \alpha = 1 + \mathbf{v}^T C^{-1}\mathbf{u}, \qquad (6.1.29)$$

6.1 Gaussian Elimination

which is easily verified (Exercise 6.1.20). The condition for nonsingularity of $C + \mathbf{u}\mathbf{v}^T$ is that $\alpha \neq 0$. Note that the matrix added to C^{-1} on the right side of (6.1.29) is also a rank-one matrix.

In order to solve a linear system of the form

$$(C + \mathbf{u}\mathbf{v}^T)\mathbf{x} = \mathbf{b}, \tag{6.1.30}$$

we do not wish to form the inverse as in (6.1.29), but only solve linear systems. Thus we use (6.1.29) to write the solution of (6.1.30) as

$$\begin{aligned}\mathbf{x} &= (C + \mathbf{u}\mathbf{v}^T)^{-1}\mathbf{b} = C^{-1}\mathbf{b} - \alpha^{-1}C^{-1}\mathbf{u}\mathbf{v}^T C^{-1}\mathbf{b} \\ &= \mathbf{y} - \alpha^{-1}(\mathbf{v}^T\mathbf{y})\mathbf{z}, \quad \alpha = 1 + \mathbf{v}^T\mathbf{z},\end{aligned} \tag{6.1.31}$$

where $\mathbf{y}$ is the solution of $C\mathbf{y} = \mathbf{b}$ and $\mathbf{z}$ the solution of $C\mathbf{z} = \mathbf{u}$. In particular, if C is a tridiagonal matrix, we can solve the system (6.1.30) by solving two tridiagonal systems and then combining those solutions as shown in (6.1.31). Thus, for the matrix of (6.1.28), the solution of $A\mathbf{x} = \mathbf{b}$ would consist of the steps:

$$\text{Solve } T\mathbf{y} = \mathbf{b}, \quad T\mathbf{z} = \mathbf{u}, \tag{6.1.32}$$

$$\text{Form } \alpha = 1 + \mathbf{v}^T\mathbf{z}, \quad \mathbf{x} = \mathbf{y} - \alpha^{-1}(\mathbf{v}^T\mathbf{y})\mathbf{z}. \tag{6.1.33}$$

Exercise 6.1.22 shows that this approach has a slightly lower operation count than (6.1.26). More importantly, it requires only a code for solving tridiagonal systems, plus the additional operations of (6.1.33).

Although we were able to convert the matrix A of (6.1.27) to the form (6.1.28), which involved only a single rank-one matrix, in many situations we wish to deal with a matrix of the form $C + R$, where R is a matrix of rank m. A rank m matrix may be written in the form $R = UV^T$, where U and V are $n \times m$ matrices. Then (6.1.29) extends (Exercise 6.1.20) to the *Sherman-Morrison-Woodbury* formula

$$(C + UV^T)^{-1} = C^{-1} - C^{-1}U(I + V^T C^{-1}U)^{-1}V^T C^{-1}. \tag{6.1.34}$$

The matrix $I + V^T C^{-1}U$ is $m \times m$, and the Sherman-Morrison formula (6.1.29) is the special case $m = 1$. We could apply (6.1.34) to solve the system $A\mathbf{x} = \mathbf{b}$, where A is given by (6.1.27), although it is slightly more efficient to use (6.1.28) and (6.1.29). To use the formula (6.1.34) we take $C = B$, $U = (c\mathbf{e}_1, d\mathbf{e}_n)$, and $V = (\mathbf{e}_n, \mathbf{e}_1)$. The details of the computation are left to Exercise 6.1.23.

Fill for the Poisson Matrix

We next give another example of fill and its consequences. The matrix (5.5.18), obtained by discretizing Poisson's equation, has five non-zero diagonals. If these non-zero diagonals were adjacent to the main diagonal so that the matrix had semibandwidth 2, then by (6.1.17), the operation count to do the LU factorization would be approximately $4N^2$, since $n = N^2$. However, the effect of the outlying diagonals in (5.5.18) is to cause considerable fill and, consequently, a much higher operation count. To see how this fill occurs, we will consider only the block 3×3 form of (5.5.18) since this clearly exhibits the fill pattern. It is convenient to do the analysis in terms of the LU decomposition of A; if we partition L and U corresponding to A we have:

$$\begin{bmatrix} T & -I & \\ -I & T & -I \\ & -I & T \end{bmatrix} = \begin{bmatrix} L_{11} & & \\ L_{21} & L_{22} & \\ & L_{32} & L_{33} \end{bmatrix} \begin{bmatrix} U_{11} & U_{12} & \\ & U_{22} & U_{23} \\ & & U_{33} \end{bmatrix}. \quad (6.1.35)$$

Equating the corresponding submatrices in (6.1.35), gives

$$L_{11}U_{11} = T, \quad (6.1.36a)$$

$$L_{21}U_{11} = -I \text{ or } L_{21} = -U_{11}^{-1}; \quad U_{12} = -L_{11}^{-1}, \quad (6.1.36b)$$

$$L_{22}U_{22} = T - L_{21}U_{12} = T - U_{11}^{-1}L_{11}^{-1} = T - T^{-1}, \quad (6.1.36c)$$

$$L_{32} = -U_{22}^{-1}, \quad U_{23} = -L_{22}^{-1}, \quad (6.1.36d)$$

$$L_{33}U_{33} = T - L_{32}U_{23} = T - U_{22}^{-1}L_{22}^{-1}. \quad (6.1.36e)$$

Here, L_{11} and U_{11} are the LU factors of the tridiagonal matrix T, and by (6.1.19), these factors have the form

$$L_{11} = \begin{bmatrix} 1 & & & \\ l_2 & 1 & & \\ & \ddots & \ddots & \\ & & l_N & 1 \end{bmatrix}, U_{11} = \begin{bmatrix} u_1 & -1 & & \\ & u_2 & \ddots & \\ & & \ddots & -1 \\ & & & u_N \end{bmatrix}, \quad (6.1.37)$$

where the -1's in U_{11} are the off-diagonal elements of T. Even though L_{11} has only two non-zero diagonals, the same is not true of L_{11}^{-1}; it is a full lower triangular matrix. To see why this is true, recall from (6.1.21) that the ith column of L_{11}^{-1} is the solution of the system

$$L_{11}\mathbf{x}_i = \mathbf{e}_i, \quad (6.1.38)$$

where $\mathbf{e}_i$ is the vector with 1 in the ith position and zero elsewhere. The solution of (6.1.38) for $i = 1$ is

$$x_1 = 1, x_2 = -l_2 x_1 = -l_2, x_3 = -l_3 x_2 = l_2 l_3, \cdots, x_N = \pm l_2 \cdots l_N.$$

Thus provided that none of the l_i is zero, which is the case if L_{11} is the factor of T, all components of the first column of L_{11}^{-1} are non-zero. Doing the analogous computation for general i, one sees that the first non-zero component of the solution is in the ith position and then each subsequent component of the solution is non-zero. It follows that each column of L_{11}^{-1} has all non-zero elements below the main diagonal. The same is true for $(U_{11}^{-1})^T$ so that U_{11}^{-1} is full above the main diagonal. It is easy to verify (Exercise 6.1.26) that the product $U_{11}^{-1}L_{11}^{-1}$ is then a completely full matrix, and therefore the factors L_{22} and U_{22} in (6.1.36c) are full below and above the main diagonal, respectively. The same is true of the factors L_{32}, L_{33}, U_{23}, and U_{33}. Thus the non-zero structure of the factor L of (6.1.35) is as shown in Figure 6.1.2.

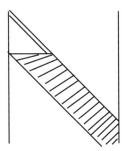

Figure 6.1.2: *Non-zero Structure of L and U^T*

U^T has the same structure and complete fill has occurred within the band, except for the first block. The same is true no matter how many blocks are in the matrix A; the 3×3 block structure of (6.1.35) was just used as an example. Thus, the amount of work to carry out the LU factorization is almost as much as if the matrix A were a full banded matrix: the sparse structure of A within the band has essentially been lost because of the fill. Since the semibandwidth of A is N, by (6.1.17) the operation count for the LU factorization is (Exercise 6.1.31) approximately N^4. For a three-dimensional problem on an $N \times N \times N$ grid, the semibandwidth will be N^2, and the operation count will be $O(N^7)$, since now $n = N^3$. If, for example, $N = 100$, then $O(10^{14})$ operations would be required. With a machine capable of an operation rate of 1 Gflop, this would require 10^5 seconds or about 28 hours.

Reordering to Eliminate Fill

One approach to circumventing this problem of fill is a reordering of the equations and unknowns. Consider a matrix with non-zero elements in only the first row, first column and main diagonal as illustrated in Figure 6.1.3(a). If Gaussian elimination is applied to this matrix, all elements will, in general, fill. On the other hand, for the matrix of Figure 6.1.3(b), no fill will occur. The matrix of Figure 6.1.3(b) may be obtained from that of 6.1.3(a) by a reordering of the unknowns and equations (Exercise 6.1.27). In general, it

228   Chapter 6 Direct Solution of Linear Equations

will not be known in advance how to do a reordering that will minimize fill, but algorithms are known that can approximate this; see the Supplementary Discussion.

(a)   (b)

Figure 6.1.3: *Arrowhead Matrices*

Domain Decomposition Reordering

We now consider a way to order the system of equations (5.5.13) for the discrete Poisson problem so that Gaussian elimination can be carried out with less fill, and therefore fewer arithmetic operations, than if we used the natural ordering. We consider for illustration a rectangular grid of 22 interior points, as shown in Figure 6.1.4. We partition this grid into three subdomains D_1, D_2 and D_3, as well as two vertical lines of grid points called the *separator set*, labeled S. Such a partitioning is an example of a *domain decomposition*. In the current context, it is also called *one-way dissection*. We next number the grid points in the first subdomain using the natural ordering, followed by the points in the second and third subdomains, and then finally those in the separator set. This is illustrated by the grid point numbers in the example of Figure 6.1.4.

●	●	●	●	●	●	●	●	●	●	●
4	5	6	20	10	11	12	22	16	17	18
●	●	●	●	●	●	●	●	●	●	●
1	2	3	19	7	8	9	21	13	14	15
D_1			S	D_2			S	D_3		

Figure 6.1.4: *Domain Decomposition*

We now order the equations and unknowns according to the grid point numbering of Figure 6.1.4. The resulting coefficient matrix is shown in Figure 6.1.5. Also shown in Figure 6.1.5 is the fill pattern that results from Gaussian elimination: the original elements of the matrix are 4 and −1, and

6.1 Gaussian Elimination

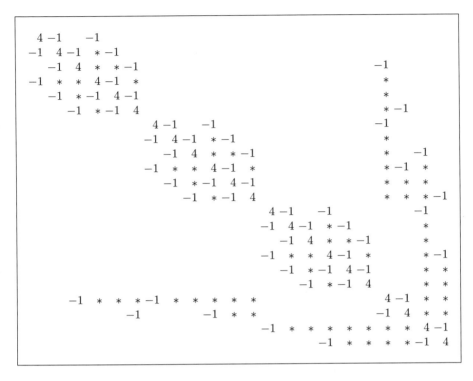

Figure 6.1.5: *Domain Decomposition Matrix and Fill Pattern*

asterisks indicate fill elements. It is left to Exercise 6.1.28 to verify the details of Figure 6.1.5. There are 76 fill elements as contrasted to 182 had we used the natural ordering and applied Gaussian elimination to the resulting banded system (Exercise 6.1.30). We note that the matrix (before fill) of Figure 6.1.5 is related by a permutation matrix to the matrix that would result from using the natural ordering (Exercise 6.1.29).

The above discussion has illustrated in a very simple case the principle of domain decomposition. More generally, if we have p subdomains and corresponding separator sets that prevent unknowns in one subdomain from being connected to unknowns in any other subdomain, then the coefficient matrix will take the *block arrowhead matrix* form,

$$A = \begin{bmatrix} A_1 & & & & B_1^T \\ & A_2 & & & B_2^T \\ & & \ddots & & \vdots \\ & & & A_p & B_p^T \\ B_1 & B_2 & \cdots & B_p & A_s \end{bmatrix}. \qquad (6.1.39)$$

Figure 6.1.5 (without the fill) is the special case $p = 3$. This domain decomposition process can be applied to more general partial differential equations, more general domains, and more general discretizations. It is of particular importance in parallel computing, as we will see later.

The various algorithms in this section have all been predicated on the assumption that a_{11} and the subsequent diagonal elements of the reduced matrix do not vanish. In practice, it is not sufficient that these divisors be non-zero; they must also be large enough in some sense or severe rounding error problems may occur. In Section 6.2 we will consider these problems and the modifications that are necessary for the elimination process to be a viable procedure.

Supplementary Discussion and Reference: 6.1

1. There are many books on numerical linear algebra and these, as well as more elementary books on numerical methods, all discuss the problem of solving linear systems of equations. For an advanced treatment and more references, see Golub and Van Loan [1989].

2. The state of the art in solving linear equations has now reached a very high level, especially for full and banded systems. Probably the best set of codes has been LINPACK, a package of FORTRAN subroutines. Recently, LINPACK has evolved to a new package, LAPACK, specifically designed for use on vector and parallel computers. See Dongarra and Bunch et al. [1979] for a discussion of LINPACK and Anderson et al. [1992] for LAPACK.

3. For a review of the history and many applications of the formulas (6.1.29) and (6.1.34), see Hager [1989]. In particular, (6.1.29) was first given by J. Sherman and W. Morrison in 1949 for the special case of changing the elements in one column of C; in this case, $\mathbf{v} = \mathbf{e}_i$ if the ith column is changed. The general formula (6.1.29) was given by M. Bartlett in 1951. Simultaneously, M. Woodbury gave the still more general formula (6.1.34) in a 1950 report although it had appeared in earlier work in the mid 1940s. Both (6.1.29) and (6.1.34) must be used with caution because of their rounding error properties.

4. Because of their special structure, systems with coefficient matrices that are Hankel matrices, Toeplitz matrices or Vandermonde matrices can be solved in $O(n^2)$ operations. (A Toeplitz matrix has constant diagonals; the $(2, -1)$ tridiagonal matrix (5.3.10) is an example.) See Golub and Van Loan [1989] for further discussion.

5. An excellent reference for further reading on direct methods for sparse linear systems is George and Liu [1981]. This book includes, in particular, detailed analyses of the storage and operation counts of one-way dissection and a more general treatment of *nested dissection* in which both horizontal and vertical separator lines are used. Also, there is a discussion and analysis of other reordering techniques that reduce the fill. For more general problems (not coming from Poisson's equation), one of the best such techniques is the *minimum degree algorithm*. Another good reference for sparse systems is Duff et al. [1986].

For general sparse linear systems, the algorithmic framework is the following:

Step 1. Use a reordering strategy to minimize fill.

Step 2. Do a symbolic factorization to determine the fill. Set up data storage accordingly.

Step 3. Do the numerical LU factorization.

Step 4. Solve the corresponding triangular systems.

The symbolic factorization in Step 2 can be done surprisingly rapidly. Once this is done, the exact fill pattern is known so that the amount of storage needed for the factorization is also known. Storage can then be allocated for only the non-zero elements in the factors.

EXERCISES 6.1

6.1.1. Verify the forward reduction of (6.1.8) to obtain (6.1.9). Then verify that $A = LU$, where L is given by (6.1.10) and U is the upper triangular matrix in (6.1.9).

6.1.2. Verify that multiplication of (6.1.2) by (6.1.11) gives (6.1.3). Then verify that (6.1.5) may be obtained as indicated by (6.1.4).

6.1.3. Verify the following statements:

   a. The determinants of the L_i of (6.1.13) are all 1.

   b. Products and inverses of unit lower triangular matrices are unit lower triangular.

   c. The inverse of L_i of (6.1.13) is the same matrix with the signs of the off-diagonal elements changed.

6.1.4. Verify by induction the following summation formulas:

$$\sum_{i=1}^{n} i = \tfrac{1}{2}n(n+1), \qquad \sum_{i=1}^{n} i^2 = \tfrac{1}{6}n(n+1)(2n+1).$$

6.1.5. Show that the following operation counts are correct for Gaussian elimination:

  a. Number of additions (multiplications) to compute new right-hand side = $n(n-1)/2$.

  b. Number of divisions to compute the multipliers $l_{ik} = n(n-1)/2$.

  c. Number of divisions in back substitution $= n$.

  d. Number of additions (multiplications) in back substitution $= n(n-1)/2$.

6.1.6. Using a Gaussian elimination code (either a package or one that you write), measure the time to solve full linear systems of sizes $n = 50$ and $n = 100$. Discuss why the larger time is not exactly a factor of 8 larger than the smaller, as suggested by the $O(n^3)$ operation count.

6.1.7. If A is a banded matrix with semibandwidth β, show that $O(\beta n)$ operations are required to modify the right-hand side and do the back substitution.

6.1.8. Let A be a matrix with bandwidth $p + q + 1$, with p diagonals below and q diagonals above the main diagonal. If $A = LU$ is the LU decomposition, show that L has bandwidth $p + 1$ and U has bandwidth $q + 1$. In particular, for tridiagonal matrices show that L and U have the form (6.1.19).

6.1.9. Verify that the product of L and U in (6.1.19) is tridiagonal.

6.1.10. Write a computer program to implement Gaussian elimination for a banded matrix with p subdiagonals and q superdiagonals. Write a separate code for tridiagonal systems using one-dimensional arrays to store the matrix.

6.1.11. Multiply two $n \times n$ tridiagonal matrices. How many arithmetic operations does this require? Is the product matrix tridiagonal?

6.1.12. If $A = LU$, show that $A^{-1} = U^{-1}L^{-1}$. Using this, describe an algorithm for computing A^{-1}, taking advantage of the triangular structure of U^{-1} and L^{-1}. How does your algorithm compare with solving the systems (6.1.21)?

6.1.13. We have shown that Gaussian elimination on a full matrix requires $O(n^3)$ operations and on a tridiagonal matrix it requires $O(n)$ operations. For what size semibandwidth will it require $O(n^2)$ operations?

6.1.14. Consider the two-point boundary-value problem on the infinite interval $(0, \infty)$:

$$y''(x) = (2x+2)/(x+2)y(x), \quad y(0) = 2, \quad y(\infty) = 0.$$

  a. Approximate a solution to this problem by replacing the boundary condition at ∞ by $y(5) = 0$ and using finite differences for $h = 1/10, 1/20$ and $1/32$.

  b. Now note that $(2x+2)/(x+2) \to 2$ as $x \to \infty$. Hence, consider the problem

  $$z''(x) = 2z(x), \quad z(0) = 2, \quad z(\infty) = 0.$$

  Solve this problem exactly and find an expression for $z'(x)$ in terms of $z(x)$. Using this expression for the value of $y'(x)$ as $x \to \infty$, solve the truncated tridiagonal system and compare the solution to the truncated system used in part **a**.

6.1 Gaussian Elimination

6.1.15. Write a program to carry out (5.4.11) and apply it to the problem of Exercise 5.4.1. Use various values of Δt and Δx and verify numerically the stability of the method. Discuss your results compared to those for Exercise 5.4.1, including the relative ease and efficiency of carrying out the two methods. Modify your program to carry out the Crank-Nicolson method (5.4.15). Discuss your results and compare this method to (5.4.11).

6.1.16. Consider the $n \times n$ complex system $(A + iB)(\mathbf{u} + i\mathbf{v}) = \mathbf{b} + i\mathbf{c}$, where $A, B, \mathbf{u}, \mathbf{v}, \mathbf{b}$, and $\mathbf{c}$ are real. Show how to write this as a real $2n \times 2n$ system. Compare the operation counts of Gaussian elimination applied to the real system and the original complex system, assuming that complex multiplication requires four real multiplications and two additions. (What does complex division require?)

6.1.17. (Uniqueness of LU Decomposition) Let A be nonsingular and suppose that $A = LU = \hat{L}\hat{U}$, where L and $\hat{L}$ are unit lower-triangular and U and $\hat{U}$ are upper triangular. Show that $L = \hat{L}$ and $U = \hat{U}$.

6.1.18. Let $A = LU$. Use Exercise 6.1.17 to give the LU decomposition of AT, where T is upper triangular. Specialize this to the case that T is a diagonal matrix. What does this imply about the multipliers in Gaussian elimination when the columns of A are scaled?

6.1.19. (Jordan elimination) Show that it is possible to eliminate the elements above the main diagonal as well as below, so that the reduced system is $D\mathbf{x} = \hat{\mathbf{b}}$ where D is diagonal. How many operations does this method require to solve a linear system?

6.1.20. Verify the formulas (6.1.29), (6.1.31), and (6.1.34).

6.1.21. Show that if Gaussian elimination is done on the matrix (5.3.39) the factor L of the LU decomposition has the same non-zero structure as the lower-triangular part of A except that the last row has, in general, all non-zero elements. Show that the operation count for this algorithm, assuming that no operations are done on elements known to be zero, is given by (6.1.26). Then write a Gaussian elimination code that achieves this operation count.

6.1.22. Again for the matrix (5.3.39), show that solving $A\mathbf{x} = \mathbf{b}$ by (6.1.32), (6.1.33) requires the following operations, assuming that the LU decomposition of T is done only once: $6n - 1$ multiplications, $6n - 1$ additions and $3n - 2$ divisions. Compare this operation count with that of Exercise 6.1.21.

6.1.23. Verify that the Sherman-Morrison-Woodbury formula (6.1.34) can be used to solve the system $A\mathbf{x} = \mathbf{b}$, where A is given by (6.1.27), by carrying out the following steps:

1. Solve $B\mathbf{y} = \mathbf{b}$, $B\mathbf{w}_1 = \mathbf{e}_1$, $B\mathbf{w}_n = \mathbf{e}_n$.

2. Form the 2 × 2 matrix

$$U^T W = \begin{pmatrix} \mathbf{e}_1^T \\ \mathbf{e}_n^T \end{pmatrix} (\mathbf{w}_1, \mathbf{w}_n) = \begin{pmatrix} \mathbf{e}_1^T \mathbf{w}_1 & \mathbf{e}_1^T \mathbf{w}_n \\ \mathbf{e}_n^T \mathbf{w}_1 & \mathbf{e}_n^T \mathbf{w}_n \end{pmatrix}.$$

3. Form the 2-vector

$$\mathbf{q} = (I + U^T W)^{-1} U^T \mathbf{y}.$$

4. Form the solution

$$\mathbf{x} = \mathbf{y} - W\mathbf{q}.$$

Show that this requires $8n - 6$ multiplications, $8n - 5$ additions and $4n - 3$ divisions.

6.1.24. Let T be a symmetric nonsingular tridiagonal matrix. Give an algorithm to find numbers $p_1, \ldots, p_n$ and $q_1, \ldots, q_n$ so that the i, j element of T^{-1} is $p_i q_j$ if $i \geq j$ and $p_j q_i$ if $i < j$. (Hint: Look at the first column of T^{-1}.)

6.1.25. Write a computer program to obtain a cubic spline by first solving the tridiagonal system (4.1.36) and then using (4.1.34). Also, write a program for evaluating this cubic spline at a given point x.

6.1.26. Let L and U be lower- and upper-triangular matrices for which $l_{ij} \neq 0$, $i \geq j$, and $u_{ij} \neq 0$, $i \leq j$. Show that all elements of UL are, in general, non-zero.

6.1.27. Consider the system

$$\begin{bmatrix} * & * & * & * & * \\ * & * & & & \\ * & & * & & \\ * & & & * & \\ * & & & & * \end{bmatrix} \begin{bmatrix} x_1 \\ x_2 \\ x_3 \\ x_4 \\ x_5 \end{bmatrix} = \begin{bmatrix} b_1 \\ b_2 \\ b_3 \\ b_4 \\ b_5 \end{bmatrix},$$

where $*$ denotes a non-zero element. Show that by the reordering of unknowns $x_1 \leftrightarrow x_5$, $x_2 \leftrightarrow x_4$, $x_3 \leftrightarrow x_3$ and the corresponding reordering of the equations, the system can be written as

$$\begin{bmatrix} * & & & & * \\ & * & & & * \\ & & * & & * \\ & & & * & * \\ * & * & * & * & * \end{bmatrix} \begin{bmatrix} x_5 \\ x_4 \\ x_3 \\ x_2 \\ x_1 \end{bmatrix} = \begin{bmatrix} b_5 \\ b_4 \\ b_3 \\ b_2 \\ b_1 \end{bmatrix}.$$

Generalize this to the case of the corresponding $n \times n$ system.

6.1.28. Show that the ordering of the grid points (and hence unknowns) of Figure 6.1.4 gives the coefficient matrix of 4's and -1's of Figure 6.1.5 for the system of equations (5.5.13). Next, apply Gaussian elimination to this matrix and show that fill develops as indicated in Figure 6.1.5.

6.1 Gaussian Elimination

6.1.29. Show that the coefficient matrix A of Figure 6.1.5 (without the fill) is related by $A = P\hat{A}P$ to the matrix $\hat{A}$ that one would obtain from the natural ordering. Here P is a permutation matrix.

6.1.30. For the grid of Figure 6.1.4, write out the 22×22 coefficient matrix for the natural ordering. Then verify, using the techniques that led to Figure 6.1.2, that the number of fill elements produced by Gaussian elimination is 182.

6.1.31. Use the fact that fill develops as in Figure 6.1.2 to show that the operation count for Gaussian elimination applied to the system (5.5.19) is $O(N^4)$.

6.1.32. Let A be the block tridiagonal matrix

$$A = \begin{bmatrix} A_1 & B_1 & & \\ C_2 & \ddots & \ddots & \\ & \ddots & & B_{p-1} \\ & & C_{p-1} & A_p \end{bmatrix}.$$

Proceed as in the derivation of (6.1.36) to develop a block LU decomposition of A where

$$L = \begin{bmatrix} I & & & \\ L_1 & I & & \\ & \ddots & \ddots & \\ & & L_{p-1}I & \end{bmatrix}, U = \begin{bmatrix} U_1 & V_1 & & \\ & \ddots & \ddots & \\ & & & V_{p-1} \\ & & & U_p \end{bmatrix}.$$

What assumptions do you need to make in order that the decomposition can be carried out mathematically?

6.1.33. Write the matrix shown in Figure 6.1.5 (without the fill) as

$$\begin{bmatrix} A_1 & & B_1^T \\ & A_2 & B_2^T \\ B_1 & B_2 & A_s \end{bmatrix} = \begin{bmatrix} A_1 & & \\ & A_2 & \\ & & A_s \end{bmatrix} + \begin{bmatrix} & & B_1^T \\ & & B_2^T \\ B_1 & B_2 & \end{bmatrix}$$

Show that the second matrix is of rank 8 and show how to apply the Sherman-Morrison-Woodbury formula. Extend this to the matrix (6.1.39).

6.1.34. Let A be an $n \times n$ matrix. Show that doing the first p steps of Gaussian elimination is equivalent to a decomposition of the form

$$A = \begin{bmatrix} L_1 \\ F \end{bmatrix} [U_1 E] + \begin{bmatrix} 0 & 0 \\ 0 & X \end{bmatrix},$$

where L_1 and U_1 are $p \times p$ lower- and upper-triangular matrices. Give expressions for E, F and X.

6.1.35. The *profile* of a matrix A is the set S of indices $(i, i_1), i = 1, \ldots, n$ of the first non-zero element in each row. For example, in the matrix

$$A = \begin{bmatrix} 2 & 1 & 2 & 3 \\ 0 & 1 & 2 & 3 \\ 2 & 0 & 1 & 2 \\ 0 & 2 & 1 & 1 \end{bmatrix},$$

$S = \{(1,1), (2,2), (3,1), (4,2)\}$. Assume that A has an LU decomposition. Show that the factor L has the same profile as A.

6.1.36. Show that the following boundary value problems have the solutions indicated. Solve these problems numerically, setting up difference equations corresponding to (5.3.41) and then using Newton's method (5.3.49). Solve for various values of the step length h and discuss the discretization error.

a. $v'' = 2v^3, v(0) = 1, v(\frac{1}{2}) = 2$. Solution: $v(x) = (1-x)^{-1}$

b. $v'' = 2v^3 + 6v' - 6(1-x)^{-1} + 2, v(0) = 0, v(\frac{1}{2}) = 1$.
Solution: $v(x) = (1-x)^{-1} - 1$.

c. $v'' = \frac{3}{2}v^2, v(0) = 4, v(1) = 1$. Solution: $v(x) = 4(1+x)^{-2}$. (Note: This solution is not unique.)

6.1.37. The *leading principal minors* of an $n \times n$ matrix A are

$$a_{11}, \det \begin{bmatrix} a_{11} & a_{12} \\ a_{21} & a_{22} \end{bmatrix}, \det \begin{bmatrix} a_{11} & a_{12} & a_{13} \\ a_{21} & a_{22} & a_{23} \\ a_{31} & a_{32} & a_{33} \end{bmatrix}, \cdots.$$

Show that A has an LU decomposition if and only if all principal minors are non-zero.

6.2 Errors in Gaussian Elimination

Since in exact arithmetic the Gaussian elimination algorithm obtains the exact solution of a linear system in finitely many operations, the only error involved in solving the system on a computer is rounding error. Properly implemented, Gaussian elimination has excellent properties with respect to rounding error. However, this does not necessarily guarantee an accurate solution of the linear system, as we will see.

Interchanges

In our discussion of Gaussian elimination in the previous section, we assumed that a_{11} and all subsequent divisors were non-zero. However, we do not need to make such an assumption provided that we revise the algorithm so as to interchange equations if necessary, as we shall now describe.

6.2 Errors in Gaussian Elimination

We assume, as usual, that the coefficient matrix A is nonsingular. Suppose that $a_{11} = 0$. Then some other element in the first column of A must be non-zero or else A is singular (see Exercise 6.2.1). If, say, $a_{k1} \neq 0$, then we interchange the first equation in the system with the kth; clearly, this does not change the solution. In the new system the $(1,1)$ coefficient is now non-zero, and the elimination process can proceed. Similarly, an interchange can be done if any computed diagonal element that is to become a divisor in the next stage should vanish. Suppose, for example, that the elimination has progressed to the point

$$\begin{matrix} a_{11} & \cdots & & & a_{1n} \\ & \ddots & & & \vdots \\ & & a_{ii}^{(i-1)} & \cdots & a_{in}^{(i-1)} \\ & & \vdots & & \vdots \\ & & a_{ni}^{(i-1)} & \cdots & a_{nn}^{(i-1)}, \end{matrix}$$

and that $a_{ii}^{(i-1)} = 0$. If all of the remaining elements below $a_{ii}^{(i-1)}$ in the ith column are also zero, this matrix is singular (see Exercise 6.2.2). Since the operations of adding a multiple of one row to another row, which produced this reduced matrix, do not affect the determinant, and an interchange of rows only changes the sign of the determinant (Theorem 2.1.1), it follows that the original matrix is also singular, contrary to assumption. Hence, at least one of the elements $a_{ki}^{(i-1)}$, $k = i+1, \ldots, n$, is non-zero, and we can interchange a row that contains a non-zero element with the ith row, thus ensuring that the new (i,i) element is non-zero. Again, this interchange of rows does not change the solution of the system. However, an interchange of rows does change the sign of the determinant of the coefficient matrix, so that if the determinant is to be computed a record must be kept of whether the number of interchanges is even or odd. In any case, in exact arithmetic we can ensure that Gaussian elimination can be carried out by interchanging rows so that no divisors are zero.

Rounding Error and Instability

There are two possibilities as to how rounding error can effect the computed solution. The first is an accumulation of rounding errors during a large number of arithmetic operations. For example, if $n = 1,000$, the operation count of the previous section shows that on the order of $n^3 = 10^9$ operations will be required; even though the error in each individual operation may be small, the total buildup could be large. We shall see later that this potential accumulation of rounding error is not as serious as one might expect.

The second possibility involves catastrophic rounding errors. If an algorithm has this unfortunate characteristic, it is called *numerically unstable* and

238   Chapter 6 Direct Solution of Linear Equations

is not suitable as a general method. Although the interchange process described previously ensures that Gaussian elimination can be carried out mathematically for any nonsingular matrix, the algorithm can give rise to catastrophic rounding errors and is numerically unstable. We shall analyze a simple 2×2 example in order to see how this can occur.

Consider the system

$$\begin{bmatrix} -10^{-5} & 1 \\ 2 & 1 \end{bmatrix} \begin{bmatrix} x_1 \\ x_2 \end{bmatrix} = \begin{bmatrix} 1 \\ 0 \end{bmatrix}, \qquad (6.2.1)$$

whose exact solution is

$$x_1 = -0.4999975 \cdots, \qquad x_2 = 0.999995.$$

Now suppose that we have a decimal computer with a word length of four digits; that is, numbers are represented in the form $0. \ast \ast \ast \ast \times 10^p$. Let us carry out Gaussian elimination on this hypothetical computer. First, we note that $a_{11} \neq 0$, and no interchange is needed. The multiplier is

$$l_{21} = \frac{-0.2 \times 10^1}{0.1 \times 10^{-4}} = -0.2 \times 10^6,$$

which is exact, and the calculation for the new a_{22} is

$$a_{22}^{(1)} = 0.1 \times 10^1 - (-0.2 \times 10^6)(0.1 \times 10^1) \qquad (6.2.2)$$
$$= 0.1 \times 10^1 + 0.2 \times 10^6 \doteq 0.2 \times 10^6.$$

The exact sum in (6.2.2) is 0.200001×10^6, but since the computer has a word length of only four digits this must be represented as 0.2000×10^6; this is the first error in the calculation.

The new b_2 is

$$b_2^{(1)} = -(-0.2 \times 10^6)(0.1 \times 10^1) = 0.2 \times 10^6. \qquad (6.2.3)$$

No rounding errors occurred in this computation, nor do any occur in the back substitution:

$$x_2 = \frac{b_2^{(1)}}{a_{22}^{(1)}} = \frac{0.2 \times 10^6}{0.2 \times 10^6} = 0.1 \times 10^1,$$

$$x_1 = \frac{0.1 \times 10^1 - 0.1 \times 10^1}{-0.1 \times 10^{-4}} = 0.$$

The computed x_2 agrees excellently with the exact x_2, but the computed x_1 has no digits of accuracy. Note that the only error made in the calculation is in $a_{22}^{(1)}$, which has an error in the sixth decimal place. Every other operation

6.2 Errors in Gaussian Elimination

was exact. How, then, can this one "small" error cause the computed x_1 to deviate so drastically from its exact value?

Backward Error Analysis

The answer lies in the principle of *backward error analysis*, one of the most important concepts in scientific computing. The basic idea of backward error analysis is to "ask not what the error is, but what problem have we really solved." We shall invoke this principle here in the following form. Note that the quantity 0.000001×10^6 that was dropped from the computed $a_{22}^{(1)}$ in (6.2.2) is the original element a_{22}. Since this is the only place that a_{22} enters the calculation, the computed solution would have been the same if a_{22} were zero. Therefore the calculation on our four-digit computer has computed the exact solution of the system

$$\begin{bmatrix} -10^{-5} & 1 \\ 2 & 0 \end{bmatrix} \begin{bmatrix} x_1 \\ x_2 \end{bmatrix} = \begin{bmatrix} 1 \\ 0 \end{bmatrix}. \tag{6.2.4}$$

Intuitively, we would expect the two systems (6.2.1) and (6.2.4) to have rather different solutions, and this is indeed the case.

But why did this occur? The culprit is the large multiplier l_{21}, which made it impossible for a_{22} to be included in the sum in (6.2.2) because of the word length of the machine. The large multiplier was due to the smallness of a_{11} relative to a_{21}, and the remedy is, again, an interchange of the equations. Indeed, if we solve the system

$$\begin{bmatrix} 2 & 1 \\ -10^{-5} & 1 \end{bmatrix} \begin{bmatrix} x_1 \\ x_2 \end{bmatrix} = \begin{bmatrix} 0 \\ 1 \end{bmatrix} \tag{6.2.5}$$

on our hypothetical four-digit computer, we obtain

$$l_{21} = \frac{-0.1 \times 10^{-4}}{0.2 \times 10^1} = -0.5 \times 10^{-5}$$

$$a_{22}^{(1)} = 0.1 \times 10^1 - (-0.5 \times 10^{-5})(1) \doteq 0.1 \times 10^1$$

$$b_2^{(1)} = 0.1 \times 10^1 - (-0.5 \times 10^{-5})(0) = 0.1 \times 10^1$$

$$x_2 = \frac{0.1 \times 10^1}{0.1 \times 10^1} = 1.0$$

$$x_1 = \frac{-(0.1 \times 10^1)(1)}{0.2 \times 10^1} = -0.5 .$$

The computed solution now agrees excellently with the exact solution.

240　　　　　　　　　　Chapter 6　Direct Solution of Linear Equations

Partial Pivoting

By a relatively simple strategy we can always arrange to keep the multipliers in the elimination process less than or equal to 1 in absolute value. This is known as *partial pivoting*: at the kth stage of the elimination process an interchange of rows is made, if necessary, to place in the main diagonal position the element of largest absolute value from the kth column on or below the main diagonal. If we include this interchange strategy in the forward reduction algorithm of Figure 6.1.1(a), the result is shown in Figure 6.2.1.

For $k = 1, \ldots, n-1$
    Find $m \geq k$ such that $|a_{mk}| = \max\{|a_{ik}| : i \geq k\}$.
    If $a_{mk} = 0$, then A is singular, and stop.
        else interchange a_{kj} and a_{mj}, $j = k, k+1, \ldots, n$.
    interchange b_k and b_m.
    For $i = k+1, k+2, \ldots, n$
        $l_{ik} = a_{ik}/a_{kk}$
        For $j = k+1, k+2, \ldots, n$
            $a_{ij} = a_{ij} - l_{ik}a_{kj}$
        $b_i = b_i - l_{ik}b_k$.

Figure 6.2.1: *Forward Reduction with Partial Pivoting*

Gaussian elimination with partial pivoting has proved to be an extremely reliable algorithm in practice. However, there are two major precautions that should be kept in mind. First, the matrix must be properly scaled before the algorithm is used. To illustrate this point, consider the system

$$\begin{bmatrix} 10 & -10^6 \\ 2 & 1 \end{bmatrix} \begin{bmatrix} x_1 \\ x_2 \end{bmatrix} = \begin{bmatrix} -10^6 \\ 0 \end{bmatrix}, \tag{6.2.6}$$

which is just (6.2.1) with the first equation multiplied by -10^6. No interchange is called for by the partial pivoting strategy since the $(1,1)$ element is already the largest in the first column. However, if we carry out the elimination on our hypothetical four-digit computer (see Exercise 6.2.5), we will encounter exactly the same problem that we did with the system (6.2.1).

The use of the partial pivoting strategy is predicated on the coefficient matrix being properly scaled so that the maximum element in each row and

6.2 Errors in Gaussian Elimination

column is the same order of magnitude. This scaling is called *equilibration* or *balancing* of the matrix. Unfortunately, there is no known foolproof general procedure for such scaling, but usually it will be clear that some rows or columns of the matrix need scaling, and this can be done before the elimination starts. For example, if we were given the system (6.2.6), we should scale the first row so that its maximum element is approximately 1. Then a_{11} will be small, and the partial pivoting strategy will cause an interchange of the first and second rows.

The second precaution regarding the partial pivoting strategy is that even with an equilibrated matrix, it can be numerically unstable. Examples for which this can happen have been given (Exercise 6.2.25), but the occurrence of such matrices in practical computations seems to be sufficiently rare that the danger can be safely ignored. (For additional remarks, see the Supplementary Discussion.)

LU with Interchanges

If row interchanges are made, the Gaussian elimination process is not equivalent to a factorization of the matrix A into the product of lower- and upper-triangular matrices; the lower-triangular matrix must be modified in the following way. Interchange of rows of a matrix can be effected by multiplication on the left by a permutation matrix (see Chapter 2). For example, multiplication of a 4×4 matrix by the permutation matrix

$$P = \begin{bmatrix} 1 & 0 & 0 & 0 \\ 0 & 0 & 0 & 1 \\ 0 & 0 & 1 & 0 \\ 0 & 1 & 0 & 0 \end{bmatrix}. \tag{6.2.7}$$

will leave the first and third rows the same and interchange the second and fourth rows (see Exercise 6.2.6). Thus the row interchanges of the coefficient matrix A that are required by the partial pivoting strategy can be represented by multiplication of A on the left by suitable permutation matrices. If P_i denotes the permutation matrix corresponding to the interchange required at the ith stage, then conceptually we are generating the triangular factorization of the matrix

$$P_{n-1}P_{n-2}\cdots P_2 P_1 A = PA = LU \tag{6.2.8}$$

rather than A itself. Thus, the factorization is $A = (P^{-1}L)U$. Since the product of permutation matrices is again a permutation matrix and the inverse of a permutation matrix is a permutation matrix (Exercise 6.2.7), the first factor is a permutation of a lower-triangular matrix, whereas the second is again upper-triangular. Note that if no interchange is required at the ith stage, the permutation matrix P_i is simply the identity matrix.

Banded Systems

Row interchanges require additional time and, in the case of banded matrices, also complicate the storage. Consider first a tridiagonal system. If an interchange is made at the first stage, the elements in the first two rows will be

$$* \quad * \quad * \quad 0 \quad \cdots$$
$$* \quad * \quad 0 \quad \cdots$$

The elimination process will then reenter a (generally) non-zero element into the $(2,3)$ position, and the reduced $(n-1) \times (n-1)$ matrix will again be tridiagonal. Thus, the effect of the interchanges will be to introduce possible non-zero elements into the second superdiagonal of the reduced triangular matrix. Then the factor U in the decomposition of A will no longer be bidiagonal but will have, in general, three non-zero diagonals. Perhaps the simplest way of handling the storage is to add an additional one-dimensional array to hold these elements in the second superdiagonal.

For a banded matrix with semibandwidth β the same kind of problem occurs. If an interchange is made at the first stage between the first and $(\beta + 1)$st rows, an additional β elements will be introduced into the first row, and these, in turn, will be propagated into rows 2 through $\beta + 1$ during the elimination process. Thus, we need to provide storage for a possible additional β superdiagonals. The simplest way to handle this is to allow for an additional $n \times \beta$ array of storage at the outset. An alternative is based on the observation that the amount of additional storage needed is no more than the amount of storage required for the non-zero subdiagonals. As the subdiagonals are eliminated, we no longer will need that storage, and the new superdiagonals elements can be stored in those positions. However, it is this subdiagonal space that is normally used to store the multipliers if their retention is desired; in this case we have no alternative but to set aside additional storage.

Diagonally Dominant and Positive Definite Matrices

Although for general nonsingular matrices it is necessary to use the partial pivoting strategy, there are some types of matrices for which it is known that no interchanges are necessary. The most important of these are diagonally dominant matrices and symmetric positive definite matrices. In both cases, it is safe to use Gaussian elimination with no interchanges at all (see the Supplementary Discussion), although for positive definite matrices accuracy is often improved slightly by using both row and column interchanges. Not needing to interchange is especially advantageous for banded matrices, and it is a fortunate fact that many banded matrices arising in applications, especially from differential equations, are either diagonally dominant or symmetric and

6.2 Errors in Gaussian Elimination

positive definite. In particular, no interchanges are needed for those tridiagonal matrices that were shown to be diagonally dominant in Chapter 5.

Ill-Conditioning

Even though the Gaussian elimination algorithm, with partial pivoting if needed, has proved to be an efficient and reliable method in practice, it may fail to compute accurate solutions of systems of equations that are "ill-conditioned." A linear system of equations is said to be *ill-conditioned* if small changes in the elements of the coefficient matrix and/or right-hand side cause large changes in the solution. In this case no numerical method can be expected to produce an accurate solution, nor, in many cases, should a solution even be attempted.

We begin with a 2×2 example. Consider the system

$$\begin{aligned} 0.832 x_1 + 0.448 x_2 &= 1.00 \\ 0.784 x_1 + 0.421 x_2 &= 0, \end{aligned} \tag{6.2.9}$$

and assume that we use a three-digit decimal computer to carry out Gaussian elimination. Since a_{11} is the largest element of the matrix no interchange is required, and the computation of the new elements $a_{22}^{(1)}$ and $b_1^{(1)}$ is

$$\begin{aligned} l_{21} &= \frac{0.784}{0.832} = 0.942 \mid 308 \cdots \doteq 0.942 \\ a_{22}^{(1)} &= 0.421 - 0.942 \times 0.448 = 0.421 - 0.422 \mid 016 \doteq -0.001 \\ b_2^{(1)} &= 0 - 1.00 \times 0.942 = -0.942, \end{aligned} \tag{6.2.10}$$

where we have indicated by the vertical bars those digits lost in the computation. Hence the computed triangular system is

$$\begin{aligned} 0.832 x_1 + 0.448 x_2 &= 1.00 \\ -0.001 x_2 &= -0.942, \end{aligned} \tag{6.2.11}$$

and the back substitution produces the approximate solution

$$x_1 = -506, \quad x_2 = 942. \tag{6.2.12}$$

But the exact solution of (6.2.9), correct to three figures, is

$$x_1 = -439, \quad x_2 = 817, \tag{6.2.13}$$

so the computed solution is incorrect by about 15%. Why has this occurred?

The first easy answer is that we have lost significance in the calculation of $a_{22}^{(1)}$. Indeed, it is clear that the computed value of $a_{22}^{(1)}$ has only one significant figure, so our final computed solution will have no more than one significant figure. But this is only the manifestation of the real problem. We invoke again the principle of backward error analysis. By carrying out a more detailed computation we can show that the computed solution (6.2.12) is the exact solution of the system

$$\begin{aligned} 0.832x_1 + 0.447974\cdots x_2 &= 1.00 \\ 0.783744\cdots x_1 + 0.420992\cdots x_2 &= 0. \end{aligned} \qquad (6.2.14)$$

The maximum percentage change between the elements of this system and the original system (6.2.9) is only 0.03%; therefore, errors in the data are magnified by a factor of about 500.

The root cause of this ill-conditioning is that the coefficient matrix of (6.2.9) is "almost singular." Geometrically, this means that the lines defined by the two equations (6.2.9) are almost parallel, as indicated in Figure 6.2.2. Consider now the system of equations

$$\begin{aligned} 0.832x_1 + 0.448x_2 &= 1.00 \\ 0.784x_1 + (0.421 + \varepsilon)x_2 &= 0. \end{aligned} \qquad (6.2.15)$$

The second equation defines a family of lines depending on the parameter ε. As ε increases from zero to approximately 0.0012, the line rotates counterclockwise and its intersection with the line defined by the first equation recedes to infinity until the two lines become exactly parallel and no solution of the linear system exists.

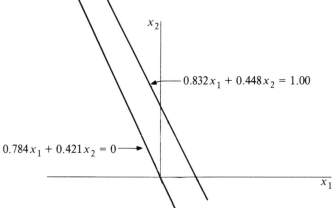

Figure 6.2.2: *Almost-Parallel Lines Defined by* (6.2.10).

For only one value of ε, say ε_0, is the coefficient matrix of (6.2.15) singular, but for infinitely many values of ε near ε_0 the matrix is almost singular. In general, the probability of a matrix being exactly singular is very small unless it was constructed in such a way that singularity is ensured. For example, we

6.2 Errors in Gaussian Elimination

saw in Section 5.3 that periodic boundary conditions can give rise to singular coefficient matrices. In many situations, however, it may not be obvious in the formulation of the problem that the resulting matrix will be singular or almost singular. This must be detected during the course of the solution and a warning issued. But, in general, it is extremely difficult to ascertain computationally if a given matrix is exactly singular. For example, if LU is the computed factorization of A and $u_{nn} = 0$, then U is singular. But u_{nn} may be contaminated by rounding error so we cannot claim that A itself is singular. Conversely, if the computed u_{nn} is not zero, this does not guarantee that A is nonsingular. The fundamental problem is the detection of zero in the presence of rounding error. However, near-singularity is easier to detect, and if that is the case, it is likely that the problem should be reformulated. For example, we may have chosen variables that are close to being dependent, and we should remove some of them or choose another set of variables.

Ramifications of Ill-Conditioning

Consider again the system (6.2.9) and suppose that (6.2.14) is the "real" system that we wish to solve but that the coefficients of this system must be measured by some physical apparatus accurate to only the third decimal place. Thus (6.2.9) is not the system that we really wish to solve but is the best approximation to it that we can make. Suppose that we can also claim that the coefficients of (6.2.9) are accurate to at least 0.05%, as indeed they are, compared with (6.2.14). Then, it is an often-heard argument that we should be able to compute the solution of the system to about the same accuracy. But we have seen that this is not true; the ill-conditioning of the coefficient matrix magnifies small errors in the coefficients by a factor of about 500 in the case of (6.2.9). Hence, no matter how accurately the system (6.2.9) is solved, we will still have the error that has come from the measurement error in the coefficients. If, for example, we need the solution of the "real" system (6.2.14) accurate to at least 1%, we need to measure the coefficients much more accurately than three decimal places.

In some cases, however, the coefficient matrix may be exact. A famous example of a class of ill-conditioned matrices is the *Hilbert matrices* (or *Hilbert segments*), in which the elements of the matrix are exact rational numbers:

$$H_n = \begin{bmatrix} 1 & 1/2 & \cdots & 1/n \\ 1/2 & & & \\ \vdots & & & \vdots \\ 1/n & & \cdots & 1/(2n-1) \end{bmatrix}. \quad (6.2.16)$$

These matrices are increasingly ill-conditioned as n increases. If for $n = 8$ the coefficients are entered in the computer as binary fractions exact to the extent possible with 27 binary digits (equivalent to about 8 decimal digits), the exact inverse of the matrix in the computer differs from the exact inverse of H_8 in the first figure!

The following is another manifestation of ill-conditioning. Suppose that $\bar{\mathbf{x}}$ is a computed solution of the system $A\mathbf{x} = \mathbf{b}$. One way to try to ascertain the accuracy of $\bar{\mathbf{x}}$ is to form the *residual vector*,

$$\mathbf{r} = \mathbf{b} - A\bar{\mathbf{x}}. \qquad (6.2.17)$$

If $\bar{\mathbf{x}}$ were the exact solution, then $\mathbf{r}$ would be zero. Thus we would expect $\mathbf{r}$ to be "small" if $\bar{\mathbf{x}}$ were a good approximation to the exact solution, and, conversely, that if $\mathbf{r}$ were small, then $\bar{\mathbf{x}}$ would be a good approximation. This is true in some cases, but if A is ill-conditioned, the magnitude of $\mathbf{r}$ can be very misleading. As an example, consider the system

$$\begin{array}{l} 0.780x_1 + 0.563x_2 = 0.217 \\ 0.913x_1 + 0.659x_2 = 0.254, \end{array} \qquad (6.2.18)$$

and the approximate solution

$$\bar{\mathbf{x}} = \begin{bmatrix} 0.341 \\ -0.087 \end{bmatrix}. \qquad (6.2.19)$$

Then, the residual vector is

$$\mathbf{r} = \begin{bmatrix} 10^{-6} \\ 0 \end{bmatrix}. \qquad (6.2.20)$$

Now consider another very different approximate solution

$$\bar{\mathbf{x}} = \begin{bmatrix} 0.999 \\ -1.001 \end{bmatrix}, \qquad (6.2.21)$$

and the corresponding residual vector

$$\mathbf{r} = \begin{bmatrix} 0.0013 \cdots \\ -0.0015 \cdots \end{bmatrix}. \qquad (6.2.22)$$

By comparing the residuals (6.2.20) and (6.2.22) we could easily conclude that (6.2.19) is the better approximate solution. However, the exact solution of (6.2.18) is $(1, -1)$, so the residuals give completely misleading information.

Determinants and Ill-conditioning

Since a matrix is singular if its determinant is zero, it is sometimes suggested that the smallness of the determinant is a measure of ill-conditioning. This is not, however, generally true as the following example shows:

$$\det \begin{bmatrix} 10^{-10} & 0 \\ 0 & 10^{-10} \end{bmatrix} = 10^{-20}, \quad \det \begin{bmatrix} 10^{10} & 0 \\ 0 & 10^{10} \end{bmatrix} = 10^{20}. \qquad (6.2.23)$$

6.2 Errors in Gaussian Elimination

The values of the two determinants are very different, but the lines defined by the two corresponding sets of equations

$$\begin{array}{ll} 10^{-10}x_1 = 0 & 10^{10}x_1 = 0 \\ 10^{-10}x_2 = 0 & 10^{10}x_2 = 0 \end{array} \quad (6.2.24)$$

are the same and are the coordinate axes. As we shall see more clearly in a moment, if the lines defined by the equations of a system are perpendicular, the system is "perfectly conditioned." Thus the magnitude of the determinant of the coefficient matrix is not a good measure of the near-singularity of the matrix. It can, however, become the basis of such a measure if the matrix is suitably scaled, as we shall now see.

For two equations it is clear that a good measure of the "almost parallelness" of the corresponding two lines is the angle between them. An essentially equivalent measure is the area of the parallelogram shown in Figure 6.2.3, in which the sides of the parallelogram are of length 1 and the height is denoted by h. The area of the parallelogram is then equal to h since the base is 1, and the angle θ between the lines defined by the two equations is related to h by $h = \sin\theta$. The area, h, varies between zero, when the lines coalesce, and 1, when they are perpendicular.

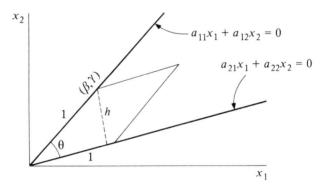

Figure 6.2.3: *The Unit Parallelogram*

From analytic geometry, the distance from the point (β, γ) to the line $a_{21}x_1 + a_{22}x_2 = 0$ is

$$h = \frac{|a_{21}\beta + a_{22}\gamma|}{\alpha_2}, \quad \alpha_2 = (a_{21}^2 + a_{22}^2)^{1/2}.$$

If we assume that $a_{11} \geq 0$, the coordinates (β, γ) are given by

$$\beta = \frac{-a_{12}}{\alpha_1}, \quad \gamma = \frac{a_{11}}{\alpha_1}, \quad \alpha_1 = (a_{11}^2 + a_{12}^2)^{1/2},$$

so that
$$h = \frac{|a_{11}a_{22} - a_{21}a_{12}|}{\alpha_1\alpha_2} = \frac{|\det A|}{\alpha_1\alpha_2}. \tag{6.2.25}$$

Hence, the area h is just the determinant divided by the product $\alpha_1\alpha_2$. This measure easily extends to n equations. Let $A = (a_{ij})$ be the coefficient matrix, and set

$$V = \frac{|\det A|}{\alpha_1\alpha_2\cdots\alpha_n} = \left|\det\begin{bmatrix} a_{11}/\alpha_1 & \cdots & a_{1n}/\alpha_1 \\ \vdots & & \vdots \\ a_{n1}/\alpha_n & \cdots & a_{nn}/\alpha_n \end{bmatrix}\right|, \tag{6.2.26}$$

where
$$\alpha_i = (a_{i1}^2 + a_{i2}^2 + \cdots + a_{in}^2)^{1/2}.$$

We have called the quantity in (6.2.26) V instead of h because it is the volume of the n-dimensional unit parallelepiped circumscribed by the lines defined by the rows of matrix A; that is,

$$\frac{1}{\alpha_i}(a_{i1}, a_{i2}, \ldots, a_{in}), \qquad i = 1, \ldots, n,$$

are the coordinates of n points in n-dimensional space located Euclidean distance 1 from the origin, and these n points define a parallelepiped whose sides are of length 1. It is intuitively clear, and can be proved rigorously, that the volume of this parallelepiped is between zero, when two or more of the edges coincide, and 1, when the edges are all mutually perpendicular. If $V = 0$, then $\det A = 0$, and the matrix is singular. If $V = 1$, then the edges are as far from being parallel as possible, and in this case the matrix is called *perfectly conditioned*.

Condition Numbers Based on Norms

We now consider another way of measuring the ill-conditioning of a matrix by means of norms (see Section 2.3 for a review of vector and matrix norms). Suppose first that $\hat{\mathbf{x}}$ is the solution of $A\mathbf{x} = \mathbf{b}$ and that $\hat{\mathbf{x}} + \Delta\mathbf{x}$ is the solution of the system with the right-hand side $\mathbf{b} + \Delta\mathbf{b}$:

$$A(\hat{\mathbf{x}} + \Delta\mathbf{x}) = \mathbf{b} + \Delta\mathbf{b}. \tag{6.2.27}$$

Since $A\hat{\mathbf{x}} = \mathbf{b}$, it follows that $A(\Delta\mathbf{x}) = \Delta\mathbf{b}$ and $\Delta\mathbf{x} = A^{-1}(\Delta\mathbf{b})$, assuming, as usual, that A is nonsingular. Thus

$$||\Delta\mathbf{x}|| \leq ||A^{-1}||\,||\Delta\mathbf{b}||, \tag{6.2.28}$$

which shows that the change in the solution due to a change in the right-hand side is bounded by $||A^{-1}||$. Thus, a small change in $\mathbf{b}$ may cause a large change

6.2 Errors in Gaussian Elimination

in $\hat{\mathbf{x}}$ if $||A^{-1}||$ is large. The notion of "large" is always relative, however, and it is more useful to deal with the relative change $||\Delta \mathbf{x}||/||\hat{\mathbf{x}}||$. From $A\hat{\mathbf{x}} = \mathbf{b}$, it follows that

$$||\mathbf{b}|| \leq ||A||\, ||\hat{\mathbf{x}}||,$$

and combining this with (6.2.28) yields

$$||\Delta \mathbf{x}||\, ||\mathbf{b}|| \leq ||A||\, ||A^{-1}||\, ||\Delta \mathbf{b}||\, ||\hat{\mathbf{x}}||,$$

or, equivalently (if $\mathbf{b} \neq 0$),

$$\frac{||\Delta \mathbf{x}||}{||\hat{\mathbf{x}}||} \leq ||A||\, ||A^{-1}|| \frac{||\Delta \mathbf{b}||}{||\mathbf{b}||}. \tag{6.2.29}$$

This inequality shows that the relative change in $\hat{\mathbf{x}}$ is bounded by the relative change in $\mathbf{b}$, multiplied by $||A||\, ||A^{-1}||$. The latter quantity is of great importance and is called the *condition number* of A (with respect to the norm being used); it will be denoted by $\text{cond}(A)$. This is the condition number for the problem of solving $A\mathbf{x} = \mathbf{b}$ or computing A^{-1}. Other problems will have different condition numbers.

Consider next the case in which the elements of A are changed so that the perturbed equations are

$$(A + \delta A)(\hat{\mathbf{x}} + \delta \mathbf{x}) = \mathbf{b}. \tag{6.2.30}$$

Thus, since $A\hat{\mathbf{x}} = \mathbf{b}$,

$$A\delta \mathbf{x} = \mathbf{b} - A\hat{\mathbf{x}} - \delta A(\hat{\mathbf{x}} + \delta \mathbf{x}) = -\delta A(\hat{\mathbf{x}} + \delta \mathbf{x}),$$

or

$$-\delta \mathbf{x} = A^{-1} \delta A(\hat{\mathbf{x}} + \delta \mathbf{x}).$$

Therefore,

$$||\delta \mathbf{x}|| \leq ||A^{-1}||\, ||\delta A||\, ||\hat{\mathbf{x}} + \delta \mathbf{x}|| = \text{cond}(A) \frac{||\delta A||}{||A||} ||\hat{\mathbf{x}} + \delta \mathbf{x}||,$$

so that

$$\frac{||\delta \mathbf{x}||}{||\hat{\mathbf{x}} + \delta \mathbf{x}||} \leq \text{cond}(A) \frac{||\delta A||}{||A||}. \tag{6.2.31}$$

Once again, the condition number plays a major role in the bound. Note that (6.2.31) expresses the change in $\hat{\mathbf{x}}$ relative to the perturbed solution, $\hat{\mathbf{x}} + \delta \mathbf{x}$, rather than $\hat{\mathbf{x}}$ itself, as in (6.2.29), although it is possible to obtain a bound relative to $\hat{\mathbf{x}}$ (Exercise 6.2.20).

The inequalities (6.2.29) and (6.2.31) need to be interpreted correctly. Note first that $\text{cond}(A) \geq 1$ (see Exercise 6.2.12). If $\text{cond}(A)$ is close to 1, then small relative changes in the data can lead to only small relative changes in the solution. In this case we say that the problem is *well-conditioned*. This also guarantees that the residual vector provides a valid estimate of the accuracy of an approximate solution $\bar{\mathbf{x}}$. From (6.2.17)

$$\mathbf{r} = A(A^{-1}\mathbf{b} - \bar{\mathbf{x}}), \qquad (6.2.32)$$

so that, if $\mathbf{e} = A^{-1}\mathbf{b} - \bar{\mathbf{x}}$ is the error in the approximate solution,

$$\mathbf{e} = A^{-1}\mathbf{r}. \qquad (6.2.33)$$

This is the fundamental relation between the residual and the error. Then

$$||\mathbf{e}|| \leq ||A^{-1}||\,||\mathbf{r}|| = \text{cond}(A)\frac{||\mathbf{r}||}{||A||}, \qquad (6.2.34)$$

so that the error is bounded by $\text{cond}(A)$ times a normalized residual vector. Note that it is necessary to normalize the residual vector somehow since we can multiply the equation $A\mathbf{x} = \mathbf{b}$ by any constant without changing the solution, but such a multiplication would change the residual by the same amount. The estimate (6.2.34) shows that if $\text{cond}(A)$ and $||\mathbf{r}||/||A||$ are both small, then the error is also small. On the other hand, from $\mathbf{r} = A\mathbf{e}$, we obtain

$$\frac{||\mathbf{r}||}{||A||} \leq ||\mathbf{e}||$$

so that if $||\mathbf{r}||/||A||$ is large, so is the error.

If the condition number is large, then small changes in the data may cause large changes in the solution, but not necessarily, depending on the particular perturbation. The practical effect of a large condition number depends on the accuracy of the data and the word length of the computer being used. If, for example, $\text{cond}(A) = 10^6$, then the equivalent of 6 decimal digits could possibly be lost. On a computer with a word length equivalent to 8 decimal digits, this could be disastrous; on the other hand if the word length were the equivalent of 14 decimal digits, it might not cause much of a problem. If the data are measured quantities, however, the computed solution may not have any meaning even if computed accurately.

Computation of the Condition Number

In general, it is very difficult to compute the condition number $||A||\,||A^{-1}||$ without knowing A^{-1}, although the packages LINPACK and LAPACK (see the Supplementary Discussion of Section 6.1) are able to estimate $\text{cond}(A)$ in the course of solving a linear system. In some cases of interest, however, it is relatively easy to compute the condition number explicitly, and we give an example of this for the $(2, -1)$ tridiagonal matrix of (5.3.10).

6.2 Errors in Gaussian Elimination

As given in Section 2.3, the l_2 norm of a symmetric matrix is its spectral radius $\rho(A)$. Thus

$$\operatorname{cond}_2(A) = ||A||_2 ||A^{-1}||_2 = \rho(A)\rho(A^{-1}). \tag{6.2.35}$$

For the $(2, -1)$ matrix of (5.3.10), we can compute explicitly (Exercise 5.3.5) the eigenvalues as

$$\lambda_k = 2 - 2\cos\frac{k\pi}{n+1} = 2 - 2\cos kh, \tag{6.2.36}$$

where we have set $h = \pi/(n+1)$. Thus, the largest eigenvalue of A is

$$\rho(A) = \lambda_n = 2 - 2\cos nh,$$

and the smallest is

$$\lambda_1 = 2 - 2\cos h > 0.$$

Since the eigenvalues of A^{-1} are $\lambda_1^{-1}, \ldots, \lambda_n^{-1}$ (Theorem 2.2.4), the spectral radius of A^{-1} is λ_1^{-1}. Thus

$$\operatorname{cond}_2(A) = \lambda_n \lambda_1^{-1} = \frac{1 - \cos nh}{1 - \cos h} = \frac{1 + \cos h}{1 - \cos h}.$$

For small h we can approximate the cosine by the first-order Taylor expansion

$$\cos h \doteq 1 - \frac{h^2}{2}.$$

Thus,

$$\operatorname{cond}_2(A) \doteq \frac{4 - h^2}{h^2} = O(h^{-2}) = O(n^2),$$

which shows that the condition number grows approximately as the square of the dimension of the matrix, or as h^{-2}. This is typical for matrices arising from boundary value problems.

Supplementary Discussion and References: 6.2

1. The interchange of rows required by partial pivoting need not be done explicitly. Instead, the interchanges may be carried out implicitly by using a permutation vector that keeps track of which rows are interchanged. Whether one should use explicit interchanges depends on the computer's "interchange" time, time required for indexing, program clarity, and other considerations.

2. In those cases in which partial pivoting is not sufficient to guarantee accuracy, we can use another strategy called *complete pivoting*, in which both rows and columns are interchanged so as to bring into the diagonal divisor position the largest element in absolute value in the remaining submatrix to be processed. This adds a significant amount of time to the Gaussian elimination process and is rarely incorporated into a standard program. See Wilkinson [1961] for further discussion.

3. In a very important paper, Wilkinson [1961] showed that the effect of rounding errors in Gaussian elimination is such that the computed solution is the exact solution of a perturbed system $(A + E)\mathbf{x} = \mathbf{b}$ (see also, for example, Golub and Van Loan [1989] and Ortega [1990] for textbook discussions). A bound on the matrix E is of the form

$$||E||_\infty \leq p(n)g(n)\varepsilon ||A||_\infty,$$

where $p(n)$ is a cubic polynomial in the size of the matrix, ε is the basic rounding error of the computer (for example, 2^{-27}), and $g(n)$ is the *growth factor* defined by

$$g = \frac{\max_{i,j,k} |a_{ij}^{(k)}|}{a}, \qquad a = \max_{i,j} |a_{ij}|,$$

where the $a_{ij}^{(k)}$ are the elements of the successive reduced matrices formed in the elimination process. The growth factor depends crucially on the interchange strategy used. With no interchanges, g may be arbitrarily large. With partial pivoting (and in exact arithmetic), $g(n)$ is bounded by 2^{n-1}, which for large n completely dominates $p(n)$. Wilkinson has exhibited matrices for which $g(n) = 2^{n-1}$ (see Exercise 6.2.25), but such matrices seem to be very rare in practice; indeed, the actual size of g has been monitored extensively by Wilkinson and others for a large number of practical problems and has seldom exceeded 10, regardless of the size of the matrix. For the complete pivoting strategy a complicated but much better bound for g was given by Wilkinson, and a long-standing conjecture was that $g(n) \leq n$. This conjecture was shown by Gould [1991] to be false if rounding error in Gaussian elimination is allowed. Subsequently Edelman and Ohlrich [1991] used Mathematica to show that the conjecture is also false in exact arithmetic. The form of the best bound for g using complete pivoting remains an open question.

For matrices that are (column) diagonally dominant, the growth factor g is bounded by 2, without any interchanges. For symmetric positive definite matrices, g is equal to 1 (see Exercise 6.2.24). This explains why for these two important classes of matrices, no interchange strategy is necessary.

6.2 Errors in Gaussian Elimination

4. The partial pivoting strategy ensures that the multipliers are all less than or equal to one in magnitude. However, for problems where the matrix is sparse it is sometimes desirable to require only that $|l_{ij}| \leq \alpha$ for some parameter $\alpha > 1$. As discussed in Section 6.1, interchanges of rows and columns can decrease the fill in sparse matrices. But interchanges for stability may conflict with interchanges to minimize fill. Generally, a compromise called *threshold pivoting* is used in which interchanges for stability are made only if the pivot element is too small, say less than .1 of the maximum element in its column. In this case, there may be several candidates for the new pivot element and the algorithm can choose the one that is best for maintaining sparsity. In a related approach for full matrices Businger [1971] discussed the following strategy. If we wish $|l_{ij}| \leq \alpha$, use partial pivoting as long as an (easily calculated) upper bound for $|a_{ij}^{(k)}|$ is less than α. If this bound fails to be less than α, then switch to complete pivoting.

5. One way to attempt to obtain an accurate solution of ill-conditioned systems – and also to detect the ill-conditioning – is *iterative refinement*, which we now describe. Let $\mathbf{x}_1$ be the computed solution of the system $A\mathbf{x} = \mathbf{b}$ and $\mathbf{r}_1 = \mathbf{b} - A\mathbf{x}_1$. If $\mathbf{x}_1$ is not the exact solution, then $\mathbf{r}_1 \neq 0$. Now solve the system $A\mathbf{z}_1 = \mathbf{r}_1$. If $\mathbf{z}_1$ were the exact solution of this system, then
$$A(\mathbf{x}_1 + \mathbf{z}_1) = A\mathbf{x}_1 + \mathbf{r}_1 = \mathbf{b},$$
so that $\mathbf{x}_1 + \mathbf{z}_1$ is the exact solution of the original system. Of course, we will not be able to compute $\mathbf{z}_1$ exactly, but we hope that $\mathbf{x}_2 = \mathbf{x}_1 + \mathbf{z}_1$ will be a better approximation to the exact solution than $\mathbf{x}_1$. For this to be the case, it is usual to compute the residual in double precision, although single precision can be useful in certain cases (see Skeel [1980] and Arioli et al. [1989]). The process can be repeated: form $\mathbf{r}_2 = \mathbf{b} - A\mathbf{x}_2$, solve $A\mathbf{z}_2 = \mathbf{r}_2$, set $\mathbf{x}_3 = \mathbf{x}_2 + \mathbf{z}_2$, and so on. One or two iterations will usually suffice to obtain an accurate solution, unless the problem is very ill-conditioned. For further discussion of iterative refinement, see Golub and Van Loan [1989].

6. For other perturbation results such as (6.2.31) see Stewart and Sun [1990].

EXERCISES 6.2

6.2.1. Suppose that the ith column of the matrix A consists of zero elements. Show that A is singular by the following different arguments:

    a. The determinant of A is zero.

b. $A\mathbf{e}_i = 0$, where $\mathbf{e}_i$ is the vector with 1 in the ith position and zeros elsewhere.

c. A has, at most, $n - 1$ linearly independent columns.

6.2.2. Let A be a matrix of the form

$$\begin{bmatrix} a_{11} & \cdots & & & a_{1n} \\ & \ddots & & & \vdots \\ & & a_{i-1,i-1} & & \\ & & a_{ii} & & \\ & & \vdots & \ddots & \\ & & a_{ni} & \cdots & a_{nn} \end{bmatrix}.$$

If $a_{ii} = a_{i+1,i} = \cdots = a_{ni} = 0$, show that A is singular.

6.2.3. Solve the following 3×3 system by Gaussian elimination by making row interchanges where needed to avoid division by zero:

$$\begin{aligned} 2x_1 + 2x_2 + 3x_3 &= 1 \\ x_1 + x_2 + 2x_3 &= 2 \\ 2x_1 + x_2 + 2x_3 &= 3 \end{aligned}$$

6.2.4. Translate the algorithm of Figure 6.2.1 into a computer program. Include back substitution.

6.2.5. Apply Gaussian elimination to the system (6.2.6) using the four-digit decimal computer of the text. Repeat the calculation after interchanging the equations.

6.2.6. a. Show that multiplication of a 4×4 matrix on the left by the permutation matrix (6.2.7) interchanges the second and fourth rows and leaves the first and third rows the same.

b. Show that multiplication on the right by the permutation matrix interchanges the second and fourth columns.

c. Give the 4×4 permutation matrix that interchanges the first and third rows and leaves the second and fourth rows the same.

6.2.7. Show that the product of two $n \times n$ permutation matrices is a permutation matrix. Show that the inverse of a permutation matrix is a permutation matrix, and that a permutation matrix is an orthogonal matrix.

6.2.8. What assumptions do you need to make so that the block LU decomposition of Exercise 6.1.32 is numerically stable?

6.2.9. A matrix H is called *Hessenberg* if $h_{ij} = 0$ when $i > j + 1$. How many operations are required to solve $H\mathbf{x} = \mathbf{b}$ by Gaussian elimination? If H is normalized so that $|h_{ij}| \leq 1$ for all i, j, and partial pivoting is used in Gaussian elimination, show that the elements of U are less than n in magnitude.

6.2 Errors in Gaussian Elimination

6.2.10. Assume that the positive definite condition $\mathbf{x}^T A \mathbf{x} > 0$ if $\mathbf{x} \neq 0$ holds for all real $\mathbf{x}$ even though A is not symmetric (such matrices are sometimes called *positive real*). Show that an LU decomposition of A exists.

6.2.11. Compute the determinant and the normalized determinant (6.2.26) for the matrix
$$A = \begin{bmatrix} 1 & 2 & 3 \\ 2 & 3 & 4 \\ 3 & 4 & 4 \end{bmatrix}.$$

6.2.12. Using properties of matrix norms, prove that $\text{cond}(A) \geq 1$.

6.2.13. Compute $\text{cond}(A)$ for the matrix in Exercise 6.2.11 using both the l_1 and l_∞ norms (see Section 2.3 for definitions of these norms).

6.2.14. Solve the system (6.2.9) for different right-hand sides. Compare the differences in these solutions to the bound (6.2.29), using the l_∞ norm.

6.2.15 Suppose that $\text{cond}_2(A) = 1$. Show that A is a scalar multiple of an orthogonal matrix.

6.2.16. Let A be the matrix of (5.3.9) in which $c_1 = c_2 = \cdots = c_n = c$. Use the fact (Theorem 2.2.3) that if A has eigenvalues λ_i, then $A + cI$ has eigenvalues $\lambda_i + c$ to show that A has eigenvalues $2 + c - 2\cos kh$, $k = 1, \ldots, n$. Use these eigenvalues to find $\text{cond}_2(A)$, and discuss how $\text{cond}_2(A)$ varies with c.

6.2.17. Let A be an $n \times n$ upper triangular matrix with 1 in every main diagonal and upper triangular position. Show that $\det A = 1$ but that the normalized determinant of (6.2.26) is $(n!)^{-1/2}$. Thus, A is very ill-conditioned. Show also that $\|A^{-1}\|_2 \doteq 2^{n-1}$.

6.2.18. Let A be the $n \times n$ matrix of (5.3.10). Show that $\det A = n + 1$ and thus the normalized determinant of (6.2.26) is $(n+1)/(5 \times \sqrt{6}^{(n-2)})$. (Hint: Let A_n be the $n \times n$ matrix. Use the cofactor expansion of Theorem 2.1.1 to establish the recurrence $\det A_n = 2\det A_{n-1} - \det A_{n-2}$.)

6.2.19. Let
$$A = \begin{pmatrix} 1.6384 & 0.8065 \\ 0.8321 & 0.4096 \end{pmatrix}, \qquad \mathbf{b} = \begin{pmatrix} 0.8319 \\ 0.4225 \end{pmatrix}.$$
Verify that $(1, -1)^T$ is the exact solution of $A\mathbf{x} = \mathbf{b}$. If $\mathbf{r} = \mathbf{b} - A\bar{\mathbf{x}}$, construct an $\bar{\mathbf{x}}$ for which $\mathbf{r} = (10^{-8}, -10^{-8})$ exactly. Find $\text{cond}_\infty(A)$. If $\mathbf{b}$ is exact, how small should the relative error in A be so that the solution can be guaranteed to have a relative error that is $\leq 10^{-8}$?

6.2.20. a. Let X be an approximate inverse of A such that if $R = I - AX$, then $\|R\| < 1$. Use Exercise 2.3.18 to show that
$$\|A^{-1}\| \leq \frac{\|X\|}{1 - \|R\|},$$

and then that
$$\|A^{-1} - X\| \le \frac{\|X\|\|R\|}{1 - \|R\|}.$$

b. Let E be such that $\|A^{-1}E\| < 1$. Show that $(A+E)^{-1}$ exists and
$$\|(A+E)^{-1} - A^{-1}\| \le \frac{\|A^{-1}E\|\|A^{-1}\|}{1 - \|A^{-1}E\|} \le \frac{\|A^{-1}\|^2\|E\|}{1 - \|A^{-1}\|\|E\|}$$
Use this result to show that if $A\mathbf{x} = \mathbf{b}$ and $(A+E)(\mathbf{x} + \delta\mathbf{x}) = \mathbf{b}$, then
$$\frac{\|\delta\mathbf{x}\|}{\|\mathbf{x}\|} \le \frac{\text{cond}(A)\alpha}{1 - \text{cond}(A)\alpha},$$
where $\alpha = \|E\|/\|A\|$.

6.2.21. Let A be a nonsingular diagonal matrix. Show that the quantity V of (6.2.26) is always equal to 1, but that $\|A\|\|A^{-1}\|$ may be arbitrarily large.

6.2.22. For the system $A\mathbf{x} = \mathbf{b}$, where A is $m \times n$ with $m \le n$ and rank $(A) = r$, show how to use Gaussian elimination with row interchanges to obtain a solution, if one exists.

6.2.23. Let $\hat{A} = A - \mathbf{e}_1\mathbf{e}_1^T$, where A is the $n \times n$ matrix of (5.3.10). Show that $\hat{A} = LL^T$, where
$$L = \begin{bmatrix} 1 & & & \\ -1 & \ddots & & \\ & \ddots & & \\ & & -1 & 1 \end{bmatrix}.$$
Conclude that $\det \hat{A} = 1$, for any n.

6.2.24. Let A be symmetric positive definite with $a = \max |a_{ij}|$. Let Gaussian elimination be carried out without interchanges and let $A^{(k)}$ be the submatrix of order $n - k + 1$ upon which the kth step will operate. Show that $A^{(k)}$ is symmetric positive definite for $k = 2, \ldots, n-1$. If $a_k = \max_{ij} |a_{ij}^{(k)}|$, show that $a_k \le a$ for $k = 2, \ldots, n-1$. (Thus, in terms of the Supplementary Discussion, the growth factor $g \le 1$.)

6.2.25. Let
$$A_5 = \begin{bmatrix} 1 & 0 & 0 & 0 & 1 \\ 1 & 1 & 0 & 0 & -1 \\ -1 & 1 & 1 & 0 & 1 \\ 1 & -1 & 1 & 1 & -1 \\ -1 & 1 & -1 & 1 & 1 \end{bmatrix},$$
and let A_n be the corresponding matrix of order n. Show that the growth factor (see the Supplementary Discussion) for Gaussian elimination with partial pivoting is 2^{n-1}.

6.2.26. Suppose that $E = \alpha A, |\alpha| < 1$. Show that the solutions of $A\mathbf{x} = \mathbf{b}$ and $(A + E)\bar{\mathbf{x}} = \mathbf{b}$ satisfy

$$\|\mathbf{x} - \bar{\mathbf{x}}\| \leq |\alpha| \|\mathbf{x}\|/(1 - |\alpha|).$$

(Note the absence of cond(A).)

6.2.27. Consider the matrices

$$A = \begin{bmatrix} 1 & -1 \\ 1 & -1.00001 \end{bmatrix}, \quad B = \begin{bmatrix} 1 & -1 \\ -1 & 1.00001 \end{bmatrix}$$

Show that the ratio of maximum to minimum eigenvalue is about 1 for A but about $4 \cdot 10^5$ for B. Show, however, that $\text{cond}_2(A) = \text{cond}_2(B)$. Conclude that the ratio of maximum to minimum eigenvalue is not a good condition number for nonsymmetric matrices. Is A well-conditioned or ill-conditioned?

6.3 Other Factorizations

So far in this chapter we have considered only Gaussian elimination and the corresponding LU factorization. But there are other factorizations of the matrix A that are sometimes very useful.

Cholesky Factorization

In the case of a symmetric positive definite matrix there is an important variant of the LU factorization called the *Cholesky factorization* (or *decomposition*):

$$A = LL^T. \tag{6.3.1}$$

Here L is a lower-triangular matrix but does not necessarily have 1's on the main diagonal as in the LU factorization. The factorization (6.3.1) is unique, if L is required to have positive diagonal elements (see Exercise 6.3.1).

The product in (6.3.1) is

$$\begin{bmatrix} l_{11} & & & & \\ \vdots & \ddots & & & \\ l_{il} & & l_{ii} & & \\ \vdots & & & \ddots & \\ l_{n1} & \cdots & & & l_{nn} \end{bmatrix} \begin{bmatrix} l_{11} & \cdots & l_{i1} & \cdots & l_{n1} \\ & \ddots & & & \\ & & l_{ii} & & \vdots \\ & & & \ddots & \\ & & & & l_{nn} \end{bmatrix}. \tag{6.3.2}$$

By equating elements of the first column of (6.3.2) with corresponding elements of A, we see that $a_{i1} = l_{i1} l_{11}$, so the first column of L is determined by

$$l_{11} = (a_{11})^{1/2}, \quad l_{i1} = \frac{a_{i1}}{l_{11}}, \quad i = 2, \ldots, n. \tag{6.3.3}$$

In general,

$$a_{ii} = \sum_{k=1}^{i} l_{ik}^2, \qquad a_{ij} = \sum_{k=1}^{j} l_{ik} l_{jk}, \qquad j < i, \qquad (6.3.4)$$

which forms the basis for determining the columns of L in sequence. Once L is computed, the solution of the linear system can proceed just as in (6.1.15) for the LU decomposition: solve $L\mathbf{y} = \mathbf{b}$ and then solve $L^T\mathbf{x} = \mathbf{y}$. The algorithm for the factorization is given in Figure 6.3.1. For the Cholesky decomposition to be carried out, it is necessary that the quantities $a_{jj} - \sum l_{jk}^2$ all be positive so that the square roots may be taken. If the coefficient matrix A is positive definite, these are indeed positive; moreover, the algorithm is numerically stable.

For $j = 1, \ldots, n$

$$l_{jj} = \left(a_{jj} - \sum_{k=1}^{j-1} l_{jk}^2\right)^{1/2}$$

For $i = j+1, \ldots, n$

$$l_{ij} = \frac{a_{ij} - \sum_{k=1}^{j-1} l_{ik} l_{jk}}{l_{jj}}$$

Figure 6.3.1: *Cholesky Factorization*

The Cholesky factorization enjoys two advantages over LU factorization. First, there are approximately half as many arithmetic operations (Exercise 6.3.2). Square roots are also required in the Cholesky factorization, although there is a variant of the algorithm which avoids these (see Exercise 6.3.3). The second advantage is that by utilizing symmetry only the lower triangular part of A needs to be stored. As with the LU factorization, the l_{ij} can be overwritten onto the corresponding portions of A as they are computed. Finally, the Cholesky factorization extends readily to banded matrices and preserves the bandwidth, just as LU factorization without interchanges (see Exercise 6.3.4). More generally, Cholesky factorization has exactly the same properties with respect to fill as LU factorization; in particular, the L in the LU factorization of a symmetric positive definite matrix and the L in Cholesky factorization have exactly the same non-zero structure (see Exercise 6.3.3).

6.3 Other Factorizations

The QR Factorization

The Cholesky factorization applies only to symmetric positive definite matrices. We next consider a factorization that applies to any matrix. Indeed, later in this section we will use this factorization for rectangular matrices, but for the moment we will assume that A is $n \times n$ and real. The *QR factorization* (or *decomposition* or *reduction*) is then

$$A = QR, \qquad (6.3.5)$$

where Q is an orthogonal matrix (see Chapter 2) and R is upper triangular. (We could denote R by U, but historically R has always been used in this context.)

We will obtain the matrix Q as a product of simpler orthogonal matrices based on the rotation matrix

$$\begin{bmatrix} \cos\theta & \sin\theta \\ -\sin\theta & \cos\theta \end{bmatrix}. \qquad (6.3.6)$$

We generalize such rotation matrices to $n \times n$ matrices of the form

$$P_{ij} = \begin{bmatrix} 1 & & & & & & & & \\ & \ddots & & & & & & & \\ & & 1 & & & & & & \\ & & & c_{ij} & & s_{ij} & & & \\ & & & & 1 & & & & \\ & & & & & \ddots & & & \\ & & & & & & 1 & & \\ & & & -s_{ij} & & c_{ij} & & & \\ & & & & & & & 1 & \\ & & & & & & & & \ddots \\ & & & & & & & & & 1 \end{bmatrix}, \qquad (6.3.7)$$

where $c_{ij} = \cos\theta_{ij}$ and $s_{ij} = \sin\theta_{ij}$ are located in the ith and jth rows and columns, as indicated. Such matrices are called *plane rotation matrices* or *Givens* or *Jacobi transformations*. Just as (6.3.6) defines a rotation of the plane, a matrix of the form (6.3.7) defines a rotation in the (i,j) plane in n-space. It is easy to show (Exercise 6.3.5) that the matrices (6.3.6) and (6.3.7) are orthogonal.

We now use the matrices P_{ij} in the following way to achieve the QR factorization (6.3.5). Let $\mathbf{a}_i$ denote the ith row of A. Then multiplication of A

by P_{12} gives the matrix

$$A_1 = P_{12}A = \begin{bmatrix} c_{12}\mathbf{a}_1 + s_{12}\mathbf{a}_2 \\ -s_{12}\mathbf{a}_1 + c_{12}\mathbf{a}_2 \\ \mathbf{a}_3 \\ \vdots \\ \mathbf{a}_n \end{bmatrix}. \quad (6.3.8)$$

Note that only the first two rows of A are changed by this multiplication. If we choose s_{12} and c_{12} so that

$$-a_{11}s_{12} + a_{21}c_{12} = 0, \quad (6.3.9)$$

then A_1 has a zero in the $(2,1)$ position, and the other elements in the first two rows of A_1 differ, in general, from those of A. We do not actually compute the angle θ_{12} to achieve (6.3.9) since we can obtain the desired sine and cosine directly by

$$c_{12} = a_{11}(a_{11}^2 + a_{21}^2)^{-1/2}, \quad s_{12} = a_{21}(a_{11}^2 + a_{21}^2)^{-1/2}. \quad (6.3.10)$$

The denominator in (6.3.10) is non-zero unless both a_{11} and a_{21} are zero; but if $a_{21} = 0$, this step can be bypassed since a zero is already in the desired position.

The transformation (6.3.8) is analogous to the first step of Gaussian elimination in which a multiple of the first row of A is subtracted from the second row to achieve a zero in the $(2,1)$ position. We next proceed, as in Gaussian elimination, to obtain zeros in the remaining positions of the first column. We form $A_2 = P_{13}A$, which modifies the first and third rows of A_1 while leaving all other rows the same; in particular, the zero produced in the first stage in the $(2,1)$ position remains unchanged. The elements c_{13} and s_{13} of P_{13} are chosen analogously to (6.3.10) so that the $(3,1)$ element of A_2 is zero. We continue in this fashion, zeroing the remaining elements in the first column one after another and then zeroing elements in the second column in the order $(3,2), (4,2), \ldots, (n,2)$, and so on. In all, we will use $(n-1) + (n-2) + \cdots + 1$ plane rotation matrices P_{ij}, and the result is that

$$PA \equiv P_{n-1,n} \cdots P_{12}A = R \quad (6.3.11)$$

is upper triangular. Since the product of orthogonal matrices is orthogonal (Exercise 2.1.14) the product P is orthogonal. Also, the inverse of an orthogonal matrix is orthogonal (Exercise 2.1.13) so that if we set $Q = P^{-1}$, then Q is orthogonal, and multiplying (6.3.11) by Q gives (6.3.5).

We next count the operations to carry out the above QR factorization. The majority of the work is in modifying the elements of the two rows that

6.3 Other Factorizations

are changed at each rotation. From (6.3.8), modification of the first two rows requires $4n$ multiplications and $2n$ additions. (For simplicity we have also counted the operations used to produce the zero in the $(2,1)$ position even though they do not need to be performed.) The same count is true for producing zeros in the remaining $n-2$ elements in the first column. Hence, the first stage requires $4n(n-1)$ multiplications and $2n(n-1)$ additions. In each of the subsequent $n-2$ stages, n decreases by 1 in this count so that the total is

$$4\sum_{k=2}^{n} k(k-1) \text{ mult } + 2\sum_{k=2}^{n} k(k-1) \text{ add } \doteq \frac{4}{3}n^3 \text{ mult } + \frac{2}{3}n^3 \text{ add}, \quad (6.3.12)$$

where we have used the summation formulas of Exercise 6.1.4. Computation of the c_{ij} and s_{ij} is also needed, but this requires only $O(n^2)$ operations. Thus, (6.3.12) shows that this QR factorization requires approximately four times the multiplications and twice the additions of LU factorization (see Section 6.1). We will discuss the relative merits of QR and LU factorization shortly, but first we show that by using other orthogonal transformations the QR factorization can be computed more economically.

Householder Transformations

A *Householder transformation* is a matrix of the form $I - 2\mathbf{ww}^T$, where $\mathbf{w}^T\mathbf{w} = 1$. It is easy to see (Exercise 6.3.6) that such matrices are symmetric and orthogonal. They are also called *elementary reflection* matrices (see Exercise 6.3.6). We now show how Householder transformations can be used to obtain the QR factorization of A. Let $\mathbf{a}_1$ be the first column of A and define (see Exercise 6.3.11)

$$\mathbf{w}_1 = \mu_1 \mathbf{u}_1, \quad \mathbf{u}_1^T = (a_{11} - s_1, a_{21}, \ldots, a_{n1}), \quad (6.3.13)$$

where

$$s_1 = \pm(\mathbf{a}_1^T \mathbf{a}_1)^{1/2}, \quad \mu_1 = (2s_1^2 - 2a_{11}s_1)^{-1/2}. \quad (6.3.14)$$

The sign of s_1 must be chosen to be opposite that of a_{11} so that there is no cancellation in the computation of μ_1; otherwise, the algorithm would be numerically unstable. The vector $\mathbf{w}_1$ satisfies

$$\mathbf{w}_1^T \mathbf{w}_1 = \mu_1^2[(a_{11} - s_1)^2 + \sum_{j=2}^{n} a_{j1}^2] = \mu_1^2(a_{11}^2 - 2a_{11}s_1 + 2s_1^2 - a_{11}^2) = 1,$$

so that $P_1 = I - 2\mathbf{w}_1\mathbf{w}_1^T$ is a Householder transformation. Moreover,

$$\mathbf{w}_1^T \mathbf{a}_1 = \mu_1[(a_{11} - s_1)a_{21} + \sum_{j=2}^{n} a_{j1}^2] = \mu_1(s_1^2 - a_{11}s_1) = \frac{1}{2\mu_1},$$

262   Chapter 6 Direct Solution of Linear Equations

so that
$$a_{11} - 2w_1 \mathbf{w}_1^T \mathbf{a}_1 = a_{11} - \frac{2(a_{11} - s_1)\mu_1}{2\mu_1} = s_1,$$

and
$$a_{i1} - 2w_i \mathbf{w}_1^T \mathbf{a}_1 = a_{i1} - \frac{2a_{i1}\mu_1}{2\mu_1} = 0, \qquad i = 2, 3, \ldots, n.$$

This shows that the first column of $P_1 A$ is

$$P_1 \mathbf{a}_1 = \mathbf{a}_1 - 2\mathbf{w}_1^T \mathbf{a}_1 \mathbf{w}_1 = (s_1, 0, \ldots, 0)^T.$$

Thus, with this one orthogonal transformation zeros have been introduced into the subdiagonal positions of the first column, as was done by $n-1$ Givens transformations.

The second stage is analogous. A Householder transformation $P_2 = I - 2\mathbf{w}_2 \mathbf{w}_2^T$ is defined by a vector $\mathbf{w}_2$ whose first component is zero and whose remaining components are as in (6.3.13, 6.3.14), using now the second column of $P_1 A$ from the main diagonal element down. The matrix $P_2 P_1 A$ then has zeros below the main diagonal in each of its first two columns. We continue in this way to zero elements below the main diagonal by Householder matrices $P_i = I - 2\mathbf{w}_i \mathbf{w}_i^T$, where $\mathbf{w}_i$ has zeros in its first $i-1$ components. Thus

$$P_{n-1} \cdots P_1 A = R,$$

where R is upper triangular. The matrices P_i are all orthogonal so that $P = P_{n-1} \cdots P_1$ and P^{-1} are also orthogonal. Therefore, $Q = P^{-1}$ is the orthogonal matrix of (6.3.5).

So far, we have discussed only the formation of the vectors $\mathbf{w}_i$ that define the Householder transformations, and we next consider the remainder of the computation. If $\mathbf{a}_1, \mathbf{a}_2, \ldots, \mathbf{a}_n$ are the columns of A, then

$$P_1 A = A - 2\mathbf{w}_1 \mathbf{w}_1^T A = A - 2\mathbf{w}_1 (\mathbf{w}_1^T \mathbf{a}_1, \mathbf{w}_1^T \mathbf{a}_2, \ldots, \mathbf{w}_1^T \mathbf{a}_n). \qquad (6.3.15)$$

Thus, the ith column of $P_1 A$ is

$$\mathbf{a}_i - 2\mathbf{w}_1^T \mathbf{a}_i \mathbf{w}_1 = \mathbf{a}_i - \gamma_1 \mathbf{u}_1^T \mathbf{a}_i \mathbf{u}_1, \qquad (6.3.16)$$

where
$$\gamma_1 = 2\mu_1^2 = (s_1^2 - s_1 a_{11})^{-1}.$$

It is more efficient computationally not to form the vector $\mathbf{w}_1$ explicitly but to work with γ_1 and $\mathbf{u}_1$ as shown in (6.3.16). Analogous computations are performed to obtain the remaining reduced matrices $P_2 P_1 A, \ldots$; the complete algorithm is summarized in Figure 6.3.2.

We next count the operations in the Householder reduction. The bulk of the work is in the formation of the new columns of the reduced matrices.

6.3 Other Factorizations

> For $k = 1$ to $n - 1$
> $$s_k = -\text{sgn}(a_{kk})\left(\sum_{l=k}^{n} a_{lk}^2\right)^{1/2}$$
> $$\mathbf{u}_k^T = (0, \ldots, 0, a_{kk} - s_k, a_{k+1,k}, \ldots, a_{nk})$$
> $$\gamma_k = (s_k^2 - s_k a_{kk})^{-1}$$
> $$a_{kk} = s_k$$
> For $j = k + 1$ to n
> $$\alpha_j = \gamma_k \mathbf{u}_k^T \mathbf{a}_j$$
> $$\mathbf{a}_j = \mathbf{a}_j - \alpha_j \mathbf{u}_k$$

Figure 6.3.2: *Householder Reduction*

Referring to the inner loop in Figure 6.3.2, at the kth stage the inner product $\mathbf{u}_k^T \mathbf{a}_j$ requires $n - k + 1$ multiplications and $n - k$ additions and the operation $\mathbf{a}_j - \alpha_j \mathbf{u}_k$ requires $n - k + 1$ additions and multiplications. Since there are $n - k$ columns to update at the kth stage, this gives approximately $2(n - k)$ additions and multiplications. Summing this over all $n - 1$ stages, we obtain

$$2\sum_{k=1}^{n-1}(n-k)^2 = \frac{n(n-1)(2n-1)}{3} \doteq \frac{2}{3}n^3 \qquad (6.3.17)$$

additions and the same count for multiplications. The number of other operations is no more than order n^2. Comparing the count (6.3.17) with (6.3.12) for the QR factorization using Givens matrices, we see that while the additions are the same, the number of multiplications has been roughly halved. The number of square roots is also reduced considerably. Hence, the Householder reduction is clearly more efficient, although there are some situations in which it may be still preferable to use Givens matrices.

Both the Givens and Householder factorizations are numerically stable, and they can be considered as alternatives to Gaussian elimination with partial pivoting. However, the operation count of even the Householder reduction is approximately twice that of Gaussian elimination, and the added expense of partial pivoting does not close this gap. Moreover, both the Givens and Householder transformations expand the bandwidth of a banded matrix (Exercise 6.3.13), just as partial pivoting does. Hence, the QR factorization is rarely used for the solution of nonsingular systems of equations, although it is a key part of some of the best algorithms for computing eigenvalues. It is also useful for dealing with rectangular matrices in least squares problems. We next indicate this application.

Application to Least Squares Problems

Recall from Section 4.2 that in the least squares problem (4.2.12), we wish to minimize over $a_0, \ldots, a_n$ the function

$$\sum_{i=1}^{m} \left[\sum_{j=0}^{n} a_j \phi_{ij} - f_i \right]^2, \quad (6.3.18)$$

where $\phi_{ij} = \phi_j(x_i)$ and where for simplicity we have taken the weights $w_i = 1$. If E is the matrix with elements ϕ_{ij}, then we can write (6.3.18) as (Exercise 6.3.12)

$$(E\mathbf{a} - \mathbf{f})^T (E\mathbf{a} - \mathbf{f}) = ||E\mathbf{a} - \mathbf{f}||_2^2. \quad (6.3.19)$$

Now suppose that we apply the QR factorization to the (generally rectangular) $m \times (n+1)$ matrix E, where $m \geq n+1$. The Householder factorization, for example, can be applied as previously described; just imagine that E is actually $m \times m$ but that we stop the process when we have completed zeroing elements in the first $n+1$ columns. Thus, we will have

$$E = Q \begin{bmatrix} R \\ 0 \end{bmatrix} \quad \text{or} \quad PE = \begin{bmatrix} R \\ 0 \end{bmatrix} = \hat{R}, \quad (6.3.20)$$

where R is an $(n+1) \times (n+1)$ upper triangular matrix and $P = Q^{-1}$. Since $||P\mathbf{x}||_2 = ||\mathbf{x}||_2$ for any vector $\mathbf{x}$ (Exercise 6.3.8), we have

$$||E\mathbf{a} - \mathbf{f}||_2 = ||P(E\mathbf{a} - \mathbf{f})||_2 = ||\hat{R}\mathbf{a} - \hat{\mathbf{f}}||_2, \quad (6.3.21)$$

where we have set $\hat{\mathbf{f}} = P\mathbf{f}$. If we partition $\hat{\mathbf{f}}$ commensurately with $\hat{R}$, we have

$$\hat{R}\mathbf{a} - \hat{\mathbf{f}} = \begin{bmatrix} R \\ 0 \end{bmatrix} \mathbf{a} - \begin{bmatrix} \hat{\mathbf{f}}_1 \\ \hat{\mathbf{f}}_2 \end{bmatrix} = \begin{bmatrix} R\mathbf{a} - \hat{\mathbf{f}}_1 \\ -\hat{\mathbf{f}}_2 \end{bmatrix}. \quad (6.3.22)$$

Thus,

$$||E\mathbf{a} - \mathbf{f}||_2^2 = ||R\mathbf{a} - \hat{\mathbf{f}}_1||_2^2 + ||\hat{\mathbf{f}}_2||_2^2, \quad (6.3.23)$$

and it is clear that $||E\mathbf{a} - \mathbf{f}||_2$ is minimized when $||R\mathbf{a} - \hat{\mathbf{f}}_1||_2$ is minimized since $\hat{\mathbf{f}}_2$ is fixed.

Now recall from Section 4.2 that the condition for (6.3.18) to have a unique minimizer is that the matrix $E^T E$ be nonsingular, which we assume. But from (6.3.20)

$$E^T E = \hat{R}^T Q^T Q \hat{R} = \hat{R}^T \hat{R} = R^T R. \quad (6.3.24)$$

6.3 Other Factorizations

Thus $R^T R$ is also nonsingular, and since R is square, R itself is nonsingular. (Note that if the signs of the diagonal elements of R are chosen to be positive, then $R^T R$ is the Cholesky factorization of $E^T E$.) Therefore, the system $R\mathbf{a} = \hat{\mathbf{f}}_1$ has a unique solution $\hat{\mathbf{a}}$. For this $\hat{\mathbf{a}}$ (6.3.23) is minimized and hence so is (6.3.18). We summarize the overall algorithm as:

Find the QR factorization (6.3.20), (6.3.25a)

Form the vector $P\mathbf{f}$ to obtain $\hat{\mathbf{f}}_1$, (6.3.25b)

Solve the system $R\mathbf{a} = \hat{\mathbf{f}}_1$. (6.3.25c)

This algorithm is much more numerically stable than forming $E^T E$ explicitly and then solving the normal equations. The number of operations, however, is somewhat higher (Exercise 6.3.15).

Fast Poisson Solvers

Let A be a symmetric matrix with eigenvalues $\lambda_1, \ldots, \lambda_n$ and corresponding orthonormal eigenvectors $\mathbf{q}_1, \ldots, \mathbf{q}_n$. Then a consequence of Theorem 2.2.8 is that A has the (spectral) factorization

$$A = QDQ^T, \qquad (6.3.26)$$

where $D = \text{diag}(\lambda_1, \ldots, \lambda_n)$ and Q is the orthogonal matrix $Q = (\mathbf{q}_1, \ldots, \mathbf{q}_n)$. Thus, the solution of the system $A\mathbf{x} = \mathbf{b}$ is

$$\mathbf{x} = A^{-1}\mathbf{b} = QD^{-1}Q^T\mathbf{b}. \qquad (6.3.27)$$

In general, this approach to solving $A\mathbf{x} = \mathbf{b}$ is not practical since we require all the eigenvalues and eigenvectors of A, and the computation of this eigensystem will be a much more formidable task than the solution of $A\mathbf{x} = \mathbf{b}$. However, in the important special case in which A is the matrix (5.5.18) obtained by discretizing Poisson's equation, the eigenvalues and eigenvectors are known explicitly (see Section 5.5). For simplicity in the following exposition, we will restrict attention to the one-dimensional Poisson problem for which the coefficient matrix A is the $(2, -1)$ matrix (5.3.10). For this matrix, the eigenvalues and eigenvectors are (Exercise 5.3.5)

$$\lambda_k = 2 - 2\cos kh, \quad k = 1, \ldots, n, \qquad (6.3.28a)$$

$$\mathbf{q}_k = c(\sin kh, \sin 2kh, \ldots, \sin nkh)^T, \quad k = 1, \ldots, n, \qquad (6.3.28b)$$

where $h = \pi/(n+1)$, and c is such that $\mathbf{q}_k^T \mathbf{q}_k = 1$.

266 Chapter 6 Direct Solution of Linear Equations

The formation of $Q^T\mathbf{b}$ in (6.3.27) requires the inner products of $\mathbf{b}$ with the vectors $\mathbf{q}_k$:

$$\mathbf{q}_k^T\mathbf{b} = c\sum_{j=1}^{n} b_j \sin jkh, \quad k = 1,\ldots,n. \quad (6.3.29)$$

Similar summations are required for the multiplication by Q. A straightforward evaluation of (6.3.29) would be too costly, even with all the eigenvalues and eigenvectors known. However, summations of the form (6.3.29) can be performed very rapidly as we now discuss.

The Fast Fourier Transform

If $i = \sqrt{-1}$, then

$$e^{\pm ix} = \pm\sin x + i\cos x, \quad (6.3.30)$$

so that trigonometric functions may be expressed in terms of complex exponential functions by

$$\sin x = \frac{1}{2}(e^{ix} - e^{-ix}), \quad \cos x = \frac{1}{2i}(e^{ix} + e^{-ix}). \quad (6.3.31)$$

Consider now the summations

$$y_k = \sum_{j=0}^{n-1} a_j w^{jk}, \quad k = 0,\ldots,n-1, \quad (6.3.32)$$

where $w = e^{2\pi i/n}$ is an nth root of unity. Assuming that the a_j in (6.3.32) are complex numbers and the w^{jk} are all known, each summation in (6.3.32) requires n complex multiplications and $n-1$ complex additions. Each complex multiplication requires four real multiplications and two real additions (but, see Exercise 6.3.20) so that approximately $8n^2$ real operations are needed to compute all n sums in (6.3.32). By means of the Fast Fourier Transform (FFT), we will be able to compute these summations in $O(n\log_2 n)$ operations.

The summations (6.3.32) may be viewed as the matrix-vector multiply $\mathbf{y} = W\mathbf{a}$, where W is the $n \times n$ matrix with entries $w^{jk}, j,k = 0,\ldots,n-1$. Hereafter, we assume for simplicity that $n = 2^q$. An FFT then can be viewed as utilizing a factorization of W of the form

$$W = W_q \ldots W_1 P^T, \quad (6.3.33)$$

where P is a permutation matrix, and each W_j has only two non-zero entries per row. Because of the latter property, $W\mathbf{a}$ can be computed in $O(n\log_2 n)$ operations, rather than $O(n^2)$.

We next indicate how the factors W_i in (6.3.33) may be obtained. Define the $m \times m$ matrix

6.3 Other Factorizations

$$B_m = \begin{bmatrix} I_{m_1} & D_{m_1} \\ I_{m_1} & -D_{m_1} \end{bmatrix}, \tag{6.3.34a}$$

where I_{m_1} and D_{m_1} are $m_1 \times m_1$ with $m_1 = m/2$, and

$$D_{m_1} = \text{diag}(1, w_m, \ldots, w_m^{m_1-1}) \tag{6.3.34b}$$

with $w_m = e^{-2i\pi/m}$. We then define

$$W_s = \text{diag}(B_m, B_m, \ldots, B_m), \tag{6.3.35}$$

where B_m is repeated r times and $r = n/m$ with $m = 2^s$. Thus, W_1 consists of $n/2$ blocks of 2×2 matrices B_2, W_2 consists of $n/4$ blocks of 4×4 matrices B_4, and so on until $W_q = B_n$. As mentioned above, each W_j has only two non-zero elements per row, one from the identity matrix and one from D_{m_1}.

We illustrate the above construction by computing the vector $\mathbf{y}$ of (6.3.32) for $n = 8$. In this case the permuted vector $P^T \mathbf{a}$ is

$$\hat{\mathbf{a}} = P^T \mathbf{a} = (a_0, a_4, a_2, a_6, a_1, a_5, a_3, a_7)^T, \tag{6.3.36}$$

as will be discussed further shortly. The matrices D_i of (6.3.34a) are

$$D_1 = 1, \quad D_2 = \text{diag}(1, w^2), \quad D_4 = \text{diag}(1, w, w^2, w^3),$$

where $w = e^{-i\pi/4}$. Then the computed vectors are (Exercise 6.3.24)

$$\begin{aligned} W_1 \hat{\mathbf{a}} &= (a_0 + a_4, a_0 - a_4, a_2 + a_6, a_2 - a_6, \\ & \quad a_1 + a_5, a_1 - a_5, a_3 + a_7, a_3 - a_7)^T, \end{aligned} \tag{6.3.37a}$$

$$\begin{aligned} W_2(W_1 \hat{\mathbf{a}}) &= (a_0 + a_4 + a_2 + a_6, a_0 - a_4 + w^2(a_7 - a_6), a_0 + a_4 - a_2 - a_6, \\ & \quad a_0 - a_4 - w^2(a_2 - a_6), a_1 + a_5 + a_3 + a_7, a_1 - a_5 + w^2(a_3 - a_7), \\ & \quad a_1 + a_5 - a_3 - a_7, a_1 - a_5 - w^2(a_3 - a_7))^T, \end{aligned} \tag{6.3.37b}$$

$$\mathbf{y} = \begin{bmatrix} a_0 + a_4 + a_2 + a_6 + a_1 + a_5 + a_3 + a_7 \\ a_0 - a_4 + w^2(a_2 - a_6) + w(a_1 - a_5) + w^3(a_3 - a_7) \\ a_0 + a_4 - a_2 - a_6 + w^2(a_1 + a_5 - a_3 - a_7) \\ a_0 - a_4 - w^2(a_2 - a_6) + w^3(a_1 - a_5) - w^5(a_3 - a_7) \\ a_0 + a_4 + a_2 + a_6 - a_1 - a_5 - a_3 - a_7 \\ a_0 - a_4 + w^2(a_2 - a_6) - w(a_1 - a_5) - w^3(a_3 - a_7) \\ a_0 + a_4 - a_2 - a_6 - w^2(a_1 + a_5 - a_3 - a_7) \\ a_0 - a_4 - w^2(a_2 - a_6) - w^3(a_1 - a_5) - w^5(a_3 - a_7) \end{bmatrix}. \tag{6.3.37c}$$

This, of course, is the same result as if the computations of (6.3.32) were performed directly (Exercise 6.3.24).

We next discuss the permuted vector $\hat{\mathbf{a}}$ of (6.3.36). Let Q_j be a $j \times j$ permutation matrix that sorts $\mathbf{x} = (x_0, x_1, \ldots)^T$ into even and odd components:

$$Q_j \mathbf{x} = (x_0, x_2, \ldots, x_1, x_3, \ldots)^T.$$

For $n = 2^q$, we then define $n \times n$ permutation matrices P_n recursively by

$$P_2 = Q_2, \quad P_4 = Q_4 \operatorname{diag}(P_2, P_2), \quad P_8 = Q_8 \operatorname{diag}(P_4, P_4), \quad (6.3.38)$$

and so on. Now $Q_j^T = Q_j$ since Q_j is its own inverse. Moreover Q_2 is the identity matrix. Thus, (6.3.38) becomes

$$P_4^T = Q_4, \quad P_8^T = \operatorname{diag}(Q_4, Q_4) Q_8,$$

and so on. Then it is easy to see that $P_8^T \mathbf{a}$ is the vector (6.3.36). The permutation matrices P_n are known as *bit-reversal* permutations.

The Fast Sine Transform

The Fast Fourier Transform discussed above is predicated on complex arithmetic. In general, the vector $\mathbf{a}$ may be complex, which will lead to a complex transformed vector $\mathbf{y}$. In our use of the FFT as the basis for a Fast Poisson Solver, all data is real. However, even if the $\mathbf{a}_j$ in (6.3.32) are real, the w^{jk} are still complex. We wish to modify the FFT so as to be able to compute sums of the form (6.3.29) entirely in real arithmetic. This may be done by the following algorithm, known as a *fast sine transform* (FST).

Assume that we wish to compute

$$y_k = \sum_{j=1}^{n} b_j \sin \frac{kj\pi}{n+1}, \quad k = 1, \ldots, n, \qquad (6.3.39)$$

where $n = 2^q - 1$. Then the FST algorithm is:

Step 1. Set $\mathbf{u} = (0, b_2, b_4, \ldots, b_{n-1}, 0, -b_{n-1}, \ldots, -b_2)^T$

$$\mathbf{v} = (b_1, b_3, \ldots, b_n, -b_n, -b_{n-2}, \ldots, -b_1)^T$$

$$-(-b_1, b_1, b_3, \ldots, b_n, -b_n, \ldots, -b_3)^T$$

Step 2. Compute the FFT $\mathbf{z}$ of $\mathbf{v} + i\mathbf{u}$.

Step 3. Set $\mathbf{w} = \frac{1}{8}(z_1 + z_1, z_2 + z_{n+1}, z_3 + z_n, \ldots, z_{n+1} + z_2)^T$

$$\mathbf{x} = \frac{1}{4}(z_1 - z_1, z_2 - z_{n+1}, z_3 - z_{n+1}, \ldots, z_{n+1} - z_2)^T$$

Step 4. $y_k = x_k + w_k / \sin \frac{k\pi}{n+1}, \quad k = 1, \ldots, n.$

6.3 Other Factorizations

The y_k computed in Step 4 are the desired sums (6.3.39).
We illustrate the above algorithm for $n = 3$. In this case

$$\mathbf{u} = (0, b_2, 0, -b_2)^T, \quad \mathbf{v} = (2b_1, b_3 - b_1, -2b_3, b_3 - b_1).$$

For the purpose of this example, we will compute the FFT $\mathbf{z}$ directly by (6.3.32) where $a_j = v_{j+1} + iu_{j+1}, y_j = z_{j+1}$ and $w = e^{-2i\pi/4} = -i$. Thus, $w^2 = -1$ and $w^3 = i$. Then $\mathbf{z}$ is the result of the matrix vector multiply

$$\mathbf{z} = \begin{bmatrix} 1 & 1 & 1 & 1 \\ 1 & -i & -1 & i \\ 1 & -1 & 1 & -1 \\ 1 & -i & -1 & -i \end{bmatrix} \begin{bmatrix} 2b_1 \\ b_3 - b_1 + ib_2 \\ -2b_3 \\ b_3 - b_1 - ib_2 \end{bmatrix} = 2 \begin{bmatrix} 0 \\ b_1 + b_2 + b_3 \\ 2b_1 - 2b_3 \\ b_1 - b_2 + b_3 \end{bmatrix}.$$

Note that $\mathbf{z}$ is real. Next, by Step 3,

$$\mathbf{w} = \frac{1}{8}(0, 4b_1 + 4b_3, 8b_1 - 8b_3, 4b_1 + 4b_3)$$

$$\mathbf{x} = \frac{1}{4}(0, 4b_2, 0, -4b_2),$$

so that by Step 4

$$y_1 = b_2 + \tfrac{1}{2}(b_1 + b_3)/\sin(\tfrac{\pi}{4}) = b_2 + \tfrac{\sqrt{2}}{2}(b_1 + b_3), \quad (6.3.40a)$$

$$y_2 = b_1 - b_3, \quad (6.3.40b)$$

$$y_3 = -b_2 + \tfrac{\sqrt{2}}{2}(b_1 + b_3). \quad (6.3.40c)$$

It is easy to verify that (6.3.40) agrees with a direct computation by (6.3.39) (see Exercise 6.3.24). In the above example, the vector $\mathbf{z}$ was real and this will be the case in general. This can lead to economies in the computation of $\mathbf{z}$.

Supplementary Discussion and References: 6.3

1. Further discussion of the QR factorization and its use in solving least squares problems may be found in Golub and Van Loan [1989], Lawson and Hanson [1974], and Stewart [1973].

2. To minimize round-off error, the computation of the sines and cosines in Givens method as given in (6.3.10) can be replaced by

   if $|a_{21}| \geq |a_{11}|$, $\quad r = a_{11}/a_{21}$, $\quad s_{12} = (1 + r^2)^{-1/2}$, $\quad c_{12} = s_{12}r$

   if $|a_{21}| < |a_{11}|$, $\quad r = a_{21}/a_{11}$, $\quad c_{12} = (1 + r^2)^{-1/2}$, $\quad s_{12} = c_{12}r$.

   Similarly, in the Householder reduction, it is possible that overflow or underflow could occur in the computation of s_k in Figure 6.3.2. This problem can be eliminated by scaling; see, for example, Stewart [1973].

3. Exercise 6.3.3 shows that the Cholesky factorization can be written in the form $A = LDL^T$. If A is positive definite, then so is D; conversely, if D is not positive definite, then neither is A. Unfortunately this type of decomposition is not necessarily numerically stable if A is not positive definite. However, we can find a decomposition of the form $A = PLDL^T P^T$, where P is a permutation matrix and D is block diagonal with 1×1 or 2×2 blocks. This is the basis for the Bunch-Kaufman algorithm (and others also) for solving symmetric systems that are not positive definite. For further discussion, see Golub and Van Loan [1989].

4. A difficult problem is to determine computationally the rank of a matrix A. As discussed in Section 2.2, the number of non-zero singular values is equal to the rank, and the singular value decomposition is the usual approach to determining the rank computationally. Another approach is the *rank-revealing* QR factorization. Here, $AP = QR$ for a permutation matrix P chosen so that R is of the form

$$R = \begin{bmatrix} R_1 & R_2 \\ & R_3 \end{bmatrix}$$

and the elements of R_3 are sufficiently small that the rank of R, and hence A, is chosen as the rank of R_1. For the necessary details of this process and some applications, see Chan and Hansen [1992].

5. The discussion of the Fast Fourier and Fast Sine Transforms follows Van Loan [1991], which contains extensive references to the literature.

6. Fast Poisson Solvers may be constructed for Neumann and periodic boundary conditions as well as Dirichlet. However, the domain of the Poisson equation must be a rectangle or other suitable region so that the problem is *separable* (that is, it can be solved, in principle, by separation of variables). The technique may also be extended to more general equations such as $a(x)u_{xx} + b(y)u_{yy} = f$ provided, again, that the problem is separable.

7. In addition to the approach outlined in the text, there are other ways to develop Fast Poisson Solvers. Cyclic reduction (see Chapter 7) can be used and the FACR algorithms combine the Fourier Transform and cyclic reduction approaches. For a review of various methods as well as their parallel properties, see Hockney and Jesshope [1988], and Swarztrauber and Sweet [1989].

6.3 Other Factorizations

EXERCISES 6.3

6.3.1. Let $A = LL^T$ be a factorization of a symmetric positive definite matrix A, where L is lower-triangular and has positive main diagonal elements. Show that if $\hat{L}$ is obtained from L by changing the sign of every element of the ith row, then $A = \hat{L}\hat{L}^T$. (This shows that the LL^T factorization is not unique although there is a unique L with positive main diagonal elements.)

6.3.2. Show that the number of additions and multiplications in Cholesky factorization is roughly half that of LU factorization.

6.3.3. Modify the Cholesky decomposition of Figure 6.3.1 to obtain the *root-free* Cholesky decomposition in which $A = LDL^T$, where L is unit lower-triangular. Use this to conclude that $U = DL^T$ is the upper-triangular matrix in the LU decomposition of A.

6.3.4. For a banded matrix A with semibandwidth β, show that the Cholesky factorization of Figure 6.3.1 can be written as

$$\text{For } j = 1, \ldots, n$$
$$q = \max(1, j - \beta)$$
$$l_{jj} = \left(a_{jj} - \sum_{k=q}^{j-1} l_{jk}^2 \right)^{1/2}$$
$$\text{For } i = j+1, \ldots, \min(j + \beta, n)$$
$$r = \max(1, i - \beta)$$
$$l_{ij} = \frac{a_{ij} - \sum_{k=r}^{j-1} l_{ik} l_{jk}}{l_{jj}}.$$

Find the operation count for this algorithm as a function of β and n and compare with the LU factorization for banded matrices.

6.3.5. Show that the matrices (6.3.6) and (6.3.7) are orthogonal.

6.3.6. Show that the matrix $P = I - 2\mathbf{w}\mathbf{w}^T$ is always symmetric and is orthogonal if and only if $\mathbf{w}^T\mathbf{w} = 1$. If $\mathbf{w}^T\mathbf{w} = 1$, show also that $P\mathbf{w} = -\mathbf{w}$, and $P\mathbf{u} = \mathbf{0}$ if $\mathbf{u}^T\mathbf{w} = 0$. Interpret this geometrically in two dimensions as a reflection of the plane across the line orthogonal to $\mathbf{w}$.

6.3.7. What are the eigenvalues, eigenvectors, and determinant of a Householder transformation $I - 2\mathbf{w}\mathbf{w}^T$? More generally, what can you say about the eigenvalues and eigenvectors of $I + \mathbf{u}\mathbf{v}^T$ for given vectors $\mathbf{u}$ and $\mathbf{v}$? (The case $\mathbf{v}^T\mathbf{u} = 0$ is special.)

6.3.8. If P is an orthogonal matrix, show that $||P\mathbf{x}||_2 = ||\mathbf{x}||_2$ for any vector $\mathbf{x}$, where $||\ ||_2$ is the Euclidean norm.

6.3.9. If $A = QR$ is a QR factorization, what is the relation of $\det R$ to $\det A$? If A is an orthogonal matrix, what can you say about R?

6.3.10. **a.** Show how to compute an orthogonal matrix such that all elements in the first row are $1/\sqrt{n}$.

b. Show how to generate a "random" orthogonal matrix Q so that
$$\sum_{j=1}^{n} q_{ij} = 0, i = 1, \ldots, n.$$

6.3.11. The vector $\mathbf{w}_1 = (w_1, \cdots, w_n)^T$ of (6.3.13) can be obtained as follows:

a. Since $P_1 = I - 2\mathbf{w}_1\mathbf{w}_1^T$ is orthogonal, use Exercise 6.3.8 to show that if $P_1\mathbf{a}_1 = s_1\mathbf{e}_1$, then $\|\mathbf{a}_1\|_2 = \|P_1\mathbf{a}_1\|_2 = |s_1|$ so that $s_1 = \pm\|\mathbf{a}_1\|_2$. (The vector $\mathbf{e}_1$ has 1 in its first component and zero elsewhere.)

b. Since P_1 is symmetric, show that $\mathbf{a}_1 = s_1 P_1 \mathbf{e}_1 = s_1(\mathbf{e}_1 - 2\mathbf{w}_1 w_1)$ so that $a_{11} = s_1(1 - 2w_1^2), a_{1j} = -2s_1 w_1 w_j$, $j > 1$. Then show that (6.3.13) and (6.3.14) follow.

c. Extend the above argument to show how to find a Householder transformation P so that $P\mathbf{p} = \mathbf{q}$, where $\mathbf{p}$ and $\mathbf{q}$ are real vectors with $\|\mathbf{p}\|_2 = \|\mathbf{q}\|_2$. What can you do if $\mathbf{p}$ and $\mathbf{q}$ are complex?

6.3.12. Verify that (6.3.18) and (6.3.19) are the same.

6.3.13. Let A be a banded matrix with semibandwidth β. Show that if $A = QR$, and Q is obtained by either Givens or Householder transformations, then in general R will have 2β non-zero diagonals above the main diagonal.

6.3.14. Let
$$A = \begin{bmatrix} 1 & 1 & 1 & 1 \\ \varepsilon & 0 & 0 & 0 \\ 0 & \varepsilon & 0 & 0 \\ 0 & 0 & \varepsilon & 0 \\ 0 & 0 & 0 & \varepsilon \end{bmatrix}, A^T A = \begin{bmatrix} 1+\varepsilon^2 & 1 & 1 & 1 \\ 1 & 1+\varepsilon^2 & 1 & 1 \\ 1 & 1 & 1+\varepsilon^2 & 1 \\ 1 & 1 & 1 & 1+\varepsilon^2 \end{bmatrix}.$$

Show that $A^T A$ has eigenvalues $4 + \varepsilon^2, \varepsilon^2, \varepsilon^2, \varepsilon^2$, and thus is nonsingular if $\varepsilon \neq 0$. Now let η be the largest positive number on the computer such that computed$(1 + \eta) = 1$. Show that if $|\varepsilon| < \sqrt{\eta}$,
$$\text{computed}(A^T A) = \begin{bmatrix} 1 & 1 & 1 & 1 \\ 1 & 1 & 1 & 1 \\ 1 & 1 & 1 & 1 \\ 1 & 1 & 1 & 1 \end{bmatrix},$$

a matrix of rank one, and consequently the numerical approximation to the normal equations will, for general $\mathbf{b}$, have no solution at all. (Note that ε need not be unrealistically small. For example, on machines with a 27-bit mantissa, $\eta \doteq 10^{-8}$, so that the above error would occur if $\varepsilon < 10^{-4}$.)

6.3 Other Factorizations

6.3.15. Modify (6.3.17) to show that the operation count of QR for an $m \times n$ matrix is approximately $2mn^2 - \frac{2}{3}n^3$. Show that this is also the high-order part of the operation count for the algorithm (6.3.25). Next, show that the operation count for minimizing (6.3.19) by forming and solving the normal equations is about $mn^2 + \frac{1}{3}n^3$. Thus, for $m \gg n$, conclude that the QR approach is approximately twice as slow.

6.3.16. Consider the least squares problem of minimizing $\|E\mathbf{a} - \mathbf{f}\|_2$ subject to the constraint $F^T\mathbf{a} = \mathbf{g}$, where $\mathbf{g}$ is a given vector. Show how the QR decomposition of F can be used in solving this problem.

6.3.17. Consider the normal equations $E^T E\mathbf{a} = E^T\mathbf{f}$ and let $\mathbf{r} = \mathbf{f} - E\mathbf{a}$, $\mathbf{g} = \mathbf{f} + \alpha\mathbf{r}$, where α is a constant. Show that if $\mathbf{b}$ is the solution of $E^T E\mathbf{b} = E^T\mathbf{g}$, then $\mathbf{b} = \mathbf{a}$. Conclude from this that the vector $\mathbf{f}$ can be changed by arbitrarily large amounts without changing the solution to the normal equations.

6.3.18. Consider the special case of (6.1.39) in which

$$A = \begin{bmatrix} A_1 & B_1^T \\ B_1 & 0 \end{bmatrix},$$

and assume that A_1 is symmetric positive definite. Show that A is not positive definite but that there is a Cholesky-like factorization of the form

$$A = \begin{bmatrix} F & 0 \\ G & H \end{bmatrix} \begin{bmatrix} F^T & G^T \\ 0 & -H^T \end{bmatrix}.$$

Give an efficient algorithm for computing this factorization.

6.3.19. Consider the special case of Exercise 6.3.18 in which $A_1 = I$ and $B_1 = E^T$. Show that the solution of the normal equations $E^T E\mathbf{a} = E^T\mathbf{f}$ can be obtained by solving the system

$$\begin{bmatrix} I & E \\ E^T & 0 \end{bmatrix} \begin{bmatrix} \mathbf{r} \\ \mathbf{a} \end{bmatrix} = \begin{bmatrix} \mathbf{f} \\ 0 \end{bmatrix}.$$

Discuss the circumstances in which you might wish to solve this expanded system in place of the normal equations.

6.3.20. Show that two complex numbers may be multiplied in three real multiplications. How many additions are required?

6.3.21. If A is an $m \times n$ matrix with $m \geq n = \text{rank } A$, the matrix $P = A(A^T A)^{-1} A^T$ is called a *projection* matrix.

  a. Show that $P^T = P$ and $P^2 = P$. Thus, $(I - 2P)^2 = I$.

  b. If $A = Q \begin{bmatrix} R \\ 0 \end{bmatrix}$, show how to construct P in terms of Q and R.

  c. What are the eigenvalues of P?

d. Show that the least squares problem min $\|A\mathbf{x}-\mathbf{b}\|_2$ can be solved by solving the system $A\mathbf{x} = P\mathbf{b}$.

6.3.22. Let $A = LL^T$ be the Cholesky factorization of A and set $\tilde{L} = LQ$, where Q is orthogonal. Show that $A = \tilde{L}\tilde{L}^T$ and that $\tilde{L}$ is lower-triangular, in general, if and only if Q is diagonal. Describe the class of diagonal orthogonal matrices.

6.3.23. Let $A = LL^T$ be the Cholesky factorization of A and let $\overline{A} = A + \alpha \mathbf{z}\mathbf{z}^T$ be a rank one modification of A. Assume that $\overline{A}$ is positive definite and $\overline{A} = \overline{L}\,\overline{L}^T$. Show that $\overline{L} = L\tilde{L}$, where $\tilde{L}$ is the Cholesky factor of $I + \alpha L^{-1}\mathbf{z}(L^{-1}\mathbf{z})^T$.

6.3.24. Verify the computations of (6.3.37). Then using the fact that (for $n = 8$) $w^4 = -1$ (and thus $w^5 = -w, w^6 = -w^2, w^7 = -w^3$), the matrix W of (6.3.33) is

$$W = \begin{bmatrix} 1 & 1 & 1 & 1 & 1 & 1 & 1 & 1 \\ 1 & w & w^2 & w^3 & -1 & -w & -w^2 & -w^3 \\ 1 & w^2 & -1 & -w^2 & 1 & w^2 & -1 & -w^2 \\ 1 & w^3 & -w^2 & w & -1 & -w^3 & w^2 & -w \\ 1 & -1 & 1 & -1 & 1 & -1 & 1 & -1 \\ 1 & -w & w^2 & -w^3 & -1 & w & -w^2 & w^3 \\ 1 & -w^2 & -1 & w^2 & 1 & -w^2 & -1 & w^2 \\ 1 & -w^3 & -w^2 & -w & -1 & w^3 & w^2 & w \end{bmatrix}$$

Then verify that $W\mathbf{a}$ agrees with (6.3.37c).

6.3.25. Compute the sums (6.3.39) for $n = 3$ and verify that they agree with (6.3.41).

Chapter 7

Parallel Direct Methods

In this chapter, we consider the implementation on parallel and vector computers of the methods of Chapter 6 for the solution of $A\mathbf{x} = \mathbf{b}$. In this and the next section we will assume that A is a full matrix. Then, in Section 7.3 we treat banded systems. Different organizations of the algorithms are often needed to utilize efficiently the particular computer architecture under consideration. Several different organizations will be considered in Section 7.2.

7.1 Basic Methods

Vectorization of LU Decomposition

As discussed in Section 6.1, the Gaussian elimination process without interchanges for the solution of $A\mathbf{x} = \mathbf{b}$ may be viewed as an LU decomposition followed by solution of the triangular systems $L\mathbf{c} = \mathbf{b}, U\mathbf{x} = \mathbf{c}$. We concentrate first on the LU decomposition, which is the time-consuming part of the process, and return to the solution of the triangular systems shortly. A pseudocode for the decomposition is given in Figure 7.1.1.

The j loop in Figure 7.1.1 subtracts multiples of the kth row of the current A from succeeding rows. These are axpy operations in which the vectors are rows of A. These axpy operations perform *updating* of the rows of A and constitute the bulk of the work in the LU decomposition. At the kth stage of the algorithm, there are $n - k$ vector operations and the vector lengths are $n - k$. Therefore, the average vector length for the updating operations is

$$\frac{(n-1)(n-1) + (n-2)(n-2) + \cdots + 1}{n - 1 + n - 2 + \cdots + 1} = O(2n/3), \qquad (7.1.1)$$

where the verification of the equality is left to Exercise 7.1.1. (As previously, we use the notation $O(cn^p)$ for the highest order term in the expression.) On

> For $k = 1$ to $n-1$
>   For $i = k+1$ to n
> $l_{ik} = a_{ik}/a_{kk}$
>   For $j = k+1$ to n
> $a_{ij} = a_{ij} - l_{ik}a_{kj}$

Figure 7.1.1: *Row-Oriented LU Decomposition*

some vector computers it may be advantageous to switch from vector to scalar arithmetic when the vector length drops below a certain threshold, although this will complicate the code.

The speed of the code will be degraded somewhat by the scalar divisions to form the multipliers at each stage. A further degradation will be caused by pivoting, if this is needed. For the partial pivoting strategy applied to the first stage, we must search the first column for the maximum element in absolute value. Although some vector computers have vector instructions to facilitate this search, if we assume that A is stored by rows the elements of the first column are not in contiguous storage locations. Once the location of the maximum element is known, the corresponding row can be swapped with the first row by vector operations or the indexing can be modified. How the pivoting strategy will be done will depend on the particular computer.

The right-hand side **b** may also be processed as part of the reduction to triangular form in order to achieve the forward substitution $\mathbf{c} = L^{-1}\mathbf{b}$. If A is stored by rows, it is useful to append b_i to the ith row, if other storage considerations permit this. Then, in Figure 7.1.1, the j loop would run to $n+1$, and the vector lengths would increase by one.

We have assumed so far that A is stored by rows. However, Fortran stores two-dimensional arrays by columns. In this case, the algorithm of Figure 7.1.1 will access the rows of A as vectors with a stride of n. As discussed in Chapter 3, this can cause bank conflicts. We can circumvent this problem by reordering the operations in Gaussian elimination as shown in Figure 7.1.2. At the kth stage, the algorithm of Figure 7.1.2 first forms the kth column of L; this can be done by a vector divide. The innermost loop, the i loop, subtracts a multiple of this kth column of L from the $n-k$ long jth column of the current A. Hence, the basic vector operation is again an axpy, but now the vectors are columns of L and columns of the current A. The right-hand side **b** can be processed in the same manner as the columns of A to effect the forward substitution.

Not counting the formation of the multiplier vector or processing the right-hand side, there are $n-k$ axpy operations at the kth stage with vector lengths

7.1 Basic Methods

> For $k = 1$ to $n - 1$
> $\quad$ For $s = k + 1$ to n
> $\quad\quad l_{sk} = a_{sk}/a_{kk}$
> $\quad$ For $j = k + 1$ to n
> $\quad\quad$ For $i = k + 1$ to n
> $\quad\quad\quad a_{ij} = a_{ij} - l_{ik}a_{kj}$

Figure 7.1.2: *Column-Oriented LU Decomposition*

of $n - k$. Hence, (7.1.1) again gives the average vector length. Only a more detailed analysis for a particular machine can determine whether the row form is more efficient than the column form, but the formation of the multipliers by a vector operation will probably show that the column form is the best. A deciding factor on which to use will be the way that A may already be stored.

The incorporation of pivoting is somewhat different in the column form. The search for the pivot element can now be done by vector operations on those machines with that capability, but the interchange of rows requires vector operations with a stride of n.

Solution of Triangular Systems

At the end of the LU decomposition, we will need to solve the triangular systems $L\mathbf{c} = \mathbf{b}, U\mathbf{x} = \mathbf{c}$. Usually, the solution of $L\mathbf{c} = \mathbf{b}$ will be incorporated into the LU decomposition, as already discussed, and we will concentrate on the solution of $U\mathbf{x} = \mathbf{c}$. However, solution of lower triangular systems will use the same ideas (see Exercise 7.1.2).

The usual back substitution algorithm is

$$x_i = (c_i - u_{ii+1}x_{i+1} - \cdots - u_{in}x_n)/u_{ii}, \quad i = n, \ldots, 1, \quad (7.1.2)$$

which is also shown in Figure 7.1.3(b). We next consider its implementation in vector operations. If U is stored by rows, as it will be if the forward reduction has been done with A stored by rows, then (7.1.2) can be carried out by inner products with vector lengths ranging from 1 to $n - 1$, together with n scalar divisions. Ignoring the divisions, the average vector length is then $O(n/2)$.

An alternative algorithm, useful if U is stored by columns, is given by the pseudocode in Figure 7.1.3(a), which is known as the *column sweep* (or *vector sum*) *algorithm*. After x_n is computed, the quantities $x_n u_{in}, i = 1, \ldots, n - 1$, are computed and subtracted from the corresponding c_i; thus, the contribution of x_n to the other solution components is made before moving to the next stage. The jth stage consists of a scalar division followed by an axpy of length $j-1$. Hence, the average vector length is again $O(n/2)$, but now the vector operations are axpys. Which of the two algorithms we use will almost certainly be dictated by the storage pattern of U, if this has been determined by the LU decomposition.

$$
\boxed{\begin{array}{l} \text{For } j = n \text{ down to } 1 \\ x_j = c_j/u_{jj} \\ \text{For } i = 1 \text{ to } j-1 \\ c_i = c_i - x_j u_{ij} \end{array}} \qquad \boxed{\begin{array}{l} \text{For } i = n \text{ down to } 1 \\ \text{For } j = i+1 \text{ to } n \\ c_i = c_i - u_{ij} x_j \\ x_i = c_i/u_{ii} \end{array}}
$$

(a) \qquad\qquad (b)

Figure 7.1.3: *Solution of* $U\mathbf{x} = \mathbf{c}$. *(a) Column Sweep (b) Inner Product*

Multiple Right-Hand Sides

We consider next the solution of the systems

$$A\mathbf{x}_i = \mathbf{b}_i, \qquad i = 1, \ldots, q \qquad (7.1.3)$$

with different right-hand sides $\mathbf{b}_i$ but the same coefficient matrix A. We can write (7.1.3) in an equivalent way as $AX = B$, where X and B are $n \times q$ matrices. The column-oriented algorithm of Figure 7.1.2 can be modified so as to process all the columns of B by considering them as additional columns of A. The vector lengths remain $n-k$ at the kth stage, but now there are $n-k+q$ vector operations. Thus, the average vector length is (Exercise 7.1.3)

$$\frac{(n-1+q)(n-1)+\cdots+(1+q)}{(n-1+q)+\cdots+(1+q)} = O\left(\frac{2n^2+3qn}{3n+6q}\right), \qquad (7.1.4)$$

which decreases slightly as a function of q, and for $q=n$ is $O(5n/9)$ compared with $O(2n/3)$ for $q=1$.

On the other hand, we can modify the row-oriented algorithm of Figure 7.1.1 by appending the rows of B to the rows of A, so that the vector lengths at the kth stage are $n-k+q$. Since there are still only $n-k$ vector operations, the average vector length is (Exercise 7.1.3)

$$\frac{(n-1)(n-1+q)+\cdots+(1+q)}{(n-1)+\cdots+1} = O(\frac{2}{3}n+q), \qquad (7.1.5)$$

7.1 Basic Methods

which is now an increasing function of q, and for $q = n$ is $O(5n/3)$, over twice that for $q = 1$. We conclude that the row-oriented algorithm, although possibly slightly inferior for $q = 1$, becomes increasingly attractive as q increases.

There remains the back substitution, and if there are sufficiently many right-hand sides, it may be beneficial to use the following scheme. Let the systems be given by $UX = C$, where X and C are $n \times q$ and $\mathbf{x}_i$ and $\mathbf{c}_i$ denote the ith rows. Then, the algorithm

$$\mathbf{x}_i = (\mathbf{c}_i - u_{ii+1}\mathbf{x}_{i+1} - \cdots - u_{in}\mathbf{x}_n)/u_{ii}, \quad i = n, n-1, \ldots, 1, \quad (7.1.6)$$

computes the rows $\mathbf{x}_n, \mathbf{x}_{n-1}, \ldots$ by using axpy operations on vectors of length q. This is just the inner product algorithm (7.1.2) applied "simultaneously" to compute the ith components of all the solution vectors "at once." Thus, the basic vector operation is an axpy and not an inner product. Whether this is preferable to the inner product algorithm applied to each right-hand side will depend on the size of q and n, as well as on the particular computer.

Cholesky Factorization

If A is symmetric positive definite, the Cholesky factorization $A = LL^T$ is given in Figure 6.3.1. We write the algorithm in a somewhat different form in Figure 7.1.4. The innermost loop, the i loop, in Figure 7.1.4 modifies columns of A by subtracting multiples of columns of L. Hence, the basic vector operation is again an axpy.

$$\begin{array}{l}
l_{11} = a_{11}^{1/2} \\
\text{For } j = 2 \text{ to } n \\
\quad \text{For } s = j \text{ to } n \\
\quad\quad l_{s,j-1} = a_{s,j-1}/l_{j-1,j-1} \\
\quad \text{For } k = 1 \text{ to } j-1 \\
\quad\quad \text{For } i = j \text{ to } n \\
\quad\quad\quad a_{ij} = a_{ij} - l_{ik}l_{jk} \\
\quad l_{jj} = a_{jj}^{1/2}
\end{array}$$

Figure 7.1.4: *Column-Oriented Cholesky Decomposition*

At the jth stage there are $j - 1$ axpys using vectors of length $n - j$. Hence, the average vector length is (Exercise 7.1.3)

$$\frac{n - 1 + 2(n-2) + \cdots + (n-1)1}{1 + 2 + \cdots + n - 1} = O\left(\frac{n}{3}\right) \quad (7.1.7)$$

which is only half that of *LU* decomposition. This is to be expected since in the Cholesky algorithm we are using the symmetry of the coefficient matrix to reduce the serial computational work by a factor of 2. On vector machines with large start-up time, the relative speed advantage of Cholesky over *LU* decomposition is much less than the nominal factor of 2 for serial computers.

QR Factorization

We next consider the factorization $A = QR$ and begin with the Householder algorithm of Figure 6.3.2, which we reproduce in Figure 7.1.5.

For $k = 1$ to $n - 1$

$$s_k = -\text{sgn}(a_{kk}) \left(\sum_{l=k}^{n} a_{lk}^2 \right)^{1/2}$$

$$\mathbf{u}_k^T = (0, \ldots, 0, a_{kk} - s_k, a_{k+1,k}, \ldots, a_{nk})$$

$$\gamma_k = (s_k^2 - s_k a_{kk})^{-1}$$

$$a_{kk} = s_k$$

For $j = k + 1$ to n

$$\alpha_j = \gamma_k \mathbf{u}_k^T \mathbf{a}_j$$

$$\mathbf{a}_j = \mathbf{a}_j - \alpha_j \mathbf{u}_k$$

Figure 7.1.5: *Column-Oriented Householder Reduction. Inner Products*

At the kth stage, the computation of s_k requires an inner product of length $n - k + 1$, and scalar operations are then needed for γ_k and $a_{kk} - s_k$. As discussed in Section 6.3, the bulk of the work in the Householder reduction is in the updating of the columns of the reduced matrices; this is the final j loop in Figure 7.1.5. This inner loop contains an inner product followed by an axpy, and the vector lengths are $n - k + 1$. Thus, the average vector length is the same as for *LU* decomposition, $O(2n/3)$, and the inner loop requires axpys of the same form. The major difference, of course, is that the inner loop also requires the inner products $\mathbf{u}_k^T \mathbf{a}_j$.

We next consider an alternative form of the Householder reduction in which the inner loop contains only axpys. If $\mathbf{a}_i$ now denotes the ith row of A, the updating (6.3.15) may also be written in the form

$$P_1 A = A - \mathbf{w}_1 \mathbf{z}^T, \qquad \mathbf{z}^T = 2\mathbf{w}_1^T A = 2\mu \sum_{i=1}^n u_i \mathbf{a}_i = 2\mu \mathbf{v}^T, \qquad (7.1.8a)$$

where $u_1, \ldots, u_n$ are the components of $\mathbf{u}_1$. Thus, the new ith row of A is

7.1 Basic Methods

$$\mathbf{a}_i - w_i \mathbf{z}^T = \mathbf{a}_i - \gamma u_i \mathbf{v}^T, \qquad (7.1.8b)$$

where the w_i are the components of $\mathbf{w}_1$. This is sometimes called the *rank one* update form since $\mathbf{w}\mathbf{z}^T$ is a rank one matrix. The overall algorithm is given in Figure 7.1.6.

For $k = 1$ to $n - 1$
$s_k = -\mathrm{sgn}(a_{kk}) \left(\sum_{l=k}^{n} \mathbf{a}_{lk}^2 \right)^{1/2}, \gamma_k = \left(s_k^2 - s_k a_{kk} \right)^{-1}$
$\mathbf{u}_k^T = (0, \ldots, 0, a_{kk} - s_k, \mathbf{a}_{k+1,k}, \ldots, a_{nk})$
$\mathbf{v}_k^T = \sum_{l=k}^{n} u_{lk} \mathbf{a}_l$
$\hat{\mathbf{v}}_k^T = \gamma_k \mathbf{v}_k^T$
For $j = k$ to n
$\mathbf{a}_j = \mathbf{a}_j - u_{jk} \hat{\mathbf{v}}_k^T$

Figure 7.1.6: *Row-Oriented Householder Reduction. Rank One Updates*

The average vector length in this algorithm is again $O(2n/3)$, and the vector operations are now all axpys. Note, however, that the computation of the s_k is now less satisfactory since the elements of the columns of A are not in contiguous locations if A is stored by rows.

Givens Reduction

We saw in Section 6.3 that the QR factorization using Givens transformations was more costly than using Householder transformations. However, on vector computers the comparison is not so clear cut.

At the first stage of the Givens reduction, after c_{12} and s_{12} are computed by scalar operations, the first two rows are modified as shown in (6.3.8). This requires a scalar-vector multiplication followed by an axpy. We know that the new (2,1) element is zero and since the Euclidean length of the first column must be preserved under multiplication by P_{12}, we also know that the new (1,1) element is $\pm(a_{11}^2 + a_{21}^2)^{1/2}$, which has already been computed. Thus, the vector lengths are $n - 1$, and to zero the elements in the first column requires $4(n - 1)$ vector operations with vector lengths of $n - 1$. Similarly, zeroing elements in the jth column will require $4(n - j)$ vector operations with vector lengths of $n - j$ so that the average vector length for the entire reduction will be (Exercise 7.1.3)

$$\frac{4(n-1)(n-1)+4(n-2)(n-2)+\cdots+4}{4(n-1)+4(n-2)+\cdots+4} = O(2n/3), \quad (7.1.9)$$

the same as LU decomposition and the Householder reduction. Figure 7.1.7 gives a Givens reduction pseudo-code, in which $\mathbf{a}_k$ is the kth row of the current A and $\hat{\mathbf{a}}_k$ is the update of $\mathbf{a}_k$. Note that in the update of $\mathbf{a}_i$ it is the current $\mathbf{a}_k$ that is used, not $\hat{\mathbf{a}}_k$.

> For $k = 1$ to n
>   For $i = k+1$ to n
>   Compute s_{ki}, c_{ki}
> $\hat{\mathbf{a}}_k = c_{ki}\mathbf{a}_k + s_{ki}\mathbf{a}_i$
> $\mathbf{a}_i = s_{ki}\mathbf{a}_k + c_{ki}\mathbf{a}_i$

Figure 7.1.7: *Row-Oriented Givens Reduction*

The Givens algorithm has twice as many vector operations as the Householder algorithms but these can be accomplished with only half the number of memory references as the Householder algorithms. Consider the first stage, consisting of zeroing the first column of A. The vectors $\mathbf{a}_1$ and $\mathbf{a}_2$ (or parts of them) are loaded into vector registers and then the necessary arithmetic is done. The updated $\mathbf{a}_1$ is left in a register so that the zeroing of the $(3,1)$ element requires only the loading of $\mathbf{a}_3$. Thus, the zeroing of the $n-1$ elements in the first column requires only a single load of each of the rows of A, for a total of n complete vector loads. This compares with $O(2n)$ loads for the Householder algorithms, and Givens will be more competitive on vector machines than a simple operation count would indicate.

Parallel LU Decomposition

We now begin consideration of parallel algorithms for the LU, Cholesky and QR factorizations. We begin with LU. We assume at first a distributed memory system with $p = n$ processors. Then one possible organization of LU decomposition is the following. Assume that the ith row of A is stored in processor i. At the first stage, the first row of A is sent to all other processors and then the computations

$$l_{i1} = a_{i1}/a_{11}, \quad a_{ij} = a_{ij} - l_{i1}a_{1j}, \quad j = 2, \ldots, n, \quad (7.1.10)$$

7.1 Basic Methods

can be done in parallel in processors $P_2, \ldots, P_n$. The computation is continued by sending the new second row of the reduced matrix from processor P_2 to processors $P_3, \ldots, P_n$, then doing the next set of calculations corresponding to (7.1.10) in parallel, and so on. Note that this approach has two major drawbacks: there is considerable communication of data between each stage, and the number of active processors decreases by one at each stage.

An alternative to storing by rows is to store column i of A in processor i. In this case, at the first stage, the multipliers l_{i1} are all computed in processor 1 and then sent to all processors. The updates

$$a_{ij} = a_{ij} - l_{i1}a_{1j}, \qquad j = 2, \ldots, n, \qquad (7.1.11)$$

are then computed in parallel in processors $2, \ldots, n$. Processor 1 ceases to be active after computing the multipliers l_{i1}, and, in general, one more processor becomes idle at each stage, which is the same load-balancing problem as in the row-oriented algorithm.

In the more realistic situation that $p \ll n$, the above load-balancing problems are mitigated to some extent. Suppose that $n = kp$, and, in the storage by rows strategy, assign the first k rows of A to processor 1, the second k to processor 2, and so on. This is called *block* or *panel* storage. Again, the first row will be sent from processor 1 to the other processors and then the computation (7.1.10) will be done but in blocks of k sets of operations in each processor. As before, processors will become idle as the computation proceeds, but the overall ratio of computation time to data transmission and idle time will be an increasing function of k.

A more attractive scheme is to interleave the storage of the rows among the processors. Assuming still that $n = kp$, rows $1, p+1, 2p+1, \ldots$ are stored in processor 1, rows $2, p+2, 2p+2, \ldots$ in processor 2, and so on, as illustrated in Figure 7.1.8. We will call this storage pattern *wrapped interleaved storage*. It greatly alleviates the problem of processors becoming idle since, for example, processor 1 will be working almost until the end of the reduction, in particular, until processing of row $(k-1)p + 1$ is complete. There will, of course, still be some imbalances in the workloads of the processors. For example, after the first stage, row 2 is complete and processor 2 will have one less row to process at the next stage than the other processors. A similar kind of imbalance will result in the likely event that n is not a multiple of the number of processors.

A similar interleaving may also be done in storage by columns where now processor 1 would hold columns $1, p+1, \ldots, (k-1)p+1$, processor 2 would hold columns $2, p+2, \ldots, (k-1)p+2$, and so on. Again, all processors would be working, with minor imbalances, almost to the end of the reduction. Comparisons of the two algorithms are given in Exercises 7.1.8 and 7.1.9. Hereafter, we use the terms "wrapped," "interleaved," and "wrapped interleaved" as synonymous.

rows $1, p+1, \ldots, (k-1)p+1$	rows $2, p+2, \ldots, (k-1)p+2$	$\cdots$	rows $p, 2p, \ldots, kp$
processor 1	processor 2		processor p

Figure 7.1.8: *Wrapped Interleaved Row Storage*

The notion of interleaved storage is motivated primarily by distributed memory systems. It can also be used on shared memory systems as a means of assigning tasks to processors. For example, with row interleaved storage, processor 1 will update rows $p+1, 2p+1, \ldots$, and so on. However, on shared memory systems, dynamic load balancing can also be achieved by using the idea of a pool of tasks. Such dynamic load balancing is less satisfactory for a distributed memory machine because of the need to reassign storage from one processor to another.

All of the above considerations apply equally well to Cholesky factorization, the main difference being that only half the matrix A needs to be stored.

Send-Ahead and Compute-Ahead

We now discuss in more detail the LU decomposition using wrapped interleaved row storage. We assume first a distributed memory system. Consider the first stage, in which zeros are introduced into the first column. If we proceed in the usual serial fashion, the multipliers l_{i1} are computed, the corresponding rows updated, and then the second stage is begun. At the beginning of the second stage, the processor holding the second row must transmit this row to all other processors. Hence, there is a delay while this transmission takes place. An obvious modification is for processor 2 to transmit the updated second row as soon as it is ready. We call this a *send-ahead* strategy. If computation and communication can be overlapped, all of the processors can be carrying out the updating of the first stage while this transmission takes place. The same will be done at each stage: at the kth stage, the $(k+1)$st row will be transmitted as soon as it is updated. How well this strategy works on a given machine will depend on the communication properties of the system.

This send-ahead strategy also has application on shared-memory systems. Here, a straightforward implementation would have a synchronization after each stage so that no processor would begin the next stage until all processors have finished the current one. During the kth stage, the processor to which row $k+1$ is assigned will have at least as much work as any other processor since it is assigned at least as many rows. Thus, other processors may have to wait until this processor is done before they can begin the next stage. The send-ahead strategy, which is now really a *compute-ahead* strategy, modifies this by marking the $(k+1)$st row "done" as soon as it has been updated. Then, as soon as other processors have completed their work on the kth stage,

7.1 Basic Methods

they may immediately start the $(k+1)$st stage if the $(k+1)$st row has been marked "done." This marking serves to carry out the necessary synchronization without unnecessary delays. We note that send-ahead and compute-ahead are also called *pipelining*.

Partial Pivoting

The use of partial pivoting to maintain numerical stability introduces additional considerations into the storage question. Suppose, first, that A is stored by column wrapped storage. Then the search for the pivot element takes place in a single processor. This has the virtue of simplicity but also the potential for all other processors to be idle while the search is done. This problem can potentially be mitigated by a compute-and-send-ahead strategy in which, at the kth stage, the pivot element in the $(k+1)$st column is determined as soon as the column has been updated. In any case, after the pivot row is determined, this information must be transmitted to all other processors. Then an interchange of rows can be done in parallel within all the processors (or the interchange can be handled implicitly by indexing).

The situation with row-wrapped storage is rather different. Now the search for the maximum element in the current column (say the kth) must take place across the processors. This can be done by a fan-in as indicated in Section 3.2 (see Exercise 3.2.3). At the end of the fan-in, a single processor will know the next pivot row and this information must be sent to all other processors. Again, we must decide whether to interchange rows physically. If we do interchange, only two processors are involved and all others may be idle during the interchange. On the other hand, if we do not interchange, then we deviate from row-wrapped storage in the subsequent calculations. Indeed, the more interchanges that are required but not done physically, the more the storage pattern will look like a random scattering of the rows to local memories.

Solution of Triangular Systems

After the LU or Cholesky decompositions, we need to solve triangular systems of equations. We will consider only upper triangular systems $U\mathbf{x} = \mathbf{c}$; the treatment of lower triangular systems is similar (Exercise 7.1.10). The basic methods will again be the column sweep and inner product algorithms, given in Figure 7.1.3.

We consider first distributed memory systems. Assuming that the triangular system has resulted from a prior LU or Cholesky decomposition, the storage of U is already determined by the storage for the decomposition. For example, if we have used the row-wrapped storage of Figure 7.1.8, then the storage of U will have the same row interleaved pattern. We will assume that the right-hand side $\mathbf{c}$ also is stored by interleaved rows so that the storage of the system is as shown in Figure 7.1.9, where $\mathbf{u}_i$ denotes the ith row of U. The column sweep algorithm can then be implemented as in Figure 7.1.10.

In this figure and the following discussion $P(i)$ denotes the processor holding the ith row (or column) of U. At the first step, c_n and u_{nn} are in the same processor. After x_n is transmitted to all processors, each processor updates the right-hand side components that it holds, and then x_{n-1} is computed in the processor holding the $(n-1)$st row. The process is now repeated to obtain x_{n-2}, then x_{n-3}, and so on. The work of updating the c_i will be fairly well balanced until the current triangular system becomes small; indeed, in the final stages, an increasing number of processors will become idle.

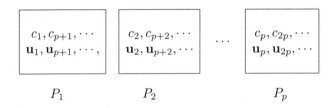

Figure 7.1.9: *Wrapped Interleaved Row Storage for $U\mathbf{x} = \mathbf{c}$*

One bad aspect of the column sweep algorithm is that the computation of x_i takes place in a single processor while all other processors are idle. Thus, it may be beneficial to try to overlap the computation of the subsequent x_i by the following compute-ahead strategy. The processor that holds the $(n-1)$st row will update c_{n-1} and then compute and transmit x_{n-1} before updating the other c_i. Hence, x_{n-1} will be available to all processors as soon as possible.

Compute $x_n = c_n/u_{nn}$ in $P(n)$
Send x_n to all processors
Compute $c_i = c_i - u_{in}x_n$, $\quad i = 1, \ldots, n-1$,
Compute $x_{n-1} = c_{n-1}/u_{n-1,n-1}$ in $P(n-1)$
Send x_{n-1} to all processors
Compute $c_i = c_i - u_{i,n-1}x_{n-1}$, $\quad i = 1, \ldots, n-2$
$\vdots$

Figure 7.1.10: *Parallel Column Sweep Algorithm for U Stored by Rows*

If U is stored by column interleaved storage, a straightforward implementation of the column sweep algorithm results in essentially a serial code. Assume that $P(n)$, the processor holding the last column of U, also holds $\mathbf{c}$. Then $P(n)$

7.1 Basic Methods

computes x_n, updates **c**, and transmits the new **c** to $P(n-1)$. $P(n-1)$ then computes x_{n-1}, updates **c**, transmits to $P(n-2)$, and so on. Thus, at any given time, only one processor is doing computation. There are ways to alleviate this problem by various compute-ahead and send-ahead strategies (see the Supplementary Discussion). We will instead consider next the inner product algorithm since it is potentially attractive when U is stored by columns.

In the inner product algorithm of Figure 7.1.3(b), the computation of x_i requires the inner product of the ith row of U, excluding the main diagonal, and the vector with components $x_{i+1}, \ldots, x_n$. We assume that $x_j \in P(j)$. Thus, processor $P(j)$ can compute $\sum u_{ij} x_j$ for all the x_j that it holds since the corresponding u_{ij} are also in $P(j)$. After these partial inner products have been computed in parallel in all processors, they can be summed by a fan-in so that the inner product is in $P(i)$. We assume that the right-hand side has been stored so that $c_i \in P(i)$, and then $P(i)$ computes x_i. We summarize this inner product algorithm in Figure 7.1.11.

For $i = n$ down to 1
All processors compute their portion of ith inner product
Fan-in partial inner products to $P(i)$
$P(i)$ computes x_i

Figure 7.1.11: *Parallel Inner Product Algorithm*

For large n and i, there is almost perfect parallelism in the computation of the partial inner products. The fan-in is less satisfactory. It will involve communication and an increasing number of processors will become idle. Moreover, in the initial stages, when i is close to n, there is very little parallelism even in the computation of the partial inner products. Some of these problems can potentially be alleviated by compute-ahead strategies. For example, during the ith stage, as soon as a processor has finished its computations required in this stage, it could begin its partial inner product for the next stage.

We note that an intrinsic difficulty in solving triangular systems on distributed memory systems is that there are only $O(n^2)$ operations over which to amortize the communication costs. This is in contrast to the factorization, in which there are $O(n^3)$ operations.

After a Cholesky decomposition, we must solve the triangular systems $L^T \mathbf{y} = \mathbf{b}$ and $L\mathbf{x} = \mathbf{y}$. If the storage of A has been by rows, then the storage of L will be the same and a column sweep algorithm analogous to Figure 7.1.10 can be done for $L\mathbf{x} = \mathbf{y}$, the only difference being that we now solve for the unknowns in the order $x_1, \ldots, x_n$ (see Exercise 7.1.10). For the solution of

$L^T \mathbf{y} = \mathbf{b}$, storage of L by rows corresponds to storage of L^T by columns. If A has been stored by columns, then L will be stored by columns and L^T by rows. Thus, in either case, solutions of triangular systems with both row and column storage will be required. For Gaussian elimination, it is usual to perform the forward substitution (the solution of $L\mathbf{y} = \mathbf{b}$) as part of the factorization. The elements of $\mathbf{b}$ can be appended to the rows of A if row storage is being used, or considered as another column of A if column storage is used.

The solution of triangular systems on shared memory machines can be carried out by either dynamic allocation of tasks or a static allocation. For example, we can use a row or column interleaved assignment scheme in which processor i is assigned the work associated with columns (or rows) $i, i+p, i+2p, \ldots$. Assuming assignment by rows, a pseudocode for each processor is given in Figure 7.1.12.

If (row n assigned to this processor)
$x_n = c_n/u_{nn}$
mark x_n as done
For $j = n - 1$ down to 1
 wait for x_{j+1} to be marked done
 If (row j assigned to this processor)
  $c_j = c_j - u_{jj+1}x_{j+1}$
  $x_j = c_j/u_{jj}$
  mark x_j as done
 For $i = 1$ to $j - 1$
  If(row i assigned to this processor)
   $c_i = c_i - u_{ij+1}x_{j+1}$

Figure 7.1.12: *Shared-Memory Column Sweep for* $U\mathbf{x} = \mathbf{c}$

In Figure 7.1.12, the necessary synchronization is achieved by marking the unknowns x_i "done" as they are computed. Processors wait until the x_i needed for the next computations are done and thus available. The marking and waiting can be implemented in different ways, depending on the system. In order to be available to other processors as soon as possible, the x_i are computed before completing the current update of the right-hand side.

Householder Reduction

Previously, we discussed the Givens and Householder orthogonal reductions for vector computers and we now wish to consider their implementation on

7.1 Basic Methods

parallel systems. We treat the Householder reduction first and assume that A is stored by interleaved column storage. We start from the vector code of Figure 7.1.5 (recall that $\mathbf{a}_i$ is the ith column of the current A).

The first stage of a parallel code on a distributed memory system might be implemented as shown in Figure 7.1.13, where the steps have been organized to achieve good load balancing. For example, while s_1 is being computed in processor 1, the other processors can begin the computation of the inner products $\mathbf{u}_1^T \mathbf{a}_j$, although they cannot be completed, of course, until s_1 is available in all processors. We also elect to compute γ_1 and $a_{11} - s_1$ in all processors rather than computing them in processor 1 and then sending them. Thus, the only communication in this stage is the broadcast of the first column and s_1 to all processors. This is a major delay in the first stage, and for the second stage it would be advantageous to do the transmission of the second column as soon as it is complete. The bulk of the work in each Householder stage is the updating of the columns, and we see that this has excellent load-balancing properties until the final stages.

---

1. Send first column of A to all processors.
2. Compute s_1 in processor 1. Start computation of the $\mathbf{u}_1^T \mathbf{a}_j$ in other processors. Send s_1 to all processors.
3. Compute γ_1 and $a_{11} - s_1$ in all processors. Continue computation of the $\mathbf{u}_1^T \mathbf{a}_j$. Replace a_{11} by s_1 in processor 1.
4. Compute $\alpha_i = \gamma_1 \mathbf{u}_1^T \mathbf{a}_i$ in all processors.
5. Compute $\mathbf{a}_i - \alpha_i \mathbf{u}_1$ in all processors.

Figure 7.1.13: *First Stage of Parallel Inner Product Form of Householder*

We next consider the rank one update form of Householder reduction as given in Figure 7.1.6. The first stage of the corresponding parallel code is shown in Figure 7.1.14. In Figure 7.1.6, the $\mathbf{a}_i$ denote the rows of A. But since we are assuming that A is stored by columns, the computation in step 4 of Figure 7.1.14 is the same as in step 4 of Figure 7.1.13. That is, the computation of $\hat{\mathbf{v}}^T$ is just γ_1 times the inner products of $\mathbf{u}_1$ and the columns of A held by each processor. Similarly, the computation in step 5 of both figures are the same. Thus, the parallel forms given in Figure 7.1.13 and 7.1.14 are identical.

If A is stored by interleaved row storage, the parallel algorithms are not nearly as satisfactory as with column storage. Consider first the inner product form. Here, we need the inner products $\mathbf{u}_j^T \mathbf{a}_i$. The partial inner products obtained from the elements already in each processor are first computed, but a fan-in across the processors is needed to complete the inner product. Then

> 1. Send first column of A to all processors.
> 2. Compute s_1 in processor 1. Start computation of $\mathbf{v}_1$ in other processors. Send s_1 to all processors.
> 3. Compute μ_1 and $a_{11} - s_1$ in all processors. Continue computation of $\mathbf{v}_1$ in all processors. Replace a_{11} by s_1 in processor 1.
> 4. Compute $\hat{\mathbf{v}}_1$ in all processors.
> 5. Compute $\mathbf{a}_j - u_j \hat{\mathbf{v}}_1^T$ in all processors.

Figure 7.1.14: *First Stage of Parallel Rank One Form of Householder*

each of these inner products must be sent to all processors. The rank-one form suffers from the same problem. Here, we must compute the vector $\mathbf{v}_j^T$, which is a linear combination of rows of A, and in order to complete this we again need to fan-in across the processors and then transmit the result to all processors. We conclude that column storage is better than row storage for the Householder reduction.

Givens Reduction

Recall from (6.3.8) that in the first step of the Givens reduction the first two rows of A are modified by

$$\hat{\mathbf{a}}_1 = c_{12}\mathbf{a}_1 + s_{12}\mathbf{a}_2, \qquad \hat{\mathbf{a}}_2 = -s_{12}\mathbf{a}_1 + c_{12}\mathbf{a}_2, \qquad (7.1.12)$$

where s_{12} and c_{12} are the sine and cosine defined by (6.3.10). This step produces a zero in the (2,1) position of A. Subsequent steps zero the remaining elements of the first column, then the subdiagonal elements of the second column, and so on. The complete row-oriented algorithm is given in Figure 7.1.7 but we wish to consider first a column-oriented algorithm.

Assume that A is stored by column interleaved storage. Then the initial steps of a parallel algorithm are shown in Figure 7.1.15. Note that there is a delay at the outset in computing the first sine-cosine pair and transmitting them to all processors. In subsequent stages, we compute and transmit the next sine-cosine pair as soon as the data are ready; for example, as soon as processor 1 has updated its (1,1) element in step 2 it can perform step 3 while the other processors continue their updates of rows 1 and 3. During the first stage, processor 2 has the additional work of computing all the sine-cosine pairs so that it will complete the first stage after the other processors. However, column 1 is now complete and processor 1 has one less column so it will begin to catch up while processor 2 falls behind on the next stage, and so on. All in all, this form of the Givens reduction has good parallel properties.

7.1 Basic Methods

> 1. Compute c_{12} and s_{12} in processor 1. Send to all processors.
> 2. Begin update of rows 1 and 2 in all processors.
> 3. Compute c_{13} and s_{13} in processor 1. Send to all processors.
> 4. Complete update of rows 1 and 2 in all processors.
> 5. Begin update of rows 1 and 3 in all processors.
> $\vdots$

Figure 7.1.15: *Column Form of Parallel Givens*

The same is not true if we use row interleaved storage in a straightforward way. At the first step only processors 1 and 2 will be active in processing rows 1 and 2; at the second step only processors 2 and 3 will be active, and so on. This problem is alleviated to some extent by another type of parallelism that is inherent in the Givens method; see the Supplementary Discussion and Exercise 7.1.12.

Multiple Right-Hand Sides

We next make a few comments on the system $AX = B$, where X and B are $n \times q$. Suppose that any of the methods of this section has produced a decomposition $A = SU$, where U is upper triangular and S will depend on the decomposition used. Then we need to solve the systems

$$SY = B, \qquad UX = Y. \tag{7.1.13}$$

Consider first the system $UX = Y$. Assume that there are q processors, that U is stored in each processor, and that the ith column of Y is stored in processor i. Then the q systems $U\mathbf{x}_i = \mathbf{y}_i$ can be solved in parallel, and the analogous computations can be done for $SY = B$. Thus, these computations can be done with perfect parallelism, and we next discuss the practicality of this approach.

As mentioned previously, there are two main cases in which multiple right-hand sides arise. The first is when $B = I, q = n$, and $X = A^{-1}$. The second is where there may be different loads to be analyzed in a structures problem or analogous situations for other problems. In this case, q may vary from small to very large. Clearly, the above approach is not viable if $q << p$, where p is the number of processors, since $p - q$ processors are left idle. In this case, we would solve the systems, with, for example, the column sweep algorithm. For example, if $q = 5$ and $p = 20$ we might allocate four processors to each system.

On the other hand, if $p << q$, then the solution of q/p systems in each processor is potentially attractive. The main difficulty is that if the reduction has been carried out by, for example, LU decomposition with interleaved row storage, then U will be distributed across the processors with the same interleaved row storage and it will be necessary to collect all of U in each processor before doing the computations. Only a more detailed analysis for a particular parallel system can ascertain if this is more efficient than solving each system $U\mathbf{x}_i = \mathbf{y}_i$ across the processors.

Supplementary Discussion and References: 7.1

1. A variation of Gaussian elimination is the Gauss-Jordan algorithm, which reduces A to a diagonal matrix, rather than upper triangular. At the kth stage, the partially reduced matrix has the form

$$\begin{pmatrix} a_{11} & & & \hat{a}_{1k} & \cdots & \hat{a}_{1n} \\ & \ddots & & \vdots & & \vdots \\ & & \hat{a}_{k-1,k-1} & & & \\ & & & \hat{a}_{kk} & \cdots & \hat{a}_{kn} \\ & & & \vdots & & \vdots \\ & & & \hat{a}_{nk} & \cdots & \hat{a}_{nn} \end{pmatrix}$$

where the carets indicate that the elements have probably been changed from their original values. Assuming that $\hat{a}_{kk} \neq 0$, a multiple of the kth row is then subtracted from all other rows so as to eliminate all entries in the kth column, except the diagonal one. The basic update step is the axpy $\mathbf{a}_j - a_{kj}\mathbf{m}_k$, where $\mathbf{a}_j$ is the current column of A and $\mathbf{m}_k$ is the vector of multipliers a_{ik}/a_{kk} with the kth component of $\mathbf{m}_k$ set to zero. Partial pivoting is easily incorporated. Peters and Wilkinson [1975] have shown that the Gauss-Jordan algorithm has essentially the same numerical stability properties as Gaussian elimination with the exception that the residual of the computed solutions may be larger for Gauss-Jordan, especially for ill-conditioned matrices. This problem is corrected by Dekker and Hoffman [1989] by using partial pivoting with column interchanges rather than row interchanges. The potential advantage of the Gauss-Jordan algorithm on vector or parallel machines is that the vector lengths remain n throughout the computation, rather than decrease in length as the reduction proceeds. However, the serial operation count is $O(n^3)$ rather than the $O(2n^3/3)$ of Gaussian elimination and the extra operations tend to cancel out the advantage of the longer vector lengths. Moreover, for banded matrices, there will be complete fill above the main diagonal in general and the elimination of this fill substantially increases the operation count over Gaussian elimination.

7.1 Basic Methods

2. The idea of interleaved storage was considered independently by O'Leary and Stewart [1985], who call it *torus assignment*, Ipsen et al. [1986], who call it *scattering*, and Geist and Heath [1986], who call it a *wrapped mapping*. O'Leary and Stewart [1985] also consider a related reflection storage.

3. The discussion of Givens reduction assumed that zeros are produced in the lower-triangular part of A in the usual order. But there are many other orderings, called *annihilation patterns*, that have the basic property that once a zero is produced, it remains zero. Many of these orderings have good parallel properties and have been proposed for use with Givens method; see, for example, Exercise 7.1.12. For further discussion and references, see Ortega [1988a].

4. The costs of partial pivoting in Gaussian elimination can be substantially reduced by *pairwise pivoting*. In the first step a_{11} is compared with a_{21} and rows 1 and 2 interchanged if $|a_{11}| < |a_{21}|$. At the next step, a_{31} is compared with the current a_{11}, and so on. Clearly, all multipliers l_{ij} satisfy $|l_{ij}| \le 1$, although the algorithm is not quite as stable as partial pivoting. See Sorenson [1985] for further discussion and analysis.

EXERCISES 7.1

7.1.1. Use the summation formulas of Exercise 6.1.4 to show that the expression in (7.1.1) is
$$\frac{1}{3}\frac{n(n-1)(2n-1)}{n(n-1)} = \frac{1}{3}(2n-1) = O(2n/3)$$

7.1.2. Formulate the inner product and column sweep algorithms for the solution of a lower triangular system of equations $L\mathbf{x} = \mathbf{b}$.

7.1.3. Use the summation formulas of Exercise 6.1.4 to verify (7.1.4), (7.1.5), (7.1.7) and (7.1.9).

7.1.4. Formulate a column-oriented rank-one update form of Householder reduction by using Householder transformations on the right: $A(I - \alpha\mathbf{w}\mathbf{w}^T)$. Thus, in place of the QR decomposition of A the sequence of transformations
$$AP_1P_2\cdots P_{n-1} = L$$
gives an LQ decomposition, where L is lower triangular and Q is orthogonal. This is equivalent to doing the QR decomposition of A^T, but without performing an explicit transposition of A.

7.1.5. Formulate a column-oriented Givens reduction and discuss its vector properties vis-a-vis the row-oriented algorithm of Figure 7.1.7.

7.1.6. Assume that a parallel system requires t time units to do arithmetic operations and αt time units to send one word of data from one to arbitrarily many processors. Assume that the number of processors $p = n$, the size of the system. Without pivoting, compute the total time for the LU decomposition using storage by rows and also storage by columns, as discussed in the text. Conclude which is best as a function of α.

7.1.7. Assume that a parallel system has the characteristics of Exercise 7.1.6 and that the comparison of the magnitude of two numbers also requires t time units. Compare the two algorithms with partial pivoting added.

7.1.8. Assume that a parallel system has the characteristics of Exercise 7.1.6, and assume that $n = kp$. Compare the number of time steps for LU decomposition, without pivoting, if (a) the first k rows of A are stored in processor 1, the second k in processor 2, and so on, and (b), if the storage of the rows is interleaved as shown in Figure 7.1.8. Repeat the analysis if partial pivoting is added.

7.1.9. Repeat Exercise 7.1.8 if A is stored by columns.

7.1.10. Discuss the parallel column sweep and inner product algorithms for a lower-triangular system.

7.1.11. Show that a straightforward implementation of the column sweep algorithm on a shared memory machine with assignment of tasks by interleaved columns has very poor load balancing. Discuss ways to improve the implementation.

7.1.12. Consider the *Sameh-Kuck* annihilation pattern illustrated below for an 8×8 matrix.

row							
2	7						
3	6	8					
4	5	7	9				
5	4	6	8	10			
6	3	5	7	9	11		
7	2	4	6	8	10	12	
8	1	3	5	7	9	11	13

Extend this pattern to general n and discuss which Givens transformations can be done in parallel. Compare with the parallelism of (6.3.11). Can you find an annihilation pattern with better parallelism?

7.2 Other Organizations of Factorization

In the previous section, we discussed some basic parallel and vector approaches to LU and other factorizations. But there are many other organizations of the factorization algorithms that have potentially desirable properties

7.2 Other Organizations of Factorization

on parallel and vector machines. In this section we consider several such possibilities.

The ijk forms of LU Decomposition

Just as for matrix multiplication, as discussed in Chapter 3, there are six different organizations of LU factorization that are obtained by permuting the indices i, j, k in the generic triple loop

$$\begin{aligned} &\text{For } \_\_\_ \\ &\quad \text{For } \_\_\_ \\ &\quad\quad \text{For } \_\_\_ \\ &\quad\quad\quad a_{ij} = a_{ij} - l_{ik}a_{kj}. \end{aligned} \quad (7.2.1)$$

The permutation of i, j, k occurs only in the For statements and the arithmetic statement remains the same in all cases. Also, we do not show the formation of the multipliers, and the actual final codes are somewhat more complicated than is suggested by (7.2.1). (See Figures 7.1.1, 7.1.2, 7.2.1, and Exercise 7.2.3.) As examples, the row-oriented algorithm of Figure 7.1.1 is the kij form, and the column-oriented algorithm of Figure 7.1.2 is the kji form. Another important organization is the jki form shown in Figure 7.2.1. This is a column-oriented algorithm and a corresponding row-oriented algorithm is the ikj form (see Exercise 7.2.3).

For $j = 2$ to n
$\quad$ For $s = j$ to n
$\quad\quad l_{sj-1} = a_{sj-1}/a_{j-1,j-1}$
$\quad$ For $k = 1$ to $j - 1$
$\quad\quad$ For $i = k + 1$ to n
$\quad\quad\quad a_{ij} = a_{ij} - l_{ik}a_{kj}$

Figure 7.2.1: *The jki Form of LU Decomposition*

The data access patterns of the kji and jki algorithms are shown in Figure 7.2.2. In the kth stage of the kji form, the kth column of L and $(k+1)$st row of U are computed and the remaining submatrix updated. The previously computed portions of L and U are not accessed and this is sometimes called a *right-looking* variant of LU decomposition since data references are to the right. In contrast, in the jth stage of the jki form, the jth columns of L and U are computed, using the columns of L previously computed. Thus, data references are to the left of the current column and this is a *left-looking* variant.

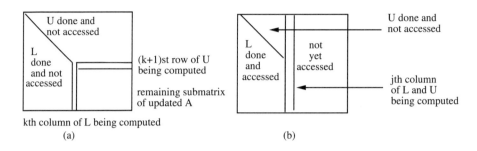

kth column of L being computed
(a)   (b)

Figure 7.2.2: *Data Access Patterns of (a) kji and (b) jki*

Another important difference between the kji and jki forms is that the kji algorithm updates the remaining submatrix as soon as possible, whereas the jki algorithm does all the updates on column j during the jth stage. Thus, the kji form is known as an *immediate update* algorithm and the jki form is a *delayed update* algorithm. On the other hand, the inner loop of both the kji and jki forms (as well as their kij and ikj row-oriented counterparts) is an axpy operation. This is in contrast to the ijk and jik forms in which the inner loop is an inner product. (See Exercise 7.2.3 for these as well as the sixth form.) The basic properties of the six ijk forms are summarized in Table 7.2.1. Note that the ijk and jik forms require access by both rows and columns so that no matter how A is stored, some vectors will have stride greater than one.

Form	Operation	Update	A access	L access
kij	Axpy	Immediate	Row	Scalar
kji	Axpy	Immediate	Column	Column
ikj	Axpy	Delayed	Row	Scalar
jki	Axpy	Delayed	Column	Column
ijk	Inner product	Delayed	Column	Row
jik	Inner product	Delayed	Column	Row

Table 7.2.1: The ijk Forms of LU Decomposition

The jki Form

The jki form has an advantage over the kji form on some vector computers with registers. At the jth stage of the LU decomposition, the steps are:

Load first column of L

Load jth column of A

7.2 Other Organizations of Factorization

Modify jth column of A; load second column of L

Modify jth column of A; load third column of L

$$\vdots$$

The advantage is that there is only one store of the jth column of A, when its modification is complete. This is in contrast with the kji form in which a column of A is loaded, modified by an axpy operation and stored. The jki form was particularly useful for the CRAY-1, for which it was not possible to both load and store vector registers simultaneously.

A problem with the jki form is that the axpy operations at the jth stage work with vectors of different lengths since they utilize the columns of L, and it is useful to reformulate the algorithm in terms of matrix-vector operations. Assume that the first $j-1$ columns of L have been computed and the first j columns of A updated. At the beginning of the next stage the jth column of L is computed and then the update of the $(j+1)$st column of A can be done in two stages. The first stage updates the first $j+1$ components of the $(j+1)$st column of A by restricting the i loop in Figure 7.2.1 to run only up to $j+1$. It is easy to see (Exercise 7.2.1) that this is equivalent to solving the triangular system of equations

$$\begin{bmatrix} 1 & & & \\ l_{21} & 1 & & \\ \vdots & & \ddots & \\ l_{j+1,1} & \cdots & l_{j+1,j} & 1 \end{bmatrix} \begin{bmatrix} a'_{1,j+1} \\ \vdots \\ a'_{j+1,j+1} \end{bmatrix} = \begin{bmatrix} a_{1,j+1} \\ \vdots \\ a_{j+1,j+1} \end{bmatrix}, \quad (7.2.2)$$

where the $a'_{i,j+1}$ denote the updated components of the $(j+1)$st column. Then the remainder of the $(j+1)$st column can be updated by

$$\begin{bmatrix} a'_{j+2,j+1} \\ \vdots \\ a'_{n,j+1} \end{bmatrix} = \begin{bmatrix} a_{j+2,j+1} \\ \vdots \\ a_{n,j+1} \end{bmatrix} - \begin{bmatrix} l_{j+2,1} & \cdots & l_{j+2,j} \\ \vdots & & \vdots \\ l_{n,1} & \cdots & l_{n,j} \end{bmatrix} \begin{bmatrix} a'_{1,j+1} \\ \vdots \\ a'_{j,j+1} \end{bmatrix}. \quad (7.2.3)$$

Note that (7.2.2) corresponds to updating the columns of A using columns of L of different lengths.

Block Algorithms

As discussed in Section 3.3, block algorithms for matrix-matrix multiplication can be more efficient for machines with cache memory or vector registers. The same is true for LU decomposition. We next describe the basic structure of a block LU decomposition algorithm corresponding to the kji form of Figure 7.1.2.

Let n_b be the block size and write A as

$$A = \begin{bmatrix} A_{11} & A_{12} \\ A_{21} & A_{22} \end{bmatrix}, \qquad (7.2.4)$$

where A_{11} is $n_b \times n_b$. At the first stage, the first step is the factorization

$$\begin{bmatrix} A_{11} \\ A_{21} \end{bmatrix} = \begin{bmatrix} L_{11} \\ L_{21} \end{bmatrix} U_{11}, \qquad (7.2.5)$$

where L_{11} is lower-triangular with unit diagonal. This is analogous to finding the first column of multipliers in the kji algorithm. It also is equivalent to performing Gaussian elimination on the first n_b columns of A. The factorization (7.2.5) can be achieved by the kji form for Gaussian elimination, using partial pivoting if necessary. We will discuss shortly how to incorporate partial pivoting into the overall block algorithm.

Next, U_{12} is computed by solving

$$L_{11} U_{12} = A_{12}, \qquad (7.2.6)$$

which is a triangular system with multiple right-hand sides. The first stage is then completed by updating A_{22}:

$$A'_{22} = A_{22} - L_{21} U_{12}. \qquad (7.2.7)$$

The basic operation here is a matrix-matrix multiplication followed by the indicated subtraction. Note that this is an immediate update, as called for by the kji paradigm.

If necessary, row interchanges are easily incorporated into the algorithm. A record is kept of the row interchanges performed in the factorization (7.2.5), and before (7.2.6) and (7.2.7) are done these same interchanges are performed on the rows of A_{12} and A_{22}.

At the end of the first stage, the matrix has the form

$$\begin{bmatrix} U_{11} & U_{12} \\ 0 & A'_{22} \end{bmatrix},$$

and now the same process is applied to the updated matrix A'_{22}. We continue in this fashion until A has been reduced to upper-triangular form.

We next describe the basic structure of a block version of the jki algorithm. Let A now be written as

$$A = \begin{bmatrix} A_{11} & A_{12} & \cdots & A_{1p} \\ \vdots & & & \vdots \\ A_{p1} & \cdots & & A_{pp} \end{bmatrix}, \qquad (7.2.8)$$

7.2 Other Organizations of Factorization

where all submatrices A_{ij} are $n_b \times n_b$ with the possible exception of those in the last block row and column (if n_b divides n, these will also be $n_b \times n_b$). The first stage of the algorithm is again (7.2.5) which, with the notation of (7.2.8), becomes

$$\begin{bmatrix} A_{11} \\ \vdots \\ A_{p1} \end{bmatrix} = \begin{bmatrix} L_{11} \\ \vdots \\ L_{p1} \end{bmatrix} U_{11}. \tag{7.2.9}$$

Next, U_{12} is obtained by solving the triangular system

$$L_{11}U_{12} = A_{12}, \tag{7.2.10}$$

and the remainder of the second block column of A is updated by

$$\begin{bmatrix} A'_{22} \\ \vdots \\ A'_{p2} \end{bmatrix} = \begin{bmatrix} A_{22} \\ \vdots \\ A_{p2} \end{bmatrix} - \begin{bmatrix} L_{21} \\ \vdots \\ L_{p1} \end{bmatrix} U_{12}. \tag{7.2.11}$$

In keeping with the delayed update paradigm of the jki form, the remainder of A is not updated at this stage.

In the second stage, the updated second block column is first factored:

$$\begin{bmatrix} A'_{22} \\ \vdots \\ A'_{p2} \end{bmatrix} = \begin{bmatrix} L_{22} \\ \vdots \\ L_{p2} \end{bmatrix} U_{22}. \tag{7.2.12}$$

Next, U_{13} and U_{23} are obtained by solving the block triangular system

$$\begin{bmatrix} L_{11} & \\ L_{21} & L_{22} \end{bmatrix} \begin{bmatrix} U_{13} \\ U_{23} \end{bmatrix} = \begin{bmatrix} A_{13} \\ A_{23} \end{bmatrix}. \tag{7.2.13}$$

Then the remainder of the third block column of A is updated by

$$\begin{bmatrix} A'_{33} \\ \vdots \\ A'_{p3} \end{bmatrix} = \begin{bmatrix} A_{33} \\ \vdots \\ A_{p3} \end{bmatrix} - \begin{bmatrix} L_{31} \\ \vdots \\ L_{p1} \end{bmatrix} U_{13} - \begin{bmatrix} L_{32} \\ \vdots \\ L_{p2} \end{bmatrix} U_{23}$$

$$= \begin{bmatrix} A_{33} \\ \vdots \\ A_{p3} \end{bmatrix} - \begin{bmatrix} L_{31} & L_{32} \\ \vdots & \vdots \\ L_{p1} & L_{p2} \end{bmatrix} \begin{bmatrix} U_{13} \\ U_{23} \end{bmatrix}. \tag{7.2.14}$$

Note that the update of A_{23} is included in the solution of (7.2.13), which is equivalent to

$$U_{13} = L_{11}^{-1} A_{13}, \qquad U_{23} = L_{22}^{-1}(A_{23} - L_{21}U_{13}),$$

where $A_{23} - L_{21}U_{13}$ is the A_{23} update. Note also that (7.2.14) contains the current update as well as the update delayed from the first stage.

The continuation of the algorithm is as follows. At the beginning of the $(k+1)$st stage we have

$$\begin{bmatrix} L_{11} & & & \\ & \ddots & & \\ \vdots & & L_{kk} & \\ \vdots & & & \\ L_{p1} & \cdots & L_{pk} & \end{bmatrix} \begin{bmatrix} U_{11} & \cdots & & U_{1k} & A_{1,k+1} & \cdots & A_{1p} \\ & \ddots & & \vdots & \vdots & & \vdots \\ & & U_{k-1,k-1} & U_{k-1,k} & & & \\ & & & A'_{kk} & & & \\ & & & \vdots & & & \\ & & & A'_{pk} & A_{p,k+1} & \cdots & A_{pp} \end{bmatrix}$$

where the indicated parts of L and U have been computed, the remainder of the kth block column of A has been updated, and the remaining block columns of A have not yet been processed. The following steps of the $(k+1)$st stage then correspond to (7.2.12), (7.2.13) and (7.2.14): the factorization, solve and update.

$$\begin{bmatrix} A'_{kk} \\ \vdots \\ A'_{pk} \end{bmatrix} = \begin{bmatrix} L_{kk} \\ \vdots \\ L_{pk} \end{bmatrix} U_{kk}, \qquad (7.2.15)$$

$$\begin{bmatrix} L_{11} & & \\ \vdots & \ddots & \\ L_{k1} & \cdots & L_{kk} \end{bmatrix} \begin{bmatrix} U_{1,k+1} \\ \vdots \\ U_{k,k+1} \end{bmatrix} = \begin{bmatrix} A_{1,k+1} \\ \vdots \\ A_{k,k+1} \end{bmatrix}, \qquad (7.2.16)$$

$$\begin{bmatrix} A'_{k+1,k+1} \\ \vdots \\ A'_{p,k+1} \end{bmatrix} = \begin{bmatrix} A_{k+1,k+1} \\ \vdots \\ A_{p,k+1} \end{bmatrix} - \begin{bmatrix} L_{k+1,1} & \cdots & L_{k+1,k} \\ \vdots & & \vdots \\ L_{p,1} & \cdots & L_{p,k} \end{bmatrix} \begin{bmatrix} U_{1,k+1} \\ \vdots \\ U_{k,k+1} \end{bmatrix}. \qquad (7.2.17)$$

Partial pivoting is incorporated into this algorithm in the analogous way as in the block kji algorithm: Interchanges required in the factorizations (7.2.9), (7.2.12) or (7.2.15) are applied to the remainder of A before final processing; that is, all necessary interchanges are applied to the $(k+1)$st block column before beginning the computations (7.2.16) and (7.2.17) of the $(k+1)$st stage.

Supplementary Discussion and References: 7.2

1. Dongarra et al.[1984] introduced in a systematic way the ijk forms of LU decomposition. Earlier, Moler [1972] advocated the kji form for serial computers with virtual memory; this form had also become standard on the CDC vector computers. Fong and Jordan [1977] observed that the jki form was quite suitable for the CRAY-1. See Ortega [1988a,b] for

7.2 Other Organizations of Factorization

a more detailed treatment of the ijk forms for both LU and Cholesky factorization as well as some other algorithms. See also Mattingly et al.[1989] for a discussion of ijk forms for Givens and Householder reductions.

2. One advantage of viewing the jki algorithm in the matrix form (7.2.2) and (7.2.3) is that we can use the BLAS to carry out the computations. Recall from Section 3.3 that the BLAS are a set of subroutines, assumed highly optimized for each particular machine, to do basic vector and matrix operations.

3. Block algorithms form the basis for LAPACK, the successor to LINPACK. For more discussion of block algorithms in the context of LAPACK, see Anderson et al.[1992] and Dongarra and Duff et al.[1990].

4. For the implementation of block algorithms on distributed memory machines, Dongarra and van de Geijn [1992] suggest the extension of the interleaved column storage of Section 7.1 to interleaved blocks of columns. They call this interleaved *panel storage*.

5. The QR factorization based on the Householder reduction can also be done in block form; see Bischof and Van Loan [1987], Schreiber and Van Loan [1989] and Puglisi [1992].

EXERCISES 7.2

7.2.1. Verify that the update of the first $j+1$ components of A in the jki LU decomposition algorithm is given by the solution of the triangular system (7.2.2).

7.2.2. Repeat the discussion of this section for Cholesky factorization, rather than LU factorization.

7.2.3. Show that the following algorithms give LU decompositions of A and that inner products are the basic operations of the ijk and jik forms.

For $i=2$ to n	For $j=2$ to n	For $i=2$ to n
For $j=2$ to i	For $s=j$ to n	For $k=1$ to $i-1$
$l_{ij-1}=a_{ij-1}/a_{j-1,j-1}$	$l_{sj-1}=a_{sj-1}/a_{j-1,j-1}$	$l_{ik}=a_{ik}/a_{kk}$
For $k=1$ to $j-1$	For $i=2$ to j	For $j=k+1$ to n
$a_{ij}=a_{ij}-l_{ik}a_{kj}$	For $k=1$ to $i-1$	$a_{ij}=a_{ij}-l_{ik}a_{kj}$
For $j=i+1$ to n	$a_{ij}=q_{ij}-l_{ik}a_{kj}$	
For $k=1$ to $i-1$	For $i=j+1$ to n	
$a_{ij}=a_{ij}-l_{ik}a_{kj}$	For $k=1$ to $j-1$	
	$a_{ij}=a_{ij}-l_{ik}a_{kj}$	
ijk form	jik form	ikj form

7.3 Banded and Tridiagonal Systems

In the previous two sections we assumed that the matrix A was full; now we treat banded systems. For simplicity, as in Section 6.1, we shall consider explicitly only symmetrically-banded systems with semibandwidth β but most of the discussion extends in a natural way to general banded systems.

LU Decomposition

We begin with LU decomposition. The pseudocodes of Figure 7.1.1 and 7.1.2 adapted to a system with semibandwidth β are shown in Figure 7.3.1. Adaptations of the other ijk forms of LU decomposition in Section 7.1 are easily given (see Exercise 7.3.1 for the jki form).

For $k = 1$ to $n - 1$
  For $i = k + 1$ to $\min(k + \beta, n)$
    $l_{ik} = a_{ik}/a_{kk}$
    For $j = k + 1$ to $\min(k + \beta, n)$
      $a_{ij} = a_{ij} - l_{ik}a_{kj}$

For $k = 1$ to $n - 1$
  For $s = k + 1$ to $\min(k + \beta, n)$
    $l_{sk} = a_{sk}/a_{kk}$
  For $j = k + 1$ to $\min(k + \beta, n)$
    For $i = k + 1$ to $\min(k + \beta, n)$
      $a_{ij} = a_{ij} - l_{ik}a_{ij}$

(a) kij (b) kji

Figure 7.3.1: *Banded LU Decomposition*

On serial computers, it is common to store a banded matrix by diagonals as discussed in Section 6.1, but this is not satisfactory for vector computers. For example, for the kij algorithm of Figure 7.3.1, the natural storage is by rows. However, we do not wish to use a full $n \times n$ two-dimensional array for the storage since much of this space would be wasted, especially when β is small. An alternative is to store the rows in a one-dimensional array as indicated in Figure 7.3.2 in which $\mathbf{a}_i$ denotes the ith row and the corresponding row length is shown. Figure 7.3.2 also provides a schematic for column storage if $\mathbf{a}_i$ is interpreted as the ith column.

The row or column updates in the last loops of Figure 7.3.1 can utilize vectors of length β until the final $\beta \times \beta$ submatrix is reached and the lengths then decrease by 1 at each subsequent stage. Thus, when β is suitably large, Figure 7.3.1 provides the basis for potentially satisfactory algorithms for vector computers, but they are increasingly inefficient as β becomes small. In particular, for tridiagonal systems, for which $\beta = 1$, they are totally inadequate. We will consider other algorithms for small bandwidth systems shortly.

7.3 Banded and Tridiagonal Systems

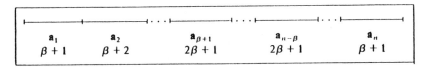

Figure 7.3.2: *Row (Column) Storage for Banded Matrix*

Parallel Implementation

We treat next the parallel implementation of LU decomposition. As in Section 7.1, we assume that we will use row or column wrapped interleaved storage, as illustrated in Figure 7.1.8. The general considerations of Section 7.1 for LU decomposition apply also to the banded case, but the major issue now is the effect of the bandwidth on the parallelism.

For the algorithms of Figure 7.3.1, the basic parallelism of the updates of the last loop is β, the same as the vector length. In particular, with row interleaved assignment, at the first stage rows $2, \ldots, \beta + 1$ will have an element in the first column to be eliminated. These β rows can then be updated in parallel in the kij form. Hence, if $\beta < p$, the number of processors, then only β processors will be used during the first stage. Thus, we need $\beta \geq p$ in order to utilize fully all the processors. Moreover, there is more potential for load imbalance with banded systems even if $\beta \geq p$. For example, suppose that we have 50 processors and $n = 10{,}032$. Then 32 processors will have 201 rows and 18 will have 200. For a full system, this would be a very small imbalance. Suppose, however, that $\beta = 75$. Then the number of rows each processor holds is still 201 or 200 but, with a straightforward implementation of the kij algorithm, only half of the processors would be active during the second part of a stage. Clearly, the worst case is when $\beta = p + 1$ so that only one processor would be active during the second part of the stage. This load imbalance problem can be alleviated to a large extent by a compute-ahead strategy. For example, during the kth stage, as soon as row $k + 1$ has been updated it will be sent to all other processors. Then, when a processor finishes its work on the kth stage it will immediately start the $(k+1)$st stage.

Triangular Systems

The treatment in Section 7.1 of the solution of triangular systems on vector computers goes over immediately to banded systems. Consider first the solution of the triangular system $U\mathbf{x} = \mathbf{c}$, where U has an upper semibandwidth of β as shown below:

$$\begin{bmatrix} u_{11} & \cdots & u_{1\beta+1} & & \\ & \ddots & & & \\ & & \ddots & u_{n-\beta,n} & \\ & & & & \vdots \\ & & & & u_{nn} \end{bmatrix} \begin{bmatrix} x_1 \\ \vdots \\ x_n \end{bmatrix} = \begin{bmatrix} c_1 \\ \vdots \\ c_n \end{bmatrix}. \quad (7.3.1)$$

If U is stored by its columns within the band, the column sweep algorithm of Figure 7.1.3 may be written as in Figure 7.3.3. In Figure 7.3.3, $\mathbf{u}_j$ consists of the elements in the jth column from the edge of the band to the main diagonal, but not including the main diagonal element. These vectors all have length β until $j \leq \beta$ at which point the vector lengths decrease to 1 as j decreases to 2. Thus, the vector lengths of the axpy operations are β for most of the computation, decreasing to 1 in the final stage.

For $j = n$ down to 2
$x_j = c_j/u_{jj}$
$\mathbf{c} = \mathbf{c} - x_j \mathbf{u}_j$
$x_1 = c_1/u_{11}$

Figure 7.3.3: *Column Sweep Algorithm*

If U is stored by rows, we would use the inner product algorithm (7.1.2), which now becomes

$$x_i = (c_i - u_{ii+1}x_{i+1} - \cdots - u_{i,i+\beta+1}x_{i+\beta+1})/u_{ii}, \quad i = n-\beta, \ldots, 1. \quad (7.3.2)$$

For $i = n, \ldots, n - \beta + 1$, (7.3.2) would be modified to reflect a row length less than $\beta + 1$. An algorithm based on (7.3.2) consists of inner products of vectors of length β, while the corresponding inner products for $i = n, \ldots, n - \beta + 1$ use vectors of lengths $1, \ldots, \beta - 1$.

As with full matrices, the column sweep algorithm may be appealing on some vector computers because of the axpy operations as opposed to the inner products in the row algorithm. Moreover, for the row-oriented kij algorithm of Figure 7.3.1, and with storage as indicated by Figure 7.3.2, the element b_i of the right-hand side of the system cannot just be appended to the ith row of A, as was the case for full systems. This points more strongly towards the use of storage by columns than is the case for full matrices. However, for small

7.3 Banded and Tridiagonal Systems

β and several right-hand sides, this is no longer the case, as will be discussed shortly.

Consider, next, distributed memory machines and assume that U is stored by interleaved row storage. The parallel column sweep algorithm given in Figure 7.1.10 is modified for banded systems as shown in Figure 7.3.4. Clearly, the parallelism in this algorithm in the beginning stages is β. If $\beta < p$, then $p - \beta$ processors are not used. Even if $\beta > p$ it is likely that there will be an imbalance in the workload unless a compute-ahead strategy is included, as previously discussed for the reduction phase. If U is stored by interleaved column storage, then the inner product algorithm in Figure 7.1.11 is easily adapted to banded systems.

Compute $x_n = c_n/u_{nn}$. Send x_n to all processors
Compute $c_i = c_i - u_{in}x_n$, $\quad i = n - \beta, \ldots, n - 1$
Compute $x_{n-1} = c_{n-1}/u_{n-1,n-1}$. Send x_{n-1} to all processors
Compute $c_i = c_i - u_{i,n-1}x_{n-1}$, $\quad i = n - \beta - 1, \ldots, n - 2$
$\vdots$

Figure 7.3.4: *Parallel Column Sweep Algorithm for Banded Systems*

Interchanges

As discussed in Section 6.2, the LU or Cholesky decompositions preserve the bandwidth (that is, the upper triangular matrix U will have the same bandwidth as A) but the row interchanges needed for partial pivoting increase the bandwidth. This has consequences somewhat more severe for vector and parallel computers than for serial ones. One way to handle the problem of increased bandwidth is simply to allow enough extra storage at the outset. Thus, if A is stored by rows, we would modify Figure 7.3.2 as shown in Figure 7.3.5, so that β extra storage locations are allowed for each row down to row $n - 2\beta$, at which point the storage requirement decreases by one for each row. In the LU decomposition using this storage pattern, we would like to use vector lengths that are no longer than necessary. Thus, for example, if there is no interchange at the first stage, then vector lengths of only β are required, as before. In general, whenever there is an interchange, we will set the vector lengths for the subsequent operations accordingly. This adds both overhead and program complexity but is probably preferable to using the maximum vector lengths of Figure 7.3.5 when there are few interchanges or when the interchanges introduce relatively few extra nonzero elements. The considerations for storage by column are similar.

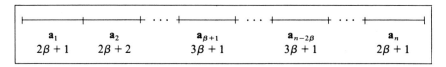

Figure 7.3.5: *Row Storage for Interchanges*

Multiple Right-Hand Sides

With multiple right-hand sides $(\mathbf{b}_1, \ldots, \mathbf{b}_q) = B$ and storage of A and B by columns, the algorithm of Figure 7.3.1(b) may be modified to process the q right-hand sides during the reduction. This adds q axpy operations during each stage of the reduction, just as with full matrices.

If A is stored by rows, we would like to append the rows of B to the rows of A. As discussed in Section 7.1 for full matrices A, this increases the vector lengths during the reduction by q. For banded systems, however, we encounter the problem illustrated below:

$$\begin{array}{llllll} a_{11} & \cdots & a_{1\beta+1} & b_{11} & \cdots & b_{1q} \\ a_{21} & \cdots & a_{2\beta+1} & a_{2\beta+2} & b_{21} & \cdots & b_{2q}. \end{array}$$

Because of the varying lengths of the rows of A, the columns of B do not align properly. If β and q are both reasonably large, we can forego the idea of appending the rows of B and operate on A and B separately.

A case of some interest is when β is small, but q is sufficiently large that vector operations of length q are efficient. In this case it may well be cost efficient to do the decomposition of A and then do vector operations for processing all right-hand sides "at once."

Cyclic Reduction

The methods discussed so far are satisfactory for banded systems as long as the bandwidth β is sufficiently large. For small β, however, the vector lengths or the degree of parallelism are too small. In contrast to matrix-vector multiplication, as discussed in Section 3.3, there are no efficient elimination algorithms based on the diagonals of A.

The previous algorithms are completely unsatisfactory in the important special case of tridiagonal matrices, for which $\beta = 1$. We next describe another

7.3 Banded and Tridiagonal Systems

approach to the solution of tridiagonal systems, which we write in the form

$$a_1 x_1 + b_1 x_2 = d_1$$
$$c_2 x_1 + a_2 x_2 + b_2 x_3 = d_2$$
$$c_3 x_2 + a_3 x_3 + b_3 x_4 = d_3 \qquad (7.3.3)$$
$$\vdots$$
$$c_n x_{n-1} + a_n x_n = d_n.$$

We multiply the first equation of (7.3.3) by c_2/a_1 and subtract it from the second equation to eliminate the coefficient of x_1 in the second equation; this is just the usual first step of Gaussian elimination. But next we multiply the third equation by b_2/a_3 and also subtract it from the second equation to eliminate the coefficient of x_3 in the second equation. This produces a new second equation of the form

$$a_2' x_2 + b_2' x_4 = d_2'.$$

We now do the same thing with equations 3, 4 and 5, subtracting multiples of equations 3 and 5 from equation 4 to produce a new equation of the form

$$c_3' x_2 + a_4' x_4 + b_4' x_6 = d_4',$$

in which the coefficients of x_3 and x_5 have been eliminated from the original fourth equation. We continue in this way, working with overlapping groups of three equations to produce new middle equations in which the odd numbered variables have been eliminated. This is illustrated schematically in Figure 7.3.6 for $n = 7$. Assuming that n is odd, we have at the end of the process a modified system

$$a_2' x_2 + b_2' x_4 = d_2'$$
$$c_2' x_2 + a_4' x_4 + b_4' x_6 = d_4' \qquad (7.3.4)$$
$$\vdots$$

involving only the variables $x_2, x_4, \ldots, x_{n-1}$.

The system (7.3.4) is tridiagonal in the variables $x_2, x_4, \ldots$, and we can repeat the process to obtain a new system containing only the variables $x_4, x_8, \ldots$. We continue in this fashion until no further reduction is possible. In particular, if $n = 2^q - 1$, the algorithm will terminate with a single final equation. This final equation is solved and then a back substitution is initiated. This is illustrated in Figure 7.3.7 for $n = 7$, which continues the schematic of Figure 7.3.6. In Figure 7.3.7, the three equations obtained from the first step, as shown in Figure 7.3.6, are combined to yield a final equation in the single unknown x_4. This is solved and then the first and last reduced equations can be solved for x_2 and x_6. The original first and last equations can then be solved for x_1 and x_7 and the original third and fifth equations may be solved for x_3 and x_5. If

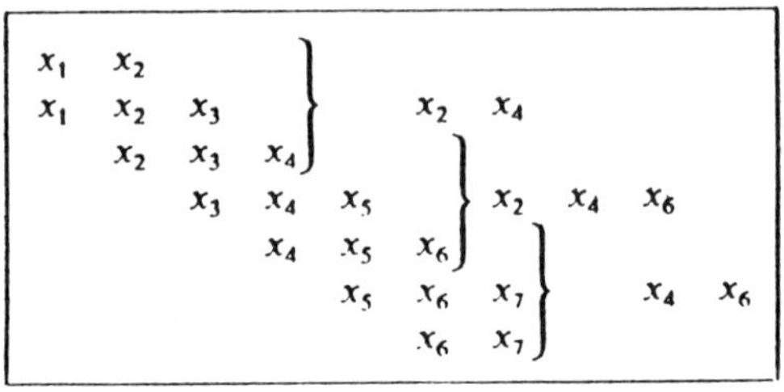

308   Chapter 7 Parallel Direct Methods

$$\begin{bmatrix} x_1 & x_2 & & & & & & & & & \\ x_1 & x_2 & x_3 & & & & x_2 & x_4 & & & \\ & x_2 & x_3 & x_4 & & & & & & & \\ & & x_3 & x_4 & x_5 & & & x_2 & x_4 & x_6 & \\ & & & x_4 & x_5 & x_6 & & & & & \\ & & & & x_5 & x_6 & x_7 & & x_4 & x_6 & \\ & & & & & x_6 & x_7 & & & & \end{bmatrix}$$

Figure 7.3.6: *Cyclic Reduction*

$n \neq 2^q - 1$, the process can be terminated in a system with a small number of variables which can be solved separately before the back substitution begins. An alternative is to add dummy equations of the form $x_i = 1$ to the system so that the total number of variables becomes $2^q - 1$, for some q.

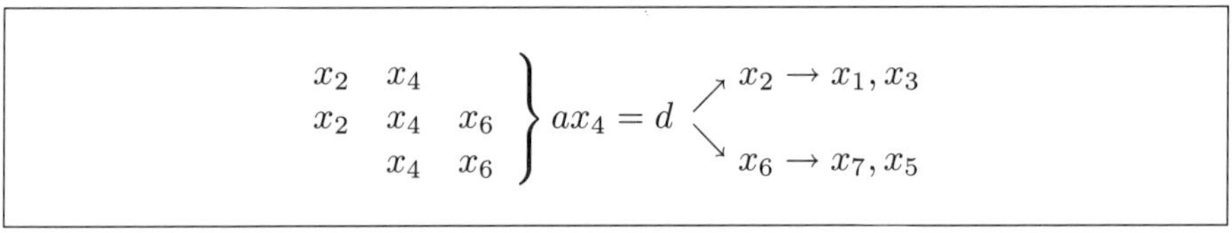

$$\left.\begin{array}{ccc} x_2 & x_4 & \\ x_2 & x_4 & x_6 \\ & x_4 & x_6 \end{array}\right\} ax_4 = d \begin{array}{c} \nearrow x_2 \to x_1, x_3 \\ \searrow x_6 \to x_7, x_5 \end{array}$$

Figure 7.3.7: *Back Substitution*

The above algorithm is known as *cyclic* or *odd-even reduction*. Although it was devised originally without parallel computing in mind, it has considerable inherent parallelism, which we now examine. Note first that the operations leading to the new equations for $x_2, x_4, \ldots, x_{n-1}$ are independent and can be done in parallel. Suppose that there are $p = (n-1)/2$ processors with the storage pattern illustrated in Figure 7.3.8. The processors could then do in parallel the operations leading to the first reduced system containing the even-numbered unknowns. At the end of this step, each processor would contain the coefficients for exactly one new equation, and before the next step the coefficients for two additional equations would have to be transmitted to other processors. One natural scheme is that each "middle" processor receives this information as indicated below:

$$P_1 \to P_2 \leftarrow P_3 \to P_4 \leftarrow P_5.$$

The odd-numbered processors are now finished and the even-numbered processors do the next elimination step. The process is continued with approximately half of the remaining processors becoming inactive after each step; the final combination of three equations to give one equation in a single unknown would

7.3 Banded and Tridiagonal Systems

be done in one processor. Note that $q - 1 = \log[(n + 1)/2]$ parallel stages are required for the reduction. The back substitution essentially works in the reverse fashion. The final equation is solved in one processor and then additional processors can be used as the back substitution proceeds.

$$\boxed{\text{rows 1, 2, 3}} \quad \boxed{\text{rows 3, 4, 5}} \quad \ldots \quad \boxed{\text{rows } n-2, n-1, n}$$

$$P_1 \qquad\qquad P_2 \qquad\qquad\qquad\qquad P_p$$

Figure 7.3.8: *Processor Assignment for Cyclic Reduction*

In the more realistic case that $p \ll n$, several rows would be distributed to each processor initially and processors would work with high parallel efficiency in the early stages. However, there may be some load imbalances depending on the relative sizes of p and n. For example, for $n = 15$ and $p = 2$, a schematic for the computation is shown in Figure 7.3.9.

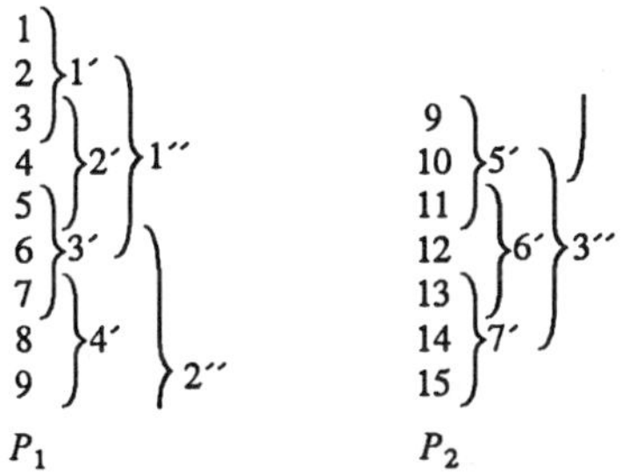

Figure 7.3.9: *Cyclic Reduction on Two Processors for $n = 15$*

At the first stage there will be seven reduction steps leading to the rows denoted by $1', \ldots, 7'$ in the new tridiagonal matrix. As shown in Figure 7.3.9, P_1 does four of them and P_2 the other three, so that there is a load imbalance on this first step. Note that row 9 must be available to both processors. On the second stage, there will be three reduction steps leading to the new rows $1'', 2'', 3''$; two of these steps are done in P_1 and the other in P_2, another load imbalance. Note that row $5'$ must be transmitted from P_2 to P_1 before this step. Finally, row $3''$ will be sent to P_1 and the final reduction performed there. The load imbalances indicated in this example will become relatively less important the larger n is.

Relation to Gaussian Elimination

We next show that the cyclic reduction algorithm is actually Gaussian elimination applied to a reordered system of equations. We reorder the unknowns as

$$x_1, x_3, \ldots, x_n, x_2, x_4, \ldots, x_{n-1}, \qquad (7.3.5)$$

which is known as the *odd/even* ordering. (It is also the red/black ordering.) With this ordering the system of equations (7.3.3) has the form

$$\begin{bmatrix} a_1 & & & & & b_1 & & & & \\ & a_3 & & & & c_2 & b_3 & & & \\ & & \ddots & & & & c_4 & \ddots & & \\ & & & \ddots & & & & \ddots & b_{n-2} & \\ & & & & a_n & & & & c_{n-1} & \\ c_1 & b_2 & & & & a_2 & & & & \\ & c_3 & & & & & a_4 & & & \\ & & \ddots & \ddots & & & & \ddots & & \\ & & & c_{n-2} & b_{n-1} & & & & a_{n-1} \end{bmatrix} \begin{bmatrix} x_1 \\ x_3 \\ \vdots \\ x_n \\ x_2 \\ \vdots \\ x_{n-1} \end{bmatrix} = \begin{bmatrix} d_1 \\ d_3 \\ \vdots \\ d_n \\ d_2 \\ \vdots \\ d_{n-1} \end{bmatrix}. \qquad (7.3.6)$$

Then the first stage of cyclic reduction is equivalent (Exercise 7.3.3) to applying Gaussian elimination to the first $(n+1)/2$ columns of (7.3.6). This eliminates the subdiagonal elements in these columns and produces the following tridiagonal system in the even numbered variables:

$$\begin{bmatrix} a_2' & b_1' & & & \\ c_1' & a_4' & \ddots & & \\ & \ddots & \ddots & b_{n-2}' \\ & & c_{n-2}' & a_{n-1}' \end{bmatrix} \begin{bmatrix} x_2 \\ \vdots \\ x_{n-1} \end{bmatrix} = \begin{bmatrix} d_2' \\ \vdots \\ d_{n-1}' \end{bmatrix}. \qquad (7.3.7)$$

The next stage of cyclic reduction is the same as the first but applied now to the system (7.3.7). Again we can view this as first performing an odd/even reordering of (7.3.7) and then Gaussian elimination on the corresponding first part of the permuted matrix. Each stage of cyclic reduction is analogous so that in the $q-1$ stages of the process, we will do $q-1$ permutations. In order to relate the cyclic reduction process back to a complete LU decomposition, we must do these permutations on the whole $n \times n$ matrix, not just the smaller tridiagonal systems. Thus, corresponding to the second stage, we do the odd-even permutation on the last $(n-1)/2$ rows and columns of the original system, and so on at each stage. Figure 7.3.9 shows the form (Exercise 7.3.4) of the reordered matrix for $n = 15$. In this figure, ignore the F_i which will be discussed later.

7.3 Banded and Tridiagonal Systems

$$\begin{bmatrix}
a_1 & & & & & & & b_1 & & & & & & & \\
 & a_3 & & & & & & & c_2 & & b_3 & & & & \\
 & & a_5 & & & & & & & b_5 & & c_4 & & & \\
 & & & a_7 & & & & & & c_6 & & & & b_7 & \\
 & & & & a_9 & & & & & & b_9 & & & c_8 & \\
 & & & & & a_{11} & & & & & c_{10} & & b_{11} & & \\
 & & & & & & a_{13} & & & & & b_{15} & & c_{12} & \\
 & & & & & & & a_{15} & & & & c_{14} & & & \\
c_1 & b_2 & & & & & & & a_2 & & & F_1 & & & \\
 & c_5 & b_6 & & & & & & & a_6 & & F_1 & & F_1 & \\
 & & c_9 & b_{10} & & & & & & & a_{10} & & & & F_1 \\
 & & & c_{13} & b_{14} & & & & & & & a_{14} & & F_1 & \\
c_3 & b_4 & & & & & & & F_1 & F_1 & & & a_4 & & F_2 \\
 & & & & & c_{11} & b_{12} & & & & F_1 & F_1 & & a_{12} & F_2 \\
 & & & & & & c_7 & b_8 & & & & F_1 & F_1 & F_2 & F_2 & a_8
\end{bmatrix}$$

Figure 7.3.10: *Cyclic Reduction Reordered Matrix for* $n = 15$

We note that if A is symmetric positive definite or diagonally dominant, these same properties hold for the reordered matrix (Exercise 7.3.5). Thus, the cyclic reduction algorithm will be numerically stable. If, however, the original matrix is such that Gaussian elimination without pivoting is numerically unstable on the reordered matrix, the same will be true for cyclic reduction.

Operation Count for Cyclic Reduction

The F_i in Figure 7.3.10 indicate elements that are zero in the original matrix but fill during Gaussian elimination. This fill in turn increases the operation count so that cyclic reduction is more expensive than Gaussian elimination applied to the original tridiagonal system. The operation count for the forward reduction phase of cyclic reduction may be obtained (Exercise 7.3.7) either directly from the algorithm or from the LU decomposition of the matrix of Figure 7.3.10. It is

$$O_F = (2n - 2q) \text{ div} + (4n - 6q + 2) \text{ mult} + (2n - 2q) \text{ add}, \quad (7.3.8)$$

where, as before, $n = 2^q - 1$. The operation counts for modifying the right-hand side and doing the back substitution are

$$O_{RHS} = (2n - 2q) \text{ mult} + (2n - 2q) \text{ add} \quad (7.3.9)$$

and

$$O_{BS} = n \text{ div} + (2n - 2q) \text{ mult} + (2n - 2q) \text{ add}. \quad (7.3.10)$$

Thus, the total operation count to solve the tridiagonal system by cyclic reduction is

$$O_{CR} = (3n - 2q) \text{ div} + (8n - 10q + 2) \text{ mult} + (6n - 2q) \text{ add} \qquad (7.3.11)$$

$$\doteq 3n \text{ div} + 8n \text{ mult} + 6n \text{ add}$$

where the last approximation is valid if $q << n$.

Comparing (7.3.11) with the operation count (6.1.18) for solving the original tridiagonal system by Gaussian elimination, we see that cyclic reduction requires about twice as many additions, almost three times as many multiplications and 1.5 as many divisions. If we count divisions the same as multiplications and additions, cyclic reduction requires approximately $17n$ operations versus $8n$ for Gaussian elimination. One consequence of this is that on p processors, the maximum speedup of cyclic reduction over Gaussian elimination can be no more than $p/2$. It will, in fact, be considerably less due to lack of perfect parallelism, communication, etc. Nevertheless, cyclic reduction, or one of its many variants, is the best known parallel algorithm for tridiagonal systems although the size of the system must be large to achieve any gain over serial Gaussian elimination.

Domain Decomposition for Banded Systems

We now reinterpret the matrix of (7.3.6) in terms of domain decomposition. Recall that we used domain decomposition in Section 6.1 as a way to reorder the discrete equations of Poisson's equation so as to reduce fill in Gaussian elimination. Here, we shall use it as a way of introducing parallelism into the solution of banded systems of equations. Such a system may or may not arise from the discretization of a differential equation, but we can always proceed as follows. We divide the unknowns into $2p - 1$ sets $D_1, \ldots, D_p$ and $S_1, \ldots, S_{p-1}$ such that each D_i has q unknowns and the unknowns in the S_i "separate" the unknowns in the D_i. We can imagine the unknowns lined up and partitioned as

$$D_1 \quad S_1 \quad D_2 \quad S_2 \quad \cdots \quad S_{p-1} \quad D_p. \qquad (7.3.12)$$

Each set S_i in (7.3.12) contains β unknowns, where β is the semibandwidth. Since the jth equation in the system $A\mathbf{x} = \mathbf{b}$ is

$$\sum_{k=j-\beta}^{j+\beta} a_{jk} x_k = b_j, \qquad (7.3.13)$$

if $x_j \in D_i$, none of the unknowns in the equation (7.3.13) are in any of the other D_k. The sets S_i separate the unknowns in the D_i so that each equation in the system contains unknowns in only one D_i.

7.3 Banded and Tridiagonal Systems

If we now renumber the unknowns so that those in the separator sets are numbered last, and write the equations in the corresponding order, we can write the system in the block arrowhead form

$$\begin{bmatrix} A_1 & & & & B_1 \\ & A_2 & & & B_2 \\ & & \ddots & & \vdots \\ & & & A_p & B_p \\ C_1 & C_2 & \cdots & C_p & A_S \end{bmatrix} \begin{bmatrix} \mathbf{x}_1 \\ \mathbf{x}_2 \\ \vdots \\ \mathbf{x}_p \\ \mathbf{x}_S \end{bmatrix} = \begin{bmatrix} \mathbf{b}_1 \\ \mathbf{b}_2 \\ \vdots \\ \mathbf{b}_p \\ \mathbf{b}_S \end{bmatrix} \qquad (7.3.14)$$

in which $A_1, \ldots, A_p$ are $q \times q$ and A_S is $s \times s$. Here, $s = \beta(p-1)$ is the number of unknowns in the separator set S, which is the union of the sets S_i. Then the B_i are $q \times s$ and the C_i are $s \times q$. We have assumed, for simplicity, that $n = pq + s$ and will return to this point shortly.

Now let

$$A_I = \text{diag}(A_1, \ldots, A_p), \quad B^T = (B_1^T, \ldots, B_p^T), \quad C = (C_1, \ldots, C_p). \qquad (7.3.15)$$

Then the system (7.3.15) can be written as

$$A_I \mathbf{x}_I + B \mathbf{x}_S = \mathbf{b}_I \qquad (7.3.16a)$$

$$C \mathbf{x}_I + A_S \mathbf{x}_S = \mathbf{b}_S, \qquad (7.3.16b)$$

where $\mathbf{x}_I^T = (\mathbf{x}_1^T, \ldots, \mathbf{x}_p^T)$ and $\mathbf{b}_I^T = (\mathbf{b}_1^T, \ldots, \mathbf{b}_p^T)$. We assume that A_I is nonsingular. Then, if we multiply (7.3.16a) by CA_I^{-1} and subtract from (7.3.16b), we obtain the equation

$$\hat{A}\mathbf{x}_S = \mathbf{b}, \qquad (7.3.17a)$$

where

$$\hat{A} = A_S - CA_I^{-1}B, \quad \hat{\mathbf{b}} = \mathbf{b}_S - CA_I^{-1}\mathbf{b}_I. \qquad (7.3.17b)$$

Note that this is just block Gaussian elimination for the system (7.3.16). The matrix $\hat{A}$ is called a *Schur complement* or *Gauss transform*. Once the system (7.3.17a) is solved for $\mathbf{x}_S$, the remaining $\mathbf{x}_i$ can be computed by solving the following systems obtained from (7.3.14):

$$A_i \mathbf{x}_i = \mathbf{b}_i - B_i \mathbf{x}_S, \quad i = 1, \ldots, p. \qquad (7.3.18)$$

We now discuss the computation in more detail. We assume that the A_i have stable LU decompositions

$$A_i = L_i U_i, \quad i = 1, \ldots, p, \qquad (7.3.19)$$

although other factorizations, or other ways to solve systems with the coefficient matrices A_i, could be used. We then solve the systems

$$L_i Y_i = B_i, \qquad L_i \mathbf{y}_i = \mathbf{b}_i, \qquad i = 1, \ldots, p, \qquad (7.3.20)$$

$$U_i Z_i = Y_i, \qquad U_i \mathbf{z}_i = \mathbf{y}_i, \qquad i = 1, \ldots, p. \qquad (7.3.21)$$

Since
$$C_i A_i^{-1} B_i = C_i (L_i U_i)^{-1} B_i = C_i U_i^{-1} Y_i = C_i Z_i$$
and, similarly, $C_i A_i^{-1} \mathbf{b}_i = C_i \mathbf{z}_i$, we can write

$$\hat{A} = A_S - \sum_{i=1}^{p} C_i A_i^{-1} B_i = A_S - \sum_{i=1}^{p} C_i Z_i \qquad (7.3.22)$$

and

$$\hat{\mathbf{b}} = \mathbf{b}_S - \sum_{i=1}^{p} C_i A_i^{-1} \mathbf{b}_i = \mathbf{b}_S - \sum_{i=1}^{p} C_i \mathbf{z}_i, \qquad (7.3.23)$$

which shows how $\hat{A}$ and $\hat{\mathbf{b}}$ are to be obtained. The overall procedure can then be summarized as shown in Figure 7.3.11.

Step 1. Form the decompositions (7.3.20) and solve the systems (7.3.20) and (7.3.21).

Step 2. Form $C_i Z_i$ and $C_i \mathbf{z}_i$, $\qquad i = 1, \ldots, p$.

Step 3. Form $\hat{A}$ and $\hat{\mathbf{b}}$ and solve $\hat{A} \mathbf{x}_S = \hat{\mathbf{b}}$.

Step 4. Form $\mathbf{c}_i = \mathbf{b}_i - B_i \mathbf{x}_S, \qquad i = 1, \ldots, p$.

Step 5. Solve $A_i \mathbf{x}_i = \mathbf{c}_i, i = 1, \ldots, p$, using the decompositions (7.3.19).

Figure 7.3.11: *Domain Decomposition Algorithm*

It is clear that Steps 1, 2, 4, and 5 are highly parallel. For example, on a distributed memory system, assume that A_i, B_i, C_i, and $\mathbf{b}_i$ are assigned to processor i. Then, the computations in each of Steps 1 and 2 are parallel and require no communication. The potential bottleneck is in Step 3. One possibility is to solve the system $\hat{A} \mathbf{x}_S = \hat{\mathbf{b}}$ in a single processor, say, P_1. In this case, the summations of (7.3.22, 23) would be done by a fan-in across the processors with the final sums arriving in P_1. Alternatively, we could solve $\hat{A} \mathbf{x}_S = \hat{\mathbf{b}}$ in parallel by Gaussian elimination with interleaved storage. Then, we would wish to do the fan-in additions of (7.3.22) in such a way that the first

7.3 Banded and Tridiagonal Systems

row of $\hat{A}$ is in processor 1, the second row in processor 2, and so on. In any case, after $\mathbf{x}_S$ has been obtained, we need to send copies of it to all processors. Then, steps 4 and 5 in Figure 7.3.11 can be carried out in parallel with no further communication.

We next summarize the changes to the domain decomposition algorithm in the important special case that A is symmetric positive definite. Since the matrix of (7.3.14), call it $\bar{A}$, arises from A by interchanges of equations and unknowns, it is related to A by $\bar{A} = PAP^T$, where P is a permutation matrix. Hence, $\bar{A}$ is symmetric positive definite (Exercise 7.3.5) and, thus, so are A_S and the A_i (Theorem 2.1.3). In step 1 of Figure 7.3.11, we would probably now use Cholesky decompositions $A_i = L_i L_i^T$. Then, since $C_i = B_i^T$ by the symmetry, we have

$$\hat{A} = A_s - \sum_{i=1}^p B_i^T A_i^{-1} B_i, \qquad \hat{\mathbf{b}} = \mathbf{b}_S - \sum_{i=1}^p B_i^T A_i^{-1} \mathbf{b}_i.$$

If we write

$$B_i^T A_i^{-1} B_i = B_i^T (L_i L_i^T)^{-1} B_i = Y_i^T Y_i,$$

where $Y_i = L_i^{-1} B_i$, then with $\mathbf{y}_i = L_i^{-1} \mathbf{b}_i$

$$\hat{A} = A_S - \sum_{i=1}^p Y_i^T Y_i, \qquad \hat{\mathbf{b}} = \mathbf{b}_S - \sum_{i=1}^p Y_i^T \mathbf{y}_i. \qquad (7.3.24)$$

We summarize the domain decomposition algorithm for symmetric positive definite matrices in Figure 7.3.12. Again, Steps 1 and 3 are highly parallel and Step 2 is the possible bottleneck.

Step 1. Do the Cholesky decompositions $A_i = L_i L_i^T$ and solve the systems $L_i Y_i = B_i, L_i \mathbf{y}_i = \mathbf{b}_i, i = 1, \ldots, p$.

Step 2. Form $\hat{A}$ and $\hat{\mathbf{b}}$ by (7.3.24) and solve $\hat{A} \mathbf{x}_S = \hat{\mathbf{b}}$.

Step 3. Form $\mathbf{c}_i = \mathbf{b}_i - B_i \mathbf{x}_S$ and solve the systems $A_i \mathbf{x}_i = \mathbf{c}_i, i = 1, \ldots, p$.

Figure 7.3.12: *Symmetric Positive Definite Domain Decomposition*

An important observation is that the matrix $\hat{A}$ is symmetric positive definite whenever A itself is. Clearly, it is symmetric. For the positive definiteness, let $\mathbf{x}_2$ be any nonzero vector of length s, and set $\mathbf{x}_1 = -A_I^{-1} B \mathbf{x}_2$. Then

$$0 < (\mathbf{x}_1^T, \mathbf{x}_2^T) \begin{bmatrix} A_I & B \\ B^T & A_S \end{bmatrix} \begin{bmatrix} \mathbf{x}_1 \\ \mathbf{x}_2 \end{bmatrix}$$

by the positive definiteness of A. Expanding this out gives

$$0 < \mathbf{x}_1^T A_I \mathbf{x}_1 + 2\mathbf{x}_1^T B \mathbf{x}_2 + \mathbf{x}_2^T A_S \mathbf{x}_2 = \mathbf{x}_2^T \hat{A} \mathbf{x}_2, \qquad (7.3.25)$$

which shows that $\hat{A}$ is positive definite.

We now relate the domain decomposition procedure to cyclic reduction for tridiagonal matrices. For tridiagonal matrices, the semibandwidth $\beta = 1$ and so each of the separator sets of S_i of (7.3.12) can be chosen to have exactly one point. If we also take each domain set D_i to have only one point, then $p = (n+1)/2$, and the matrix (7.3.14) is just the matrix of (7.3.6), corresponding to the first step of cyclic reduction.

Structure of the Reordered Matrix and Complexity

The algorithm of Figure 7.3.11 applies to any matrix of the form (7.3.14) whenever the A_i are suitable for an LU decomposition. However, when (7.3.14) arises by reordering of a banded system, there is additional useful structure.

We write the original banded matrix in partitioned form

$$\begin{bmatrix} A_{11} & B_{11} & & & & & & \\ C_{11} & A_{12} & B_{12} & & & & & \\ & C_{12} & A_{21} & B_{21} & & & & \\ & & C_{21} & A_{22} & B_{22} & & & \\ & & & C_{22} & A_{31} & B_{31} & & \\ & & & & C_{31} & A_{32} & B_{32} & \\ & & & & & C_{32} & A_{41} & \end{bmatrix} \begin{bmatrix} \mathbf{x}_{11} \\ \mathbf{x}_{12} \\ \mathbf{x}_{21} \\ \mathbf{x}_{22} \\ \mathbf{x}_{31} \\ \mathbf{x}_{32} \\ \mathbf{x}_{41} \end{bmatrix} = \begin{bmatrix} \mathbf{b}_{11} \\ \mathbf{b}_{12} \\ \mathbf{b}_{21} \\ \mathbf{b}_{22} \\ \mathbf{b}_{31} \\ \mathbf{b}_{32} \\ \mathbf{b}_{41} \end{bmatrix}, \qquad (7.3.26)$$

where we assume for illustration purposes that the number of subdomains is $p = 4$. In (7.3.26), the unknowns in the vectors $\mathbf{x}_{11}, \mathbf{x}_{21}, \mathbf{x}_{31}$ and $\mathbf{x}_{41}$ correspond to the domains D_1, D_2, D_3 and D_4, and those in $\mathbf{x}_{12}, \mathbf{x}_{22}$ and $\mathbf{x}_{32}$ correspond to the separator sets. Thus, the matrices A_{12}, A_{22} and A_{32} are all $\beta \times \beta$ and the matrices A_{11}, A_{21}, A_{31} and A_{41} are $m \times m$, where $m = (n - 3\beta)/4$. (We assume for simplicity that m is an integer; if not, the size of A_{41} will be smaller.) The dimensions of the B_{ij} and C_{ij} are then determined accordingly: B_{11} and C_{12} are $m \times \beta$, B_{12} and C_{11} are $\beta \times m$, and so on.

The domain decomposition reordering of (7.3.26) gives the permuted system

$$\begin{bmatrix} A_{11} & & & & B_{11} & & \\ & A_{21} & & & C_{12} & B_{21} & \\ & & A_{31} & & & C_{22} & B_{31} \\ & & & A_{41} & & & C_{32} \\ C_{11} & B_{12} & & & A_{12} & & \\ & C_{21} & B_{22} & & & A_{22} & \\ & & C_{31} & B_{32} & & & A_{32} \end{bmatrix} \begin{bmatrix} \mathbf{x}_{11} \\ \mathbf{x}_{21} \\ \mathbf{x}_{31} \\ \mathbf{x}_{41} \\ \mathbf{x}_{12} \\ \mathbf{x}_{22} \\ \mathbf{x}_{32} \end{bmatrix} = \begin{bmatrix} \mathbf{b}_{11} \\ \mathbf{b}_{21} \\ \mathbf{b}_{31} \\ \mathbf{b}_{41} \\ \mathbf{b}_{12} \\ \mathbf{b}_{22} \\ \mathbf{b}_{32} \end{bmatrix}. \qquad (7.3.27)$$

7.3 Banded and Tridiagonal Systems

(By multiplying out (7.3.26) and (7.3.27), it is clear that the individual equations are exactly the same.) The domain decomposition method is then mathematically equivalent to the solution of (7.3.27) by LU decomposition.

To relate (7.3.27) back to (7.3.14), we note that $A_i = A_{i1}, i = 1, \ldots, 4, A_S = \mathrm{diag}(A_{12}, A_{22}, A_{32}), B_1 = (B_{11}, 0, 0), B_2 = (C_{12}, B_{21}, 0)$, and so on. The matrix $\hat{A}$ of (7.3.17) is then

$$\hat{A} = \begin{bmatrix} A'_{12} & B'_{21} & 0 \\ C'_{12} & A'_{22} & B'_{31} \\ 0 & C'_{22} & A'_{32} \end{bmatrix}, \qquad (7.3.28)$$

where, for example,

$$C'_{12} = -C_{21} A_{21}^{-1} C_{12}, \quad C'_{22} = -C_{31} A_{31}^{-1} C_{22}, \qquad (7.3.29)$$

and the B'_{i1} are given by similar expressions. The C'_{i2} and B'_{i1} in (7.3.28) are full $\beta \times \beta$ matrices, as are the A'_{i2}. Thus, the semibandwidth of (7.3.28) is $2\beta - 1$ (see Exercise 7.3.8). These properties persist for larger p: $\hat{A}$ is block tridiagonal with semibandwidth $2\beta - 1$.

As in cyclic reduction, it is possible to apply the domain decomposition method again to the reduced system with the coefficient matrix $\hat{A}$. However, since the semibandwidth of $\hat{A}$ is $2\beta - 1$ it may be more effective to use just a banded algorithm. Note that in the case $\beta = 1$ for a tridiagonal matrix, $2\beta - 1 = 1$ so that the reduced system is again tridiagonal. It was this property that was used in the cyclic reduction algorithm.

It is possible to show that for large n and small β, the domain decomposition algorithm requires approximately four times as many multiplications and additions and twice as many divisions as Gaussian elimination applied to the original banded system. As with cyclic reduction, this increased operation count may be viewed as a consequence of fill that occurs when Gaussian elimination is applied to the reordered system (7.3.27). Note the seeming paradox that a domain decomposition reordering was applied in Section 6.1 to reduce fill in Gaussian elimination on the Poisson system (5.5.18).

As a consequence of the increased operation count, the speedup of the domain decomposition algorithm for banded systems will be limited to $p/4$ on p processors compared with Gaussian elimination on the original system. Nevertheless, it and its variants are the best parallel algorithms known at this time for narrow banded systems.

Supplementary Discussion and References: 7.3

1. The recognition that the degree of vectorization or parallelism of LU or Cholesky decomposition depended on the bandwidth of the system

motivated considerable early research on narrow banded systems, especially tridiagonal systems. Stone [1973] gave an algorithm for tridiagonal systems based on "recursive doubling." Lambiotte and Voigt [1975] modified this algorithm, as did Stone [1975], but these methods have not been widely used.

2. The cyclic reduction algorithm, or its variants, has been one of the most preferred methods for tridiagonal systems on both parallel and vector computers. It was originally proposed for general tridiagonal systems by G. Golub and developed by R. Hockney for special block tridiagonal systems (see Hockney [1965]). Heller [1976] showed, under certain assumptions, that during the cyclic reduction process the off-diagonal elements decrease in size relative to the diagonal elements at a quadratic rate; this allows termination before the full $\log n$ steps have been performed. It was recognized by G. Golub (see Lambiotte and Voigt [1975]) that cyclic reduction is just Gaussian elimination applied to PAP^T for a suitable permutation matrix P, as shown in the text. Thus, if A is symmetric positive definite, so is PAP^T and cyclic reduction is numerically stable. However, it is still necessary to handle the right-hand side carefully in order to retain the numerical stability; see Golub and Van Loan [1989]. The cyclic reduction algorithm discussed in the text loses parallelism in the final stages of the reduction and in the beginning stages of the back substitution. A more parallel revision, which essentially eliminates the back substitution, has been given by Hockney and Jesshope [1988]. (See, also, Swarztrauber and Sweet [1989].) For further discussion of cyclic reduction and some of its variants see Hockney and Jesshope [1988] and Johnsson [1987].

3. The domain decomposition method for banded systems is closely related to a method proposed by Johnsson [1985]. Other similar methods have been given by Lawrie and Sameh [1984] and Meier [1985] for banded systems and by Wang [1981] for tridiagonal systems. For a summary of these methods and further comments on tridiagonal and banded systems, see Ortega [1988a]. The idea of domain decomposition has a long history in the engineering community, independent of parallel methods, where it is called *substructuring*. The original motivation was to break up a large problem into smaller parts that could be solved separately, and then the final solution is obtained from solutions of these subproblems.

4. The fact that the domain decomposition method for banded systems increases the operation count considerably is disappointing but it may be unavoidable. Cleary [1989] showed that any reordering of the coefficient matrix that gives a degree of parallelism p must suffer fill in at least $p-2$ blocks and thus has a higher operation count.

7.3 Banded and Tridiagonal Systems

EXERCISES 7.3

7.3.1. Corresponding to Figure 7.3.1 for the kij and kji forms, give a pseudo code for banded matrices for the jki form of LU decomposition.

7.3.2. Discuss the vector and parallel properties of Cholesky decomposition for banded symmetric positive definite matrices along the lines of the LU discussion in the text.

7.3.3. Verify that the first stage of cyclic reduction corresponds to applying Gaussian elimination to the first $(n+1)/2$ columns of (7.3.6).

7.3.4. Verify that the whole cyclic reduction algorithm for $n = 15$ is equivalent to doing Gaussian elimination on a system with the coefficient matrix of Figure 7.3.9 (without the F_i). What is the right-hand side of the reordered system?

7.3.5. Show that the coefficient matrix $\hat{A}$ of (7.3.6) can be written as $\hat{A} = P^T A P$, where A is the original tridiagonal matrix and P is a permutation matrix. Conclude from this that $\hat{A}$ is symmetric positive definite if A is. Do the same for the matrix of Figure 7.3.10, without the F_i, and also conclude that this matrix is diagonally dominant if the tridiagonal matrix is.

7.3.6. Show that Gaussian elimination applied to the matrix of (7.3.6) for $n = 15$ produces fill elements as indicated by the F's in Figure 7.3.10.

7.3.7. Verify the operation counts (7.3.8)–(7.3.11) for cyclic reduction.

7.3.8. Show that the matrices C'_{i2} of (7.3.29) (as well as the B'_{i1} of (7.3.28)) are, in general, full $\beta \times \beta$ matrices and that, therefore, the semibandwidth of (7.3.28) is $2\beta - 1$.

7.3.9. (Two-Way Gaussian Elimination) For a system $A\mathbf{x} = \mathbf{b}$ with an $n \times n$ tridiagonal matrix A and n assumed to be odd, start Gaussian elimination simultaneously at the first and last equations and continue working towards the middle, eliminating the subdiagonal elements from the top and superdiagonal elements from the bottom. At the end of the elimination steps, there will be a single equation with a single unknown. Back substitution can then be carried out, obtaining two more unknowns at each stage. Write out this method in detail and discuss its parallel properties. Can you extend this approach to banded systems?

7.3.10. a. Let A be the coefficient matrix of (7.3.16) and assume that A_I has an LU decomposition. Show that the matrix $\hat{A}$ of (7.3.17) is exactly the reduced matrix that would be produced by Gaussian elimination applied to the columns of A corresponding to A_I.

b. Consider the matrix
$$A = \begin{bmatrix} C & B \\ B^T & 0 \end{bmatrix},$$

where C is $r \times r$ and symmetric positive definite. Is A positive definite? Apply Gaussian elimination to the first r columns of A and show that 0 is replaced by the Schur complement $-B^T C^{-1} B$.

7.3.11. Show that any banded matrix A can be written as a block tridiagonal matrix

$$A = \begin{bmatrix} A_1 & B_1 & & \\ C_1 & \ddots & \ddots & \\ & \ddots & & B_{2m-1} \\ & & C_{2m-1} & A_{2m} \end{bmatrix}.$$

If A is $n \times n$ with semibandwidth β, what is the maximum size of m?

7.3.12. Show how to write

$$\begin{bmatrix} A_1 & B \\ B_1^T & A_2 \end{bmatrix} = \begin{bmatrix} \hat{A}_1 & 0 \\ 0 & \hat{A}_2 \end{bmatrix} + \begin{bmatrix} L \\ H \end{bmatrix} (L^T H^T).$$

Use this together with the Sherman-Morrison-Woodbury formula (6.1.34) to develop an algorithm that solves $A\mathbf{x} = \mathbf{b}$ by solving two systems of roughly half the size independently.

Chapter 8

Iterative Methods

We have seen in Chapter 6 that when we apply Gaussian elimination to a large sparse system of equations, such as those arising from discretizing Poisson's equation, fill occurs and increases the operation count and the storage requirements. An alternative approach is to use iterative methods. Generally, these methods have more modest storage requirements than direct methods and may also be faster, depending on the iterative method and the problem. They usually also have better vectorization and parallelization properties. In this chapter we shall consider some iterative methods based on "relaxation." Some of these methods should usually not be used in practice, but they provide a useful framework for the introduction of iterative methods and also will be used as parts of the more sophisticated multigrid and conjugate gradient type methods, discussed in Section 8.3 and Chapter 9.

8.1 Relaxation-Type Methods

We consider the linear system $A\mathbf{x} = \mathbf{b}$ and assume that the diagonal elements of A are non-zero:

$$a_{ii} \neq 0, \quad i = 1, \ldots, n. \tag{8.1.1}$$

Perhaps the simplest iterative procedure is *Jacobi's method*. Assume that an initial approximation $\mathbf{x}^0$ to the solution is chosen. Then the next iterate is given by

$$x_i^{(1)} = \frac{1}{a_{ii}}(b_i - \sum_{j \neq i} a_{ij} x_j^{(0)}), \quad i = 1, \ldots, n. \tag{8.1.2}$$

It will be useful to write this in matrix-vector notation, and we let $D = \text{diag}(a_{11}, \ldots, a_{nn})$ and $B = D - A$. Then it is easy to verify that (8.1.2)

may be written as
$$\mathbf{x}^1 = D^{-1}(\mathbf{b} + B\mathbf{x}^0),$$
and the entire sequence of Jacobi iterates is defined by
$$\mathbf{x}^{k+1} = D^{-1}(\mathbf{b} + B\mathbf{x}^k), \qquad k = 0, 1, \ldots, \tag{8.1.3}$$
where the superscripts denote iteration number.

The Gauss-Seidel Method

A closely related iteration is derived from the following observation. After $x_1^{(1)}$ is computed in (8.1.2) it is available to use in the computation of $x_2^{(1)}$, and it is natural to use this new value rather than the original estimate $x_1^{(0)}$. If we use new values as soon as they are available, then (8.1.2) becomes

$$x_i^{(1)} = \frac{1}{a_{ii}}(b_i - \sum_{j<i} a_{ij} x_j^{(1)} - \sum_{j>i} a_{ij} x_j^{(0)}), \qquad i = 1, \ldots, n, \tag{8.1.4}$$

which is the first step in the *Gauss-Seidel iteration*. To write this iteration in matrix-vector form, let $-L$ and $-U$ denote the strictly lower- and upper-triangular parts of A; that is, both L and U have zero main diagonals and

$$A = D - L - U. \tag{8.1.5}$$

If we multiply (8.1.4) through by a_{ii}, then it is easy to verify that the n equations in (8.1.4) can be written as

$$D\mathbf{x}^1 - L\mathbf{x}^1 = \mathbf{b} + U\mathbf{x}^0. \tag{8.1.6}$$

Since $D - L$ is a lower-triangular matrix with non-zero diagonal elements, it is nonsingular. Then the entire sequence of Gauss-Seidel iterates is defined by

$$\mathbf{x}^{k+1} = (D - L)^{-1}(U\mathbf{x}^k + \mathbf{b}), \qquad k = 0, 1, \ldots. \tag{8.1.7}$$

The matrix-vector representations (8.1.3) and (8.1.7) of the Jacobi and Gauss-Seidel iterations are useful for theoretical purposes, but the actual computations would usually be done using the componentwise representations (8.1.2) and (8.1.4).

In these iterative methods, storage is needed in general only for the non-zero elements of A, the right hand side $\mathbf{b}$, and the current and previous iterates $\mathbf{x}^k$ and $\mathbf{x}^{k+1}$. (In the Gauss-Seidel method, one vector suffices for both $\mathbf{x}^k$ and $\mathbf{x}^{k+1}$ since new components of $\mathbf{x}^{k+1}$ may be written into the corresponding positions of $\mathbf{x}^k$ as soon as they are completed.) As opposed to a direct method such as Gaussian elimination, there is no need to allow additional storage for fill, since the original matrix is not modified.

8.1 Relaxation-Type Methods

Application to Laplace's Equation

We next consider the application of these iterative methods to Laplace's equation on a square. The difference equations for this problem were given by (5.5.13) (with $f_{ij} = 0$) in the form

$$-u_{i+1,j} - u_{i-1,j} - u_{i,j+1} - u_{i,j-1} + 4u_{ij} = 0, \quad i,j = 1,\ldots,N. \quad (8.1.8)$$

Here, the unknowns are the $u_{ij}, i,j = 1, \ldots, N$, and the remaining values of the u_{ij} are assumed known from the boundary conditions. Given initial approximations $u_{ij}^{(0)}$, a Jacobi step applied to (8.1.8) is

$$u_{ij}^{(1)} = \frac{1}{4}(u_{i+1,j}^{(0)} + u_{i-1,j}^{(0)} + u_{i,j+1}^{(0)} + u_{i,j-1}^{(0)}),$$

so that the new Jacobi approximation at the (i,j) grid point is simply the average of the previous approximations at the four surrounding grid points. It is for this reason that the Jacobi method is sometimes known as the *method of simultaneous displacements*. Note that for the Jacobi method the order in which the equations are processed is immaterial, although storage considerations might make some orderings preferable. For the Gauss-Seidel method, however, each different ordering of the equations actually corresponds to a different iterative process. If we order the grid points left to right and bottom to top, as was done in Section 5.5, then a typical Gauss-Seidel step is

$$u_{ij}^{(1)} = \frac{1}{4}(u_{i-1,j}^{(1)} + u_{i,j-1}^{(1)} + u_{i+1,j}^{(0)} + u_{i,j+1}^{(0)}).$$

This new approximation at the (i,j) grid point is again an average of the approximations at the four surrounding grid points, but now using two old values and two new values. The difference between the two methods is shown schematically in Figure 8.1.1. Because the Jacobi and Gauss-Seidel method give a new iterate by averaging current values at neighboring points, they were known historically as *relaxation methods*.

Convergence

We next consider the question of the convergence of iterative methods. Both the Jacobi and Gauss-Seidel methods can be written in the form

$$\mathbf{x}^{k+1} = H\mathbf{x}^k + \mathbf{d}, \quad k = 0, 1, \ldots. \quad (8.1.9)$$

In particular, $H = D^{-1}B$ and $\mathbf{d} = D^{-1}\mathbf{b}$ for the Jacobi process, whereas $H = (D-L)^{-1}U$ and $\mathbf{d} = (D-L)^{-1}\mathbf{b}$ for Gauss-Seidel. Now assume that $\mathbf{x}^*$ is the exact solution of the system $A\mathbf{x} = \mathbf{b}$. For the Jacobi method we then have

$$(D-B)\mathbf{x}^* = \mathbf{b} \quad \text{or} \quad \mathbf{x}^* = D^{-1}B\mathbf{x}^* + D^{-1}\mathbf{b},$$

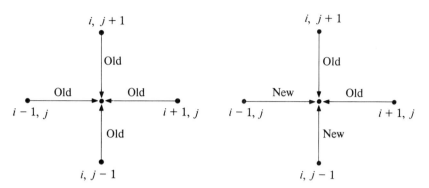

Figure 8.1.1: *Jacobi and Gauss-Seidel Updates*

and for the Gauss-Seidel method

$$(D - L - U)\mathbf{x}^* \quad \text{or} \quad \mathbf{x}^* = (D - L)^{-1}U\mathbf{x}^* + (D - L)^{-1}\mathbf{b}.$$

Thus, in both cases

$$\mathbf{x}^* = H\mathbf{x}^* + \mathbf{d}. \tag{8.1.10}$$

If we subtract (8.1.10) from (8.1.9), we obtain

$$\mathbf{e}^{k+1} = H\mathbf{e}^k, \quad k = 0, 1, \ldots, \tag{8.1.11}$$

where $\mathbf{e}^k = \mathbf{x}^k - \mathbf{x}^*$ is the error at the kth step.

Iterative methods of the form (8.1.9) are called *stationary one-step* methods and (8.1.11) is the basic error relation for such methods. We can analyze the errors in the following way. Assume that H has n linearly independent eigenvectors $\mathbf{v}_1, \ldots, \mathbf{v}_n$ with corresponding eigenvalues $\lambda_1, \ldots, \lambda_n$. (See Chapter 2 for the linear algebra results used in this section.) The initial error $\mathbf{e}^0$ can then be expressed as some linear combination of the eigenvectors:

$$\mathbf{e}^0 = c_1\mathbf{v}_1 + c_2\mathbf{v}_2 + \cdots + c_n\mathbf{v}_n. \tag{8.1.12}$$

Thus,

$$\mathbf{e}^k = H^k\mathbf{e}^0 = c_1\lambda_1^k\mathbf{v}_1 + c_2\lambda_2^k\mathbf{v}_2 + \cdots + c_n\lambda_n^k\mathbf{v}_n. \tag{8.1.13}$$

In order that $\mathbf{e}^k \to 0$ as $k \to \infty$ for any $\mathbf{x}^0$ (and, hence, any c_i in (8.1.12)), we must have $|\lambda_i| < 1, i = 1, \ldots, n$; that is, the spectral radius, $\rho(H)$, must be less than one. Moreover, it is $\rho(H)$ that determines the *asymptotic* rate of convergence, the rate of convergence as k becomes large. Suppose that $|\lambda_n| = \rho(H)$, and $|\lambda_n| > |\lambda_i|, i = 1, \ldots, n-1$. Then the term $\lambda_n^k\mathbf{v}_n$ in (8.1.13) will be the most slowly converging and for large k, each additional iteration

8.1 Relaxation-Type Methods

reduces the error by approximately the factor $|\lambda_n|$. Thus, $\rho(H)$ is called the *asymptotic convergence factor*. These results are true also when H does not have n linearly independent eigenvectors, but the proofs are more difficult (see the Supplementary Discussion). We state this basic convergence theorem as:

THEOREM 8.1.1 *If (8.1.10) holds, the iterates (8.1.9) converge to the solution* $\mathbf{x}^*$ *for any starting vector* $\mathbf{x}^0$ *if and only if* $\rho(H) < 1$. *Moreover, the smaller* $\rho(H)$ *the faster is the asymptotic rate of convergence.*

Theorem 8.1.1 is the basic theoretical result for iterative methods of the form (8.1.9), but it does not immediately tell us if a particular iterative method is convergent; for this, we need to ascertain if the spectral radius of the iteration matrix for the method is less than 1. In general, this is a difficult problem. But for some iterative methods and for certain classes of matrices it is relatively easy to determine that the convergence criterion is satisfied. The following are examples of this for the Jacobi and Gauss-Seidel methods.

THEOREM 8.1.2 *Assume that the matrix A is strictly diagonally dominant:*

$$|a_{ii}| > \sum_{j \neq i} |a_{ij}|, \qquad i = 1, \ldots, n. \tag{8.1.14}$$

Then both the Jacobi and Gauss-Seidel iterations converge to the unique solution of $A\mathbf{x} = \mathbf{b}$ for any starting vector $\mathbf{x}^0$.

The proof of this theorem is very simple for the Jacobi method. Since $H = D^{-1}B$, the condition (8.1.14) implies that the sums of the absolute values of the elements in each row of H are less than 1. Hence $\|H\|_\infty < 1$, and therefore all eigenvalues of H are less than 1 in absolute value. Thus Theorem 8.1.1 applies. The proof for the Gauss-Seidel method is a little harder. Let λ be any eigenvalue of H and $\mathbf{v}$ a corresponding eigenvector. Then

$$\lambda \mathbf{v} = H\mathbf{v} = (D - L)^{-1} U\mathbf{v},$$

or

$$\lambda(D - L)\mathbf{v} = U\mathbf{v}. \tag{8.1.15}$$

Let

$$|v_k| = \max\{|v_i| : i = 1, \ldots, n\}. \tag{8.1.16}$$

The kth equation of (8.1.15) is

$$\lambda\left(a_{kk}v_k + \sum_{j<k} a_{kj}v_j\right) = -\sum_{j>k} a_{kj}v_j, \tag{8.1.17}$$

and we set
$$\alpha = \sum_{j<k} \frac{a_{kj}v_j}{a_{kk}v_k}, \qquad \beta = \sum_{j>k} \frac{a_{kj}v_j}{a_{kk}v_k}.$$

Then (8.1.17) can be written as

$$\lambda(1+\alpha) = -\beta,$$

so that

$$|\lambda| \le \frac{|\beta|}{|1+\alpha|} \le \frac{|\beta|}{1-|\alpha|} < 1,$$

by (8.1.14) and (8.1.16). Thus, we have shown that $\rho(H) < 1$ and Theorem 8.1.1 applies.

The condition of strict diagonal dominance is a rather stringent one and does not apply to the difference equations (8.1.8) for Laplace's equation: in most rows of the coefficient matrix there are four coefficients of absolute value 1 in the off-diagonal positions, so that strict inequality does not hold in (8.1.14). However, we showed in Section 5.5 that the coefficient matrix for (8.1.8) is irreducibly diagonally dominant and this is a sufficient condition for both the Jacobi and Gauss-Seidel iterations to converge (see the Supplementary Discussion).

We also showed in Section 5.5 that the coefficient matrix of the equations (8.1.8) is symmetric positive-definite. Indeed, for many discrete analogs of elliptic partial differential equations, the coefficient matrix will be symmetric and positive-definite. In this case the Gauss-Seidel iteration will always converge, although positive-definiteness is not, in general, sufficient for the Jacobi method to converge. We state the following theorem without proof:

THEOREM 8.1.3 *Assume that the matrix $A = D - B$ is symmetric and positive definite. Then the Gauss-Seidel iterates converge to the unique solution of $A\mathbf{x} = \mathbf{b}$ for any starting vector $\mathbf{x}^0$. The Jacobi iterates converge for any $\mathbf{x}^0$ if and only if the matrix $D + B$ is also positive definite.*

The SOR Method

Even when the Jacobi and Gauss-Seidel methods are convergent, the rate of convergence may be so slow as to preclude their usefulness. For example, for equation (8.1.8) with $N = 44$, the error in each iteration of the Gauss-Seidel method will decrease asymptotically only by a factor of about 0.995. Moreover, the Jacobi method is about twice as slow on this problem, and the rate of convergence of both methods becomes worse as N increases.

8.1 Relaxation-Type Methods

In certain cases it is possible to accelerate considerably the Gauss-Seidel method. Given the current approximation $\mathbf{x}^k$, we first compute the Gauss-Seidel iterate

$$\hat{x}_i^{(k+1)} = \frac{1}{a_{ii}}(b_i - \sum_{j<i} a_{ij} x_j^{(k+1)} - \sum_{j>i} a_{ij} x_j^{(k)}) \tag{8.1.18}$$

as an intermediate value, and then take the final value of the new approximation to the ith component to be

$$x_i^{(k+1)} = x_i^{(k)} + \omega(\hat{x}_i^{(k+1)} - x_i^{(k)}). \tag{8.1.19}$$

Here ω is a parameter that has been introduced to accelerate the convergence.

We can rewrite (8.1.18) and (8.1.19) in the following way. First substitute (8.1.18) into (8.1.19):

$$x_i^{(k+1)} = (1-\omega)x_i^{(k)} + \frac{\omega}{a_{ii}}(b_i - \sum_{j<i} a_{ij} x_j^{(k+1)} - \sum_{j>i} a_{ij} x_j^{(k)}), \tag{8.1.20}$$

and then rearrange this equation into the form

$$a_{ii} x_i^{(k+1)} + \omega \sum_{j<i} a_{ij} x_j^{(k+1)} = (1-\omega) a_{ii} x_i^{(k)} - \omega \sum_{j>i} a_{ij} x_j^{(k)} + \omega b_i.$$

This relationship between the new iterates $x_i^{(k+1)}$ and the old $x_i^{(k)}$ holds for $i = 1, \ldots, n$, and using (8.1.5) we can write it in matrix-vector form as

$$D\mathbf{x}^{k+1} - \omega L \mathbf{x}^{k+1} = (1-\omega) D \mathbf{x}^k + \omega U \mathbf{x}^k + \omega \mathbf{b}.$$

Since the matrix $D - \omega L$ is again lower triangular and, by assumption, has non-zero diagonal elements, it is nonsingular, so we may write

$$\mathbf{x}^{k+1} = (D - \omega L)^{-1}[(1-\omega)D + \omega U]\mathbf{x}^k + \omega(D - \omega L)^{-1}\mathbf{b}. \tag{8.1.21}$$

This defines the *successive overrelaxation (SOR) method*, although, as with Gauss-Seidel, the componentwise prescription (8.1.18), (8.1.19) would usually be used for the actual computation. Note that if $\omega = 1$, (8.1.21) reduces to the Gauss-Seidel iteration.

For real values of the parameter ω a necessary condition for the SOR iteration (8.1.21) to be convergent is that $0 < \omega < 2$ (see Exercise 8.1.5). In general, a choice of ω in this range will *not* give convergence, but in the important case that the coefficient matrix A is symmetric and positive definite, we have the following extension of Theorem 8.1.3, which we also state without proof:

THEOREM 8.1.4 (Ostrowski's Theorem) *If A is symmetric and positive definite, then for any $\omega \in (0, 2)$ and any starting vector $\mathbf{x}^0$, the SOR iterates (8.1.21) converge to the solution of $A\mathbf{x} = \mathbf{b}$.*

We would like to be able to choose the parameter ω so as to optimize the rate of convergence of the iteration (8.1.21). In general this is a very difficult problem, and we will attempt to summarize, without proofs, a few facts that are known about its solution. For a class of matrices that are called *consistently ordered with property A*, there is a rather complete theory that relates the rate of convergence of the SOR method to that of the Jacobi method and gives important insights into how to choose the optimum value of ω. We will not define this class of matrices precisely; suffice it to say that it includes the matrix (5.5.18) of equations (8.1.8) as well as many other matrices.

The fundamental result that holds for this class of matrices is a relationship between the eigenvalues of the SOR iteration matrix

$$H_\omega = (D - \omega L)^{-1}[(1 - \omega)D + \omega U] \qquad (8.1.22)$$

and the eigenvalues μ_i of the Jacobi iteration matrix $J = D^{-1}(L+U)$. Under the assumption that the μ_i are all real and less than 1 in absolute value, it can be shown that the optimum value of ω, denoted by ω_0, is given in terms of the spectral radius, $\rho(J)$, of J by

$$\omega_0 = \frac{2}{1 + \sqrt{1 - \rho^2}}, \qquad \rho = \rho(J), \qquad (8.1.23)$$

and is always between 1 and 2 (hence the term "over relaxation"). The corresponding value of the spectral radius of H_ω is

$$\rho(H_{\omega_0}) = \omega_0 - 1, \qquad (8.1.24)$$

and it is this quantity that governs the ultimate rate of convergence of the method. Moreover, we can ascertain the behavior of $\rho(H_\omega)$ as a function of ω, as is shown in Figure 8.1.2.

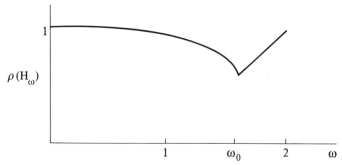

Figure 8.1.2: $\rho(H_\omega)$ as a Function of ω

We can obtain an idea of the acceleration of convergence that is possible by considering the equations (8.1.8). For this problem the eigenvalues of the

8.1 Relaxation-Type Methods

Jacobi iteration matrix J can be computed explicitly, and the largest is

$$\rho(J) = \cos \pi h, \qquad h = \frac{1}{N+1}. \tag{8.1.25}$$

If we put this in (8.1.23), we obtain

$$\omega_0 = \frac{2}{1 + \sqrt{1 - \cos^2 \pi h}}, \qquad \rho(H_{\omega_0}) = \frac{1 - \sqrt{1 - \cos^2 \pi h}}{1 + \sqrt{1 - \cos^2 \pi h}}. \tag{8.1.26}$$

If, again for illustration, we take $N = 44$, then

$$\rho(J) \doteq 0.9976, \quad \rho(H_1) \doteq 0.995, \quad \omega_0 \doteq 1.87, \quad \rho(H_{\omega_0}) \doteq 0.87. \tag{8.1.27}$$

This shows that, asymptotically, the error in Jacobi's method will decrease by a factor of 0.9976 at each step, and that of the Gauss-Seidel method by a factor of $0.995 = (.9976)^2$, which is twice as fast. But the error in the SOR method will decrease by a factor of $0.87 = (.995)^{30}$, so that SOR is about thirty times as fast as the Gauss-Seidel method. Moreover, the improvement increases as N increases (see Exercises 8.1.8 and 8.1.9).

The preceding discussion indicates that dramatic improvements in the rate of convergence of the Gauss-Seidel method are possible. However, a number of caveats are in order. First of all, many – perhaps most – large sparse matrices that arise in practice do not enjoy being "consistently ordered with property A," and the preceding theory will not hold. It is still possible that introduction of the parameter ω into the Gauss-Seidel method will produce a substantial increase in the rate of convergence, but this will not necessarily be known in advance, nor will we know how to choose a good value of ω. Even if the coefficient matrix is "consistently ordered with property A," it still may be difficult to obtain a good estimate of ω_0. It was possible to compute explicitly the quantities of (8.1.27) only because of the very special nature of the equations (8.1.8), which allowed an exact computation of $\rho(J)$. In general this will not be possible, and to use (8.1.23) will require estimating $\rho(J)$, which is itself a difficult problem. Thus, even in those cases where the preceding theory holds, it may be necessary to use an approximation process to obtain a suitable value of ω. In particular, there are "adaptive methods" that help to approximate a good value of ω as the SOR iteration proceeds (see the Supplementary Discussion). Finally, for most problems the multigrid method to be discussed in Section 8.3, and the conjugate gradient methods of Chapter 9 will usually be preferred to the methods of this section.

Supplementary Discussion and References: 8.1

1. The Jacobi and Gauss-Seidel iterations are classical methods that go back to the last century. The basic theory of the SOR method was developed

by Young [1950], [1954]. For extensive discussions of the Jacobi, Gauss-Seidel, and SOR methods and their many variants, see Varga [1962] and Young [1971]. For ways to compute adaptively the ω in SOR, see Hageman and Young [1981] and Young and Mai [1990].

2. The basic convergence Theorem 8.1.1 is equivalent to powers of a matrix tending to zero. The condition that $H^k \mathbf{e}^0 \to 0$ as $k \to \infty$ for any $\mathbf{e}^0$ is equivalent to $H^k \to 0$ as $k \to \infty$. If $H = PJP^{-1}$, where J is the Jordan canonical form of H, then $H^k = PJ^kP^{-1}$ and $H^k \to 0$ if and only if $J^k \to 0$ as $k \to \infty$. If J is diagonal, then the result is essentially what was shown in the text. Otherwise one needs to analyze powers of a Jordan block. If E is a matrix of 1's on the first super diagonal, it is easy to see that for $k \geq n$

$$(\lambda I + E)^k = \begin{bmatrix} \lambda^k & k\lambda^{k-1} & \binom{k}{2}\lambda^{k-2} & \cdots & \binom{k}{n-1}\lambda^{k-n+1} \\ & & & & \vdots \\ & & \ddots & \ddots & \\ & & & & k\lambda^{k-1} \\ & & & & \lambda^k \end{bmatrix},$$

which shows that the powers of a Jordan block tend to zero if and only if $|\lambda| < 1$. Although $\rho(H) < 1$ is a necessary and sufficient condition that the iterates (8.1.9) converge for any $\mathbf{x}^0$, it is not necessarily the case that the relation $||\mathbf{e}^{k+1}|| < ||\mathbf{e}^k||$ for the error vectors will hold in some usual norm (see Exercise 8.1.10).

3. Theorem 8.1.2 can be extended to show that if A is irreducibly diagonally dominant, then both the Jacobi and Gauss-Seidel iterates converge to the unique solution of $A\mathbf{x} = \mathbf{b}$ for any starting vector $\mathbf{x}^0$. (See, for example, Ortega [1990] or Varga [1962].)

4. Theorem 8.1.4 is a special case of a more general convergence theorem: if A is symmetric positive definite and $A = P - Q$, where P is nonsingular and $\mathbf{x}^T(P+Q)\mathbf{x} > 0$ for all $\mathbf{x} \neq 0$, then $\rho(P^{-1}Q) < 1$. This is sometimes referred to as the Householder-John Theorem but it is actually due to Weissinger; see Ortega [1990], where it is called the P-regular splitting theorem. A converse of Theorem 8.1.4 also holds: If A is symmetric with positive diagonal elements and SOR converges for some $\omega \in (0,2)$ and every $\mathbf{x}^0$, then A is positive definite. If this converse is added, Theorem 8.1.4 is known as the Ostrowski-Reich Theorem.

8.1 Relaxation-Type Methods

EXERCISES 8.1

8.1.1. Apply the Jacobi and Gauss-Seidel iterations to the system of $A\mathbf{x} = \mathbf{b}$, where

$$A = \begin{bmatrix} 3 & 1 & 1 \\ 1 & 3 & 1 \\ 1 & 1 & 3 \end{bmatrix}, \quad \mathbf{b} = \begin{bmatrix} 1 \\ 2 \\ 3 \end{bmatrix}.$$

Use the starting approximation $\mathbf{x}^0 = (1, 1, 1)$ and take enough steps of the iterative processes for the pattern of convergence to become clear.

8.1.2. Write computer programs for the Jacobi and Gauss-Seidel methods. Test them on the problem of Exercise 8.1.1.

8.1.3. Write out in detail the Jacobi and Gauss-Seidel iterations for the equations (8.1.8) for $N = 3$.

8.1.4. Consider the *Helmholtz equation* $u_{xx} + u_{yy} + cu = 0$ with u prescribed on the boundary of a square domain. Derive the difference equations corresponding to (8.1.8). If c is a negative constant, show that the resulting coefficient matrix is strictly diagonally dominant.

8.1.5. (Kahan's Lemma) For the matrix of (8.1.22), show that $\det(D - \omega L)^{-1} = \det D^{-1}$, and then that $\det H_\omega = (1-\omega)^n$. Using the fact that the determinant of a matrix is the product of its eigenvalues, conclude that if ω is real and $\omega \leq 0$ or $\omega \geq 2$, than at least one eigenvalue of H_ω must be greater than or equal to 1 in magnitude.

8.1.6. Carry out several steps of the SOR iteration for the problem of Exercise 8.1.1. Use the values $\omega = 0.6$ and $\omega = 1.4$ and compare the rates of convergence to the Gauss-Seidel iteration.

8.1.7. Write a computer program to apply the SOR iteration to (8.1.8).

8.1.8. Use the relations (8.1.25) and (8.1.26) to compute $\rho(J)$, ω_0, and $\rho(H_{\omega_0})$ for (8.1.8) for $N = 99$ and $N = 999$.

8.1.9. Expand $\cos \pi h$ in a Taylor series to show that, by (8.1.25), $\rho(J) = 1 - O(h^2) = 1 - O(N^{-2})$. Similarly, show that $\sqrt{1 - \cos^2 \pi h} = O(h)$ and then from (8.1.26), that

$$\rho(H_{\omega_0}) = \frac{1 - O(h)}{1 + O(h)} = [1 - O(h)][1 - O(h)] = 1 - O(h) = 1 - O(N^{-1}).$$

8.1.10. Let $H = \begin{bmatrix} 0.5 & \alpha \\ 0 & 0 \end{bmatrix}$. Compute H^k and show that if $\mathbf{e}^0 = (0, 1)^T$, then $\mathbf{e}^k = H^k \mathbf{e}^0 = (\alpha 2^{-k+1}, 0)^T$. Thus, if $\alpha = 2^p$, then $\|\mathbf{e}^k\|_2 \geq \|\mathbf{e}^0\|_2$ for $k \leq p+1$.

8.1.11. **a.** Let
$$A = \begin{bmatrix} 1 & \alpha & \alpha \\ \alpha & 1 & \alpha \\ \alpha & \alpha & 1 \end{bmatrix}.$$

Show that A is positive definite when $-1 < 2\alpha < 2$ but that the conditions of Theorem 8.1.3 for the convergence of the Jacobi iteration are satisfied only for $-1 < 2\alpha < 1$. Hence, conclude that for $1 \leq 2\alpha < 2$, the Gauss-Seidel iteration is convergent but Jacobi is not.

b. For the matrix
$$A = \begin{bmatrix} 1 & -2 & 2 \\ -1 & 1 & -1 \\ -2 & -2 & 1 \end{bmatrix},$$

show that the Jacobi iteration is convergent but Gauss-Seidel is not.

8.1.12. For the $(2, -1)$ matrix A of (5.3.10), use Exercise 5.3.5 to show that the eigenvalues of the Jacobi iteration matrix are

$$\cos \frac{k\pi}{N+1}, \quad k = 1, \ldots, N,$$

and the eigenvectors are the same as those of A. Conclude that the Jacobi iteration is convergent.

8.1.13. Give an example of a sequence of matrices $B_0, B_1, \ldots$ such that $\rho(B_i) = 0$ for all i but the vectors

$$\mathbf{e}^{k+1} = B_k \mathbf{e}^k, \quad k = 0, 1, \ldots$$

do not converge to zero.

8.1.14. Assume that the Jacobi iteration matrix $H = D^{-1}B$ has real eigenvalues $\mu_1 \leq \cdots \leq \mu_n$ and consider the iteration

$$\mathbf{x}^{k+1} = \mathbf{x}^k + \omega(H\mathbf{x}^k + \mathbf{d} - \mathbf{x}^k)$$

where $\mathbf{d} = D^{-1}\mathbf{b}$.

a. What is the iteration matrix H_ω?

b. If $\rho(H) \geq 1$, can you choose ω so that $\rho(H_\omega) < 1$?

c. What value of ω as a function of $\mu_1, \ldots, \mu_n$ minimizes $\rho(H_\omega)$?

d. If $-\mu_1 = \mu_n$, show that $\omega = 1$ is optimal.

8.1.15. Let
$$A = \begin{bmatrix} 1 & \alpha \\ -\alpha & 1 \end{bmatrix}.$$

a. Under what conditions on α will Jacobi and Gauss-Seidel converge?

b. Under what conditions on α and ω will SOR converge?

8.2 Parallel and Vector Implementations

We next discuss the implementation of the methods of Section 8.1 on parallel and vector computers. For Jacobi's method, it is clear from (8.1.8) that, in principle, all of the $x_j^{(1)}$ can be computed in parallel; it is for this reason that Jacobi's method is considered to be a prototypical parallel method. However, various details need to be considered in implementing it on vector or parallel computers. We discuss next some of the simplest issues for parallel systems, for the moment remaining with the discrete Laplace equation (8.1.8).

Suppose first that the parallel system consists of a mesh connected array of $p = q^2$ processors P_{ij} arranged in a two-dimensional lattice with each connected to its north, south, east, and west neighbors, as illustrated in Figure 8.2.1. Suppose that $N^2 = mp$. Then it is natural to assign m unknowns to each processor and there are a variety of ways to accomplish this. One of the simplest is to imagine the array of processors overlaid on the grid points, as illustrated in Figure 8.2.1 for four unknowns per processor. We assume that those processors that contain interior points adjacent to a boundary hold the corresponding known boundary values.

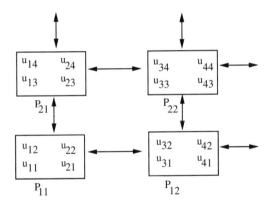

Figure 8.2.1: *Grid Point and Unknown Assignments*

On a distributed memory system, at the end of each iteration the new iterates at certain grid points will need to be transmitted to adjacent processors. We will next discuss this in more detail. A limiting case of the above arrangement is when there are $p = N^2$ processors, and each processor contains exactly one unknown. In this case, the work in an interior processor will proceed as

$$\begin{array}{l} \text{compute } u_{ij}^{k+1} \text{ in processor } P_{ij} \\ \text{send } u_{ij}^{k+1} \text{ to processors } P_{i+1,j}, P_{i-1,j}, P_{i,j-1}, P_{i,j+1}. \end{array} \quad (8.2.1)$$

Thus, at each stage a compute step, using (8.1.8) in the case of Laplace's equation, is followed by transmissions of the updated iterates to the processors that need them for the next iteration. Although this arrangement exhibits the perfect parallelism of Jacobi's method, the requirement of N^2 processors is generally unrealistic, and, moreover, an exorbitant amount of time will be spent in communication.

Assume, then, that $N^2 = mp$ and each processor holds m unknowns. This is illustrated in Figure 8.2.1 for $m = 4$, but a more realistic case might be $N = 100$ and $p = 20$, which implies that $m = 500$. Each processor will proceed to compute iterates for the unknowns it holds. For example, for the situation of Figure 8.2.1 the work in processor P_{22} would be

$$\begin{aligned} &\text{compute } u_{33}^{k+1}, u_{43}^{k+1}, \text{ send to } P_{12} \\ &\text{compute } u_{34}^{k+1}, \text{ send } u_{33}^{k+1}, u_{34}^{k+1} \text{ to } P_{21} \\ &\vdots \end{aligned} \qquad (8.2.2)$$

with analogous work proceeding in parallel in the other processors.

We denote by *edge values* those values of u_{ij} that are needed by other processors at the next iteration and must be transmitted. This is illustrated in Figure 8.2.2 for a single processor. A compute-and-send-ahead strategy should be used in which the edge values are computed first and then sent. While this communication is proceeding, the interior unknowns are then updated. If there are $m = q^2$ unknowns in a processor, then all m must be updated but at most $4q$ are transmitted. Thus, if we have a fixed number of processors, then the larger the problem, the higher will be the ratio of computation time to communication. In particular, the number of interior grid points for which no data communication is required grows quadratically as a function of the number of unknowns, whereas the number of grid points for which communication is required grows only linearly. This is known as the *volume-to-surface* effect. Hence, for sufficiently large problems, the data communication time becomes increasingly negligible relative to the computation time.

Synchronous and Asynchronous Methods

In the computation scheme indicated by (8.2.2), it is necessary to add a synchronization mechanism to ensure that the Jacobi iteration is carried out correctly. For example, the computation of u_{22}^{k+1} cannot be completed until u_{32}^k and u_{23}^k have been received from the neighboring processors. As discussed in Section 3.2, the time to carry out the synchronization, as well as the time that a processor must wait for its data to be ready, adds an overhead to the computation.

For certain iterative methods, such as Jacobi's method, a potentially attractive alternative is to allow the processors to proceed asynchronously, without any synchronization. In this case, some iterates may be computed incorrectly. For example, in the situation of Figure 8.2.1 suppose that at the time u_{12}^{k+1} is

8.2 Parallel and Vector Implementations

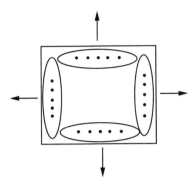

Figure 8.2.2: *Communication of Edge Values*

to be computed, u_{13}^k has not yet been received from processor P_{12}. Then the previous value, u_{13}^{k-1}, will be used in the computation of u_{12}^{k+1}, and this new iterate is not the Jacobi iterate. However, this need not be overly detrimental to the overall iteration and it may be cost-effective in some cases to allow iterative methods to proceed asynchronously. The Jacobi method implemented in this fashion will be called an *asynchronous* method.

Convergence Tests

In carrying out any iterative method, it is necessary to test for convergence of the iterates. If $\{\mathbf{x}^k\}$ is the sequence of iterates, two common tests are

$$\|\mathbf{x}^{k+1} - \mathbf{x}^k\| \le \varepsilon \quad \text{or} \quad \|\mathbf{x}^{k+1} - \mathbf{x}^k\| \le \varepsilon\|\mathbf{x}^k\|, \tag{8.2.3}$$

where $\|\ \|$ denotes a vector norm. It is also relatively common to use the residual $\mathbf{r}^k = \mathbf{b} - A\mathbf{x}^k$ and require instead of (8.2.3) that

$$\|\mathbf{r}^k\| \le \varepsilon \quad \text{or} \quad \|\mathbf{r}^k\| \le \varepsilon\|\mathbf{r}^0\|. \tag{8.2.4}$$

We next consider the implementation of the tests (8.2.3) for Jacobi's method in the situation of Figure 8.2.1. Two useful norms are the l_2 and l_∞ norms defined for a vector $\mathbf{x}$ by (see Chapter 2)

$$\|\mathbf{x}\|_2 = \left(\sum_{i=1}^n x_i^2\right)^{\frac{1}{2}}, \qquad \|\mathbf{x}\|_\infty = \max_{1 \le i \le n} |x_i|. \tag{8.2.5}$$

The use of the l_2 norm requires the computation of an inner product across the processors, which was discussed in Section 3.3. On the other hand, the first test of (8.2.3) can be implemented in the l_∞ norm in the following way. Each processor tests its unknowns at each iteration to see if they all satisfy

$$|u_{ij}^{k+1} - u_{ij}^k| \le \varepsilon.$$

If so, a "flag" is set indicating that the processor's iterates have converged, and then some mechanism is used to examine the flags of all processors. If they are all set, the iteration has converged and is stopped; if not, the iteration proceeds. Thus, this procedure is very much like a synchronization mechanism. Note that the vector norm is not actually computed in order to carry out the convergence test.

The second test of (8.2.3) is a relative error test and is useful when the magnitude of the solution is not known a priori so as to guide in the choice of ε. However, this test requires the computation of a norm across the processors followed by a transmission of the norm back to the processors and is more costly to implement.

Vectorization of the Jacobi Iteration

For many problems, the iteration matrix H of the Jacobi iteration is diagonally sparse but does not have constant diagonals as with the Poisson equation. We illustrate this by the important "generalized Poisson" equation

$$(au_x)_x + (bu_y)_y = f, \tag{8.2.6}$$

where a and b are given positive functions of x and y (not necessarily continuous). Again, for simplicity, we shall assume that the domain of (8.2.6) is the unit square and the solution u is prescribed on the boundary. A standard finite difference discretization of (8.2.6) is (see the Supplementary Discussion of Section 5.3 for the corresponding one-dimensional discretization)

$$a_{i-\frac{1}{2},j}u_{i-1,j} + a_{i+\frac{1}{2},j}u_{i+1,j} + b_{i,j-\frac{1}{2}}u_{i,j-1}$$
$$+ b_{i,j+\frac{1}{2}}u_{i,j+1} - c_{ij}u_{ij} = h^2 f_{ij}, \tag{8.2.7}$$

where

$$c_{ij} = a_{i-\frac{1}{2},j} + a_{i+\frac{1}{2},j} + b_{i,j-\frac{1}{2}} + b_{i,j+\frac{1}{2}}. \tag{8.2.8}$$

In (8.2.7) the coefficients a and b are evaluated at the midpoints of the intervals between grid points. For example, $a_{i-\frac{1}{2},j} = a(ih - \frac{1}{2}h, jh)$ and similarly for the others. This assures symmetry of the coefficient matrix, as seen below. Note that (8.2.7) reduces to (5.5.13) if $a = b = 1$.

The equations (8.2.7) can be written in the form $A\mathbf{x} = \mathbf{b}$, where $\mathbf{x}$ is the vector of unknowns u_{ij} given by

$$\mathbf{x}^T = (u_{11}, \ldots, u_{N1}, u_{12}, \ldots, u_{N2}, \ldots, u_{1N}, \ldots, u_{NN}), \tag{8.2.9}$$

8.2 Parallel and Vector Implementations

b is the vector containing boundary values and $-h^2 f_{ij}$, and

$$A = \begin{bmatrix} T_1 & B_1 & & & \\ B_1 & \ddots & \ddots & & \\ & \ddots & \ddots & B_{N-1} \\ & & B_{N-1} & T_N \end{bmatrix}, \quad (8.2.10)$$

where

$$T_j = \begin{bmatrix} c_{1j} & -a_{\frac{1}{2},j} & & & \\ -a_{\frac{1}{2},j} & c_{2j} & -a_{\frac{3}{2},j} & & \\ & \ddots & \ddots & & \\ & & \ddots & & -a_{N-\frac{1}{2},j} \\ & & & -a_{N-\frac{1}{2},j} & c_{Nj} \end{bmatrix} \quad (8.2.11)$$

and

$$B_j = \mathrm{diag}(-b_{i,j+\frac{1}{2}}, \ldots, -b_{N,j+\frac{1}{2}}).$$

The matrix A of (8.2.10) has five nonzero diagonals. The Jacobi iteration matrix H has the same structure, except that its main diagonal is zero. We assume that H is stored by diagonals and that the multiplication $H\mathbf{x}^k$ in the Jacobi iteration is carried out by the multiplication by diagonals procedure of Section 3.3. Thus, the vector lengths in carrying out the Jacobi iteration are $O(N^2)$. More precisely, the main diagonal of H is of length N^2, the close off-diagonals are of length $N^2 - 1$, and the far off-diagonals are of length $N^2 - N$. Hence, the storage of the matrix H requires $2N^2 - N - 1$ words since, by symmetry, we need only store the upper triangular part. There is wasted storage of $N-1$ words for the zeros on the close-in diagonal, but this is necessary to obtain the long vector lengths. The above method of implementing the Jacobi iteration in terms of long vector operations is efficient for those problems with variable coefficients for which the equivalent of the matrix H must be stored in any case. For Poisson's equation, H consists of only 0's and $\frac{1}{4}$'s and there is no need to store the matrix.

Block Methods

Let a given matrix A be partitioned into submatrices as

$$A = \begin{bmatrix} A_{11} & \cdots & A_{1q} \\ \vdots & & \vdots \\ A_{q1} & \cdots & A_{qq} \end{bmatrix}, \quad (8.2.12)$$

where each A_{ii} is assumed to be nonsingular. Then the corresponding *block Jacobi method* for the solution of $Ax = b$ is

$$A_{ii}\mathbf{x}_i^{k+1} = -\sum_{j \neq i} A_{ij}\mathbf{x}_j^k + \mathbf{b}_i, \qquad i = 1, \ldots, q, \qquad k = 0, 1, \ldots, \qquad (8.2.13)$$

where $\mathbf{x}$ and $\mathbf{b}$ are partitioned commensurately with A. Thus, to carry out one block Jacobi iteration requires the solution of the q systems of (8.2.13) with coefficient matrices A_{ii}. Note that in the special case that each A_{ij} is 1×1, (8.2.13) reduces to the (point) Jacobi method previously discussed. It is known that in certain cases block methods require fewer iterations than point methods.

As an example of a block Jacobi method, consider again the discrete Poisson equation (5.5.19). In this case, $q = N, A_{ii} = T, A_{i,i+1} = A_{i+1,i} = -I$, and all other A_{ij} are zero. Thus (8.2.13) becomes

$$T\mathbf{x}_i^{k+1} = \mathbf{x}_{i+1}^k + \mathbf{x}_{i-1}^k + \mathbf{b}_i, \qquad i = 1, \ldots, N, \qquad (8.2.14)$$

where we assume that $\mathbf{x}_0^k$ and $\mathbf{x}_{N+1}^k$ contain the boundary values along the bottom and top boundaries, respectively, and the vectors $\mathbf{b}_i$ contain the boundary values on the sides of the region as well as the $h^2 f$ values. The effect of (8.2.14) is to update all unknowns on each row of grid points simultaneously using the values of the unknowns from the previous iteration at the adjacent grid lines. Thus, (8.2.14) is known as a *line Jacobi method*. We would implement (8.2.14) by doing a Cholesky or LU decomposition of T once and for all at the outset, and then use the factors to solve the systems of (8.2.14). Thus, if $T = LU$, then (8.2.14) is carried out as shown in Figure 8.2.3.

$$\begin{array}{ll} \mathbf{d}_i^k = \mathbf{x}_{i+1}^k + \mathbf{x}_{i-1}^k + \mathbf{b}_i, & i = 1, \ldots, N \\ \text{Solve } L\mathbf{y}_i^k = \mathbf{d}_i^k, & i = 1, \ldots, N \\ \text{Solve } U\mathbf{x}_i^{k+1} = \mathbf{y}_i^k, & i = 1, \ldots, N \end{array}$$

Figure 8.2.3: *Line Jacobi Method*

The systems for the $\mathbf{y}_i^k$ are independent and can be solved in parallel; similarly for the $\mathbf{x}_i^{k+1}$ once the $\mathbf{y}_i^k$ are known. Thus, for example, if there are $p = N$ processors, we can assign one pair of systems to each processor of the linear array illustrated in Figure 8.2.4. After each iteration the new iterates $\mathbf{x}_i^{k+1}$ must be transmitted to each of the neighboring processors P_{i-1}

8.2 Parallel and Vector Implementations

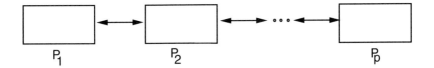

Figure 8.2.4: *A Linear Array for Parallel Line Jacobi Method*

and P_{i+1}. In the likely case that $N \gg p$, several systems would be assigned to each processor and there would be relatively less communication.

Block methods may be run asynchronously, just as previously discussed for point methods, but now there is the potential for whole blocks of unknowns to fail to be updated in time for the next iteration, depending upon how the data transmission is accomplished.

Vectorization of Block Methods

We consider next the implementation of the line Jacobi method (8.2.14) on vector computers. Assume that T has been factored into LU, where L is a lower bidiagonal matrix with unit diagonal. Then the forward and back substitutions of Figure 8.2.3 can be accomplished by "vectorizing across the systems" as illustrated in Figure 8.2.5.

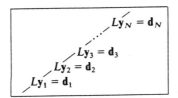

Figure 8.2.5: *Vectorizing Across Systems*

A pseudocode is given in Figure 8.2.6, in which it is assumed that $L1$ is a one-dimensional array that holds the elements of L in the first diagonal below the main diagonal, and y is a two-dimensional array such that $y(\ ,j)$ is the N-long vector of the jth components of all of the solution vectors. Similarly, $d(\ ,j)$ is the N-long vector of the jth components of the right-hand sides. Thus, the vector lengths in Figure 8.2.6 are all N. The back substitution $U\mathbf{x} = \mathbf{y}$ can be handled in an analogous way.

At each iteration it is also necessary to determine the new right-hand sides as given by (8.2.14). These can also be computed using vectors across the $\mathbf{x}_i$; indeed, this storage of the $\mathbf{x}_i$ is what is produced by the back substitution since the code analogous to Figure 8.2.6 will give $x(\ ,j)$, the vectors of the jth components. Thus, the computation of the next set of right-hand sides would

$$\boxed{\begin{array}{l} y(\ ,1) = d(\ ,1) \\ \text{For } j = 2 \text{ to } N \\ \quad y(\ ,j) = -L1(j)^* y(\ ,j-1) + d(\ ,j) \end{array}}$$

Figure 8.2.6: *Vector Forward Substitution for Several Systems*

be accomplished by

$$\begin{array}{l} \text{for } j = 1 \text{ to } N \\ \quad d(\ ,j) = x(\ ,j)_{-1} + x(\ ,j)_{+1} + b(\ ,j) \end{array}$$

where $x(\ ,j)_{\pm 1}$ denotes the vector $x(\ ,j)$ shifted up or down one position.

This same idea of "vectorizing across systems" may be applied in many other situations provided that the structures of all entities involved in the vectorization are identical.

The Gauss-Seidel Iteration

Although the Gauss-Seidel iteration is similar in some respects to the Jacobi iteration, it is much harder to carry out in parallel. By (8.1.7), a Gauss-Seidel step requires solution of a lower triangular system of equations, a problem for which the previous parallel methods are only moderately successful, in general. However, the Gauss-Seidel (or SOR) iteration will, in practice, only be used for large sparse systems of equations, and in this case the previous methods for triangular systems may be totally unsatisfactory.

One approach to successful parallel implementation of the SOR method is to reorder the equations in such a way that the solution of the lower triangular system can be done efficiently in parallel. An important classical ordering of the grid points is the *red-black* (or *checkerboard*) ordering illustrated in Figure 8.2.7. The grid points are first divided into two classes, red and black, and then ordered within each class. Figure 8.2.7 shows the ordering left to right, bottom to top within each class. The unknowns at the grid points are ordered analogously so that, for example, u at $R1$ will be the first unknown, u at $R2$ the second, and so on until the red points are exhausted; the ordering then continues with the black points. This is illustrated in Figure 8.2.8 by writing out the equations for $N = 2$ for Poisson's equation.

For general N, at any interior point, red grid points will be coupled only with black grid points and conversely, as illustrated in Figures 8.2.8 and 8.2.9. Thus, when we write the equations with all of the unknowns at the red points

8.2 Parallel and Vector Implementations

```
•R6  •B6  •R7  •B7  •R8
•B3  •R4  •B4  •R5  •B5
•R1  •B1  •R2  •B2  •R3
```

Figure 8.2.7: *The Red-Black Ordering of Grid Points*

$\dot{B}2 \quad \dot{R}2$

$\dot{R}1 \quad \dot{B}1$

$$\begin{bmatrix} 4 & 0 & -1 & -1 \\ 0 & 4 & -1 & -1 \\ -1 & -1 & 4 & 0 \\ -1 & -1 & 0 & 4 \end{bmatrix} \begin{bmatrix} u_{R1} \\ u_{R2} \\ u_{B1} \\ u_{B2} \end{bmatrix} = \begin{bmatrix} b_1 \\ b_2 \\ b_3 \\ b_4 \end{bmatrix}$$

(a) grid (b) equations

Figure 8.2.8: *Red-Black Equations for Four Unknowns*

ordered first, the system will be of the form

$$\begin{bmatrix} D_R & C \\ C^T & D_B \end{bmatrix} \begin{bmatrix} \mathbf{u}_R \\ \mathbf{u}_B \end{bmatrix} = \begin{bmatrix} \mathbf{b}_1 \\ \mathbf{b}_2 \end{bmatrix}. \qquad (8.2.15)$$

Here $D_R = 4I_R$ and $D_B = 4I_B$, where I_R is the identity matrix of size equal to the number of red interior grid points and similarly for I_B for the black points. The matrix C defines the interactions between the red and black unknowns. The special case of (8.2.15) for $N = 2$ has been illustrated in Figure 8.2.8. Note that the coefficient matrix of (8.2.15) must be symmetric since it is of the form PAP^T, where P is a permutation matrix and A is the symmetric matrix of the equations in the natural ordering.

Now consider the Gauss-Seidel iteration applied to (8.2.15):

$$\begin{bmatrix} D_R & 0 \\ C^T & D_B \end{bmatrix} \begin{bmatrix} \mathbf{u}_R^{k+1} \\ \mathbf{u}_B^{k+1} \end{bmatrix} = \begin{bmatrix} \mathbf{b}_1 \\ \mathbf{b}_2 \end{bmatrix} - \begin{bmatrix} 0 & C \\ 0 & 0 \end{bmatrix} \begin{bmatrix} \mathbf{u}_R^k \\ \mathbf{u}_B^k \end{bmatrix}, \qquad (8.2.16)$$

which uncouples into the two separate parts

$$\mathbf{u}_R^{k+1} = D_R^{-1}(\mathbf{b}_1 - C\mathbf{u}_B^k), \qquad \mathbf{u}_B^{k+1} = D_B^{-1}(\mathbf{b}_2 - C^T\mathbf{u}_R^{k+1}). \qquad (8.2.17)$$

Thus, although (8.2.16) still requires the solution of a lower triangular system of equations to advance one iteration, because D_R and D_B are diagonal, the

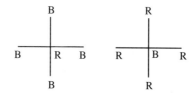

Figure 8.2.9: *Red-Black Stencils*

solution of this system reduces to the matrix-vector multiplications indicated by (8.2.17).

In the case of Poisson's equation, the multiplications of (8.2.17) are particularly easy to carry out since the elements of C are all 1. For more general differential equations like (8.2.6) the red/black ordering of the grid points can again be used (Exercise 8.2.5) and the resulting equations written as (8.2.15). Hence, (8.2.17) is still valid and C will be diagonally sparse so that matrix-vector multiplication by diagonals may be used.

For the SOR iteration the parameter ω is easily inserted. If $\hat{\mathbf{u}}_R^{k+1}$ denotes the Gauss-Seidel values at the red points, then the SOR iterate is

$$\mathbf{u}_R^{k+1} = \mathbf{u}_R^k + \omega(\hat{\mathbf{u}}_R^{k+1} - \mathbf{u}_R^k). \tag{8.2.18}$$

Using these updated values at the red points, the new values at the black points are obtained in an analogous way. The new Gauss-Seidel iterate for the red points is the same as the Jacobi iterate for these points, and similarly for the black points. Thus, the SOR iteration is implemented by using two Jacobi sweeps through the grid, each on roughly half of the grid points.

Parallel Implementation

Suppose first that we have $p = N^2/2$ processors arranged in a mesh-connected array. We assign one red and one black interior grid point to these processors as shown in Figure 8.2.10. The processors holding grid points adjacent to the boundary will be assumed to hold the corresponding boundary values. In order to do one Gauss-Seidel iteration, all processors will first update their red unknowns in parallel and then their black unknowns. As noted previously, these are just the computations for Jacobi iterations. The relaxation parameter ω will be inserted after each half step to yield the SOR values; for the red points, this is the computation (8.2.18) and similarly for the black points. In the more realistic case that $p << N^2$, for example, if $N^2 = 2mp$, then m red points and m black points would be assigned to each processor. If p is not an even multiple of N^2, then we will assign as close as possible the same number of red and black points to each processor.

8.2 Parallel and Vector Implementations

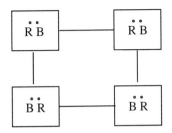

Figure 8.2.10: *Red-Black Processor Assignments*

The above SOR iteration involves the same considerations as in the Jacobi method concerning communication on distributed memory machines, convergence testing, and synchronization. The synchronization here would be done after each Jacobi step; that is, all new black values would be computed and passed to the processors that need them before the computation on the red unknown would proceed, and conversely. Alternatively, just as with the Jacobi iteration, the iteration may be carried out asynchronously, and we will call this the *asynchronous red-black SOR iteration*.

Multicoloring

The implementation of the SOR iteration by means of the red-black ordering of grid points is limited to rather simple partial differential equations, such as Poisson's equation, and rather simple discretizations, as we will now illustrate.

Consider the equation

$$u_{xx} + u_{yy} + au_{xy} = 0 \qquad (8.2.19)$$

again on the unit square, where a is a constant. This is just Laplace's equation with an additional term. The standard finite difference approximation to u_{xy} is

$$u_{xy} \doteq \frac{1}{4h^2}[u_{i+1,j+1} - u_{i-1,j+1} - u_{i+1,j-1} + u_{i-1,j-1}], \qquad (8.2.20)$$

which combined with the previous approximation (8.1.8) for Laplace's equation gives the system of equations

$$u_{i+1,j} + u_{i-1,j} + u_{i,j+1} + u_{i,j-1} - 4u_{ij}$$
$$+ \frac{a}{4}[u_{i+1,j+1} - u_{i-1,j+1} - u_{i+1,j-1} + u_{i-1,j-1}] = 0, \quad (8.2.21)$$

for $i, j = 1, \ldots, N$. For a red-black ordering of the grid points, as in Figure 8.2.7, it is easy to see (Exercise 8.2.7) that the system (8.2.21) may be written in the form (8.2.15) as was done for Poisson's equation. However, the matrices D_R

and D_B in (8.2.15) are no longer diagonal. The problem is that the unknown at the i,j grid point is now coupled with its eight nearest neighbors, some of which have the same color as the center point. This is illustrated in Figure 8.2.11.

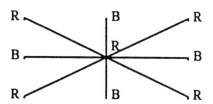

Figure 8.2.11: *Red-Black Ordering for* (8.2.21)

The solution to this problem is to introduce more colors, as illustrated in Figure 8.2.12. Now, if we write the equations for all the red points first, all the black points next, and so on, the system takes the form

$$\begin{bmatrix} D_1 & B_{12} & B_{13} & B_{14} \\ B_{21} & D_2 & B_{23} & B_{24} \\ B_{31} & B_{32} & D_3 & B_{34} \\ B_{41} & B_{42} & B_{43} & D_4 \end{bmatrix} \begin{bmatrix} \mathbf{u}_R \\ \mathbf{u}_B \\ \mathbf{u}_G \\ \mathbf{u}_W \end{bmatrix} = \begin{bmatrix} \mathbf{b}_1 \\ \mathbf{b}_2 \\ \mathbf{b}_3 \\ \mathbf{b}_4 \end{bmatrix}, \qquad (8.2.22)$$

where the diagonal blocks D_i are diagonal. The Gauss-Seidel iteration can then be written as

$$D_1 \mathbf{u}_R^{k+1} = -B_{12} \mathbf{u}_B^k - B_{13} \mathbf{u}_G^k - B_{14} \mathbf{u}_W^k + \mathbf{b}_1, \qquad (8.2.23a)$$

$$D_2 \mathbf{u}_B^{k+1} = -B_{21} \mathbf{u}_R^{k+1} - B_{23} \mathbf{u}_G^k - B_{24} \mathbf{u}_W^k + \mathbf{b}_2, \qquad (8.2.23b)$$

and similarly for the other two colors. Since the D_i are diagonal, the solution of the triangular system to carry out a Gauss-Seidel iteration has again reduced to matrix multiplications.

The four-color ordering of Figure 8.2.12 was based on the coupling of grid points illustrated in Figure 8.2.11, and we will call such a pattern a *stencil*. A stencil shows the connections of a grid point to its neighbors and depends on both the differential equation and the discretization. The determination of the number of colors needed is simplified if the stencil is the same at all points of the grid. Then the criterion for a successful coloring is that when the stencil is put at each point of the grid, the center point has a color different from that of all other points to which it is connected. It is this "local uncoupling" of the unknowns that allows a matrix representation of the problem that in general

8.2 Parallel and Vector Implementations

```
•G  •W  •R  •B  •G  •W  •R  •B
•R  •B  •G  •W  •R  •B  •G  •W
•G  •W  •R  •B  •G  •W  •R  •B
•R  •B  •G  •W  •R  •B  •G  •W
```

Figure 8.2.12: *A Four-Color Ordering*

has the form

$$\begin{bmatrix} D_1 & B_{12} & \cdots & & B_{1c} \\ B_{21} & & & & \vdots \\ \vdots & & \ddots & & B_{c-1,c} \\ B_{c1} & \cdots & & B_{c,c-1} & D_c \end{bmatrix} \begin{bmatrix} \mathbf{u}_1 \\ \mathbf{u}_2 \\ \vdots \\ \mathbf{u}_c \end{bmatrix} = \begin{bmatrix} \mathbf{b}_1 \\ \mathbf{b}_2 \\ \vdots \\ \mathbf{b}_c \end{bmatrix}, \qquad (8.2.24)$$

in which the diagonal blocks D_i are diagonal. Note that $c = 4$ in (8.2.22) and $c = 2$ for the red-black ordering. With the system in the form (8.2.24), the Gauss-Seidel iteration can be carried out, analogously to (8.2.23), by

$$\mathbf{u}_i^{k+1} = D_i^{-1} \left[\mathbf{b}_i - \sum_{j<i} B_{ij} \mathbf{u}_j^{k+1} - \sum_{j>i} B_{ij} \mathbf{u}_j^k \right], \qquad i = 1, \ldots, c, \qquad (8.2.25)$$

and, again, the solution of the triangular system is reduced to matrix-vector multiplications.

In general, one wishes to use the minimum number of colors that are necessary to achieve the matrix form (8.2.24). For arbitrary stencils that can vary at each point of the grid and arbitrary grids this is a difficult problem. However, if the stencil is repeated at each grid point it is generally evident how to achieve the minimum number of colors. But the coloring pattern need not be unique, even when the minimum number of colors is used. Exercise 8.2.9 gives other four-color orderings, different from that of Figure 8.2.12, for the stencil of Figure 8.2.11.

For systems of partial differential equations the number of colors will generally be multiplied by the number of equations. Thus, for example, if the grid stencil requires three colors and there are two partial differential equations, then six colors will be required to achieve (8.2.24).

Parallel and Vector Implementation

The implementation of the Gauss-Seidel and SOR iterations using multicolor orderings is similar to that using the red/black ordering. For simplicity, we consider again only a square grid with N^2 interior grid points. We begin with a distributed memory mesh-connected system with p processors, and assume that $N^2 = cmp$, where c is the number of colors. Thus, each processor will hold mc unknowns. This is illustrated in Figure 8.2.13 for $m = 1$ and $c = 4$, using the color pattern of Figure 8.2.12. The processors first all update their red points and then transmit the new values to the processors that will need them. In Figure 8.2.13 we have used the stencil of Figure 8.2.11 to show a sample communication pattern. Thus, the updated red value in processor P_{12} must be transmitted to processors P_{11}, P_{21}, and P_{22}, but not to processors P_{13} and P_{23}. The black values are then all updated, followed by the green and then the white. Between each stage there is communication and synchronization; alternatively, the iteration could be run asynchronously. For each set of colored values, we do only Jacobi operations, plus the SOR modification using ω. Hence, the iteration is implemented, in essence, by four Jacobi sweeps of the grid, each updating $O(N^2/4)$ points or, in general, c sweeps if c colors are used. Our assumption that $N^2 = cmp$ is based on the tacit assumption that updating at each grid point requires the same amount of work so that the distribution of mc unknowns to each processor balances the overall workload.

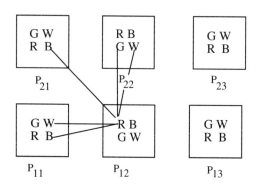

Figure 8.2.13: *Processor Assignment for Four Colors*

On vector machines, the critical thing is to implement efficiently the matrix-vector multiplies of (8.2.25). As with the red/black ordering, for problems like (8.2.21) the matrices B_{ij} will be diagonally sparse and multiplication by diagonals may be used. However, in general, the more colors used, the shorter the vector lengths will be.

Block and Line SOR

As with the Jacobi iteration, we can apply the Gauss-Seidel iteration to a partitioned matrix

$$A = \begin{bmatrix} A_{11} & \cdots & A_{1q} \\ \vdots & & \vdots \\ A_{q1} & \cdots & A_{qq} \end{bmatrix} \qquad (8.2.26)$$

to obtain the *block Gauss-Seidel method*

$$A_{ii}\mathbf{x}_i^{k+1} = -\sum_{j<i} A_{ij}\mathbf{x}_j^{k+1} - \sum_{j>i} A_{ij}\mathbf{x}_j^k + \mathbf{b}_i, \qquad i = 1, \ldots, q. \qquad (8.2.27)$$

Here $\mathbf{x}^k$ and $\mathbf{b}$ are partitioned commensurately with A, and it is assumed that the A_{ii} are all nonsingular. The parameter ω is introduced analogously to (8.2.18) by

$$\mathbf{x}_i^{k+1} = \mathbf{x}_i^k + \omega(\hat{\mathbf{x}}_i^{k+1} - \mathbf{x}_i^k), \qquad (8.2.28)$$

where $\hat{\mathbf{x}}_i^{k+1}$ is the Gauss-Seidel iterate produced by (8.2.27); this gives the *block SOR iteration*. As an example of these block methods, consider again the discrete Poisson equation (5.5.19). In this case, one form of (8.2.27) is

$$T\mathbf{x}_i^{k+1} = \mathbf{x}_{i-1}^{k+1} + \mathbf{x}_{i+1}^k + \mathbf{b}_i, \qquad i = 1, \ldots, N. \qquad (8.2.29)$$

As with the corresponding line Jacobi method (8.2.14), the new iterates on each line of the grid are updated simultaneously, the difference now being that these new iterates are used as soon as we move to the next line. Thus, (8.2.29) is called a *line Gauss-Seidel method* and the corresponding SOR iteration is generally called a *successive line overrelaxation (SLOR or LSOR) method*.

Next, we briefly discuss the parallel and vector implementation of the line Gauss-Seidel method (8.2.29). If the equations are written in the natural order, the parallelization of (8.2.29) has the same difficulty as the point Gauss-Seidel method, and again one approach is a reordering of the equations. This time, however, we use a red-black ordering by line as illustrated in Figure 8.2.14. This is sometimes called a "zebra ordering" (replace red by white). With the ordering of Figure 8.2.14, the equations for the unknowns on the ith line couple red points with only black points on the adjacent lines. If we denote the vector of unknowns on the ith red line by $\mathbf{u}_{Ri}$, and similarly for the black unknowns, then a line Gauss-Seidel step for updating the ith red line is

$$T\mathbf{u}_{Ri}^{k+1} = \mathbf{u}_{Bi-1}^k + \mathbf{u}_{Bi}^k + \mathbf{b}_i, \qquad (8.2.30)$$

and similarly for updating the black points. The tridiagonal systems (8.2.30) can now be solved in parallel or by vectorizing across the systems just as was done with the line Jacobi method. The only difference is that now the vector lengths are $O(N/2)$ rather than $O(N)$. For more general equations and/or

$$\begin{array}{ccccc} B & B & B & \cdots & B \\ R & R & R & \cdots & R \\ B & B & B & \cdots & B \\ R & R & R & \cdots & R \end{array}$$

Figure 8.2.14: *Red-Black Ordering by Lines*

discretizations the same principle applies, although additional colors may be needed (see Exercise 8.2.10).

The Diagonal Ordering

We now return to the natural ordering as shown in Figure 5.5.3 and consider another approach to obtaining parallelism. We illustrate this by the 4 × 4 grid shown in Figure 8.2.15.

$$\begin{array}{cccc} \bullet & \bullet & \bullet & \bullet \\ 10 & 13 & 15 & 16 \\ \bullet & \bullet & \bullet & \bullet \\ 6 & 9 & 12 & 14 \\ \bullet & \bullet & \bullet & \bullet \\ 3 & 5 & 8 & 11 \\ \bullet & \bullet & \bullet & \bullet \\ 1 & 2 & 4 & 7 \end{array}$$

Figure 8.2.15: *Diagonal Ordering of Grid Points*

If we are carrying out the Gauss-Seidel iteration in the natural ordering for the five-point stencil, we would first update left to right the unknowns corresponding to grid points on the first row, then on the second row and so on. But the update at the grid point labeled 3 in Figure 8.2.15 depends only on the new value at grid point 1, and not on the new values on the rest of the first row. Thus, we can update at both points 2 and 3 in parallel. The same is true along the next diagonal of grid points: we can update at points 4, 5 and 6 in parallel as soon as 2 and 3 have been updated, and so on. This is sometimes called the *"compute when you can"* principle.

On an $N \times N$ grid the lengths of the diagonals of grid points are $1, 2, 3, \ldots,$ $N-1, N, N-1, \ldots, 1$, so that there are varying degrees of parallelism corresponding to the different diagonals. However, greater parallelism is available since we can begin the second iteration at grid point 1 as soon as the unknowns

8.2 Parallel and Vector Implementations

at points 2 and 3 have been updated. Thus, the unknowns at grid points 1, 4, 5, 6 may be done in parallel, then those at points 2, 3, 7, 8, 9, 10, and so on. As soon as points 2 and 3 have been updated a second time, a third iteration may be started, and then a fourth, etc. This is sometimes called a "wave front" computation, which is illustrated schematically in Figure 8.2.16. In this figure are depicted the diagonals currently being processed in iterations 1 up to 5, and one can imagine waves of a computation passing over the grid.

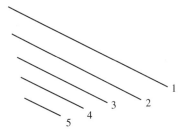

Figure 8.2.16: *Wave Front Computation*

The same basic idea may be used for three-dimensional problems but now one works with "diagonal planes," rather than lines. For example, for a $3 \times 3 \times 3$ grid the first plane is a single point, there are 3 points on the second plane and these can be updated in parallel, 6 points on the third plane, and so on. On a $N \times N \times N$ grid the maximum number of points on a plane will be $O(N^2)$.

Vectorization of the Diagonal Ordering

We return now to the two-dimensional problem and the diagonal ordering illustrated by Figure 8.2.15. If we reorder the equations and unknowns according to this ordering of the grid points, the coefficient matrix takes the form illustrated in Figure 8.2.17 (Exercise 8.2.11).

This matrix is of the form (8.2.24) and thus we may consider the diagonal ordering as a multicolor ordering if we assign separate colors to each diagonal. The Gauss-Seidel iteration may then be carried out as in (8.2.25) and the key aspect of this is the multiplication by the off-diagonal blocks. These off-diagonal blocks are shown in Figure 8.2.17. The (3,2) block has diagonals of length 2, the (4,3) block has diagonals of length 3, and so on. These diagonals become the vectors in the matrix-vector multiplications of (8.2.25) and for an $N \times N$ grid are of length $1, 2, \ldots, N-1, \ldots, 1$; that is, they are exactly one less than the lengths of the diagonals in the grid, and mirror the parallelism previously discussed. Thus, the diagonal ordering yields a rather modest parallelism or vectorization compared with the red/black ordering in which the vector lengths are of $O(\frac{1}{2}N^2)$. However, we will see in the next chapter that

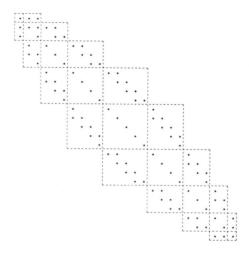

Figure 8.2.17: *Matrix for Diagonal Ordering (Two Dimensions)*

the diagonal ordering is useful in certain circumstances in conjunction with the conjugate gradient method.

Rate of Convergence

Reorderings may effect the rate of convergence of the Gauss-Seidel (and SOR) method. The natural and diagonal orderings have the same rate of convergence since exactly the same iterates are produced in both cases, as previously discussed. The iterates produced by the natural and red/black orderings are not the same. But, provided that the matrices D_R and D_B in (8.2.15) are diagonal, a consequence of the theory discussed at the end of Section 8.1 for "consistently ordered" matrices is that the asymptotic rate of convergence of the SOR iteration is exactly the same for both the natural and red/black orderings. The number of iterations to satisfy a convergence criterion may be different, however, and it seems to be the case that the red/black ordering many times provides slightly faster convergence. The effect of multicolor orderings on the rate of convergence is not yet completely understood. (See the Supplementary Discussion.)

Supplementary Discussion and References: 8.2

1. We have only considered in the text the simplest kind of partitioning of the grid points to the processors in a parallel system. More difficult problems include irregular regions, nonuniform grids, and dynamically changing grids. For these and other questions, see for example, Berger

and Bokhari [1987], and Morison and Otto [1987]. The latter paper considers mappings of irregular domains to hypercubes (or other mesh-connected arrays) by *scattered decomposition*.

2. The idea of asynchronous iterations dates back at least to the "chaotic relaxation" of Chazan and Miranker [1969]. A detailed study of asynchronous iterations, including experiments, was done by Baudet [1978].

3. Block Jacobi and SOR methods are classical and a detailed treatment of line methods and their rate of convergence is given in Varga [1962]. More recent work on block methods includes Parter and Steurwalt [1980,1982].

4. It was recognized in the early days of parallel computing (see, for example, Lambiotte [1975]) that the red-black ordering for the discrete Poisson equation allowed a high degree of parallelism in the SOR method. Some years later, several authors independently advocated using more than two colors for more complicated situations; the discussion in the text follows Adams and Ortega [1982]. The effect of multicoloring on the rate of convergence is still not completely clear, although Adams and Jordan [1985] have shown that in some situations the asymptotic rate of convergence using certain multicolor orderings is identical to that using the natural ordering. See also Adams et al. [1988] for an analysis of the rate of convergence of four-color orderings applied to a nine-point discretization of Poisson's equation, and Adams [1986] for further discussion of multicolor orderings.

5. O'Leary [1984] has proposed other types of multicolor orderings, one of which is illustrated below.

$$\begin{array}{cccccccccc} 3 & 3 & 1 & 1 & 3 & 3 & 1 & 2 & 3 & 3 \\ 3 & 3 & 2 & 2 & 3 & 3 & 2 & 2 & 3 & 1 \\ 3 & 1 & 2 & 2 & 1 & 1 & 2 & 2 & 1 & 1 \\ 1 & 1 & 2 & 3 & 1 & 1 & 3 & 3 & 1 & 1 \\ 1 & 1 & 3 & 3 & 1 & 2 & 3 & 3 & 2 & 2 \end{array}$$

Here, the nodes are grouped in blocks of five, except at the boundaries. First, all points labeled 1 are ordered, followed by all points labeled 2, then all points labeled 3. The resulting system has the form (8.2.24) with $c = 3$, but now the D_i are block diagonal matrices with blocks that are 5×5 or less. A block SOR iteration can then be carried out with block Jacobi sweeps that involve solving 5×5, or smaller, systems.

6. O'Leary and White [1985] consider *multisplittings* of A of the form

$$A = B_i - C_i, \qquad i = 1, \ldots, k, \qquad \sum_{i=1}^{k} D_i = I,$$

$$H = \sum_{i=1}^{k} D_i B_i^{-1} C_i, \qquad \mathbf{d} = \left(\sum_{i=1}^{k} D_i B_i^{-1} \right) \mathbf{b},$$

to give the iteration $\mathbf{x}^{k+1} = H\mathbf{x}^k + \mathbf{d}$. A number of standard methods, including the SOR method, can be cast in this form, which leads to another approach to parallelism.

7. The diagonal ordering is used in Young [1971] as an example of obtaining consistently ordered matrices. It was advocated by Hayes [1978] as an approach to vectorization. It has been used by several people since then; see, for example, Ashcraft and Grimes [1988] and van der Vorst [1989a], [1989b]. Elman and Golub [1990] consider a related "torus" ordering in which the diagonals wrap around.

EXERCISES 8.2

8.2.1. For a system of equations in three unknowns, write out the six possible re-orderings of these equations obtained by interchanging equations and corresponding unknowns.

8.2.2. Assume a distributed memory system of p processors in which arithmetic computation requires α time units per result and transmittal of q words of data between any two processors requires $s + \beta q$ time units. Assume that $N^2 = mp$ so that m unknowns are assigned to each processor. Ascertain the time for one Jacobi iteration for the discrete Poisson equation.

8.2.3. Assuming the parallel system of Exercise 8.2.2 with $p = N$, estimate the time for the line Jacobi iteration (8.2.14).

8.2.4. Write out the equations (8.2.15) explicitly for $N = 3$ and $N = 4$ for the discrete Poisson problem.

8.2.5. Show that the system of equations (8.2.7) can be written in the red/black form (8.2.15), where D_R and D_B are again diagonal.

8.2.6. Show that the matrix representation of the SOR iteration for the system (8.2.15) is

$$\begin{bmatrix} D_R & 0 \\ \omega C^T & D_B \end{bmatrix} \begin{bmatrix} \mathbf{u}_R^{k+1} \\ \mathbf{u}_B^{k+1} \end{bmatrix} = \begin{bmatrix} \omega \mathbf{b}_1 \\ \omega \mathbf{b}_2 \end{bmatrix} + \begin{bmatrix} (1-\omega)D_R & -\omega C \\ 0 & (1-\omega)D_B \end{bmatrix} \begin{bmatrix} \mathbf{u}_R^k \\ \mathbf{u}_B^k \end{bmatrix}.$$

8.2.7. Write the equations (8.2.21) in the form (8.2.15) using the red-black ordering and show the structure of the matrices D_R and D_B. Write out the equations explicitly for $N = 4$.

8.2.8. Write out the system (8.2.22) for the equations (8.2.21) using the four-color ordering of Figure 8.2.12 with $N = 4$.

8.3 The Multigrid Method

8.2.9. For the nine point stencil of Figure 8.2.11 show that the three four-color orderings

$$\begin{array}{ccc} GORB & OBOB & GOGO \\ RBGO & GRGR & RBRB \\ GORB & BOBO & GOGO \\ RBGO & RGRG & RBRB \end{array}$$

are the only ones (up to permutations of the colors) that decouple the stencil points.

8.2.10. Show that the indicated three-coloring by lines is suitable for a line SOR method for the stencil indicated.

```
G G G ··· G
R R R ··· R
B B B ··· B
G G G ··· G
R R R ··· R
```

8.2.11 Show that if the unknowns and equations of the discrete Poisson problem are ordered according to Figure 8.2.15, then the coefficient matrix has the structure shown in Figure 8.2.17.

8.2.12. If A is partitioned as in (8.2.12) and

$$\sum_{j=1}^{q} \|A_{ii}^{-1} A_{ij}\| < 1, \quad i = 1, \ldots, q,$$

in some norm, show that the block Jacobi method (8.2.13) converges.

8.2.13. Show that the discretization (8.2.20) of u_{xy} is second-order accurate.

8.3 The Multigrid Method

Suppose that the system $Ax = b$ has arisen from the discretization (8.2.7) of the two-dimensional Poisson-type equation (8.2.6) on the unit square and with Dirichlet boundary conditions. One approach to obtain a starting approximation for an iterative method, say the SOR method, is to solve the problem on a coarser grid. For example, in Figure 8.3.1, the points marked x are points of a grid with spacing $2h$, as opposed to spacing h on the original grid. If there are N^2 grid points in the original grid (called the *fine* grid) and N is odd, then there will be $(N-1)^2/4$ coarse grid points; this is illustrated in Figure 8.3.1 for $N = 5$.

Since the coarse grid problem is much smaller, it will be easier to solve. We can use our iterative method to obtain an approximate solution on the

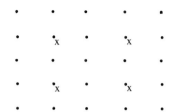

Figure 8.3.1: *Coarse Grid and Fine Grid*

coarse grid, and then use linear interpolation (to be discussed more thoroughly shortly) to obtain approximate values at the other fine grid points. This provides the starting approximation for the iterative method on the fine grid. The hope, of course, is that the work expended to obtain an approximate solution on the coarse grid is more than repaid by having a better approximation to start the iteration on the fine grid.

The same idea can be used to obtain a starting approximation for the coarse grid problem, and so on for increasingly coarser grids. In the case that $N = 2^q - 1$ for some integer q, the coarse grids will have $(2^i - 1)^2$ points for $i = q - 1, q - 2, \ldots, 1$. Thus, we can begin by solving a problem with a single unknown. The next finer grid will have $(2^2 - 1)^2 = 9$ grid points and a starting approximation will be obtained by interpolation using the values from the previous coarse grid and the boundary values. This problem with 9 unknowns is then solved approximately and these values used to obtain the starting approximation for the next finer grid, and so on. This process, which is sometimes called *nested iteration*, is summarized in Figure 8.3.2.

Solve problem on coarsest grid
Interpolate solution to next finer grid
Approximate solution on this grid
Interpolate to next finer grid
$\vdots$
Interpolate to finest grid
Solve fine grid problem

Figure 8.3.2: *Starting Approximation Using Coarse Grids*

8.3 The Multigrid Method

Interpolation

It will be useful for what follows to develop the details of this process and we begin with linear interpolation. Consider first the corresponding one-dimensional problem

$$u''(x) = f(x), \quad 0 \leq x \leq 1, \quad u(0) = \alpha, \quad u(1) = \beta. \tag{8.3.1}$$

In Figure 8.3.3 is shown a coarse and a fine grid for this problem, in which the coarse grid points are labeled c_1, c_2 and c_3.

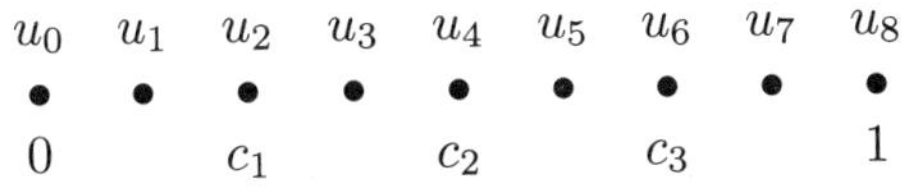

$$\begin{array}{ccccccccc}
u_0 & u_1 & u_2 & u_3 & u_4 & u_5 & u_6 & u_7 & u_8 \\
\bullet & \bullet & \bullet & \bullet & \bullet & \bullet & \bullet & \bullet & \bullet \\
0 & & c_1 & & c_2 & & c_3 & & 1
\end{array}$$

Figure 8.3.3: *Fine and Coarse One-Dimensional Grids*

If the u_{2i} are known at the coarse points, then values at the intermediate fine grid points can be obtained by the linear interpolations

$$u_{2i+1} = \frac{1}{2}(u_{2i+2} + u_{2i}), \quad i = 0, \ldots, (N-1)/2, \tag{8.3.2}$$

where N is the number of interior grid points, assumed to be odd. (In Figure 8.3.3, $N = 7$.) At the coarse grid points, the values are just reproduced. Thus, in matrix-vector form

$$\begin{bmatrix} u_1 \\ u_2 \\ u_3 \\ u_4 \\ u_5 \\ u_6 \\ u_7 \end{bmatrix} = \frac{1}{2} \begin{bmatrix} 1 & 0 & 0 \\ 2 & 0 & 0 \\ 1 & 1 & 0 \\ 0 & 2 & 0 \\ 0 & 1 & 1 \\ 0 & 0 & 2 \\ 0 & 0 & 1 \end{bmatrix} \begin{bmatrix} u_2 \\ u_4 \\ u_6 \end{bmatrix} + \frac{1}{2} \begin{bmatrix} u_0 \\ 0 \\ 0 \\ 0 \\ 0 \\ 0 \\ u_8 \end{bmatrix}. \tag{8.3.3}$$

Note that (8.3.3) is an *affine* transformation (a linear transformation plus a fixed vector) in which the boundary values reside in a separate vector. In key parts of the multigrid method to be discussed shortly, these boundary values will be zero and (8.3.3) is then a linear transformation.

For the corresponding interpolation process in two dimensions, consider the portion of the grid illustrated in Figure 8.3.4, where the coarse grid points are indicated by x. We will approximate values at the four fine grid points on the edges by the one-dimensional linear interpolations

$$u_{i-1,j} = \tfrac{1}{2}(u_{i-1,j+1} + u_{i-1,j-1}), \quad u_{i+1,j} = \tfrac{1}{2}(u_{i+1,j+1} + u_{i+1,j-1}), \tag{8.3.4a}$$

$$\begin{array}{ccc} x & \bullet & x \\ i-1,j+1 & i,j+1 & i+1,j+1 \\ \bullet & \bullet & \bullet \\ i-1,j & i,j & i+1,j \\ x & \bullet & x \\ i-1,j-1 & i,j-1 & i+1,j-1 \end{array}$$

Figure 8.3.4: *Two-Dimensional Coarse and Fine Grids*

$$u_{i,j+1} = \tfrac{1}{2}(u_{i-1,j+1} + u_{i+1,j+1}), \quad u_{i,j-1} = \tfrac{1}{2}(u_{i-1,j-1} + u_{i+1,j-1}). \quad (8.3.4\text{b})$$

The value at the center point can then be approximated by averaging the values in (8.3.4a) or (8.3.4b) to give

$$u_{ij} = (u_{i-1,j+1} + u_{i+1,j+1} + u_{i-1,j-1} + u_{i+1,j-1})/4. \tag{8.3.4c}$$

The matrix-vector form of the interpolation formulas (8.3.4) is illustrated in (8.3.5) for the case of $N = 5$, a 5×5 grid of interior grid points as shown in Figure 8.3.5. (See Exercise 8.3.1.) In this figure, the integers denote the indices of x_i and y_j, and the x points are the coarse grid. Again, as with (8.3.3), (8.3.5) is an affine transformation with the boundary values in the fixed vector, and is a linear transformation if the boundary values are zero.

```
 •   •   •   •   •   •   •
06  16  26  36  46  56  66
 •   •   •   •   •   •   •
05  15  25  35  45  55  65
 •   •   x   •   x   •   •
04  14  24  34  44  54  64
 •   •   •   •   •   •   •
03  13  23  33  43  53  63
 •   •   x   •   x   •   •
02  12  22  32  42  52  62
 •   •   •   •   •   •   •
01  11  21  31  41  51  61
 •   •   •   •   •   •   •
00  10  20  30  40  50  60
```

Figure 8.3.5: *A Fine and Coarse Grid, Including Boundary Points*

$$\begin{bmatrix} u_{11} \\ u_{21} \\ u_{31} \\ u_{41} \\ u_{51} \\ u_{12} \\ u_{22} \\ u_{32} \\ u_{42} \\ u_{52} \\ u_{13} \\ \vdots \end{bmatrix} = \frac{1}{4} \begin{bmatrix} 1 & 0 & 0 & 0 \\ 2 & 0 & 0 & 0 \\ 1 & 1 & 0 & 0 \\ 0 & 2 & 0 & 0 \\ 0 & 1 & 0 & 0 \\ 2 & 0 & 0 & 0 \\ 4 & 0 & 0 & 0 \\ 2 & 2 & 0 & 0 \\ 0 & 4 & 0 & 0 \\ 0 & 2 & 0 & 0 \\ 1 & 0 & 1 & 0 \\ \vdots \end{bmatrix} \begin{bmatrix} u_{22} \\ u_{42} \\ u_{24} \\ u_{44} \end{bmatrix} + \frac{1}{4} \begin{bmatrix} u_{00} + u_{20} + u_{02} \\ 2u_{20} \\ u_{20} + u_{40} \\ 2u_{40} \\ u_{40} + u_{60} + u_{62} \\ \vdots \end{bmatrix} \quad (8.3.5)$$

A Simple Multigrid Method

The procedure that we have described for obtaining an initial approximation to the solution on the fine grid is not yet a multigrid (MG) method. In a MG method, of which there are numerous variants, the basic idea is to continue to use all of the coarse grids throughout the solution process. To fix our ideas, we next describe one simple procedure and then discuss some variations of it. In the multigrid context, interpolation is also called *prolongation* and, hereafter, we will use these terms interchangeably.

Let the grids be denoted by $G_1, \ldots, G_q$, where G_1 is the coarsest grid and G_q is the fine grid. For definiteness, we will assume for now that the iterative method is Gauss-Seidel. Let $\mathbf{u}_q$ be an approximate solution on the fine grid; $\mathbf{u}_q$ may have been obtained by the previously discussed nested iteration process. We first form the residual $\mathbf{r}_q = \mathbf{b} - A\mathbf{u}_q$, and then restrict $\mathbf{r}_q$ to the grid G_{q-1} to obtain the vector $\tilde{\mathbf{r}}_{q-1}$. The simplest such restriction is *injection* in which the values of $\mathbf{r}_q$ at the grid points of G_{q-1} are taken to be the elements of $\tilde{\mathbf{r}}_{q-1}$. Other restrictions are sometimes desirable and will be discussed later. We next take m Gauss-Seidel iterations on the equation

$$A_{q-1}\mathbf{e}_{q-1} = \tilde{\mathbf{r}}_{q-1} \quad (8.3.6)$$

to obtain an approximate solution $\mathbf{e}_{q-1}$ of (8.3.6). Here, A_{q-1} is the coefficient matrix corresponding to G_{q-1}; its formation will be discussed shortly.

We will continue this process on the coarser grids but let us examine first why we are using (8.3.6). Suppose that we could solve the error equation $A\mathbf{e} = \mathbf{r}_q$ on the fine grid. Then

$$A(\mathbf{u}_q + \mathbf{e}) = \mathbf{b} - \mathbf{r}_q + \mathbf{r}_q = \mathbf{b}, \quad (8.3.7)$$

so that $\mathbf{u}_q + \mathbf{e}$ is the desired solution of the system $A\mathbf{u} = \mathbf{b}$. Thus, $\mathbf{e}$ may be considered a correction to the approximate solution $\mathbf{u}_q$. If $\mathbf{e}_{q-1}$, as obtained from (8.3.6), is prolongated (interpolated back) to G_q to give $\mathbf{e}_q$ on the fine grid, then $\mathbf{e}_q$ is an approximation to $\mathbf{e}$ and may be considered as an approximate correction to $\mathbf{u}_q$.

We now continue the process on the coarser grids. Rather than prolongating $\mathbf{e}_{q-1}$ to G_q, we form the residual $\mathbf{r}_{q-1} = \tilde{\mathbf{r}}_{q-1} - A_{q-1}\mathbf{e}_{q-1}$ of the error equation on G_{q-1}, restrict this residual to G_{q-2} to obtain $\tilde{\mathbf{r}}_{q-2}$, take m Gauss-Seidel iterations on the equation $A_{q-2}\mathbf{e} = \tilde{\mathbf{r}}_{q-2}$ to obtain $\mathbf{e}_{q-2}$, and continue in this fashion until $\mathbf{e}_1$ is obtained on the coarsest grid G_1. We are then ready to begin the correction process. This is similar to how the initial approximation for the finest grid is computed by nested iteration, but we now work with corrections. First, $\mathbf{e}_1$ is prolongated to G_2 and added to $\mathbf{e}_2$ to obtain $\hat{\mathbf{e}}_2$. Then, $\hat{\mathbf{e}}_2$ is refined by m Gauss-Seidel iterations on the error equation $A_2\mathbf{e} = \tilde{\mathbf{r}}_2$ to obtain $\tilde{\mathbf{e}}_2$. Next, $\tilde{\mathbf{e}}_2$ is prolongated to G_3, and the process continued until finally $\tilde{\mathbf{e}}_{q-1}$ is prolongated to the finest grid and added to $\mathbf{u}_q$ to obtain a new approximation to the solution of $A\mathbf{u} = \mathbf{b}$. We note that, under the assumption that the original differential equation has Dirichlet boundary conditions, the errors are all zero at grid points corresponding to boundary values. Thus, the interpolation processes for the various error vectors are linear, rather than affine, transformations.

The process we have just described is known as a *V-cycle* and is depicted in Figure 8.3.6(a), where the bottom point of the V represents the coarsest grid and the top points the fine grid. A common variation of the V-cycle is to do two correction cycles at each level before returning to the next higher level; this is the W-cycle shown in Figure 8.3.6(b). In this case, there are two correction cycles beginning at level 3 and within each of these cycles there are two correction cycles beginning at level 2. Hereafter, we will restrict attention to the V-cycle.

Figure 8.3.6: *(a) V-cycle (b) W-cycle*

We note that it is necessary to work with the residual equation in the V-cycle rather than the original equation. If we used the latter in the ascending part of the V-cycle, we would just be repeating the nested iteration process that led to the initial approximation on the fine grid. By using the residual equation, we retain all approximations already made and then correct them. The V-cycle can be summarized as shown in Figure 8.3.7, where "smooth"

8.3 The Multigrid Method

means taking a certain number of steps of the Gauss-Seidel or other suitable iteration.

Smooth current $\mathbf{u}_q$ on G_q to give $\tilde{\mathbf{u}}_q$.
Form $\mathbf{r}_q = \mathbf{b} - A_q \tilde{\mathbf{u}}_q$
For $i = q$ down to 3
  Restrict $\mathbf{r}_i$ onto G_{i-1} to give $\tilde{\mathbf{r}}_{i-1}$
  Smooth on $A_{i-1}\mathbf{e} = \tilde{\mathbf{r}}_{i-1}$ starting with 0 to give $\mathbf{e}_{i-1}$
  Form $\mathbf{r}_{i-1} = \tilde{\mathbf{r}}_{i-1} - A_{i-1}\mathbf{e}_{i-1}$
Restrict $\mathbf{r}_2$ to G_1 to give $\tilde{\mathbf{r}}_1$
Solve exactly $A_1 \mathbf{e} = \tilde{\mathbf{r}}_1$ to give $\tilde{\mathbf{e}}_1$
For $i = 1, \ldots, q-2$
  Prolongate $\tilde{\mathbf{e}}_i$ to G_{i+1} and add to $\mathbf{e}_{i+1}$ to obtain $\hat{\mathbf{e}}_{i+1}$
  Smooth on $A_{i+1}\mathbf{e} = \tilde{\mathbf{r}}_{i+1}$ starting from $\hat{\mathbf{e}}_{i+1}$ to obtain $\tilde{\mathbf{e}}_{i+1}$
Prolongate $\tilde{\mathbf{e}}_{q-1}$ to get $\hat{\mathbf{e}}_q$
Add $\hat{\mathbf{e}}_q$ to $\tilde{\mathbf{u}}_q$ to obtain new $\mathbf{u}_q$ on G_q

---

Figure 8.3.7: *A V-cycle*

A Detailed Example

We next give a detailed example that will not only illustrate the process but also begin to give some insight as to why the multigrid method works. For this example, we will use the Jacobi method for the relaxation process rather than the Gauss-Seidel process. For simplicity and concreteness, we will work with the one-dimensional equation (8.3.1). The coefficient matrix of the corresponding discrete system $A_h \mathbf{u} = \mathbf{b}$ is the $(2, -1)$ matrix given by (5.3.10) but divided by h^2; it is important to retain the factor h^{-2}. By Exercise 5.3.5, A_h has the eigenvalues (see Exercise 8.3.2)

$$\lambda_k^h = h^{-2}(2 - 2\cos k\eta) = 4h^{-2} \sin^2 \frac{k\eta}{2}, \quad k = 1, \ldots, N, \tag{8.3.8}$$

where $\eta = \pi/(N+1)$ and the second equality follows from the identity $1 - \cos 2\theta = 2 \sin^2 \theta$. The corresponding eigenvectors are

$$\mathbf{v}_k^h = (\sin k\eta, \sin 2k\eta, \ldots, \sin Nk\eta)^T, \quad k = 1, \ldots, N. \tag{8.3.9}$$

By Exercise 8.1.12, the Jacobi iteration matrix, J, then has eigenvalues

$$\mu_k^h = \cos k\eta, \quad k = 1, \ldots, N, \tag{8.3.10}$$

and the same eigenvectors, $\mathbf{v}_k^h$, as A_h. Note that the kth eigenvector corresponds to the function

$$v_k(x) = \sin k\pi x, \quad 0 \le x \le 1, \qquad (8.3.11)$$

which is an *eigenfunction* of the differential operator of (8.3.1):

$$\frac{d^2}{dx^2} v_k(x) = -k^2 \pi^2 v_k(x), \qquad k = 1, 2, \ldots . \qquad (8.3.12)$$

The components of $\mathbf{v}_k^h$ are then the function values

$$v_k(\frac{j}{N+1}), \qquad j = 1, \ldots, N. \qquad (8.3.13)$$

The larger k, the more oscillatory the function $v_k(x)$ is; in particular, $v_1(x)$ consists of a single half sine wave and $v_k(x)$ consists of $k/2$ full sine waves. For this reason, the eigenvectors $\mathbf{v}_k^h$ for small k are called the "smooth" eigenvectors and those with large k the "oscillatory" ones. Clearly, there is no sharp dividing line between smooth and oscillatory, and the functions v_k of (8.3.11) continue to become more oscillatory as k increases.

As discussed in Section 8.1 in general, the error vectors $\mathbf{e}^k$ of the Jacobi iteration satisfy

$$\mathbf{e}^k = J^k \mathbf{e}^0 = \sum_{i=1}^{N} c_i (\mu_i^h)^k \mathbf{v}_i^h, \qquad (8.3.14)$$

where $\mathbf{e}^0 = \sum c_i \mathbf{v}_i^h$ is the expansion of the initial error in terms of the eigenvectors of J. The largest eigenvalues in magnitude occur for $k = 1$ and $k = N$:

$$\mu_1^h = -\mu_N^h = \cos \frac{\pi}{N+1}.$$

Therefore, the errors in the first few directions $\mathbf{v}_1^h, \mathbf{v}_2^h, \ldots$ and $\mathbf{v}_N^h, \mathbf{v}_{N-1}^h, \ldots$ diminish the most slowly while the errors in the directions $\mathbf{v}_i^h$ for $i \doteq N/2$ diminish most rapidly. For example, for $N = 1000$

$$\mu_1^2 = \cos \frac{\pi}{1001} \doteq .99999,$$

so that powers of μ_1^h go to zero very slowly. On the other hand

$$\mu_{500}^h \doteq 0,$$

so that powers of μ_{500}^h go to zero very quickly.

We would like to modify the Jacobi iteration so that errors in the directions corresponding to the most oscillatory modes go to zero the most rapidly. We

8.3 The Multigrid Method

can achieve this with a *damped* (underrelaxed) Jacobi iteration in which the iteration matrix is

$$J_\omega = (1-\omega)I + \omega J \tag{8.3.15}$$

for some $\omega < 1$. By Theorem 2.2.3, the eigenvalues of J_ω are

$$\mu_{k,\omega}^h = 1 - \omega + \omega \mu_k^h. \tag{8.3.16}$$

In Figure 8.3.8, the lines $1 - \omega + \omega\mu$ are plotted as functions of ω for values of the parameter μ. For a given value of ω, the eigenvalue $\mu_{j,\omega}^h$ lies on the intersection of the vertical line through ω and the line that takes on the value μ_j^h at $\omega = 1$.

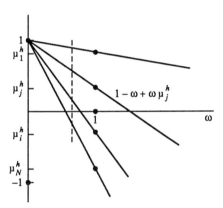

Figure 8.3.8: *Eigenvalues of the Damped Jacobi Method*

It is clear from Figure 8.3.8 that, since $\mu_1^h = -\mu_N^h$, $\rho(H_\omega)$ is minimized when $\omega = 1$: for $\omega > 1$, $\mu_{N,\omega}^h > \mu_N^h$ and for $\omega < 1$, $\mu_{1,\omega}^h > \mu_1^h$. However, by choosing $\omega < 1$, we can move the negative eigenvalues closer to zero, at the expense of increasing the positive eigenvalues. For example, as illustrated in Figure 8.3.8, $\omega = \frac{2}{3}$ decreases the magnitude of the largest negative eigenvalues considerably while increasing the largest positive ones only slightly. The effect of this is that the errors in the directions $\mathbf{v}_N^h, \mathbf{v}_{N-1}^h, \ldots$ go to zero more rapidly and there will be a larger number of directions for which this is true. The errors in the directions $\mathbf{v}_1^h, \mathbf{v}_2^h \ldots$ go to zero more slowly but it is exactly these errors that the multigrid process will handle, as we shall now see.

Assume that we have iterated a few times on the fine grid and that we have a current approximation $\mathbf{u}^h$ for which the error $\mathbf{e}$ is largely in the direction of the eigenvectors $\mathbf{v}_1^h, \ldots, \mathbf{v}_m^h$. We then restrict the current residual to the next coarser grid, which has grid spacing $2h$. The coefficient matrix, A_{2h}, for this coarser grid is again the $(2,-1)$ matrix divided now by $(2h)^2$ and of size $N_1 \times N_1$ where, if N is odd, $N_1 = (N-1)/2$. The eigenvalues and eigenvectors of A_{2h} have the same form as in (8.3.8) and (8.3.9) with h replaced by $2h$ and η replaced by $\eta_1 = \pi/(N_1 + 1)$. Thus (Exercise 8.3.4)

$$\lambda_k^{2h} = 4(2h)^{-2}\sin^2\frac{k\eta_1}{2} = h^{-2}\sin^2\frac{k\eta_1}{2}, \quad k = 1, \ldots, N_1, \tag{8.3.17}$$

$$\begin{aligned}\mathbf{v}_k^{2h} &= (\sin k\eta_1, \sin 2k\eta_1, \ldots, \sin N_1 k\eta_1) \\ &= (\sin 2k\eta, \sin 4k\eta, \ldots, \sin(N-1)k\eta), \quad k = 1, \ldots, N_1.\end{aligned} \tag{8.3.18}$$

For future use, we note the following relations between the eigenvalues λ_k^h and λ_k^{2h}, which follow from $\eta_1 = 2\eta$ and the identity $\sin 2\theta = 2\sin\theta\cos\theta$:

$$\lambda_k^{2h} = h^{-2}\sin^2 k\eta = 4h^{-2}\sin^2\frac{k\eta}{2}\cos^2\frac{k\eta}{2} = \lambda_k^h \cos^2\frac{k\eta}{2}, \tag{8.3.19}$$

$k = 1, \ldots, N_1$. Similarly, the eigenvalues of the Jacobi iteration matrix have the same form as (8.3.10) but with N replaced by N_1 so that

$$\mu_k^{2h} = \cos k\eta_1 = \cos 2k\eta, \quad k = 1, \ldots, N_1. \tag{8.3.20}$$

The largest of these in magnitude is

$$\max\left|\mu_k^{2h}\right| = \cos 2\eta < \cos\eta, \tag{8.3.21}$$

where the last quantity is the spectral radius of J itself.

The next step is to restrict the residual $\mathbf{r}^h$ to the coarse grid. As mentioned earlier, the simplest restriction is injection, in which $\mathbf{r}^{2h}$ contains every other component of $\mathbf{r}^h$. We will also use what is called *full weighting restriction*. In this case, the restriction $\mathbf{w}^{2h}$ of a vector $\mathbf{w}^h$ on the fine grid is defined by

$$\mathbf{w}^{2h} = R\mathbf{w}^h, \tag{8.3.22}$$

where R is the $N \times N$ matrix

$$R = \frac{1}{4}\begin{bmatrix} 1 & 2 & 1 & & & & \\ & & 1 & 2 & 1 & & \\ & & & & 1 & 2 & 1 \\ & & & & & & \ddots \end{bmatrix}. \tag{8.3.23}$$

Thus, a typical component of $\mathbf{w}^{2h}$ is defined by

$$w_{2i}^{2h} = \frac{1}{4}(w_{i-1}^h + 2w_i^h + w_{i+1}^h). \tag{8.3.24}$$

Note that there is a close connection between the restriction R of (8.3.23) and the prolongation matrix of (8.3.3). In fact, denoting the latter matrix by P, we have

$$2R = P^T. \tag{8.3.25}$$

8.3 The Multigrid Method

Moreover, an easy computation (Exercise 8.3.5) shows that

$$A_{2h} = RA_h P. \tag{8.3.26}$$

This indicates a way, known as the *Galerkin technique*, to obtain the coefficient matrices on coarser grids directly from those on the finer grids by means of the restriction and prolongation operators. Note that, in general, whenever $R = cP^T$ for some constant c and A_h is symmetric and positive definite, then $RA_h P$ will be symmetric, and also positive definite provided that R and P are of full rank (Exercise 8.3.6).

We use the restriction operator R to obtain $\tilde{\mathbf{r}}^{2h} = R\mathbf{r}^h$, and this provides the right-hand side for the correction equation on the coarse grid:

$$A_{2h}\mathbf{e}^{2h} = \tilde{\mathbf{r}}^{2h}. \tag{8.3.27}$$

Now suppose that we could solve (8.3.27) exactly. We then interpolate $\mathbf{e}^{2h}$ back to the fine grid using the prolongation operator P of (8.3.3) to obtain

$$\hat{\mathbf{e}}^h = P\mathbf{e}^{2h}. \tag{8.3.28}$$

Then we correct the current approximate solution $\mathbf{u}^0$ on the fine grid by

$$\mathbf{u}^1 = \mathbf{u}^0 + \hat{\mathbf{e}}^h. \tag{8.3.29}$$

Although it is unreasonable, in general, to solve the coarse grid equation (8.3.27) exactly, this *two-grid* procedure serves as a useful theoretical model. It can be analyzed in detail for our model one-dimensional problem (see the Supplementary Discussion), and we mention here briefly some of the results of that analysis.

Assume that the initial error is expanded in a linear combination

$$\mathbf{e}^0 = \sum_{k=1}^{N} a_k \mathbf{v}_k^h \tag{8.3.30}$$

of the eigenvectors $\mathbf{v}_k^h$ of A_h. Then, assuming that damped Jacobi with $\omega = \frac{1}{2}$ is used as the smoother, the error after the two-grid correction is

$$\mathbf{e}^1 = \sum_{k=1}^{N} (a_k + a_{N-k+1}) s_k^2 c_k^2 \mathbf{v}_k^h, \tag{8.3.31}$$

where

$$c_k^2 = \cos^2 \frac{k\eta}{2}, \qquad s_k^2 = \sin^2 \frac{k\eta}{2}. \tag{8.3.32}$$

Note that by (8.3.19) c_k^2 is the ratio of the eigenvalues of A_h and A_{2h}. Now consider, for example, the coefficient in (8.3.31) of $\mathbf{v}_1^h$. This will be small because $s_1^2 \doteq 0$. The term $a_1 s_1^2 c_1^2$ comes from the coarse grid correction and

$a_N s_1^2 c_1^2$ comes from the smoothing. Thus, the two parts of the multigrid process, the smoothing and the coarse grid correction, play complementary roles in decreasing the error.

Computational Considerations

We continue now with a more detailed description of the multigrid process for the one-dimensional model problem. In particular, we will expand on the general form of the V-cycle given in Figure 8.3.7. We will assume that $N = 2^{q-1}$, and that the unknowns and right-hand sides are stored in one-dimensional arrays as indicated in Figure 8.3.9. Consider first the storage of the solutions. The current approximate solution on the fine grid requires $2^q - 1$ storage locations, which are the left-most positions in the array of Figure 8.3.9. The solution on the next coarser grid, which is the correction vector $\tilde{\mathbf{e}}_{q-1}$, requires $2^{q-1} - 1$ storage locations, as shown in Figure 8.3.9. And so on down to the final coarse grid, which we assume for now contains a single point. The right-hand sides will have corresponding storage in a second one-dimensional array: the right-hand side $\mathbf{b}$ of the equations on the fine grid will require $2^q - 1$ storage locations, the right-hand side, $\tilde{\mathbf{r}}_{q-1}$, of the correction equation on the next coarser grid will require $2^{q-1} - 1$ locations, and so on down to the right-hand side $\tilde{\mathbf{r}}_1$ on the coarsest grid, which requires only a single location.

$2^q - 1$	$2^{q-1} - 1$	$2^{q-2} - 1$	$\cdots$	3	1
q	$q-1$	$q-2$		2	1

Figure 8.3.9: *Storage for Solutions or Right-Hand Sides*

We next describe the computations. The first step in the V-cycle is to do smoothing steps on the fine grid. In the following, we will denote A_h by A_q, A_{2h} by A_{q-1}, and so on, so that A_i is the coefficient matrix on the ith grid G_i. We assume that the smoother is the damped Jacobi iteration so that if one smoothing step is taken starting from $\mathbf{u}_q$, the next approximate solution is

$$\tilde{\mathbf{u}}_q = (1 - \omega)\mathbf{u}_q + \omega D_q^{-1} B_q \mathbf{u}_q + \omega D_q^{-1}\mathbf{b}, \qquad (8.3.33)$$

where $A_q = D_q - B_q$ is the splitting of A_q into its diagonal and off-diagonal parts. The vector $\tilde{\mathbf{u}}_q$ replaces $\mathbf{u}_q$ in storage. Next, the residual

$$\mathbf{r}_q = \mathbf{b} - A_q \tilde{\mathbf{u}}_q \qquad (8.3.34)$$

is computed and restricted to the grid G_{q-1}:

$$\tilde{\mathbf{r}}_{q-1} = R_q \mathbf{r}_q. \qquad (8.3.35)$$

If we wish, steps (8.3.34) and (8.3.35) can be combined into

$$\tilde{\mathbf{r}}_{q-1} = R_q(\mathbf{b} - A_q \tilde{\mathbf{u}}_q), \qquad (8.3.36)$$

8.3 The Multigrid Method

and $\tilde{\mathbf{r}}_{q-1}$ computed using just a few temporary storage locations. Upon completion of (8.3.36), $\tilde{\mathbf{r}}_{q-1}$ is stored in the $2^{q-1} - 1$ locations of the $q - 1$ level of the right-hand side array. Next, at the $q - 1$ level a smoothing step is taken on $A_{q-1}\mathbf{e} = \tilde{\mathbf{r}}_{q-1}$, starting from a zero initial guess, to obtain

$$\mathbf{e}_{q-1} = \omega D_{q-1}^{-1} \tilde{\mathbf{r}}_{q-1}. \tag{8.3.37}$$

The vector $\mathbf{e}_{q-1}$ is retained in the $2^{q-1} - 1$ locations at level $q-1$ of the solution storage vector. Then the residual is computed and restricted by

$$\tilde{\mathbf{r}}_{q-2} = R_{q-1}\tilde{\mathbf{r}}_{q-1} - R_{q-1}A_{q-1}\mathbf{e}_{q-1}, \tag{8.3.38}$$

and $\tilde{\mathbf{r}}_{q-2}$ is stored in the $2^{q-2} - 1$ locations at level $q - 2$ of the right-hand side vector. The process now repeats at each level:

$$\mathbf{e}_i = \omega D_i^{-1}\tilde{\mathbf{r}}_i, \qquad \tilde{\mathbf{r}}_{i-1} = R_i(\tilde{\mathbf{r}}_i - A_i\mathbf{e}_i), \quad i = q - 2, \ldots, 2. \tag{8.3.39}$$

At the coarsest level the equation $A_1\mathbf{e} = \tilde{\mathbf{r}}_1$ is solved exactly to give $\tilde{\mathbf{e}}_1$. At this point, the successive solutions $\mathbf{u}_q, \mathbf{e}_{q-1}, \ldots, \mathbf{e}_2, \mathbf{e}_1$ occupy the solution storage array. The ascending portion of the V-cycle now begins, and at each level $i = 2, \ldots, q - 1$ can be written as

$$\hat{\mathbf{e}}_i = \mathbf{e}_i + P_i\tilde{\mathbf{e}}_{i-1}, \quad \tilde{\mathbf{e}}_i = (1 - \omega)\hat{\mathbf{e}}_i + \omega D_i^{-1}B_i\hat{\mathbf{e}}_i + \omega D^{-1}\tilde{\mathbf{r}}_i. \tag{8.3.40}$$

The first computation in (8.3.40) prolongates $\tilde{\mathbf{e}}_{i-1}$ to level i and adds to the current correction vector $\mathbf{e}_i$ to give $\hat{\mathbf{e}}_i$. Then a smoothing step is taken starting from $\hat{\mathbf{e}}_i$ and the result $\tilde{\mathbf{e}}_i$ replaces $\hat{\mathbf{e}}_i$ in the solution array. At the termination of (8.3.40), we have $\tilde{\mathbf{e}}_{q-1}$ which is prolongated to $\hat{\mathbf{e}}_q$ and added to $\tilde{\mathbf{u}}_q$ to obtain the new approximation on the fine grid:

$$\mathbf{u}_q^1 = \tilde{\mathbf{u}}_q + \tilde{\mathbf{e}}_q. \tag{8.3.41}$$

This finishes the V-cycle.

The two storage arrays for the solutions and the right-hand sides require a total space of

$$2(2^q - 1 + 2^{q-1} - 1 + \cdots + 2^1 - 1) = 2(2^{q+1} - q) \le 4n. \tag{8.3.42}$$

The smoothing step (8.3.40) also requires temporary storage of the components of $\tilde{\mathbf{e}}_i$ while they are being computed, as is always the case with a Jacobi iteration (Note that (8.3.37) does not require additional storage). This temporary storage can be the positions of the right-hand side array occupied by $\tilde{\mathbf{r}}_i$, so that no storage is required beyond (8.3.42). For the Jacobi iteration itself on $A_q\mathbf{u} = \mathbf{b}$, we would require $3n$ storage locations, and for Gauss-Seidel (or SOR), $2n$ locations. Thus, the storage needed for the multigrid method applied to the one-dimensional model problem is no more than twice as much as for the basic iterative methods themselves. For higher-dimensional problems, the situation is even better. For the Poisson equation in two dimensions, each coarse grid requires $\frac{1}{4}$ of the storage of the next finer grid and in three dimensions, this ratio becomes 2^{-3}. Thus, the extra storage required for the coarse grid solutions for a d-dimensional problem is

$$S_C = S(2^{-d} + 2^{-2d} + \cdots + 2^{-(q-1)d}) = \frac{S(2^{-dq} - 1)}{2^{-d} - 1} < \frac{S}{1 - 2^{-d}}, \quad (8.3.43)$$

where S is the storage on the fine grid. Thus, for $d = 2$, $S_C < \frac{4}{3}S$ and for $d = 3, S_C < \frac{8}{7}S$. The same is true for storage of the right-hand sides. Note that the above storage counts do not include the matrices A_i. For the Poisson model problem (in any number of dimensions) storage for these matrices is not required. For more general problems, it will be required but, again, the total additional storage required for the coarse grid matrices $A_1, \ldots, A_{q-1}$ will usually be less than that required for A_q itself.

A similar thing occurs for the computational work. It is usual to measure this in terms of the *work unit* W, defined to be the cost of a residual computation (or a relaxation sweep, which is an almost equivalent amount of work) on the fine grid. The work on each grid is roughly proportional to the number of unknowns so that for a d-dimensional problem the work on each level is about 2^{-d} the work on the next finer level. Thus, the total work is approximately

$$[2 + 3(2^{-d} + \cdots + 2^{-(q-1)d})]W = [2 + \frac{3(2^{-dq} - 1)}{2^{-d} - 1}]W, \quad (8.3.44)$$

an expression analogous to (8.3.43). The factor 3 in (8.3.44) accounts for the fact that on each coarse grid there are two relaxation sweeps plus one residual evaluation, whereas on the finest grid, there is one relaxation and one residual evaluation. For $d = 1$, (8.3.44) becomes approximately $8W$, whereas for $d = 2$ and 3 it is 6 and 5, respectively.

A Parallel Multigrid Algorithm

We next consider how to carry out the multigrid V-cycle on a parallel computer. For concreteness, we will assume a distributed memory machine but many of the considerations apply also to shared memory. Again, we focus on the one-dimensional model problem although we will make comments on more general situations.

We assume that the unknowns on the fine grid will be distributed to p processors in a linear array, as indicated in Figure 8.3.10. Each processor's memory will also hold the coarse grid information corresponding to its portion of the fine grid. Thus, the storage arrays shown in Figure 8.3.9 for the solutions and the right-hand sides will be distributed across the processors. For example,

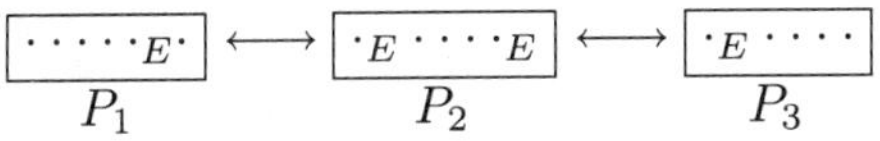

8.3 The Multigrid Method

$$\boxed{\cdots\cdot E\cdot} \longleftrightarrow \boxed{\cdot E\cdots E} \longleftrightarrow \boxed{\cdot E\cdots}$$
$$\;\;P_1 \qquad\qquad\;\; P_2 \qquad\qquad\;\; P_3$$

Figure 8.3.10: *Communication of Edge Values*

approximately one *pth* of the solution vectors on all levels will be held in processor 1, the next *pth* in processor 2, and so on, and similarly for the right-hand sides. We still assume that there are $N = 2^q - 1$ grid points on the fine grid and this number may not distribute evenly over the processors. For example, if $N = 2047 (q = 11)$ and $p = 32$, then each processor will hold 64 fine grid unknowns except the last processor, which will hold 63. Thus, there will be a slight load imbalance. A similar situation holds on the coarser grids, except for the very coarsest ones. Again for the example of 32 processors, consider the grid which contains only 31 points. Each processor except the last will contain one of these points. The next coarser grid will consist of 15 points, which will be held by processors 2, 4, ..., 30. As the grid keeps coarsening, the distribution of the points to the processors is shown in Figure 8.3.11.

> 31 grid points: processors 1, 2, 3, ..., 31
> 15 grid points: processors 2, 4, ..., 30
> 7 grid points: processors 4, 8, ..., 28
> 3 grid points: processors 8, 16, 24
> 1 grid point: processor 16

Figure 8.3.11: *Distribution of Coarse Grids to Processors*

We consider next the computations. The smoothing step (8.3.33) is done within each processor for the unknowns assigned to that processor. As discussed in Section 8.1, the Jacobi iteration is perfectly parallel and use of the underrelaxation factor ω does not change this. Communication is required, however: the edge unknowns will need data from a neighboring processor as shown in Figure 8.3.10. Note that only communication to nearest neighbors is required and for a one-dimensional problem, a linear array is adequate. For two- or three-dimensional problems, mesh connected arrays would be suitable.

The smoothing step is followed by the residual computations (8.3.34). Here, the multiplication $A_q \tilde{\mathbf{u}}_q$ will require communication of edge values between neighboring processors. These edge values have been modified by the previous smoothing step and the best strategy is to smooth and transmit these edge values before smoothing the interior unknowns. Thus, the edge values needed for the residual calculation will be available with minimal delay. This compute-and-send-ahead strategy should be used throughout the multigrid process whenever possible if the system allows overlap of computation and communication. The restriction operation (8.3.35) does not require communication of edge values if injection is used but does require it for full-weighting. As indicated by (8.3.36), the computation of the residual and its restriction may be combined if it is not desirable to store the residual separately.

Analogous considerations apply at each level until the coarsest levels are reached; these will be discussed shortly. On the return up the V-cycle, the computations (8.3.40) are needed. The prolongation is similar to the full weighting restriction in that edge values need to be communicated. For example, if we are interpolating at the right edge point in processor P_1, the value of $\tilde{e}_{i-1}$ at the left edge point in processor P_2 will need to be communicated. We should arrange to send-ahead these needed edge values as much as possible.

We next consider the coarsest grids and, again as an example, assume that we have 32 processors and $N = 2047$. Then, as discussed previously, for the coarse grid with 31 points each processor except the last will contain the information for a single grid point. Thirty-one processors will participate in the smoothing and restriction steps (8.3.37) and (8.3.38) but then at the next coarse grid only 15 processors will participate, and so on down until at the coarsest (one point) grid only one processor is active. In addition to the load balancing becoming worse as processors drop out, there are additional inefficiencies due to the fact that there is very little computation in even the active processors and the compute-and-send-ahead strategies are no longer useful. Analogous considerations apply as we move back up the V-cycle from the coarsest grid.

There are two natural alternatives on these coarsest grids. One is to solve the coarse grid equations exactly by a direct method when the number of grid points in a coarse grid drops below the number of processors. A potential direct method is cyclic reduction but it will also suffer the same loss of active processors as does the multigrid method. The other alternative is to do several smoothing steps on that coarse grid for which the number of grid points is approximately equal to the number of processors. The idea here is to obtain an approximate solution of these coarse grid equations by the relaxation method. Note that if we do this at the level that each processor contains at most one grid point, the smoothing steps will be dominated by communication and it may be better to stop the V-cycle sooner so as to have more computation over which to amortize the communication costs.

8.3 The Multigrid Method

The Gauss-Seidel Iteration as a Smoother

The previous discussion has been based on the damped Jacobi iteration as the smoother, but the Gauss-Seidel iteration is usually used in practice. As a smoother, it does not pay to use SOR ($\omega > 1$) or even damped Gauss-Seidel ($\omega < 1$). For the parallel implementation, we use the red/black ordering, or a multicolor ordering if necessary. The Gauss-Seidel iterations can then be implemented as discussed in Section 8.2, and the rest of the multigrid algorithm remains as before. An interesting aspect of the red/black ordering is that it has been found to be slightly superior to the natural ordering even on serial machines.

Supplementary Discussion and References: 8.3

The multigrid method has become increasingly popular over the last 15 years and has been applied to a variety of problems. For an introduction to the multigrid method see Briggs [1987], and for a more advanced treatment see Hackbrush [1985]. This last reference contains, in particular, the analysis of the two-grid procedure leading to (8.3.31).

EXERCISES 8.3

8.3.1. Verify that the interpolation formula (8.3.4) can be written in the form (8.3.5) for $N = 5$, as in Figure 8.3.5.

8.3.2. Verify that (8.3.8) gives the eigenvalues of A_h.

8.3.3. Verify that Figure 8.3.8 represents the eigenvalues (8.3.16) of the matrix J_ω of (8.3.15).

8.3.4. Verify the relations (8.3.17) - (8.3.20).

8.3.5. Verify the relation (8.3.26).

8.3.6. Assume that A_h is symmetric positive and A_{2h} is given by (8.3.26) with R and P matrices of full rank that satisfy (8.3.25). Show that A_{2h} is symmetric positive definite.

Chapter 9

Conjugate Gradient-Type Methods

9.1 The Conjugate Gradient Method

Minimization Methods

A large number of iterative methods for solving linear systems of equations can be derived as minimization methods. If A is symmetric and positive definite, then the quadratic function

$$q(\mathbf{x}) = \tfrac{1}{2}\mathbf{x}^T A \mathbf{x} - \mathbf{x}^T \mathbf{b} + c \qquad (9.1.1)$$

has a unique minimizer which is the solution of $A\mathbf{x} = \mathbf{b}$ (Exercise 9.1.1). Thus, methods that attempt to minimize (9.1.1) are also methods to solve $A\mathbf{x} = \mathbf{b}$.

Many minimization methods for (9.1.1) can be written in the form

$$\mathbf{x}_{k+1} = \mathbf{x}_k - \alpha_k \mathbf{p}_k, \qquad k = 0, 1, \ldots . \qquad (9.1.2)$$

(In this chapter, we will use subscripts to indicate the iteration number since we will frequently be using transposes.) Given the *direction vector* $\mathbf{p}_k$, one way to choose the scalar α_k is to minimize q along $\mathbf{x}_k - \alpha \mathbf{p}_k$ so that

$$q(\mathbf{x}_k - \alpha_k \mathbf{p}_k) = \min_{\alpha} q(\mathbf{x}_k - \alpha \mathbf{p}_k). \qquad (9.1.3)$$

For fixed $\mathbf{x}_k$ and $\mathbf{p}_k$, $q(\mathbf{x}_k - \alpha \mathbf{p}_k)$ is a quadratic function of α and may be minimized explicitly (Exercise 9.1.2) to give

$$\alpha_k = -(\mathbf{p}_k, \mathbf{r}_k)/(\mathbf{p}_k, A\mathbf{p}_k), \qquad (9.1.4)$$

where $\mathbf{r}_k = \mathbf{b} - A\mathbf{x}_k$. In (9.1.4), and henceforth, we use the notation $(\mathbf{u}, \mathbf{v})$ to denote the inner product $\mathbf{u}^T \mathbf{v}$.

Although there are many other ways to choose the α_k, we will use only (9.1.4) and concentrate on different choices of the direction vectors $\mathbf{p}_k$. One simple choice is $\mathbf{p}_k = \mathbf{r}_k$, which gives the *method of steepest descent*:

$$\mathbf{x}_{k+1} = \mathbf{x}_k - \alpha_k(\mathbf{b} - A\mathbf{x}_k), \qquad k = 0, 1, \ldots . \tag{9.1.5}$$

This is also known as *Richardson's method* and is closely related to Jacobi's method (Exercise 9.1.3). As with Jacobi's method, the convergence of (9.1.5) is usually very slow.

Another simple strategy is to take $\mathbf{p}_k$ as one of the unit vectors $\mathbf{e}_i$, which has a 1 in position i and is zero elsewhere. Then, if $\mathbf{p}_0 = \mathbf{e}_1, \mathbf{p}_1 = \mathbf{e}_2, \ldots, \mathbf{p}_{n-1} = \mathbf{e}_n$, and the α_k are chosen by (9.1.4), n steps of (9.1.2) are equivalent to one Gauss-Seidel iteration on the system $A\mathbf{x} = \mathbf{b}$ (Exercise 9.1.4). In the context of minimization, the Gauss-Seidel method is sometimes known as the method of *univariate relaxation* since at each iteration (9.1.2) only a single variable is changed when $\mathbf{p}_k$ is one of the directions $\mathbf{e}_i$.

For any method of the form (9.1.2), we can add a relaxation parameter ω to the minimization principle so that α_k is chosen to be

$$\alpha_k = \omega \hat{\alpha}_k, \tag{9.1.6}$$

where $\hat{\alpha}_k$ is now the value of α that minimizes q in the direction $\mathbf{p}_k$. This is illustrated in Figure 9.1.1 for an $\omega > 1$. Thus with an $\omega \neq 1$, the new $\mathbf{x}_{k+1}$ no longer minimizes q in the direction $\mathbf{p}_k$.

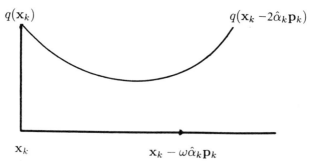

Figure 9.1.1: *Overrelaxation*

It is clear from Figure 9.1.1 that $q(\mathbf{x}_k - \omega\hat{\alpha}_k\mathbf{p}_k) < q(\mathbf{x}_k)$ for $\omega > 0$ until a value of ω is reached for which $q(\mathbf{x}_k - \omega\hat{\alpha}_k\mathbf{p}_k) = q(\mathbf{x}_k)$. By the symmetry of a quadratic function in a single variable about its minimum, it follows that this value of ω is 2, as illustrated in the figure. Thus $q(\mathbf{x}_{k+1}) < q(\mathbf{x}_k)$ if $0 < \omega < 2$, and $q(\mathbf{x}_{k+1}) \geq q(\mathbf{x}_k)$ otherwise (Exercise 9.1.5). It is this descent property for $0 < \omega < 2$ that is the basis of Theorem 8.1.4 for the convergence of the SOR iteration.

9.1 The Conjugate Gradient Method

Conjugate Direction Methods

An extremely interesting and important class of methods, called *conjugate direction* methods, arises when there are n non-zero direction vectors $\mathbf{p}_0, \ldots, \mathbf{p}_{n-1}$ that satisfy

$$\mathbf{p}_i^T A \mathbf{p}_j = 0, \qquad i \neq j. \tag{9.1.7}$$

Such vectors are orthogonal with respect to the inner product $(\mathbf{x}, \mathbf{y})_A \equiv \mathbf{x}^T A \mathbf{y}$ defined by A, and are thus called *A-orthogonal*; they are also called *conjugate* with respect to A.

The basic property of conjugate direction methods is given by the following result.

THEOREM 9.1.1 *If A is a real $n \times n$ symmetric positive definite matrix and $\mathbf{p}_0, \ldots, \mathbf{p}_{n-1}$ are nonzero vectors that satisfy (9.1.7), then for any $\mathbf{x}_0$ the iterates (9.1.2), where α_k is chosen by (9.1.4), converge to the solution of $A\mathbf{x} = \mathbf{b}$ in no more than n steps.*

Note that Theorem 9.1.1 guarantees not only that the iterates converge but also that, in the absence of rounding error, they do so in a finite number of steps, no more than n. Thus, conjugate direction methods are really direct methods, but their utility still lies in being treated as iterative methods as will be discussed further shortly.

To see why Theorem 9.1.1 should be true, consider first a quadratic function

$$q(\mathbf{x}) = \frac{1}{2} \sum_{i=1}^{n} a_{ii} x_i^2 - \mathbf{b}^T \mathbf{x} + c \tag{9.1.8}$$

for which the matrix A is diagonal. It is easy to see (Exercise 9.1.6) that univariate relaxation minimizes the function (9.1.8) in n steps. But $\mathbf{e}_1, \ldots, \mathbf{e}_n$ constitute a set of conjugate directions with respect to a diagonal matrix and thus, in this case, univariate relaxation is equivalent to the conjugate direction method in which $\mathbf{p}_{i-1} = \mathbf{e}_i, i = 1, \ldots, n$.

In general, if P is the matrix with columns $\mathbf{p}_0, \ldots, \mathbf{p}_{n-1}$, then (9.1.7) is equivalent to $P^T A P = D$, where D is a diagonal matrix. Hence, with the change of variable $\mathbf{x} = P\mathbf{y}$, the quadratic form (9.1.1) becomes

$$\frac{1}{2}(P\mathbf{y})^T A P \mathbf{y} - \mathbf{b}^T P \mathbf{y} = \frac{1}{2}\mathbf{y}^T D \mathbf{y} - (P^T \mathbf{b})^T \mathbf{y}, \tag{9.1.9}$$

and thus is of the form (9.1.8) in the $\mathbf{y}$ variables. It is easy to see that the conjugate direction method in the original $\mathbf{x}$ variables is equivalent to univariate relaxation in the $\mathbf{y}$ variables (Exercise 9.1.7) and it follows from the previous discussion that the quadratic function will be minimized in at most n steps. It is also possible to give a direct proof of Theorem 9.1.1 (see Exercise 9.1.8).

In order to utilize a conjugate direction method we must, of course, have the vectors $\mathbf{p}_j$ that satisfy (9.1.7). One classical set of conjugate vectors consists of the eigenvectors of A. If $\mathbf{x}_1, \ldots, \mathbf{x}_n$ are n orthogonal eigenvectors with corresponding eigenvalues $\lambda_1, \ldots, \lambda_n$, then

$$\mathbf{x}_i^T A \mathbf{x}_j = \lambda_j \mathbf{x}_i^T \mathbf{x}_j = 0, \qquad i \neq j,$$

so that the $\mathbf{x}_i$ are conjugate with respect to A. This does not provide a practical approach to a conjugate direction method, however, since finding eigenvectors of A is a much more substantial problem than solving the linear system. Another possibility is to orthogonalize a set of linearly independent vectors $\mathbf{y}_1, \ldots, \mathbf{y}_n$ with respect to the inner product $(\mathbf{x}, \mathbf{y})_A = \mathbf{x}^T A \mathbf{y}$. This, however, is also too costly.

The Conjugate Gradient Method

The most efficient way, in general, to obtain a set of conjugate direction vectors to be used in solving $A\mathbf{x} = \mathbf{b}$ is the *conjugate gradient method*, which generates the direction vectors in conjunction with carrying out the conjugate direction method. The basic algorithm is given in Figure 9.1.2, in which $\mathbf{r}_k \equiv \mathbf{b} - A\mathbf{x}_k$ is the residual at the kth step.

Choose $\mathbf{x}_0$, set $\mathbf{p}_0 = \mathbf{r}_0 = \mathbf{b} - A\mathbf{x}_0$.

For $k = 0, 1, \ldots$

$\alpha_k = -(\mathbf{r}_k, \mathbf{r}_k)/(\mathbf{p}_k, A\mathbf{p}_k) \quad (= -(\mathbf{p}_k, \mathbf{r}_k)/(\mathbf{p}_k, A\mathbf{p}_k))$ \hfill (9.1.10a)

$\mathbf{x}_{k+1} = \mathbf{x}_k - \alpha_k \mathbf{p}_k$ \hfill (9.1.10b)

$\mathbf{r}_{k+1} = \mathbf{r}_k + \alpha_k A\mathbf{p}_k$ \hfill (9.1.10c)

if $\|\mathbf{r}_{k+1}\|_2^2 \geq \varepsilon$, continue \hfill (9.1.10d)

$\beta_k = (\mathbf{r}_{k+1}, \mathbf{r}_{k+1})/(\mathbf{r}_k, \mathbf{r}_k) \quad (= -(\mathbf{p}_k, A\mathbf{r}_{k+1})/(\mathbf{p}_k, A\mathbf{p}_k))$ \hfill (9.1.10e)

$\mathbf{p}_{k+1} = \mathbf{r}_{k+1} + \beta_k \mathbf{p}_k$ \hfill (9.1.10f)

Figure 9.1.2: *Conjugate Gradient Algorithm*

In (9.1.10a) and (9.1.10e) we have given two equivalent forms of α_k and β_k. The second form of α_k is (9.1.4) and the second form of β_k is the condition that $\mathbf{p}_{k+1}$ be A-orthogonal to $\mathbf{p}_k$. The first forms given are most suitable for computation since only the inner products $(\mathbf{r}_k, \mathbf{r}_k)$ and $(\mathbf{p}_k, A\mathbf{p}_k)$ are required.

9.1 The Conjugate Gradient Method

The step (9.1.10c) computes the next residual recursively rather than directly since
$$\mathbf{r}_{k+1} = \mathbf{b} - A\mathbf{x}_{k+1} = \mathbf{b} - A(\mathbf{x}_k - \alpha_k \mathbf{p}_k).$$
Because $A\mathbf{p}_k$ is already known, this saves the multiplication $A\mathbf{x}_{k+1}$.

We have used in (9.1.10d) one of many possible convergence tests. The advantage of (9.1.10d) is that it requires virtually no extra work since $(\mathbf{r}_{k+1}, \mathbf{r}_{k+1})$ will be needed on the next step if convergence has not occurred. In contrast, a test on $\|\mathbf{x}_{k+1} - \mathbf{x}_k\| = \|\alpha_k \mathbf{p}_k\|$ requires splitting the axpy operation of (9.1.10b) as well as a separate computation of the norm. In many situations, we would normalize the convergence test by $\|\mathbf{r}_{k+1}\|^2 \geq \varepsilon \|\mathbf{r}_0\|_2^2$ or $\|\mathbf{r}_{k+1}\|_2^2 \geq \varepsilon \|\mathbf{b}\|_2^2$.

The final two steps of (9.1.10) compute the new direction vector. The next theorem, which we state without proof, guarantees that these direction vectors are indeed conjugate.

THEOREM 9.1.2 *Assume that A is an $n \times n$ symmetric positive definite matrix and that $\hat{\mathbf{x}}$ is the solution of $A\mathbf{x} = \mathbf{b}$. Then the vectors $\mathbf{p}_k$ generated by (9.1.10f) are conjugate and $\mathbf{p}_k \neq 0$ unless $\mathbf{x}_k = \hat{\mathbf{x}}$. Hence, $\mathbf{x}_m = \hat{\mathbf{x}}$ for some $m \leq n$.*

As indicated in Theorem 9.1.2, it is possible that a conjugate gradient iterate $\mathbf{x}_k$ will equal the solution before $k = n$. Two important cases in which this must be true are given in the following theorem.

THEOREM 9.1.3 *The conjugate gradient iterates will converge to the solution in no more than m steps if either of the following conditions hold.*

a) $\mathbf{p}_0 = \mathbf{r}_0$ is a linear combination of m eigenvectors of A.

b) A has only m distinct eigenvalues.

We will show that Theorem 9.1.3 is true in the case $m = 1$. The extension to general m is left to Exercise 9.1.10. If $\mathbf{p}_0 = \mathbf{r}_0$ is an eigenvector of A, then $A\mathbf{p}_0 = \lambda \mathbf{p}_0$ and $A^{-1}\mathbf{p}_0 = \lambda^{-1}\mathbf{p}_0$. Since $A\hat{\mathbf{x}} - A\mathbf{x}_0 = \mathbf{r}_0$, the error satisfies

$$\hat{\mathbf{x}} - \mathbf{x}_0 = A^{-1}\mathbf{r}_0 = \lambda^{-1}\mathbf{p}_0.$$

Thus, the error is in the direction $\mathbf{p}_0$ so that the first conjugate gradient step, which minimizes the error in that direction, obtains the exact solution. If A has only one distinct eigenvalue λ and $\mathbf{p}_0 = \mathbf{r}_0$ is any linear combination of eigenvectors of A, then $A\mathbf{p}_0 = \lambda \mathbf{p}_0$ so that the above argument also applies.

Krylov Subspaces and Error Estimates

An important property of the conjugate gradient iteration is given in the following theorem, stated without proof. Recall from Chapter 2 that $\|\mathbf{y}\|_A^2 = \mathbf{y}^T A \mathbf{y}$ and span$(\mathbf{y}_1, \ldots, \mathbf{y}_m)$ is the set of all linear combinations of $\mathbf{y}_1, \ldots, \mathbf{y}_m$.

THEOREM 9.1.4 *The kth conjugate gradient iterate* $\mathbf{x}_k$ *minimizes* $\|\hat{\mathbf{x}} - \mathbf{x}\|_A$ *over*

$$S_{k-1} = \mathbf{x}_0 + \operatorname{span}(\mathbf{p}_0, A\mathbf{p}_0, \ldots, A^{k-1}\mathbf{p}_0). \qquad (9.1.11)$$

A subspace generated by powers of a matrix operating on a single vector, as in (9.1.11), is called a *Krylov subspace*. Such subspaces are very important in the study of many iterative methods, especially conjugate gradient methods. The notation in (9.1.11) of a vector plus a subspace denotes an *affine* or translated subspace in which elements are of the form

$$\mathbf{x}_0 + \sum_{i=0}^{k-1} \eta_i A^i \mathbf{p}_0. \qquad (9.1.12)$$

It is possible to use Theorem 9.1.4, and the analysis done in its proof, to show that the error in the kth conjugate gradient iterate satisfies

$$\|\hat{\mathbf{x}} - \mathbf{x}_k\|_A \leq 2\gamma^k \|\hat{\mathbf{x}} - \mathbf{x}_0\|_A, \qquad (9.1.13)$$

where

$$\gamma = (\sqrt{\kappa} - 1)/(\sqrt{\kappa} + 1), \qquad (9.1.14)$$

and $\kappa = \|A\|_2 \|A^{-1}\|_2$ is the l_2 condition number of A. To rewrite (9.1.13) in terms of the l_2 norm, we first recall (see Chapter 2) that $\kappa = \lambda_n/\lambda_1$, where λ_1 and λ_n are the smallest and largest eigenvalues of A, and that $\lambda_n = \|A\|_2, \lambda_1^{-1} = \|A^{-1}\|_2$. Then by (2.2.13)

$$\|\mathbf{x}\|_2^2 / \|A^{-1}\|_2 = \lambda_1 \mathbf{x}^T \mathbf{x} \leq \mathbf{x}^T A \mathbf{x} \leq \lambda_n \mathbf{x}^T \mathbf{x} = \|A\|_2 \|\mathbf{x}\|_2^2.$$

Applying this to (9.1.13) yields

$$\|\hat{\mathbf{x}} - \mathbf{x}_k\|_2^2 \leq \|A^{-1}\|_2 \|\hat{\mathbf{x}} - \mathbf{x}_k\|_A^2 \leq (2\gamma^k)^2 \|A^{-1}\|_2 \|\hat{\mathbf{x}} - \mathbf{x}_0\|_A^2$$
$$\leq (2\gamma^k)^2 \|A^{-1}\|_2 \|A\|_2 \|\hat{\mathbf{x}} - \mathbf{x}_0\|_2^2$$

or

$$\|\hat{\mathbf{x}} - \mathbf{x}_k\|_2 \leq 2\sqrt{\kappa} \gamma^k \|\hat{\mathbf{x}} - \mathbf{x}_0\|_2. \qquad (9.1.15)$$

Since $\gamma < 1$, the right-hand side of (9.1.15) decreases as k increases. This does not necessarily imply that the errors actually decrease at each step but this important property is indeed true, as stated in the next theorem.

THEOREM 9.1.5 *The conjugate gradient iterates satisfy*

$$\|\hat{\mathbf{x}} - \mathbf{x}_k\|_2 < \|\hat{\mathbf{x}} - \mathbf{x}_{k-1}\|_2 \qquad (9.1.16)$$

unless $\mathbf{x}_{k-1} = \hat{\mathbf{x}}$.

9.1 The Conjugate Gradient Method

Parallel and Vector Implementation

The main computational requirements for the conjugate gradient algorithm are the evaluation of $A\mathbf{p}_k$, the inner products $(\mathbf{r}_k, \mathbf{r}_k)$ and $(\mathbf{p}_k, A\mathbf{p}_k)$, and three axpys to obtain $\mathbf{x}_{k+1}, \mathbf{r}_{k+1}$ and $\mathbf{p}_{k+1}$. For any substantial problem, the evaluation of $A\mathbf{p}_k$ will be dominant and it is critical that this computation be done efficiently. For example, if A is a diagonally sparse matrix, on vector machines A could be stored by diagonals, and the matrix-vector multiplication-by-diagonals technique of Section 3.3 be used. On distributed memory machines, the assignment of the components $\mathbf{x}_k$ to the processors also determines the assignment of $\mathbf{p}_k$ and $\mathbf{r}_k$. For example, for the discrete Poisson equation the assignment of a grid point to a processor implies that the components of $\mathbf{x}_k, \mathbf{p}_k$, and $\mathbf{r}_k$ associated with that grid point are all assigned to the same processor. Then, the multiplication $A\mathbf{p}_k$ follows exactly the discussion of Jacobi's method in Section 8.2, with $\mathbf{p}_k$ now playing the role of the unknowns. The inner products are done by a fan-in, and α_k and β_k must then be broadcast to the other processors. The axpy operations for $\mathbf{x}_{k+1}, \mathbf{p}_{k+1}$, and $\mathbf{r}_{k+1}$ are particularly efficient since each processor just updates the components it is assigned and no communication is required. Thus, the key aspect in implementing the conjugate gradient algorithm efficiently on parallel machines is the multiplication $A\mathbf{p}_k$, which depends on the structure of A. However, even if this can be done efficiently, the necessary inner products and broadcast of α_k and β_k will degrade the parallel efficiency somewhat.

Domain Decomposition

One of the advantages of the conjugate gradient method is that the coefficient matrix A does not need to be known explicitly as long as the product $A\mathbf{p}$ can be evaluated. We will illustrate this by means of the domain decomposition technique discussed in Sections 6.1 and 7.3.

Consider a system with the coefficient matrix of (6.1.39):

$$\begin{bmatrix} A_1 & & & & B_1 \\ & A_2 & & & \vdots \\ & & \ddots & & \\ & & & A_p & B_p \\ B_1^T & \cdots & & B_p^T & A_S \end{bmatrix} \begin{bmatrix} \mathbf{x}_1 \\ \vdots \\ \mathbf{x}_p \\ \mathbf{x}_S \end{bmatrix} = \begin{bmatrix} \mathbf{b}_1 \\ \vdots \\ \mathbf{b}_p \\ \mathbf{b}_S \end{bmatrix}. \qquad (9.1.17)$$

In order to carry out Step 2 of the domain decomposition algorithm of Figure 7.3.12 we need to solve the Schur complement system $\hat{A}\mathbf{x}_S = \hat{\mathbf{b}}$, where

$$\hat{A} = A_S - \sum_{i=1}^{p} B_i^T A_i^{-1} B_i, \qquad \hat{\mathbf{b}} = \mathbf{b}_S - \sum_{i=1}^{p} B_i^T A_i^{-1} \mathbf{b}_i. \qquad (9.1.18)$$

Recall that we showed in Section 7.3 that $\hat{A}$ is symmetric positive definite if the coefficient matrix of (9.1.17) is positive definite.

In many problems, the matrices A_i and B_i will be sparse, but that need not be the case for the $B_i^T A_i^{-1} B_i$ and, hence, $\hat{A}$. In order to utilize the original sparsity, we will apply the conjugate gradient method to $\hat{A}\mathbf{x}_S = \hat{\mathbf{b}}$ without explicitly forming $\hat{A}$. Let $\mathbf{p}$ be any direction vector. Then we can compute $\hat{A}\mathbf{p}$ without forming $\hat{A}$ by:

Step 1. Form $\mathbf{y}_i = B_i\mathbf{p}, \quad i = 1, \ldots, p.$ (9.1.19a)

Step 2. Solve the systems $A_i\mathbf{z}_i = \mathbf{y}_i, \quad i = 1, \ldots, p.$ (9.1.19b)

Step 3. Form $\mathbf{w}_i = B_i^T\mathbf{z}_i, \quad i = 1, \ldots, p.$ (9.1.19c)

Step 4. Form $\hat{A}\mathbf{p} = A_S\mathbf{p} - \sum_{i=1}^{p}\mathbf{w}_i.$ (9.1.19d)

The formation of the vector $\hat{\mathbf{b}}$ is done in a similar way.

The solution of the systems in (9.1.19b) might be accomplished by using the Cholesky decompositions $A_i = L_i L_i^T$. In any case, steps 1 to 3 in (9.1.19) can be distributed across p processors and done in parallel with no communication. Only Step 4, as well as the computation of the inner products in the conjugate gradient method, will require fan-ins and communication.

Supplementary Discussion and References: 9.1

1. Minimization methods are discussed in a variety of books. See, for example, Dennis and Schnabel [1983], which includes problems with constraints. See Ortega [1988a] for further development of the conjugate gradient algorithm along the lines in the text, including proofs of Theorems 9.1.2, 9.1.4 and 9.1.5.

2. The conjugate gradient method was developed by Hestenes and Stiefel [1952], who analyzed its basic properties. Since, in exact arithmetic, the conjugate gradient method converges to the exact solution in n steps, it can be viewed as a direct method and an alternative to Cholesky factorization. But rounding error destroys this finite convergence property, and it was soon realized that the method was not a viable alternative to factorization for dense systems. Reid [1971], however, observed that for certain large sparse problems, such as arise from discretized partial differential equations, conjugate gradient methods gave sufficiently good convergence in far less than n iterations. This spurred a revival of interest in the method, especially when preconditioning is used, as will be

9.1 The Conjugate Gradient Method

discussed in the next section. An annotated bibliography of papers concerning the conjugate gradient method in the period 1948-1976 is given in Golub and O'Leary [1989].

3. Theorem 9.1.5 states that the 2-norms of the error vectors, $\hat{\mathbf{x}} - \mathbf{x}_k$, decrease monotonically. The same is not necessarily true for $\|\mathbf{r}_k\|_2$, the 2-norms of the residuals. Indeed, Hestenes and Stiefel [1952] give an example in which the norms of the residuals increase at every step except the last! Although this is an extreme case, it is not unusual to see $\|\mathbf{r}_k\|_2$ fluctuate as the iteration proceeds. However, in the norm defined by $\|\mathbf{r}\|^2 = \mathbf{r}^T A^{-1} \mathbf{r}$, the residuals do decrease at each iteration. This follows from

$$(\mathbf{b} - A\mathbf{x}_k)^T A^{-1}(\mathbf{b} - A\mathbf{x}_k) = (\hat{\mathbf{x}} - \mathbf{x}_k) A (\hat{\mathbf{x}} - \mathbf{x}_k) = \|\hat{\mathbf{x}} - \mathbf{x}_k\|_A^2,$$

and the fact that $\|\hat{\mathbf{x}} - \mathbf{x}_{k+1}\|_A^2 \leq \|\hat{\mathbf{x}} - \mathbf{x}_k\|_A^2$ by Theorem 9.1.4.

4. Block conjugate gradient algorithms can be formulated and have the potential of requiring less communication on parallel machines. See O'Leary [1987] for further discussion and a parallel implementation.

5. Theoretical results on the rate of convergence of the conjugate gradient method are given in van der Sluis and van der Vorst [1986], and Axelsson and Lindskog [1986]. For studies on the effect of rounding error on the conjugate gradient iterates, see Greenbaum and Strakos [1992] and van der Vorst [1990].

6. The conjugate gradient iterates satisfy a three-term recurrence relation (sometimes known as the *Rutishauser form*)

$$\mathbf{x}_{k+1} = \left(1 + \frac{\alpha_k \beta_k}{\alpha_{k-1}}\right) \mathbf{x}_k - \alpha_k (\mathbf{b} - A\mathbf{x}_k) - \frac{\alpha_k \beta_k}{\alpha_{k-1}} \mathbf{x}_{k-1},$$

where the α's and β's are as given in (9.1.10). Other forms of the recurrence relation may be given; see, for example, Hageman and Young [1981] and Concus, Golub and O'Leary [1976].

7. There is a close relationship between the conjugate gradient algorithm and the Lanczos method for computing eigenvalues of a symmetric matrix. In the latter, a sequence of $i \times i$ tridiagonal matrices T_i is generated such that as i increases, the eigenvalues of T_i more closely approximate the larger and smaller eigenvalues of A. Let R_i and P_i be matrices with

columns $r_0, \ldots, r_{i-1}$ and $p_0, \ldots, p_{i-1}$, respectively, and let

$$B_i = \begin{bmatrix} 1 & -\beta_0 & & \\ & 1 & & \\ & & \ddots & \\ & & & -\beta_{i-2} \\ & & & 1 \end{bmatrix}.$$

The relation (9.1.10f) for $k = 0, \ldots, i-2$ may be written as $R_i = P_i B_i$. Thus,

$$R_i^T A R_i = B_i^T P_i^T A P_i B_i$$

is tridiagonal since $P_i^T A P_i$ is diagonal. If $D = \text{diag}(\|r_0\|, \ldots, \|r_{i-1}\|)$, then

$$T_i = D_i^{-1} R_i^T A R_i D_i^{-1}$$

is the $i \times i$ tridiagonal matrix generated by the Lanczos process. For more on the Lanczos method, see Golub and Van Loan [1989].

EXERCISES 9.1

9.1.1. Show that the gradient vector for (9.1.1) is $\nabla q(\mathbf{x}) = A\mathbf{x} - \mathbf{b}$. If A is symmetric positive definite, show that the unique solution of $A\mathbf{x} = \mathbf{b}$ is the unique minimizer of $\mathbf{q}$.

9.1.2. For the function q of (9.1.1), show that

$$q(\mathbf{x} - \alpha\mathbf{p}) = \tfrac{1}{2}\mathbf{p}^T A \mathbf{p} \alpha^2 + \mathbf{p}^T(\mathbf{b} - A\mathbf{x})\alpha - \tfrac{1}{2}\mathbf{x}^T(2\mathbf{b} - A\mathbf{x}).$$

For $\mathbf{x} = \mathbf{x}_k$ and $\mathbf{p} = \mathbf{p}_k$, minimize this function of α to obtain (9.1.4).

9.1.3. If $\alpha_k \equiv 1$ and A is such that its main diagonal is I, show that (9.1.5) is Jacobi's method.

9.1.4. Let $\mathbf{p}_0 = \mathbf{e}_1, \mathbf{p}_1 = \mathbf{e}_2, \ldots, \mathbf{p}_{n-1} = \mathbf{e}_n$, where $\mathbf{e}_i$ is the vector with 1 in the ith component and zero elsewhere. Show that n steps of (9.1.2) with α_k given by (9.1.4) is equivalent to one Gauss-Seidel iteration on the system $A\mathbf{x} = \mathbf{b}$.

9.1.5. If $\hat{\alpha}_k$ is the minimizer of $q(\mathbf{x}_k - \alpha\mathbf{p}_k)$, verify that $q(\mathbf{x}_k - \omega\hat{\alpha}_k\mathbf{p}_k) < q(\mathbf{x}_k)$ if and only if $0 < \omega < 2$.

9.1.6. Verify that univariate relaxation minimizes the quadratic function (9.1.8) in n steps.

9.1.7. Let $P = (\mathbf{p}_0, \ldots, \mathbf{p}_{n-1})$, where the $\mathbf{p}_i$ are conjugate with respect to A. Show that $P^{-1}\mathbf{p}_{i-1} = \mathbf{e}_i$, and thus a conjugate direction step $\mathbf{x}_i = \mathbf{x}_{i-1} - \alpha_{i-1}\mathbf{p}_{i-1}$ in the variables $\mathbf{y} = P^{-1}\mathbf{x}$ is

$$\mathbf{y}_i = \mathbf{y}_{i-1} - \alpha_{i-1}\mathbf{e}_i.$$

9.1.8. Prove Theorem 9.1.1 as follows. First, show that

$$(A\mathbf{x}_{k+1} - \mathbf{b})^T \mathbf{p}_j = (A\mathbf{x}_k - \mathbf{b})\mathbf{p}_j$$

if $j < k$, and 0 if $j = k$. Use this to conclude that $(A\mathbf{x}_n - \mathbf{b})^T \mathbf{p}_j = 0, j = 0, \ldots, n-1$, and thus, since the $\mathbf{p}_j$ are linearly independent, $A\mathbf{x}_n - \mathbf{b} = 0$.

9.1.9. Let

$$A = \begin{bmatrix} A_I & B \\ B^T & A_S \end{bmatrix}$$

be symmetric positive definite with minimum and maximum eigenvalues $0 < \lambda_1 \le \lambda_n$. If $\mathbf{x}_1 = A_I^{-1} B \mathbf{x}_2$, use (7.3.25) to show that $\mathbf{x}_2^T \hat{A} \mathbf{x}_2 \ge \lambda_1 \mathbf{x}_2^T \mathbf{x}_2$ for any $\mathbf{x}_2$, where $\hat{A} = A_S - B^T A_I^{-1} B$. Conclude that all eigenvalues of $\hat{A}$ are at least as large as λ_1. Show also that

$$\mathbf{x}_2^T \hat{A} \mathbf{x}_2 = (0, \mathbf{x}_2)^T A \begin{bmatrix} 0 \\ \mathbf{x}_2 \end{bmatrix} \le \lambda_n \mathbf{x}_2^T \mathbf{x}_2,$$

and conclude that $\text{cond}_2(\hat{A}) \le \text{cond}_2(A)$.

9.1.10. If $\mathbf{p}_0 = \mathbf{r}_0 = \sum_{i=1}^m c_i \mathbf{v}_i$, where $\mathbf{v}_1, \ldots, \mathbf{v}_m$ are linearly independent eigenvectors of A, show that the conjugate gradient method converges in at most m iterations. (Hint: Show that the spaces (9.1.11) satisfy $S_{m-1} = S_m = \cdots = S_{n-1}$. Thus, Theorem 9.1.4 ensures that $\mathbf{x}_m$ is already the solution.) Next, if A has only m distinct eigenvalues, show that $A\mathbf{p}_0$ may be written as a linear combination of m eigenvectors of A, and thus $\mathbf{x}_m$ is again the solution.

9.2 Preconditioning

The result of the previous section (Theorem 9.1.1) that conjugate direction methods converge in at most n steps is of only theoretical interest, and we wish to view the conjugate gradient algorithm as an iterative method. An estimate of the rate of convergence was given by (9.1.15):

$$\|\mathbf{x}_k - \hat{\mathbf{x}}\|_2 \le 2\sqrt{\kappa}\gamma^k \|\mathbf{x}_0 - \hat{\mathbf{x}}\|_2, \tag{9.2.1}$$

where $\hat{\mathbf{x}}$ is the exact solution of the system, $\gamma = (\sqrt{\kappa} - 1)/(\sqrt{\kappa} + 1)$ and $\kappa = \text{cond}(A)$ is the condition number of A in the l_2 norm. Note that $\gamma = 0$ when $\kappa = 1$, and $\gamma \to 1$ as $\kappa \to \infty$. Hence, the larger κ, the closer γ will be to 1 and the slower will be the rate of convergence.

This leads to the idea of increasing the rate of convergence by *preconditioning* A with the *congruence transformation*

$$\hat{A} = SAS^T, \tag{9.2.2}$$

where S is a nonsingular matrix chosen so that $\text{cond}(\hat{A}) < \text{cond}(A)$. The system to be solved is then, in principle,

$$\hat{A}\hat{\mathbf{x}} = \hat{\mathbf{b}}, \qquad (9.2.3)$$

where $\hat{\mathbf{x}} = S^{-T}\mathbf{x}$ and $\hat{\mathbf{b}} = S\mathbf{b}$.

The simplest preconditioning is with a diagonal matrix S. In particular, if $S = \text{diag}(a_{11}^{-\frac{1}{2}}, \ldots, a_{nn}^{-\frac{1}{2}})$, then the main diagonal elements of $\hat{A}$ are all equal to 1. This is called *diagonal scaling* and is often a useful preconditioning. However, if the main diagonal elements of A are identical, as in the Poisson matrix (5.5.18), then diagonal scaling does not change the condition number (Exercise 9.2.9). In the remainder of this section, we will examine several other types of preconditioners. As opposed to the use of a diagonal matrix, these more complicated preconditioning matrices will usually destroy the sparsity of the matrix A. For this reason, as well as to avoid the work necessary to carry out (9.2.2), we will formulate the conjugate gradient method so as to work with the original matrix A, although the rate of convergence of the conjugate gradient iterates will be that for the system (9.2.3)

We apply the conjugate gradient algorithm of Figure 9.1.2 to the system (9.2.3) to obtain iterates $\hat{\mathbf{x}}_k$, and define $\mathbf{x}_k = S^T\hat{\mathbf{x}}_k$. Then it can be shown (Exercise 9.2.1) that these $\mathbf{x}_k$ satisfy the *preconditioned conjugate gradient* (PCG) *algorithm* shown in Figure 9.2.1.

Choose $\mathbf{x}_0$. Set $\mathbf{r}_0 = \mathbf{b} - A\mathbf{x}_0$. Solve $M\tilde{\mathbf{r}}_0 = \mathbf{r}_0$. Set $\mathbf{p}_0 = \tilde{\mathbf{r}}_0$.
For $k = 0, 1, \ldots$

$$\alpha_k = -(\tilde{\mathbf{r}}_k, \mathbf{r}_k)/(\mathbf{p}_k, A\mathbf{p}_k) \qquad (9.2.4\text{a})$$

$$\mathbf{x}_{k+1} = \mathbf{x}_k - \alpha_k \mathbf{p}_k \qquad (9.2.4\text{b})$$

$$\mathbf{r}_{k+1} = \mathbf{r}_k + \alpha_k A\mathbf{p}_k \qquad (9.2.4\text{c})$$

Solve $M\tilde{\mathbf{r}}_{k+1} = \mathbf{r}_{k+1}$ \qquad (9.2.4\text{d})

Test for convergence \qquad (9.2.4\text{e})

$$\beta_k = (\tilde{\mathbf{r}}_{k+1}, \mathbf{r}_{k+1})/(\tilde{\mathbf{r}}_k, \mathbf{r}_k) \qquad (9.2.4\text{f})$$

$$\mathbf{p}_{k+1} = \tilde{\mathbf{r}}_{k+1} + \beta_k \mathbf{p}_k \qquad (9.2.4\text{g})$$

Figure 9.2.1: *Preconditioned Conjugate Gradient Algorithm*

9.2 Preconditioning

In (9.2.4d), M is the matrix

$$M = (S^T S)^{-1}. \qquad (9.2.5)$$

If $M = I$, (9.2.4) reduces to the original conjugate gradient algorithm. We do not, in general, obtain the matrix M by choosing S and then forming (9.2.5). Rather we will choose M directly and usually we will never need S. Note that we must choose M to be symmetric positive definite since $S^T S$ is symmetric positive definite.

There are two general criteria for choosing M. On the basis of Theorem 9.1.3, the conjugate gradient method will converge in no more than m iterations if A has only m distinct eigenvalues. It is unlikely that this will be the case in practical problems but sometimes clusters of eigenvalues can be achieved by proper choice of the matrix M. Then the effect on the rate of convergence will be almost the same as having multiple eigenvalues.

The second criterion is to choose M so that $\hat{A}$ of (9.2.2) has a smaller condition number than A. By (9.2.2) and (9.2.5) $\hat{A}$ satisfies

$$S^T \hat{A} S^{-T} = S^T S A = M^{-1} A. \qquad (9.2.6)$$

Therefore, the matrix $M^{-1} A$ is similar to $\hat{A}$ so that

$$\text{cond}(\hat{A}) = \lambda_{\max}(M^{-1} A) / \lambda_{\min}(M^{-1} A), \qquad (9.2.7)$$

the ratio of the maximum and minimum eigenvalues of $M^{-1} A$. Thus, to reduce the condition number of $\hat{A}$ as much as possible, we wish to choose M so that the ratio of the largest and smallest eigenvalues of $M^{-1} A$ is as small as possible. In principle, we can make $\text{cond}(\hat{A}) = 1$ by choosing $M = A$, but this is not a practical choice. Since the auxiliary system $M\tilde{\mathbf{r}} = \mathbf{r}$ must be solved at each conjugate gradient iteration, it is critical that this system be "easy" to solve. On the other hand, in order that the preconditioning be effective, we want M^{-1} to be a "good" approximation to A^{-1} in the sense of minimizing the ratio of the largest and smallest eigenvalues of $M^{-1} A$. Clearly, the two requirements are in conflict since the more closely M^{-1} approximates A^{-1}, the more likely that the system $M\hat{\mathbf{r}} = \mathbf{r}$ will be nearly as difficult to solve as the system $A\mathbf{x} = \mathbf{b}$.

We shall consider several approaches to obtaining the matrix M after we give a short discussion of the convergence test (9.2.4e). If we use the same test, $\|\mathbf{r}_{k+1}\|_2^2 \geq \varepsilon$, as in the conjugate gradient algorithm, (9.1.10d), we have to calculate $(\mathbf{r}_{k+1}, \mathbf{r}_{k+1})$ separately since it no longer appears in the algorithm. What does appear is $(\tilde{\mathbf{r}}_{k+1}, \mathbf{r}_{k+1}) = (M^{-1} \mathbf{r}_{k+1}, \mathbf{r}_{k+1})$, which is an inner product defined by M^{-1}. We can use this quantity for the convergence test, but it may indicate convergence even though $\|\mathbf{r}_{k+1}\|_2$ is larger than we would like. Therefore, it is prudent to make an auxiliary check on $\|\mathbf{r}_{k+1}\|_2$ once the other test has been passed. In this case, the convergence test (9.2.4e) would take the form

$$\text{If } (\tilde{\mathbf{r}}_{k+1}, \mathbf{r}_{k+1}) \geq \varepsilon, \text{ continue}$$
$$\text{Else if } (\mathbf{r}_{k+1}, \mathbf{r}_{k+1}) \geq \varepsilon, \text{ continue.} \quad (9.2.8)$$

The use of (9.2.8) may cause more iterations than using just $(\mathbf{r}_{k+1}, \mathbf{r}_{k+1})$, but this is usually not the case.

Subsidiary Iteration Preconditioning

We now consider a rather general approach to preconditioning. It is based on the following series representation of A^{-1}, which we state without proof (see Exercise 9.2.2). Recall that the spectral radius $\rho(H)$ is the largest eigenvalue in absolute value of H.

THEOREM 9.2.1 *Let A be an $n \times n$ nonsingular matrix and $A = P - Q$ a splitting of A such that P is nonsingular. If $H = P^{-1}Q$ and $\rho(H) < 1$, then*

$$A^{-1} = \left(\sum_{k=0}^{\infty} H^k\right) P^{-1}. \quad (9.2.9)$$

On the basis of (9.2.9), we can consider the matrices defined by

$$M = P(I + H + \cdots + H^{m-1})^{-1}, \quad (9.2.10a)$$

$$M^{-1} = (I + \cdots + H^{m-1})P^{-1} \quad (9.2.10b)$$

to be approximations to A and A^{-1}, respectively. Thus, the solution of the auxiliary system $M\tilde{\mathbf{r}} = \mathbf{r}$ is effected by

$$\tilde{\mathbf{r}} = M^{-1}\mathbf{r} = (I + H + \cdots + H^{m-1})P^{-1}\mathbf{r}. \quad (9.2.11)$$

This is not, however, to be carried out by forming the matrix H and the indicated truncated series. Rather, we recognize (Exercise 9.2.3) that $\tilde{\mathbf{r}}$, as given by (9.2.11), is the result of applying m steps of the iterative method

$$P\mathbf{r}_{i+1} = Q\mathbf{r}_i + \mathbf{r}, \quad i = 0, 1, \ldots, m-1, \quad \mathbf{r}_0 = 0, \quad (9.2.12)$$

and setting $\tilde{\mathbf{r}} = \mathbf{r}_m$. We call (9.2.12) the *subsidiary* iterative method.

We next give some specific preconditioners based on this approach. These preconditioners are also called *truncated series preconditioners* on the basis of (9.2.10). For the Jacobi method (8.1.3), P is the main diagonal of A and (9.2.12) amounts to taking m Jacobi iterations. This gives the *m-step Jacobi PCG Method* in which (9.2.4d) becomes

$$\text{Do } m \text{ Jacobi iterations on } A\mathbf{r} = \mathbf{r}_{k+1} \text{ to obtain } \tilde{\mathbf{r}}_{k+1}. \quad (9.2.13)$$

9.2 Preconditioning

The solution of $M\tilde{\mathbf{r}}_0 = \mathbf{r}_0$ would be obtained in the same way. In (9.2.13) the starting value for the Jacobi iteration is zero, as in (9.2.12). The special case of $m = 1$ is equivalent to $M = D$, the main diagonal of A. By (9.2.5), $S = D^{-1/2} = \text{diag}(a_{11}^{-1/2}, \ldots, a_{nn}^{-1/2})$ so that the preconditioned matrix is $\hat{A} = D^{-1/2} A D^{-1/2}$. Thus, one-step Jacobi preconditioning is equivalent to diagonal scaling.

Mechanistically, we can use the Gauss-Seidel and SOR iterations to carry out (9.2.12). However, these methods do not yield a symmetric and positive definite matrix M, as required by (9.2.5). Indeed, for the Gauss-Seidel method and $m = 1$, $M = P = D - L$, where D is the main diagonal of A and $-L$ the strictly lower triangular part. Hence, M is not symmetric for $m = 1$ and this property persists for all $m > 1$ and for the SOR method (Exercise 9.2.4). This difficulty can be overcome by use of the *SSOR method* discussed below.

We first describe the *symmetric Gauss-Seidel (SGS)* method. If $A = D - L - U$ is the splitting of A into its diagonal, lower and upper triangular parts, the Gauss-Seidel iterate at the kth stage, which we denote now by $\mathbf{x}_{k+1/2}$, is

$$\mathbf{x}_{k+1/2} = (D - L)^{-1} U \mathbf{x}_k + (D - L)^{-1} \mathbf{b}. \tag{9.2.14}$$

We then proceed through the equations in reverse order using the values of $\mathbf{x}_{k+1/2}$ just computed: for $i = n, n - 1, \ldots, 1$

$$x_i^{(k+1)} = \frac{1}{a_{ii}} \left(-\sum_{j>i} a_{ij} x_j^{(k+1)} - \sum_{j<i} a_{ij} x_j^{(k+1/2)} + b_i \right). \tag{9.2.15}$$

In terms of the matrices D, L and U, this can be written as

$$D\mathbf{x}_{k+1} = U\mathbf{x}_{k+1} + L\mathbf{x}_{k+1/2} + \mathbf{b},$$

or

$$\mathbf{x}_{k+1} = (D - U)^{-1} L \mathbf{x}_{k+1/2} + (D - U)^{-1} \mathbf{b}. \tag{9.2.16}$$

Thus, the roles of L and U reverse on this step. One SGS iteration is then the combination of (9.2.14) and (9.2.16):

$$\mathbf{x}_{k+1} = (D - U)^{-1} L (D - L)^{-1} U \mathbf{x}_k + \hat{\mathbf{d}}, \tag{9.2.17a}$$

where

$$\hat{\mathbf{d}} = (D - U)^{-1} L (D - L)^{-1} \mathbf{b} + (D - U)^{-1} \mathbf{b}. \tag{9.2.17b}$$

For the SSOR iteration, one simply inserts the relaxation parameter ω, as in (8.1.19), on both steps; the matrix representation of the SSOR iteration is given in Exercise 9.2.5. Computationally, the SGS iteration is carried out by the explicit formulas (8.1.18) and (9.2.15), and the rather formidable expressions

(9.2.17) are only the matrix representations of this process. In terms of a discrete partial differential equation, the SGS or SSOR iterations amount to sweeping through the grid points on one step, and then sweeping through again in exactly the reverse order on the second step.

It remains to show that the matrix M of (9.2.10) is symmetric positive definite for the Jacobi and SGS iterations. A basic theorem in this regard is the following, which we state without proof. (Although the proof of symmetry is easy and left to Exercise 9.2.6.)

THEOREM 9.2.2 *Let $A = P - Q$ be symmetric positive definite with P symmetric and nonsingular. With $H = P^{-1}Q$, the matrix M^{-1} of (9.2.10) is symmetric and for any integer $m \geq 1$:*

1. *If m is odd, M^{-1} is positive definite if and only if P is positive definite.*
2. *If m is even, M^{-1} is positive definite if and only if $P + Q$ is positive definite.*

We next discuss the seemingly strange condition of Theorem 9.2.2 that the matrix $P + Q$ be positive definite. For Jacobi's method, $P = D$, which is the diagonal of A, and $Q = D - A$. Then the requirement that $P + Q$ be positive definite is just the necessary and sufficient condition that the Jacobi iteration be convergent, as stated in Theorem 8.1.3. This result for Jacobi's method is a special case of the following more general convergence theorem for iterations defined by the splitting $A = P - Q$. Recall (Theorem 8.1.1) that the necessary and sufficient condition for the corresponding iteration $\mathbf{x}_{k+1} = P^{-1}Q\mathbf{x}_k + \mathbf{d}$ to be convergent for all $\mathbf{x}^0$ is that the spectral radius $\rho(P^{-1}Q) < 1$.

THEOREM 9.2.3 *Let $A = P - Q$ and assume that A and P are symmetric and positive definite. Then $\rho(P^{-1}Q) < 1$ if and only if $P + Q$ is positive definite.*

On the basis of Theorem 9.2.3 we can restate Theorem 9.2.2 in the following way.

THEOREM 9.2.4 *Let $A = P - Q$ be symmetric and positive definite and P symmetric and nonsingular. With $H = P^{-1}Q$, the matrix M^{-1} of (9.2.10) is symmetric and positive definite for all $m \geq 1$ if and only if P is positive definite and $\rho(H) < 1$.*

As noted above, the matrix P must be symmetric in order that M be symmetric for all m. Then, by Theorem 9.2.4, for M to be positive definite requires that P be positive definite and that $\rho(H) < 1$, this latter condition implying that the iterative method defined by the splitting $P - Q$ be convergent. In

9.2 Preconditioning

particular, since Jacobi's method can fail to be convergent for positive definite matrices A, the corresponding M need not be positive definite for even m.

On the other hand, for the SGS iteration, we can show that the matrix M is always positive definite, provided that A is symmetric positive definite. By the symmetry of A, we can take $U = L^T$ in (9.2.17). The middle two terms of the iteration matrix of (9.2.17a) can be written as

$$L(D-L)^{-1} = LD^{-1}(I-LD^{-1})^{-1} = (I-LD^{-1})^{-1}LD^{-1} = D(D-L)^{-1}LD^{-1}$$

since matrices of the form $(I-C)^{-1}$ and C commute (Exercise 9.2.7). Then we can write the iteration matrix of (9.2.17a) as

$$H = (D-L^T)^{-1}D(D-L)^{-1}LD^{-1}L^T = P^{-1}Q, \qquad (9.2.18)$$

where

$$P = (D-L)D^{-1}(D-L^T), \qquad Q = LD^{-1}L^T. \qquad (9.2.19)$$

By assumption, D is positive definite so that P is symmetric positive definite and Q is symmetric positive semidefinite (Exercise 9.2.8). It follows that $P+Q$ is positive definite (Exercise 2.2.17), and then Theorem 9.2.2 applies to show that M is positive definite for any m.

Parallel and Vector Implementation

We next discuss the parallel and vector implementation of the preconditioned methods considered so far. We have given in the previous section the implementation of the basic conjugate gradient method, and in Section 8.2 the implementation of the Jacobi iteration. For the m-step Jacobi PCG method, it is then just necessary to combine these two (Exercise 9.2.12).

For SGS or SSOR preconditioning, the Jacobi iterations would be replaced by Gauss-Seidel iterations. We saw in Section 8.2 that in order to do Gauss-Seidel iterations with a high degree of parallelism we would use the red-black ordering of the grid points. For more general problems for which the red-black ordering does not suffice, the multicolor orderings of Section 8.2 could be used. On a distributed memory system, for either Jacobi or SSOR preconditioning, we distribute the unknowns to the processors as discussed in Section 8.2 and indicated in Figure 8.2.1. We also distribute the vectors **p** and **r** in the conjugate gradient iteration in the same fashion. The multiplication $A\mathbf{p}_k$ and the computation of the inner products and axpys in the conjugate gradient iteration is done as discussed in Section 9.1, and the preconditioning is carried out as in Section 8.2 for the Jacobi and SSOR iterations.

We noted in Section 8.2 that use of the red/black ordering for the Gauss-Seidel or SOR iterations does not degrade the rate of convergence. Unfortunately, the same is not true when SGS or SSOR are used with the red/black (or multicolor) orderings as preconditioners for the conjugate gradient method.

In this case, it may be better, at least on vector machines, to use the diagonal orderings discussed in Section 8.2. Since these are equivalent to the natural ordering, there is no loss in the rate of convergence although the vector lengths are now considerably shorter than with the red/black ordering. To illustrate the advantage of the diagonal ordering, Table 9.2.1 shows the iterations (I), time and megaflop rates for a three-dimensional Poisson-type equation on an $N \times N \times N$ grid. The machine is a single processor of a CRAY 2 and times are in seconds. For small problem sizes, the vector lengths using the red/black ordering are much longer than for the diagonal ordering, and the megaflop rates much higher. For $N = 21$ (9300 equations), the times are about equal. For larger problem sizes, the advantage is increasingly to the diagonal ordering because of its much better rate of convergence. However, on distributed memory machines with a large number of processors, the red/black ordering has advantages in communication and may be preferable.

Table 9.2.1: *Diagonal vs. Red/Black Ordering*

	Diagonal			Red/Black		
N	I	Time	Mflops	I	Time	Mflops
7	5	.0033	19	7	.0014	64
21	9	.060	50	21	.065	108
61	17	1.78	78	59	3.87	124

A potential problem with the diagonal ordering on vector machines is that for the reordered matrix of Figure 8.2.17, we no longer have the long diagonals of the naturally ordered matrix to use in the matrix-vector multiply in the conjugate gradient method. In particular, there are breaks in the diagonals. On the other hand, we can leave the matrix in the naturally ordered form and still carry out the updates according to the diagonal ordering paradigm, but the vectors now have stride $N-1$. This may cause memory bank conflicts on some vector machines. Thus, in the natural ordering, the data is arranged optimally for the matrix-vector multiply but not for the preconditioning and vice-versa for the diagonal ordering. However, we can circumvent this problem by means of the "Eisenstat trick" (see the Supplementary Discussion), which eliminates the need for the matrix-vector multiply. Then we can use the diagonal ordering so as to have the best vectors for the preconditioning. This was the approach used in obtaining the results of Table 9.2.1.

Incomplete Cholesky Preconditioning

We now consider another important approach to preconditioning which is based on *incomplete factorizations* of the matrix A. If LL^T is the Cholesky

9.2 Preconditioning

factorization of the symmetric positive definite matrix A and A is sparse, then the factor L is generally much less sparse than A because of fill-in. By an *incomplete Cholesky factorization* we shall mean a relation of the form

$$A = LL^T + R, \qquad (9.2.20)$$

where L is again lower triangular but $R \neq 0$. One way to obtain such incomplete factorizations is to suppress the fill-in, or part of it, that occurs during the Cholesky factorization. A simple example is the following. The matrix

$$L = 2^{-1/2} \begin{bmatrix} 2 & & \\ 1 & 3^{1/2} & \\ 1 & -3^{-1/2} & 2^{3/2}3^{-1/2} \end{bmatrix} \qquad (9.2.21)$$

is the Cholesky factor of

$$A = \begin{bmatrix} 2 & 1 & 1 \\ 1 & 2 & 0 \\ 1 & 0 & 2 \end{bmatrix}, \qquad (9.2.22)$$

and the (3,2) element has filled in. If we wish an incomplete factorization that has zeros in the same positions as A, we could set to zero those elements in the Cholesky factor in positions corresponding to zero elements in A. Thus, an incomplete factor for (9.2.22) would be the same as the L of (9.2.20) except the (3,2) element would be zero. This approach has the disadvantage that all of the work of the Cholesky decomposition would be done, and it is precisely this work, as well as the need for additional storage, that we wish to avoid. Hence, we desire not to compute any elements of L in positions corresponding to zero elements of A. This leads to the following *incomplete Cholesky no-fill factorization* [IC(0)] principle:

$$\begin{aligned} &\text{if } a_{ij} \neq 0, \text{ do the Cholesky calculation of } l_{ij}, \\ &\text{if } a_{ij} = 0, \text{ set } l_{ij} = 0. \end{aligned} \qquad (9.2.23)$$

An algorithm for (9.2.23) is given in Figure 9.2.2. Note that when the "if" statement in Figure 9.2.2 is removed, this is a complete factorization.

If the incomplete Cholesky algorithm of Figure 9.2.2 is applied to the matrix (9.2.22), then the incomplete factor is

$$L = 2^{-1/2} \begin{bmatrix} 2 & & \\ 1 & 3^{1/2} & \\ 1 & 0 & 3^{1/2} \end{bmatrix}.$$

Note that the (3,3) element has now changed, compared with (9.2.21), as a consequence of the (3,2) element being set to zero during the factorization.

$$l_{11} = a_{11}^{1/2}$$
For $i = 2$ to n
  For $j = 1$ to $i - 1$
    If $a_{ij} = 0$ then $l_{ij} = 0$ else
    $l_{ij} = \left(a_{ij} - \sum_{k=1}^{j-1} l_{ik} l_{jk}\right)/l_{jj}$
  $l_{ii} = \left(a_{ii} - \sum_{k=1}^{i-1} l_{ik}^2\right)^{1/2}$

Figure 9.2.2: *Incomplete Cholesky Factorization*

The no-fill principle can be relaxed to allow a certain amount of fill, and there are many such possibilities. For example, if A is a diagonally sparse matrix, one common approach is to allow the incomplete factor L to have a few more nonzero diagonals than A itself (counting only the number of diagonals of A on one side of the main diagonal, of course). Another common approach is to use a *threshold* ε so that if the computed l_{ij} satisfies $|l_{ij}| \leq \varepsilon$, then l_{ij} is set to zero. Thus, only sufficiently large elements are retained in the incomplete factor.

Assuming that an incomplete decomposition has been carried out, we then take the preconditioning matrix M of the PCG algorithm (9.2.10) to be

$$M = LL^T. \tag{9.2.24}$$

Since L is nonsingular, M is symmetric and positive definite. The solution of the systems $M\tilde{\mathbf{r}} = \mathbf{r}$ in the PCG algorithm is then carried out by solving the triangular systems

$$L\mathbf{x} = \mathbf{r}, \qquad L^T \tilde{\mathbf{r}} = \mathbf{x}. \tag{9.2.25}$$

Note that the incomplete factorization will be done once and for all at the beginning of the PCG iteration and the factor L saved for use in solving the systems (9.2.25) at each step.

The above choice of M defines the general class of *incomplete Cholesky preconditioned conjugate gradient* methods, denoted as *ICCG* methods. The way that the incomplete Cholesky factorization is obtained further defines the method. Thus, for example, if the incomplete factorization is obtained by the no-fill principle (9.2.23), the corresponding conjugate gradient method is denoted by ICCG(0).

It is sometimes convenient to consider the root-free form of incomplete Cholesky decomposition in which (9.2.20) is replaced by

$$A = LDL^T + R, \tag{9.2.26}$$

9.2 Preconditioning

where D is a positive diagonal matrix and L now has 1's on its main diagonal. With $M = LDL^T$,

$$L\mathbf{z} = \mathbf{r}, \qquad D\mathbf{x} = \mathbf{z}, \qquad L^T \tilde{\mathbf{r}} = \mathbf{x}, \qquad (9.2.27)$$

then replace the forward and back substitutions (9.2.25).

A difficulty with the ICCG method is that an incomplete Cholesky decomposition of a symmetric positive definite matrix cannot necessarily be carried out, in contrast to a complete decomposition. It may fail because the square root of a negative number is required to compute a diagonal element, or, in case of the root-free form LDL^T, the matrix D may not be positive definite. There have been several suggested remedies for this. Suppose that at the ith stage, $l_{ii}^2 \leq 0$, which would require the square root of a negative number or would mean that L is singular if $l_{ii}^2 = 0$. One of the simplest ideas is to replace l_{ii}^2 by some positive number; one particular choice for this number is the previous diagonal element $l_{i-1,i-1}^2$. Another strategy is to replace l_{ii}^2 by $(\sum_{j<i} |l_{ij}|)^2$, which ensures that the new ith row of L is diagonally dominant. (See the Supplementary Discussion for other approaches.)

Parallel and Vector ICCG

For incomplete Cholesky (IC) preconditioning, we must solve the triangular systems of (9.2.25) or (9.2.27). For a no-fill incomplete factorization, the factors L in either (9.2.25) or (9.2.27) have exactly the same non-zero structure as the lower triangular part of A and thus exactly the same non-zero structure as used in carrying out an SGS step. It follows that we can also use red/black or multicolor orderings to solve the triangular systems for IC preconditioning. To illustrate the technique, assume that A is written in the red/black form

$$A = \begin{bmatrix} D_1 & C^T \\ C & D_2 \end{bmatrix}, \qquad (9.2.28)$$

where D_1 and D_2 are diagonal. The root-free form of a complete Cholesky factorization of A is then

$$\begin{bmatrix} L_1 & \\ L_3 & L_2 \end{bmatrix} \begin{bmatrix} \hat{D}_1 & \\ & \hat{D}_2 \end{bmatrix} \begin{bmatrix} L_1^T & L_3^T \\ & L_2^T \end{bmatrix} = \begin{bmatrix} D_1 & C^T \\ C & D_2 \end{bmatrix}, \qquad (9.2.29)$$

where L_1 and L_2 are lower triangular with 1's on the main diagonal, and the $\hat{D}_i$ are diagonal. Equating terms in (9.2.29) gives

$$L_1 \hat{D}_1 L_1^T = D_1, \qquad L_3 \hat{D}_1 L_1^T = C, \qquad L_3 \hat{D}_1 L_3^T + L_2 \hat{D}_2 L_2^T = D_2. \qquad (9.2.30)$$

The first relation of (9.2.30) implies (Exercise 9.2.11) that

$$L_1 = I, \qquad \hat{D}_1 = D_1, \qquad L_3 = C D_1^{-1}. \qquad (9.2.31)$$

Now consider an IC no-fill factorization of (9.2.28) in which we denote the incomplete factorization also by

$$M = \begin{bmatrix} L_1 & \\ L_3 & L_2 \end{bmatrix} \begin{bmatrix} \hat{D}_1 & \\ & \hat{D}_2 \end{bmatrix} \begin{bmatrix} L_1^T & L_3^T \\ & L_2^T \end{bmatrix}. \qquad (9.2.32)$$

Since (9.2.31) shows that a complete Cholesky factorization gives L_1 and L_3 with no fill, L_1, L_3, and $\hat{D}_1$ as given by (9.2.31) are also the correct entries for the incomplete factorization (9.2.32). Moreover, L_2 of the incomplete factorization must be the identity since no fill is allowed; thus

$$\hat{D}_2 = \bar{D}_2 \equiv \text{diagonal part of } (D_2 - CD_1^{-1}C^T). \qquad (9.2.33)$$

Therefore, the IC factors of (9.2.32) are given by

$$L = \begin{bmatrix} I & \\ CD_1^{-1} & I \end{bmatrix}, \quad D = \begin{bmatrix} D_1 & \\ & \bar{D}_2 \end{bmatrix}. \qquad (9.2.34)$$

If we partition the vectors $\mathbf{r}, \mathbf{x}$ and $\mathbf{z}$ of the systems (9.2.27) commensurately with A, then (9.2.27) becomes

$$\begin{bmatrix} I & \\ CD_1^{-1} & I \end{bmatrix} \begin{bmatrix} \mathbf{z}_1 \\ \mathbf{z}_2 \end{bmatrix} = \begin{bmatrix} \mathbf{r}_1 \\ \mathbf{r}_2 \end{bmatrix}, \quad \begin{bmatrix} D_1 & \\ & \bar{D}_2 \end{bmatrix} \begin{bmatrix} \mathbf{x}_1 \\ \mathbf{x}_2 \end{bmatrix} = \begin{bmatrix} \mathbf{z}_1 \\ \mathbf{z}_2 \end{bmatrix},$$

$$\begin{bmatrix} I & D_1^{-1}C^T \\ & I \end{bmatrix} \begin{bmatrix} \tilde{\mathbf{r}}_1 \\ \tilde{\mathbf{r}}_2 \end{bmatrix} = \begin{bmatrix} \mathbf{x}_1 \\ \mathbf{x}_2 \end{bmatrix}. \qquad (9.2.35)$$

These systems may be solved to give

$$\begin{bmatrix} \mathbf{z}_1 \\ \mathbf{z}_2 \end{bmatrix} = \begin{bmatrix} \mathbf{r}_1 \\ \mathbf{r}_2 - CD_1^{-1}\mathbf{r}_1 \end{bmatrix}, \quad \begin{bmatrix} \mathbf{x}_1 \\ \mathbf{x}_2 \end{bmatrix} = \begin{bmatrix} D_1^{-1}\mathbf{z}_1 \\ \bar{D}_2^{-1}\mathbf{z}_2 \end{bmatrix},$$

$$\begin{bmatrix} \tilde{\mathbf{r}}_1 \\ \tilde{\mathbf{r}}_2 \end{bmatrix} = \begin{bmatrix} \mathbf{x}_1 - D_1^{-1}C^T\mathbf{x}_2 \\ \mathbf{x}_2 \end{bmatrix},$$

or

$$\tilde{\mathbf{r}}_2 = \bar{D}_2^{-1}(\mathbf{r}_2 - CD_1^{-1}\mathbf{r}_1), \quad \tilde{\mathbf{r}}_1 = D_1^{-1}C^T\tilde{\mathbf{r}}_2. \qquad (9.2.36)$$

Thus, the formation of $\tilde{\mathbf{r}}$ requires only multiplication of vectors by diagonal matrices or by C and C^T. Therefore, the solution of the triangular systems has been reduced to essentially the same operations used in a SGS step. The above approach may also be used with multicolor orderings.

Unfortunately, just as with SGS or SSOR preconditioning, the use of red-black or multicolor orderings will degrade the rate of convergence of the ICCG method, as compared with the natural ordering, and these orderings will be more useful on distributed memory machines than vector machines.

9.2 Preconditioning

In the case of the red/black ordering, the incomplete factorization itself was obtained with very little computation, but in general the algorithm of Figure 9.2.1 must be carried out. In some cases, this can be done with a reasonable degree of parallelization or vectorization, but in any case it is done only once at the beginning of the ICCG method.

Supplementary Discussion and References: 9.2

1. Although the idea of preconditioning the conjugate gradient method goes back to Hestenes [1956], most of the research on preconditioning has been done since the potential value of the conjugate gradient method as an iterative method was recognized in the early 1970s. A seminal paper on preconditioning was that of Concus, Golub, and O'Leary [1976]. See also Axelsson [1976].

2. The truncated series approach to preconditioning on vector and parallel computers was initiated by Dubois et al. [1979] for the case that the matrix P of (9.2.11) is the main diagonal of A, that is, for Jacobi's method as the subsidiary iteration. Adams [1982, 1985] studied truncated series preconditioning in the context of m steps of a subsidiary iterative method including, in particular, the Jacobi and SSOR iterations. Theorem 9.2.2 is based on this work; see also Ortega [1988a] for proofs of Theorems 9.2.2 – 9.2.4.

3. The matrix M^{-1} in (9.2.10) is formed from a particular polynomial in the matrix H. This suggests considering more general polynomials in H and defining

$$M^{-1} = (\alpha_0 I + \alpha_1 H + \cdots + \alpha_{m-1} H^{m-1}) P^{-1}.$$

This idea of *polynomial preconditioning* was studied by Johnson et al. [1983] for the case that P is the main diagonal of A, and by Adams [1985] for more general P, including that of SSOR. See also Saad [1985] for an analysis of polynomial preconditioning methods.

4. Eisenstat [1981] observed that if $M = CC^T$, the matrix-vector multiplication in the PCG algorithm can be replaced by multiplication by $K = C + C^T - A$. See also Ortega [1988a] for a discussion. In certain cases, for example, SGS preconditioning, K is a diagonal matrix. A related idea is presented by Bank and Douglas [1985]. An analysis and comparison of the Bank-Douglas and Eisenstat procedures is given in Ortega [1988c].

5. The idea of incomplete factorization dates back at least to Varga [1960]. Meijerink and van der Vorst [1977] considered incomplete factorizations

for preconditioning the conjugate gradient method and gave a detailed treatment of the ICCG method. They also showed that no-fill incomplete Cholesky factorization could be carried out for a symmetric M-matrix. (An M-*matrix* has nonpositive off-diagonal elements and an inverse with all nonnegative elements.) This was extended to H-matrices by Manteuffel [1980]. (A matrix A with positive diagonal elements is an H-*matrix* if the matrix $\hat{A}$ with elements $\hat{a}_{ij} = -|a_{ij}|, i \neq j$, and $\hat{a}_{ii} = a_{ii}$ is an M matrix.) This paper also discusses another remedy, called the *shifting strategy*, for the problem of an incomplete factorization not existing. Define $\hat{A} = \alpha I + A$. Clearly, for some choice of $\alpha > 0$, $\hat{A}$ is strictly diagonally dominant. It can be shown that a strictly diagonally dominant matrix with positive diagonal elements and nonpositive off-diagonal elements is an M-matrix. It follows that $\hat{A}$ is an H matrix, and an incomplete Cholesky factorization can be carried out.

6. Kershaw [1978] suggested the strategy of replacing the current l_{ii} in an incomplete decomposition by $\sum |l_{ij}|$ to ensure diagonal dominance. A related idea was given by Gustafsson [1978], who called it the modified ICCG method (MICCG). The modified method can be formulated in terms of a parameter α that ranges from zero (unmodified) to one. The rate of convergence will many times be optimal for an α slightly less than one. See Ashcraft and Grimes [1988] for further discussion and experimental results.

7. The use of multicoloring in ICCG was suggested by Schreiber and Tang [1982], and further developed in Poole and Ortega [1987]. These papers, as well as Ashcraft and Grimes [1988], pointed out that use of red/black or multicolor orderings could degrade the rate of convergence. Then Duff and Meurant [1989] reported on an extensive set of experiments using incomplete Cholesky preconditioning for a Poisson-type equation on a 30×30 grid in two dimensions. They considered 16 different reordering strategies of which only six gave rates of convergence comparable to the natural ordering on all problems, and three of these orderings are equivalent to the natural ordering. See also Ortega [1991] for a review.

8. Another use of the red/black ordering is in formulating the Reduced System Conjugate Gradient (RSCG) method (see Hageman and Young [1981] for further discussion). Assume that in the red/black system (9.2.28) the main diagonal has been scaled so that D_1 and D_2 are identity matrices. Then the *reduced system* (the Schur complement system) is
$$(I - CC^T)\mathbf{u}_B = \mathbf{b}_B - C\mathbf{b}_R. \qquad (9.2.37)$$
After $\mathbf{u}_B$ is computed, $\mathbf{u}_R$ can be obtained by
$$\mathbf{u}_R = -C^T\mathbf{u}_B + \mathbf{b}_R.$$

As shown in Section 7.3, $I - CC^T$ is positive definite and thus the conjugate gradient method can be applied to (9.2.37). One does not need to form the matrix $I - CC^T$ explicitly but can carry out the matrix-vector multiplication by first multiplying by C^T and then by C. One advantage of the RSCG method is that we are now working with vectors of half-length, compared with the original system. Another advantage is that SGS preconditioning of the original system is automatically built into (9.2.37). This follows from the observation that the SGS preconditioned matrix can be computed explicitly as

$$\begin{bmatrix} I & 0 \\ C & I \end{bmatrix}^{-1} \begin{bmatrix} I & C^T \\ C & I \end{bmatrix} \begin{bmatrix} I & 0 \\ C & I \end{bmatrix}^{-T} = \begin{bmatrix} I & 0 \\ 0 & I - CC^T \end{bmatrix}.$$

(See Keyes and Gropp [1987] for a more general result due to S. Eisenstat.) Thus, the rate of convergence of the conjugate gradient method applied to (9.2.37) will be exactly that of the SGS preconditioned conjugate gradient method on the original system. (Note that this implies a slower rate of convergence than SGS preconditioning applied to the naturally-ordered system.)

9. There is also the question of preconditioning the reduced system (9.2.37). Various approaches have been considered for this. For example, Mansfield [1991] uses a damped Jacobi iteration as a preconditioner; this has the advantage that the matrix $I - CC^T$ does not have to be formed explicitly. On the other hand, Elman and Golub [1990, 1991] show that if the original system (9.2.28) has arisen from a five-point discretization of an elliptic equation $\nabla(K\nabla u) = f$, then the matrix $I - CC^T$ of (9.2.37) is a nine-diagonal matrix and so is still very sparse. In this case, one can consider forming $I - CC^T$ explicitly and then doing, for example, incomplete Cholesky preconditioning.

10. A number of other preconditioners have been studied. For example, the multigrid method of Section 8.3 and the fast Poisson solvers of Section 6.3 have been used as preconditioners. Another general approach is by use of "block" methods; see, for example, Concus, Golub and Meurant [1985].

11. For further textbook discussions of parallel and vector aspects of preconditioning, see Ortega [1988a] and Dongarra and Duff et al. [1990].

EXERCISES 9.2

9.2.1. Apply the conjugate gradient algorithm to the system (9.2.3) to obtain iterates $\hat{\mathbf{x}}_k$. Define $\mathbf{x}_k = S^T \hat{\mathbf{x}}_k$ and show that these $\mathbf{x}_k$ satisfy (9.2.4) with M given by (9.2.5).

9.2.2. Use the Neumann expansion of Exercise 2.3.18 to prove Theorem 9.2.1 by noting that $A = P - Q = P(I - H)$.

9.2.3. Show that the $\tilde{\mathbf{r}}$ of (9.2.11) is given by $\mathbf{r}_m$ of (9.2.12).

9.2.4. For the SOR method, show that the matrix M of (9.2.10) is, in general, not symmetric for all $m \geq 1$ and all $0 < \omega < 2$.

9.2.5. Corresponding to (9.2.17), show that the matrix representation of the SSOR iteration is

$$\mathbf{x}_{k+1} = (D - \omega U)^{-1}[(1-\omega)D + \omega L](D - \omega L)^{-1}[(1-\omega)D + \omega U]\mathbf{x}_k + \hat{\mathbf{d}},$$

where

$$\hat{\mathbf{d}} = \omega(D - \omega U)^{-1}\{[(1-\omega)D + \omega L](D - \omega L)^{-1} + I\}\mathbf{b}.$$

9.2.6. If $A = P - Q$, show that the matrix of M^{-1} of (9.2.10b) can be written as

$$M^{-1} = P^{-1} + P^{-1}QP^{-1} + P^{-1}QP^{-1}QP^{-1} + \cdots + P^{-1}QP^{-1}\ldots P^{-1}QP^{-1}.$$

Conclude that if A and P are symmetric, then so is M^{-1} (and, hence, M).

9.2.7. Show that if $I - C$ is nonsingular, then $(I - C)^{-1}C = C(I - C)^{-1}$.

9.2.8. Let B and C be $n \times n$ matrices with B symmetric positive definite. Show that $C^T BC$ is symmetric positive semidefinite and positive definite if and only if C is nonsingular.

9.2.9. If A has identical main diagonal elements, as in (5.5.18), show that A and the diagonally scaled matrix $\hat{A}$ have the same l_2 condition numbers.

9.2.10. Consider the two-point boundary value problem $(a(x)u'(x))' = 0$ with $u(0) = 0, u(1) = 1$ and $a(x) = 1$ for $0 \leq x < \frac{1}{2}, a(x) = 1000$ for $\frac{1}{2} \leq x \leq 1$. Solve the corresponding discrete system by the conjugate gradient method with and without diagonal scaling. Discuss the efficiency of diagonal scaling as a preconditioner in this case. Compute the eigenvalues of the original and preconditioned matrices and compare them and the condition numbers.

9.2.11. Show that if L is a lower triangular matrix such that LL^T is diagonal, then L is diagonal. Conclude from this that the first relation of (9.2.30) implies that $L = I$ and $\hat{D}_1 = D_1$.

9.2.12. Give the necessary details for the parallel or vector implementation of the m-step Jacobi PCG method.

9.2.13. Show that if the algorithm of Figure 9.2.1 is applied to a tridiagonal matrix, the result is a complete Cholesky decomposition. More generally, show that if the algorithm is applied to a banded matrix that has no zeros within the band, the result is again a complete Cholesky decomposition.

9.3 Nonsymmetric and Nonlinear Problems

The conjugate gradient algorithm requires that the coefficient matrix A be symmetric and positive definite and cannot be applied directly to the system

$$A\mathbf{x} = \mathbf{b} \qquad (9.3.1)$$

with a nonsymmetric coefficient matrix. It is necessary either to convert (9.3.1) to a system with a symmetric coefficient matrix or to generalize the conjugate gradient algorithm in some suitable way. We will discuss both approaches and also consider a nonlinear problem. In the latter case, we will see how the conjugate gradient method can be used to solve the linear systems of Newton's method.

The Normal Equation Approach

The normal equations (see Section 4.2 and Exercise 9.3.1) corresponding to (9.3.1) are

$$A^T A \mathbf{x} = A^T \mathbf{b}. \qquad (9.3.2)$$

If A is nonsingular, as we assume, then $A^T A$ is symmetric positive definite and the conjugate gradient algorithm can be applied to (9.3.2). A related approach is to solve the system

$$AA^T \mathbf{y} = \mathbf{b} \qquad (9.3.3)$$

by the conjugate gradient method and then recover $\mathbf{x}$ by $\mathbf{x} = A^T \mathbf{y}$.

The solution of either (9.3.2) or (9.3.3) has two disadvantages. First, multiplications by both A and A^T are required at each step of the conjugate gradient method ($A^T A$ is rarely formed explicitly). Secondly, the condition number of $A^T A$ (or AA^T) is the square of that of A (see Exercise 9.3.2). Nevertheless, these normal equation algorithms are sometimes useful. (See the Supplementary Discussion for further remarks.)

For reasons that will be discussed shortly, we denote the conjugate gradient method applied to (9.3.2) as CGNR and the conjugate gradient method applied to (9.3.3) as CGNE. This latter algorithm is also sometimes known as *Craig's method*. These algorithms are summarized in Figure 9.3.1 (see Exercise 9.3.3) in terms of the residual $\mathbf{r} = \mathbf{b} - A\mathbf{x}$ of (9.3.1).

The Conjugate Residual Method

We next consider another approach, which has a close connection with the normal equation methods. If A is symmetric positive definite, the conjugate gradient algorithm arises from minimizing the quadratic function

$$q_1(\mathbf{x}) = \mathbf{x}^T A \mathbf{x} - 2\mathbf{b}^T \mathbf{x} + c. \qquad (9.3.4)$$

(a) CGNR	(b) CGNE
$r_0 = b - Ax_0, p_0 = A^T r_0$ For $i = 0, 1, \ldots$ until convergence $\alpha_i = -\dfrac{(A^T r_i, A^T r_i)}{(Ap_i, Ap_i)}$ $x_{i+1} = x_i - \alpha_i p_i$ $r_{i+1} = r_i + \alpha_i Ap_i$ $\beta_i = \dfrac{(A^T r_{i+1}, A^T r_{i+1})}{(A^T r_i, A^T r_{i+1})}$ $p_{i+1} = A^T r_i + \beta_i p_i$	$r_0 = b - Ax_0, p_0 = A^T r_0$ For $i = 0, 1, \ldots$ until convergence $\alpha_i = -\dfrac{(r_i, r_i)}{(p_i, p_i)}$ $x_{i+1} = x_i - \alpha_i p_i$ $r_{i+1} = r_i + \alpha_i Ap_i$ $\beta_i = \dfrac{(r_{i+1}, r_{i+1})}{(r_i, r_i)}$ $p_{i+1} = A^T r_i + \beta_i p_i$

Figure 9.3.1: *Normal Equation Conjugate Gradient Algorithms*

For nonsymmetric A, we can not obtain the solution of (9.3.1) by minimization of (9.3.4) but we can always attempt to minimize the residual. Thus, we define

$$q_2(x) = \|b - Ax\|_2^2 = (b - Ax)^T(b - Ax) = x^T A^T A x - 2b^T A x + b^T b. \quad (9.3.5)$$

Minimization of (9.3.5) by the conjugate gradient algorithm is then equivalent to applying the conjugate gradient algorithm to the normal equations (9.3.2). This is the reason for the nomenclature CGNR - the R denotes minimum residual.

Next, suppose that we attempt to minimize the error $\|x - A^{-1}b\|_2$ rather than the residual. Since

$$\|x - A^{-1}b\|_2^2 = x^T x - 2b^T A^{-T} x + b^T A^{-T} A^{-1} b, \quad (9.3.6)$$

the change of variable $x = A^T y$ converts (9.3.6) to

$$y^T A A^T y - 2b^T y + b^T A^{-T} A^{-1} b. \quad (9.3.7)$$

Minimization of (9.3.7) by the conjugate gradient algorithm is then equivalent to the CGNE algorithm of Figure 9.3.1(b); here, the E denotes minimum error.

Before continuing with the nonsymmetric case, we first consider the difference between minimizing (9.3.4) and (9.3.5) when A is symmetric positive definite. In this case, if $\hat{x}$ is the solution of $Ax = b$ and $e = x - \hat{x}$ is the error for some x, then, using $b = A\hat{x}$ and taking $c = b^T \hat{x}$, we can rewrite q_1 as

$$q_1(x) = x^T A x - 2x^T A \hat{x} + \hat{x}^T A \hat{x} = e^T A e = \|e\|_A^2. \quad (9.3.8)$$

9.3 Nonsymmetric and Nonlinear Problems

Also, by the basic relation $\mathbf{r} = A\mathbf{e}$, we can rewrite q_2 as

$$q_2(\mathbf{x}) = \|\mathbf{r}\|_2^2 = \|A\mathbf{e}\|_2^2 = \|\mathbf{e}\|_{A^2}^2. \tag{9.3.9}$$

Thus, in both cases, we are attempting to minimize $\|\mathbf{e}\|$ but in different norms, and this gives different algorithms. The *conjugate residual* algorithm minimizes (9.3.9) and may be written as shown in Figure 9.3.2. This algorithm may also be derived by applying the preconditioned conjugate gradient algorithm with preconditioning by $M = A^{-1}$ to the equation $A^2\mathbf{x} = A\mathbf{b}$ (see Exercise 9.3.4).

$\mathbf{p}_0 = \mathbf{r}_0 = \mathbf{b} - A\mathbf{x}_0$
For $i = 0, 1, \ldots$ until convergence
$\quad \alpha_i = -(\mathbf{r}_i, A\mathbf{r}_i)/(A\mathbf{p}_i, A\mathbf{p}_i)$
$\quad \mathbf{x}_{i+1} = \mathbf{x}_i - \alpha_i \mathbf{p}_i$
$\quad \mathbf{r}_{i+1} = \mathbf{r}_i + \alpha_i A\mathbf{p}_i$
$\quad \beta_i = (\mathbf{r}_{i+1}, A\mathbf{r}_{i+1})/(\mathbf{r}_i, A\mathbf{r}_i)$
$\quad \mathbf{p}_{i+1} = \mathbf{r}_{i+1} - \beta_i \mathbf{p}_i$
$\quad A\mathbf{p}_{i+1} = A\mathbf{r}_{i+1} + \beta_i A\mathbf{p}_i$

Figure 9.3.2: *The Conjugate Residual Algorithm*

In the conjugate residual algorithm of Figure 9.3.2, the last step updates $A\mathbf{p}_i$ by an axpy operation so that matrix-vector multiplication is needed only for $A\mathbf{r}_{i+1}$. Because of the preconditioning by A^{-1}, the rate of convergence of this algorithm is determined by cond(A), as in the conjugate gradient algorithm, rather than by cond$(A^T A)$ as in the normal equation approach (Exercise 9.3.4).

The Generalized Conjugate Residual Algorithm

It is tempting to consider the conjugate residual algorithm of Figure 9.3.2 for nonsymmetric matrices. One obvious problem is that β_i may not be well-defined since the denominator may be zero. This problem does not exist for nonsymmetric positive definite matrices ($\mathbf{x}^T A \mathbf{x} > 0$ for all real $\mathbf{x} \neq 0$) but there is a more fundamental problem: we lose the orthogonality of the $\mathbf{p}_i$. Indeed, the definition of β_i in Figure 9.3.2 does not even guarantee that $\mathbf{p}_{i+1}$ is $A^T A$ orthogonal to $\mathbf{p}_i$ when A is nonsymmetric. To recover this property, we can redefine β_i as

$$\beta_i = -(A\mathbf{p}_i, A\mathbf{r}_{i+1})/(A\mathbf{p}_i, A\mathbf{p}_i), \tag{9.3.10}$$

which guarantees that $(A\mathbf{p}_{i+1}, A\mathbf{p}_i) = 0$ (see Exercise 9.3.5). However, this does not imply that $\mathbf{p}_{i+1}$ is $A^T A$ orthogonal to all previous $\mathbf{p}_i$. In order to guarantee this, we must redefine the $\mathbf{p}_i$ as in Figure 9.3.3.

$$\mathbf{p}_0 = \mathbf{r}_0 = \mathbf{b} - A\mathbf{x}_0$$
For $i = 0, 1, \ldots$ until convergence
$$\alpha_i = -(\mathbf{r}_i, A\mathbf{p}_i)/(A\mathbf{p}_i, A\mathbf{p}_i)$$
$$\mathbf{x}_{i+1} = \mathbf{x}_i - \alpha_i \mathbf{p}_i$$
$$\mathbf{r}_{i+1} = \mathbf{r}_i + \alpha_i A\mathbf{p}_i$$
For $j = 0, \ldots, i$
$$\beta_j^{(i)} = -(A\mathbf{r}_{i+1}, A\mathbf{p}_i)/(A\mathbf{p}_j, A\mathbf{p}_j)$$
$$\mathbf{p}_{i+1} = \mathbf{r}_{i+1} + \sum_{j=0}^{i} \beta_j^{(i)} \mathbf{p}_j$$
$$A\mathbf{p}_{i+1} = A\mathbf{r}_{i+1} + \sum_{j=0}^{i} \beta_j^{(i)} \mathbf{p}_j$$

Figure 9.3.3: *Generalized Conjugate Residual Algorithm*

The definition of $\mathbf{p}_{i+1}$ in Figure 9.3.3 ensures that $(A\mathbf{p}_{i+1}, A\mathbf{p}_j) = 0$ for $j = 0, \ldots, i$. This, in turn, guarantees that the iterates $\mathbf{x}_i$ converge to the solution in no more than n steps, in exact arithmetic. However, it is necessary to save all $\mathbf{p}_j$ as well as to perform an increasing amount of computation as i increases. Both the storage and the computation become prohibitive and this algorithm is not viable in practice.

Restart and Truncation

There are two standard approaches to reducing the amount of work and storage in the generalized conjugate residual algorithm. The first is called *truncation* in which we retain only k prior vectors $\mathbf{p}_i, \ldots, \mathbf{p}_{i-k+1}$. We then define

$$\mathbf{p}_{i+1} = \mathbf{r}_{i+1} + \sum_{j=i-k+1}^{i} \beta_j^{(i)} \mathbf{p}_j, \qquad (9.3.11)$$

where $\beta_j^{(i)}$ is as in Figure 9.3.3. The method so modified is known as *Orthomin(k)*. If A is symmetric positive definite, Orthomin(1) is just the conjugate residual method.

The second approach is to *restart* after $k+1$ steps. Thus, $\mathbf{p}_i$ is defined as in Figure 9.3.3 for $i = 0, \ldots, k$, and then the process is restarted with $\mathbf{p}_{k+1}$. This is the *GCR(k) method*, and requires less work per iteration than Orthomin(k)

9.3 Nonsymmetric and Nonlinear Problems

since fewer orthogonalizations are performed, on average. Both truncation and restarting destroy the finite convergence property.

Note that GCR(1) ≠ Orthomin(1). However, GCR(0) = Orthomin(0), a method also called the *minimum residual algorithm*, in which no prior p_i is retained. In this case, $p_{i+1} = r_{i+1}$, and the direction vectors are the current residuals.

The Biconjugate Gradient Method

We next consider a rather different approach based upon generating two sequences of vectors $\{v_i\}$ and $\{w_i\}$ that are *biorthogonal*:

$$w_i^T v_j = 0, \quad i \neq j. \tag{9.3.12}$$

We first give one form of an algorithm based on this idea, the *biconjugate gradient* (BCG) method shown in Figure 9.3.4.

$r_0 = b - Ax_0, \beta_0 = \|r_0\|_2, v_0 = w_0 = 0, v_1 = w_1 = r_0/\beta_0, d_0 = 1$

For $i = 1, 2, \ldots$ until convergence

$$d_i = w_i^T v_i \tag{9.3.13a}$$

$$\alpha_i = (w_i^T A v_i)/d_i \tag{9.3.13b}$$

$$\beta_i = d_i/d_{i-1} \tag{9.3.13c}$$

$$v_{i+1} = Av_i - \alpha_i v_i - \beta_i v_{i-1} \tag{9.3.13d}$$

$$w_{i+1} = A^T w_i - \alpha_i w_i - \beta_i w_{i-1} \tag{9.3.13e}$$

$$y_i = \beta_0 T_i^{-1} e_1 \tag{9.3.13f}$$

$$\|r_i\|_2 = \|v_{i+1}\|_2 |e_i^T y_i| \tag{9.3.13g}$$

After convergence

$$x_k = x_0 + V_k y_k \tag{9.3.13h}$$

Figure 9.3.4: Biconjugate Gradient Algorithm

The heart of the BCG algorithm consists of the three-term recurrence relations (9.3.13d,e) for computing the v_j and w_j. The v_j are called the *right Lanczos vectors* and the w_j the *left Lanczos vectors*. The coefficients α's and

β's are defined so that the orthogonality relations (9.3.12) hold. The α_i and β_i determine the tridiagonal matrices

$$
T_i = \begin{bmatrix} \alpha_1 & \beta_2 & & \\ 1 & \alpha_2 & \ddots & \\ & \ddots & \ddots & \beta_i \\ & & 1 & \alpha_i \end{bmatrix}, \tag{9.3.14}
$$

and the vector $\mathbf{y}_i$ of (9.3.13f) is then the solution of $T_i \mathbf{y}_i = \beta_0 \mathbf{e}_1$.

We define the $n \times i$ matrices V_i and W_i by

$$V_i = (\mathbf{v}_1, \ldots, \mathbf{v}_i), \qquad W_i = (\mathbf{w}_1, \ldots, \mathbf{w}_i). \tag{9.3.15}$$

Then the orthogonality relations (9.3.12), together with (9.3.13a), can be expressed by

$$W_i^T V_i = D_i = \text{diag}(d_1, \ldots, d_i). \tag{9.3.16}$$

In terms of the matrices V_i, W_i, and T_i the recurrence relations (9.3.13d,e) may be written as

$$AV_i = V_i T_i + \mathbf{v}_{i+1} \mathbf{e}_i^T, \quad A^T W_i = W_i T_i + \mathbf{w}_{i+1} \mathbf{e}_i^T. \tag{9.3.17}$$

By (9.3.16), multiplying the first of these relations by W_i^T gives

$$W_i^T A V_i = W_i^T V_i T_i + W_i^T \mathbf{v}_{i+1} \mathbf{e}_i^T = D_i T_i \tag{9.3.18}$$

since $W_i^T \mathbf{v}_{i+1} = 0$ by (9.3.12).

Next, we explain the computation (9.3.13g). Using (9.3.13h) and (9.3.17), we have

$$\begin{aligned} \mathbf{r}_i &= \mathbf{b} - A\mathbf{x}_i = \mathbf{b} - A\mathbf{x}_0 - AV_i \mathbf{y}_i \\ &= \mathbf{r}_0 - V_i T_i \mathbf{y}_i - \mathbf{v}_{i+1} \mathbf{e}_i^T \mathbf{y}_i = -\mathbf{e}_i^T \mathbf{y}_i \mathbf{v}_{i+1} \end{aligned} \tag{9.3.19}$$

since

$$V_i T_i \mathbf{y}_i = \mathbf{v}_i \beta_0 \mathbf{e}_1 = \beta_0 \mathbf{v}_1 = \mathbf{r}_0. \tag{9.3.20}$$

As a consequence of (9.3.19), we can efficiently compute $\|\mathbf{r}_i\|_2$ in terms of $\mathbf{v}_{i+1}$ and the last component of $\mathbf{y}_i$.

Note that the approximations $\mathbf{x}_i$ to the solution are not used in the BCG algorithm. Hence, there is no need to compute them until convergence, as determined by $\|\mathbf{r}_i\|_2$. If this occurs at the kth iteration, then $\mathbf{x}_k$ is obtained in (9.3.13h) as $\mathbf{x}_0$ plus a linear combination of $\mathbf{v}_1, \ldots, \mathbf{v}_k$. The particular linear combination is obtained by the criterion that the residual $\mathbf{r}_i = \mathbf{b} - A\mathbf{x}_i$ be orthogonal to $\mathbf{w}_1, \ldots, \mathbf{w}_i$. If $V_i \mathbf{y}_i$ is the linear combination, then

$$\mathbf{r}_i = \mathbf{r}_0 - AV_i \mathbf{y}_i \tag{9.3.21}$$

9.3 Nonsymmetric and Nonlinear Problems

and the orthogonality criterion may be written as

$$W_i^T(\mathbf{r}_0 - AV_i\mathbf{y}_i) = 0$$

or

$$W_i^T AV_i \mathbf{y}_i = W_i^T \mathbf{r}_0. \quad (9.3.22)$$

By (9.3.18), $W_i^T AV_i = D_i T_i$ and since $\mathbf{r}_0 = \beta_0 \mathbf{v}_1$, by (9.3.12) and (9.3.13a), $W_i^T \mathbf{r}_0 = d_1 \beta_0 \mathbf{e}_1$. Thus, (9.3.22) is equivalent to $T_i \mathbf{y}_i = \beta_0 \mathbf{e}_1$ and this is how the prescription (9.3.13f) for $\mathbf{y}_i$ arises.

The BCG algorithm has several disadvantages. Multiplications by both A and A^T are required and the solutions of the tridiagonal systems (9.3.13f) are a bottleneck for parallel computation. Moreover, the norms $\|\mathbf{r}_i\|_2$ can sometimes behave very erratically and the method can break down. This break-down occurs if some $d_i = 0$, and the method can be unstable if $d_i \neq 0$ but very small. Potential remedies for these difficulties are given in the Supplementary Discussion.

The Arnoldi Process

In the BCG algorithm, we are in essence reducing A to a tridiagonal matrix. In the Arnoldi process, only a single sequence of vectors is generated and the matrix A is reduced to Hessenberg form rather than tridiagonal. The algorithm for generating these vectors is shown in Figure 9.3.5.

$\mathbf{v}_1$ given, $\|\mathbf{v}_1\|_2 = 1$

For $j = 1$ to n

$$h_{ij} = (A\mathbf{v}_i, \mathbf{v}_j), \quad i = 1, \ldots, j \quad (9.3.23a)$$

$$\hat{\mathbf{v}}_{j+i} = A\mathbf{v}_j - \sum_{i=1}^{j} h_{ij}\mathbf{v}_i \quad (9.3.23b)$$

$$h_{j+1,j} = \|\hat{\mathbf{v}}_{j+1}\|_2 \quad (9.3.23c)$$

$$\mathbf{v}_{j+1} = \hat{\mathbf{v}}_{j+1}/h_{j+1,j}. \quad (9.3.23d)$$

Figure 9.3.5: The Arnoldi Process

The $\mathbf{v}_j$ generated by the Arnoldi process are othonormal, as we now show by induction. Assume that $\mathbf{v}_1, \ldots, \mathbf{v}_j$ are orthonormal. Then, for $i \leq j$

$$\mathbf{v}_i^T \hat{\mathbf{v}}_{j+1} = \mathbf{v}_i^T (A\mathbf{v}_j - \sum_{l=1}^{j} h_{lj}\mathbf{v}_l) = h_{ij} - h_{ij} = 0.$$

Thus, $\hat{\mathbf{v}}_{j+1}$ is orthogonal to all previous $\mathbf{v}_i$ and (9.3.23d) normalizes $\hat{\mathbf{v}}_{j+1}$ to length 1.

Corresponding to (9.3.17), we can rewrite (9.3.23b,c,d) as

$$AV_j = V_j H_j + h_{j+1,j} \mathbf{v}_{j+1} \mathbf{e}_j^T \tag{9.3.24}$$

where V_j is the matrix with columns $\mathbf{v}_1, \ldots, \mathbf{v}_j$, and H_j is an upper Hessenberg matrix with elements h_{kl}. By the orthonormality of the $\mathbf{v}$'s, we have $V_j^T \mathbf{v}_{j+1} = 0$ and $V_j^T V_j = I$ so that (9.3.24) gives

$$V_j^T AV_j = H_j, \tag{9.3.25}$$

which corresponds to (9.3.18). This is the reduction to Hessenberg form that the Arnoldi process achieves.

The FOM, IOM and GMRES Methods

The Arnoldi process forms the basis for various methods for solving $Ax = \mathbf{b}$. The first is the *full orthogonalization method* (FOM). Analogous to (9.3.13f,h) of the BCG algorithm, at the ith stage we have H_i and V_i, and then a new approximation $\mathbf{x}_i$ to the solution is obtained from

$$H_i \mathbf{y}_i = \beta_0 \mathbf{e}_1 \tag{9.3.26}$$

$$\mathbf{x}_i = \mathbf{x}_0 + V_i \mathbf{y}_i. \tag{9.3.27}$$

Using (9.3.27) and (9.3.24), the computation (9.3.19) may be replaced by

$$\mathbf{r}_i = \mathbf{r}_0 - V_i H_i \mathbf{y}_i - h_{i+1,i} \mathbf{v}_{i+1} \mathbf{e}_i^T \mathbf{y}_i = -h_{i+1,i} \mathbf{e}_i^T \mathbf{y}_i \mathbf{v}_{i+1}. \tag{9.3.28}$$

Thus, since $\|\mathbf{v}_{i+1}\|_2 = 1$,

$$\|\mathbf{r}_i\|_2 = |h_{i+1,1} \mathbf{e}_i^T \mathbf{y}_i|, \tag{9.3.29}$$

and the $\|\mathbf{r}_i\|_2$ can be efficiently monitored for convergence without computing the iterates $\mathbf{x}_i$. As in the Generalized Conjugate Residual method, the FOM method suffers from having to retain all the $\mathbf{v}_i$ and this storage plus the computations of the Arnoldi process become prohibitively expensive. Again, one may consider a truncated version, called the *incomplete orthogonalization method* (IOM), in which only the previous k vectors $\mathbf{v}_i, \ldots, \mathbf{v}_{i-k+1}$ are retained and $\mathbf{v}_{i+1}$ is made orthogonal to these. The Hessenberg matrices H_i are then banded, with bandwidth k.

In the *generalized minimum residual* (GMRES) method, the Arnoldi process is again used to generate $\mathbf{v}_1, \mathbf{v}_2, \ldots$, and $\mathbf{x}_i$ is obtained as in (9.3.27). But the vector $\mathbf{y}_i$ is computed in a different manner. Let

$$\overline{H}_i = \begin{bmatrix} H_i \\ h_{i+1,i} \mathbf{e}_i^T \end{bmatrix} \tag{9.3.30}$$

9.3 Nonsymmetric and Nonlinear Problems

be the $(i+1) \times i$ Hessenberg matrix whose $(i+1)$st row is zero except for $h_{i+1,i}$. Then, $\mathbf{y}_i$ is defined as the solution of the minimization problem

$$\min_{\mathbf{y}} \|\overline{H}_i \mathbf{y} - \beta_0 \mathbf{e}_1\|_2. \tag{9.3.31}$$

As discussed in Section 6.3, this type of minimization problem can be solved by the QR factorization as follows. Let

$$Q_i \overline{H}_i = \overline{R}_i = \begin{bmatrix} R_i \\ 0 \end{bmatrix},$$

where Q_i is an $(i+1) \times (i+1)$ orthogonal matrix and R_i is an $i \times i$ upper triangular matrix. To obtain $\mathbf{y}_i$ we solve the system

$$R_i \mathbf{y} = \beta_0 \mathbf{q}, \tag{9.3.32}$$

where $\mathbf{q}$ is the first column of Q_i with the last element deleted. Note that since $\overline{H}_i$ is a Hessenberg matrix, the formation of Q_i may be done efficiently by Givens transformations.

To contrast this way of obtaining $\mathbf{y}_i$ with that of (9.3.26), note that the vector $\overline{H}_i \mathbf{y} - \beta_0 \mathbf{e}_1$ of (9.3.31) is

$$\begin{bmatrix} H_i \mathbf{y} - \beta_0 \mathbf{e}_1 \\ h_{i+1,i} \mathbf{e}_i^T \mathbf{y} \end{bmatrix}.$$

If (9.3.26) is used to obtain $\mathbf{y}_i$, then $H_i \mathbf{y}_i - \beta_0 \mathbf{e}_1 = 0$ and $h_{i+1,i} \mathbf{e}_i^T \mathbf{y}_i$ is determined by $\mathbf{y}_i$. On the other hand, the next iterate $\mathbf{x}_i$ is given by (9.3.27) so that

$$\mathbf{r}_i = \mathbf{r}_0 - A V_i \mathbf{y}_i = \beta_0 \mathbf{v}_1 - A V_i \mathbf{y}_i.$$

By (9.3.24) and (9.3.30)

$$A V_i \mathbf{y}_i = V_i H_i \mathbf{y}_i + h_{i+1,i} \mathbf{v}_{i+1} \mathbf{e}_i^T \mathbf{y}_i = V_{i+1} \overline{H}_i \mathbf{y}_i,$$

which gives

$$\mathbf{r}_i = \beta_0 \mathbf{v}_1 - V_{i+1} \overline{H}_i \mathbf{y}_i = V_{i+1}(\beta_0 \mathbf{e}_1 - \overline{H}_i \mathbf{y}_i).$$

Since the columns of V_{i+1} are orthonormal, we then have

$$\|\mathbf{r}_i\|_2 = \|V_{i+1}(\beta_0 \mathbf{e}_1 - \overline{H}_i \mathbf{y}_i)\|_2 = \|\beta_0 \mathbf{e}_1 - \overline{H}_i \mathbf{y}_i\|_2. \tag{9.3.33}$$

Therefore, (9.3.31) minimizes the residual at the ith stage, and in that sense is a better choice than choosing $\mathbf{y}_i$ to satisfy (9.3.26). In particular, one expects to obtain smaller residuals at each stage than in the FOM method.

As in (9.3.29), it is possible to compute the norm of the residual at each stage in an economical fashion without computing the next approximate solution $\mathbf{x}_i$. By the QR process for (9.3.31) we have from (9.3.33) that

$$\|\mathbf{r}_i\|_2 = \|Q_i(\beta_0 \mathbf{e}_1 - \overline{H}_i \mathbf{y}_i)\|_2 = \|\beta_0 \mathbf{q}_1 - \overline{R}_i \mathbf{y}_i\|_2, \tag{9.3.34}$$

where $\mathbf{q}_1$ is the first column of Q_i. Since $\mathbf{y}_i$ is the solution of the system (9.3.32), (9.3.34) reduces to

$$\|\mathbf{r}_i\|_2 = \beta_0 \times |\text{ last component of } \mathbf{q}_1|, \qquad (9.3.35)$$

and is thus easily obtained as $\mathbf{y}_i$ is computed.

The GMRES method suffers the same difficulty as the FOM method in that all $\mathbf{v}_i$ need to be retained. Consequently, it is usual to use a restarted version, denoted by GMRES(k), in which the process is restarted every k steps.

Preconditioning

The GMRES method, as well as the other methods, should usually be used with preconditioning. Since A is nonsymmetric, the preconditioning $\hat{A} = S^T A S$ in the symmetric case is replaced by

$$\hat{A} = S_1 A S_2, \qquad (9.3.36)$$

where S_1 and S_2 are nonsingular. If $S_2 = I$, then (9.3.36) is called *left preconditioning*; if $S_1 = I$, it is *right preconditioning*. If $M = (S_2 S_1)^{-1}$, then corresponding to (9.2.6), we have

$$M^{-1}A = S_2 S_1 A S_2 S_2^{-1} = S_2 \hat{A} S_2^{-1},$$

so that $M^{-1}A$ is similar to $\hat{A}$.

As in the symmetric case, one does not carry out (9.3.36) explicitly but, rather, incorporates solutions of auxiliary systems with the matrix M or the matrices S_1^{-1} and S_2^{-1} into the algorithm. For example, for right preconditioning, the matrix $\hat{A}$ of (9.3.36) is AM^{-1}. To incorporate this preconditioning into the Arnoldi process for the GMRES algorithm, the multiplication $A\mathbf{v}_i$ in (9.3.23) would be replaced by $AM^{-1}\mathbf{v}_i$. To perform this, one would first solve the system $M\mathbf{y} = \mathbf{v}_i$, and then do the multiplication $A\mathbf{y}$.

One of the most common preconditioners for nonsymmetric matrices is *incomplete LU (ILU)* factorization, which corresponds to incomplete Cholesky factorization for symmetric matrices. Figure 9.3.6 gives a pseudocode for a no-fill factorization.

In Figure 9.3.6, if $a_{ki} = 0$, then l_{ik} is set to zero and the update of the ith row of the matrix is bypassed; thus, no elements in the ith row are changed. If $a_{ki} \neq 0$, then updates in the ith row are made only to those elements that are already non-zero so that no-fill is allowed. Therefore, at the end of the ILU factorization, we have a lower triangular matrix L, with 1's on its main diagonal, and an upper triangular matrix U such that both L and U have non-zero off-diagonal elements in only the same positions as A itself. The preconditioning matrices are then $S_1 = L^{-1}, S_2 = U^{-1}$ and $M = LU$. As with incomplete Cholesky factorization, a common modification of the no-fill

9.3 Nonsymmetric and Nonlinear Problems

> For $k = 1$ to n
> $\quad$ For $i = k+1$ to n
> $\quad\quad$ If $(a_{ki} = 0)$ then $l_{ik} = 0$
> $\quad\quad$ else
> $\quad\quad\quad l_{ik} = a_{ik}/a_{kk}$
> $\quad\quad\quad$ For $j = k+1$ to n
> $\quad\quad\quad\quad$ If $(a_{ij} \neq 0)$ then
> $\quad\quad\quad\quad\quad a_{ij} = a_{ij} - l_{ik}a_{kj}$

Figure 9.3.6: *No-fill ILU Factorization*

principle is to compute l_{ik} and then set it to zero if $|l_{ik}| \leq \varepsilon$, for some threshold parameter ε. As with symmetric matrices, preconditioning of nonsymmetric matrices can also be achieved by the use of subsidiary iterative methods such as Jacobi and SSOR.

Nonlinear Problems

We complete this section by considering the nonlinear partial differential equation

$$u_{xx} + u_{yy} = f(u) \tag{9.3.37}$$

on the unit square as domain and with Dirichlet boundary conditions. Here, f is a given nonlinear function of a single variable. In Section 5.3 we treated corresponding nonlinear boundary value problems for ordinary differential equations and the same general ideas apply to (9.3.37). We discretize the partial derivatives in (9.3.37) as in Section 5.5 to obtain the discrete equations

$$-u_{i,j-1} - u_{i,j+1} - u_{i-1,j} - u_{i+1,j} - 4u_{ij} = -h^2 f(u_{ij}) \tag{9.3.38}$$

for $i, j = 1, \ldots, N$. This is a system of $n = N^2$ nonlinear equations in the unknowns u_{ij}. Using the natural row-wise ordering of the grid points, in matrix-vector form (9.3.38) is

$$\mathbf{F}(\mathbf{v}) = A\mathbf{v} + \mathbf{g}(\mathbf{v}) = 0, \tag{9.3.39}$$

where A is the Poisson matrix (5.5.18) and

$$\mathbf{g}(\mathbf{v}) = h^2(f(v_1), f(v_2), \ldots, f(v_n))^T + \mathbf{b}. \tag{9.3.40}$$

Here, $\mathbf{b}$ contains the known boundary values, and $\mathbf{v}$ is the vector of unknowns

$$\mathbf{v} = (v_1, \ldots, v_n)^T = (u_{11}, \ldots, u_{1N}, u_{2N}, \ldots, u_{2N}, \ldots, u_{NN})^T. \tag{9.3.41}$$

As in Section 5.3, we may consider applying Newton's method to (9.3.39). The Jacobian matrix is
$$\mathbf{F}'(\mathbf{v}) = A + \mathbf{g}'(\mathbf{v}), \tag{9.3.42}$$
where
$$\mathbf{g}'(\mathbf{v}) = h^2 \text{diag}(f'(v_1), \ldots, f'(v_n)). \tag{9.3.43}$$
We will assume that
$$f'(u) \geq 0 \tag{9.3.44}$$
so that $\mathbf{g}'(\mathbf{v})$ is a diagonal matrix with nonnegative elements. Since the Poisson matrix A is symmetric positive definite it follows from Exercise 2.2.17 that the $\mathbf{F}'(\mathbf{v})$ is symmetric positive definite. Moreover, $\mathbf{F}'(\mathbf{v})$ is also irreducibly diagonally dominant (Exercise 9.3.6). Some particular choices for f which arise in applications and for which (9.3.44) holds are $f(u) = e^u$ and $f(u) = u^p$, where p is odd.

In principle, we may carry out the Newton iteration

Solve $[A + \mathbf{g}'(\mathbf{v}_k)]\mathbf{y}_k = -[A\mathbf{v}_k + \mathbf{g}(\mathbf{v}_k)]$ for $\mathbf{y}_k$, (9.3.45a)

Set $\mathbf{v}_{k+1} = \mathbf{v}_k + \mathbf{y}_k$, (9.3.45b)

for $k = 0, 1, \ldots$. Under the above assumptions, $A + \mathbf{g}'(\mathbf{v}_k)$ is always nonsingular and is easy to form. However, solution of (9.3.45a) by a direct method such as Gaussian elimination has the same difficulties with fill as discussed in Section 6.1 for the Poisson matrix, especially for differential equations in three space dimensions. Therefore, it is natural to consider an iterative method for the equation (9.3.45a). Any of the previous methods (for example, SOR or multigrid) may be used but we will restrict attention to the conjugate gradient method. Thus, we will have an *outer iteration* (Newton's method) and an *inner iteration* (conjugate gradient).

Since the inner iteration will not be carried to the limit, the systems (9.3.45a) will be solved only approximately at each outer iteration. Thus, we will have only an *inexact* or *truncated* Newton iteration. It is usually not necessary to solve the systems (9.3.45a) very accurately in the early stages when $\mathbf{v}_k$ is not particularly close to the solution. However, as $\mathbf{v}_k$ approaches the solution, more accuracy in (9.3.45a) will lead to a rate of convergence more closely approximating the quadratic convergence of Newton's method. This leads to the following convergence test for the inner iteration:

$$\|\mathbf{r}_i\| \leq \min(\varepsilon, \|F(\mathbf{v}_k)\|^2), \tag{9.3.46}$$

where $\mathbf{r}_i$ is the residual of the conjugate gradient iteration applied to the Newton system (9.3.45a). The parameter ε in (9.3.46) may be chosen to be some moderately small number such as $\varepsilon = 10^{-2}$ so as to obtain modest accuracy in the solution of the systems (9.3.45a) in the early stages of the

9.3 Nonsymmetric and Nonlinear Problems

outer iteration. Then, when $\mathbf{v}_k$ is close to the solution, $\|F(\mathbf{v}_k)\|$ will be small and convergence of the outer iteration will be close to quadratic. We state the following theorem without proof.

THEOREM 9.3.1 *If the residuals of the inner iteration satisfy (9.3.46), if the function f of (9.3.37) is twice continuously differentiable, and if the iterates (9.3.45) converge, then they converge quadratically:*

$$\|\mathbf{v}_{k+1} - \hat{\mathbf{v}}\| \leq c\|\mathbf{v}_k - \hat{\mathbf{v}}\|^2,$$

where $\hat{\mathbf{v}}$ is the exact solution of (9.3.39).

The condition (9.3.46) becomes an increasingly stringent convergence test for the inner iterations as $\mathbf{v}_k$ approaches the solution. For example, if $\|F(\mathbf{v}_k)\| = 10^{-6}$, then we require that $\|\mathbf{r}_k\| \leq 10^{-12}$. We can relax the condition (9.3.46) somewhat and still obtain fast, but not necessarily quadratic, convergence in the outer iteration.

In order to solve approximately the systems (9.3.45a) on parallel or vector machines, we may use any of the techniques discussed in Sections 9.1 and 9.2. In particular, we would incorporate some type of preconditioning such as incomplete Cholesky factorization or a truncated series. Note that the coefficient matrix of (9.3.45a) will change at each outer iteration, and it would be expensive to recompute an incomplete factorization each time. Usually, an incomplete factorization done at the first outer iteration will be sufficient for subsequent iterations although an additional incomplete factorization could be done after a few outer iterations if necessary.

Supplementary Discussion and References: 9.3

1. The use of the normal equations (9.3.2) were considered by Hestenes and Stiefel [1952], and (9.3.3) by Craig [1955]. See Nachtigal, Reddy and Trefethen [1992] for a comparison of the normal equation approach with other methods.

2. Several ways of generalizing the conjugate method to nonsymmetric matrices were proposed starting in the mid 1970s. The goal was to retain as many of the desirable properties of the CG method as possible but to work with the original system rather than the normal equations. Among the early papers, Concus and Golub [1976] proposed a generalized CG method based on the three-term recurrence relation given in the Supplementary Discussion of Section 9.1. A similar method was given by Widlund [1978]. Vinsome [1976] described the Orthomin method and Fletcher [1976] revived the BCG method, which had essentially been given in Lanczos [1952].

3. Young and Jea [1980] considered two other methods, Orthodir and Orthores, in the same framework as Orthomin. Orthores was based on the three-term CG recurrence relation. Orthodir and Orthomin are generalizations of the conjugate residual method, first considered by E. Stiefel in 1955 for symmetric positive definite systems. Another generalization was given by Axelsson [1980]. A precise formulation of the Generalized Conjugate Residual method, as well as a convergence analysis, was given in Eisenstat et al. [1983]. A comprehensive survey of all of these methods up to the early 1980s is Elman [1982]. We note that for the CR method for symmetric positive definite A, Chandra [1978] showed that the CR and CG residuals and errors satisfy the relations

$$\|\mathbf{r}_i^{CR}\|_2 \leq \|\mathbf{r}_i^{CG}\|_2, \quad \|\hat{\mathbf{x}} - \mathbf{x}_i^{CG}\|_A \leq \|\hat{\mathbf{x}} - \mathbf{x}_i^{CR}\|_A.$$

Thus, CR does better at decreasing the residual, while CG does better at decreasing the A-norm of the error. Also, the 2-norms of the CR residuals are monotone decreasing whereas this is not true for the CG residuals.

4. Saad [1981] adapted the method of Arnoldi [1951] to the solution of linear systems, as discussed in the text. Saad [1982] showed how this method, as well as others, could be viewed as an *oblique projection* method. Let $V_i = (\mathbf{v}_1, \ldots, \mathbf{v}_i)$ and $W_i = (\mathbf{w}_1, \ldots, \mathbf{w}_i)$. If $\mathbf{x}_i = \mathbf{x}_0 + \mathbf{z}_i$, then the requirement that $\mathbf{r}_i$ be orthogonal to $\mathbf{w}_1, \ldots, \mathbf{w}_i$ is equivalent to

$$0 = W_i^T \mathbf{r}_i = W_i^T \mathbf{r}_0 - W_i^T A \mathbf{z}_i. \qquad (9.3.47)$$

This "obliquely projects" the equations onto span$(\mathbf{w}_1, \ldots, \mathbf{w}_i)$. If we also require that $\mathbf{z}_i$ be a linear combination of $\mathbf{v}_1, \ldots, \mathbf{v}_i$ so that $\mathbf{z}_i = V_i \mathbf{y}_i$, then the "obliquely projected" equations become

$$W_i^T A V_i \mathbf{y}_i = W_i^T \mathbf{r}_0. \qquad (9.3.48)$$

These ideas were further developed in Saad [1984], Saad and Schultz [1985], and Saad and Schultz [1986], in which the important GMRES method was developed and analyzed. Also, in the *minimum residual* method of Paige and Saunders [1975] for A symmetric but indefinite $W_i = AV_i$ in (9.3.47) so that (9.3.48) becomes

$$V_i^T A^2 V_i \mathbf{y}_i = V_i^T A \mathbf{r}_0.$$

5. The three-term recurrence relation satisfied by the CG iterates when A is symmetric positive definite no longer holds in general when A is nonsymmetric. (Faber and Manteuffel [1984] have given very restrictive conditions under which it does hold.) Consequently, it becomes necessary

9.3 Nonsymmetric and Nonlinear Problems

in methods like GMRES to orthogonalize the current direction vector against all previous directions. This becomes prohibitively expensive in both computation and storage, which is the motivation of the various restart and/or truncation strategies. Inevitably, restart or truncation methods lose the finite convergence property. This is not necessarily damaging in practice, although there is no satisfactory convergence theory and, in fact, the restarted or truncated methods may not converge.

6. The BCG algorithm of Figure 9.3.4 breaks down if the quantity $d_i = \mathbf{w}_i^T \mathbf{v}_i$ of (9.3.13a) is zero. This can happen if either $\mathbf{v}_i = 0$ or $\mathbf{w}_i = 0$, in which case the iteration can be restarted with a new $\mathbf{v}_{i+1}$ or $\mathbf{w}_{i+1}$. However, if $d_i = 0$ with neither $\mathbf{v}_i$ nor $\mathbf{w}_i = 0$, this is called a *serious breakdown* and a simple restart does not suffice. Parlett et al.[1985] attempted to remedy this problem with a "look-ahead" procedure that "jumped ahead" to the computation of $\mathbf{v}_{i+2}$ and $\mathbf{w}_{i+2}$. A somewhat different look-ahead procedure that jumps ahead an arbitrary number of steps was given by Freund et al. [1993]. The QMR algorithm of Freund and Nachtigal [1991] is based on this look-ahead procedure.

7. Another disadvantage of the BCG algorithm is that multiplications are required by both A and A^T. The conjugate gradient squared (CGS) algorithm of Sonneveld [1988] avoids the multiplications by A^T. Still another difficulty with the BCG method is a sometimes very erratic behavior of the residuals. This is also true for the CGS algorithm; moreover, the residual $\mathbf{r}_i$ computed by the algorithm may be small although the actual residual $\mathbf{b} - A\mathbf{x}_i$ is large. An algorithm that attempts to remove these problems is the BiCGSTAB method of van der Vorst [1992]; a modification of this algorithm was proposed by Gutknecht [1993]. For a review of these and other "transpose-free" algorithms, including a transpose-free version of QMR, see Freund, Golub and Nachtigal [1992]. See also Nachtigal, Reichel and Trefethen [1992] for a "hybrid" algorithm based on the GMRES method. As of now, it is not clear which of these many conjugate-gradient type algorithms for nonsymmetric systems is the "method of choice." This is an active research area and still other algorithms can be expected to emerge.

8. Theorem 9.3.1 is a special case of a more general result in Dembo et al. [1982]. For further discussion of some aspects of approximate Newton outer iterations combined with inner conjugate gradient iterations, see Averick and Ortega [1991].

EXERCISES 9.3

9.3.1. Let A be a $n \times m$ matrix and define

$$q(\mathbf{x}) = \|A\mathbf{x} - \mathbf{b}\|_2^2$$

for some vector $\mathbf{b}$. Show that the gradient equation $\nabla q(\mathbf{x}) = 0$ is (9.3.2).

9.3.2. Assume that A is $n \times n$ and nonsingular. Show that $\mathrm{cond}_2(A^T A) = \mathrm{cond}_2(AA^T) = [\mathrm{cond}_2(A)]^2$.

9.3.3. Show that the CG algorithm of Figure 9.1.2 applied to the systems (9.3.2) and (9.3.3), using $\mathbf{r} = \mathbf{b} - A\mathbf{x}$, gives the algorithms of Figure 9.3.1. (For (9.3.3), first do the algorithm in terms of the variable $\mathbf{y}$, and then make the substitutions $A^T \mathbf{y} \to \mathbf{x}, A^T \mathbf{p} \to \mathbf{p}$.)

9.3.4. Apply the PCG algorithm of Figure 9.2.1 with preconditioning matrix $M = A^{-1}$ to the system $A^2 \mathbf{x} = A\mathbf{b}$ to obtain the CR algorithm of Figure 9.3.2. Conclude that the rate of convergence of the CR algorithm depends on $\mathrm{cond}_2(A)$, not $\mathrm{cond}_2(A^2)$.

9.3.5. If β_i is defined by (9.3.10) and $\mathbf{p}_{i+1} = \mathbf{r}_{i+1} - \beta_i \mathbf{p}_i$, show that $(A\mathbf{p}_{i+1}, A\mathbf{p}_i) = 0$.

9.3.6. Show that if A is an irreducibly diagonally dominant matrix with positive main diagonal elements and D is a diagonal matrix with nonnegative elements, then $A + D$ is irreducibly diagonally dominant.

9.3.7. The *convection-diffusion* equation is

$$u_{xx} + u_{yy} + au_x + bu_y + cu = f$$

where a, b, c and f are given functions of x and y. Assuming that the domain is the unit square and Dirichlet boundary conditions are given, discretize this equation by finite difference approximations, using both central and one-sided differences for the first derivatives. Conclude that the resulting coefficient matrix is nonsymmetric. (Under what conditions is it diagonally dominant?) Discuss how to carry out the GMRES(k) algorithm on this problem for both vector and parallel machines.

Bibliography

L. Adams [1982]. *Iterative Algorithms for Large Sparse Linear Systems on Parallel Computers.* Ph.D. thesis, Applied Mathematics, University of Virginia, Charlottesville, VA.

L. Adams [1985]. M-Step Preconditioned Conjugate Gradient Methods. *SIAM J. Sci. Stat. Comput.*, 6: pp. 452–463.

L. Adams [1986]. Reordering Computations for Parallel Execution. *Commun. Appl. Numer. Math*, 2: pp.263–271.

L. Adams and H. Jordan [1985]. Is SOR Color-Blind? *SIAM J. Sci. Stat. Comput.*, 7: pp. 490–506.

L. Adams, R. LeVeque, and D. Young [1988]. Analysis of the SOR Iteration for the 9-Point Laplacian. *SIAM J. Numer. Anal.*, 25: pp. 1156–1180.

L. Adams and J. Ortega [1982]. A Multi-Color SOR Method for Parallel Computation. *Proc. Int. Conf. Parallel Processing, 1982*: pp. 53–56.

A. Aho, R. Sethi, and J. Ullman [1988]. *Compilers.* Addison-Wesley, Reading, MA.

E. Allgower and K. Georg [1990]. *Numerical Continuation Methods.* Springer-Verlag, New York.

G. Amdahl [1967]. The Validity of the Single Processor Approach to Achieving Large Scale Computing Capabilities. *AFIPS Conf. Proc.*, 30: pp. 483–485.

W. Ames [1977]. *Numerical Methods for Partial Differential Equations.* Academic Press, New York.

G. Andrews and F. Schneider [1983]. Concepts and Notations for Concurrent Programming. *Comput. Surveys*, 15: pp. 3–43.

E. Anderson, Z. Bai, C. Bischof, J. Demmel, J. Dongarra, J. DuCroz, A. Greenbaum, S. Hammarling, A. McKenney, S. Ostrouchov, and D. Sorensen [1992]. *LAPACK User's Guide.* SIAM, Philadelphia.

E. Anderson and Y. Saad [1989]. Solving Sparse Triangular Linear Systems on Parallel Computers. *Int. J. High Speed Comput.*, 1: pp. 73–96.

M. Arioli, J. Demmel, and I. Duff [1989]. Solving Sparse Linear Systems with Sparse Backward Error. *SIAM J. Mat. Anal. Appl.*, 10: pp. 165–190.

W. Arnoldi [1951]. The Principle of Minimized Iteration in the Solution of Matrix Eigenvalue Problems. *Quart. Appl. Math.*, 9: pp. 17–29.

U. Ascher, R. Matteij, and R. Russell [1988]. *Numerical Solution of Boundary Value Problems for Ordinary Differential Equations*. Prentice-Hall, Englewood Cliffs, NJ.

S. Ashby, T. Manteuffel, and P. Saylor [1990]. A Taxonomy for Conjugate Gradient Methods. *SIAM J. Numer. Anal.*, 27: pp. 856–869.

C. Ashcraft and R. Grimes [1988]. On Vectorizing Incomplete Factorization and SSOR Preconditioning. *SIAM J. Sci. Stat. Comput.*, 9: pp. 121–151.

B. Averick and J. Ortega [1991]. Solution of Nonlinear Poisson-Type Equations. *Appl. Numer. Math.*, 8: pp. 443–455.

T. Axelrod [1986]. Effects of Synchronization Barriers on Multiprocessor Performance. *Parallel Comput.*, 3: pp. 129–140.

O. Axelsson [1976]. A Class of Iterative Methods for Finite Element Equations. *Comput. Meth. Appl. Mech. Eng.*, 9: pp. 123–137.

O. Axelsson [1980]. Conjugate Gradient Type Methods for Unsymmetric and Inconsistent Systems of Linear Equations. *Lin. Alg. Appl.*, 29: pp. 1–16.

O. Axelsson and V. Barker [1984]. *Finite Element Solution of Boundary Value Problems*. Academic Press, Orlando.

O. Axelsson and G. Lindskog [1986]. On the Rate of Convergence of the Preconditioned Conjugate Gradient Method. *Numerische Mathematik*, 48: pp. 499–523.

D. Bailey, K. Lee, and H. Simon [1990]. Using Strassen's Algorithm to Accelerate the Solution of Linear Systems. *J. Supercomputing*, 4: pp. 358–371.

R. Bank and C. Douglas [1985]. An Efficient Implementation for SSOR and Incomplete Factorization Preconditionings. *Appl. Numer. Math.*, 1: pp. 489–492.

G. Baudet [1978]. Asynchronous Iterative Methods for Multiprocessors. *J. ACM*, 25: pp. 226–244.

E. Becker, G. Carey, and J. Oden [1981]. *Finite Elements, An Introduction.* Prentice-Hall, Englewood Cliffs, NJ.

M. Berger and S. Bokhari [1987]. A Partitioning Strategy for Non-Uniform Problems on Multiprocessors. *IEEE Trans. Comput.*, TC36: pp. 570–580.

C. Bischof and C. Van Loan [1987]. The WY Representation for Products of Householder Products. *SIAM J. Sci. Stat. Comput.*, 8: pp. s2 – s13.

S. Bokhari [1981]. On the Mapping Problem. *IEEE Trans. Comput.*, C-30: pp. 107–214.

A. Brandt [1977]. Multigrid Adaptive Solutions to Boundary Value Problems. *Math. Comp.*, 31: pp. 333–390.

K. Brenan, S. Campbell, and L. Petzold [1989]. *Numerical Solution of Initial-Value Problems in Differential-Algebraic Equations.* American Elsevier, New York.

W. Briggs [1987]. *A Multigrid Tutorial.* SIAM, Philadelphia.

P. Brown, G. Byrne, and A. Hindmarsh [1989]. VODE: A Variable Coefficient ODE Solver. *SIAM J. Sci. Stat. Comput.*, 10: pp. 1038–1051.

P. Businger [1971]. Monitoring the Numerical Stability of Gaussian Elimination. *Numerische Mathematik*, 16: pp. 360–361.

J. Butcher [1987]. *The Numerical Analysis of Ordinary Differential Equations.* Wiley, New York.

G. Carey and J. Oden [1984]. *Finite Elements, Computational Aspects.* Prentice-Hall, Englewood Cliffs, NJ.

T. Chan and P. Hansen [1992]. Some Applications of the Rank Revealing QR Factorization. *SIAM J. Sci. Stat. Comput.*, 13: pp. 727–741.

R. Chandra [1978]. *Conjugate Gradient Methods for Partial Differential Equations.* Ph.D. thesis, Computer Science, Yale University, New Haven, CT.

B. Char [1991]. Computer Algebra as a Toolbox for Program Generation and Manipulation, in Griewank and Corliss [1991], pp. 53-60.

B. Char, K. Geddes, G. Gonnet, B. Leong, M. Monagan, and S. Watt [1992]. *First Leaves: A Tutorial Introduction to MAPLE V.* Springer-Verlag, New York.

D. Chazan and W. Miranker [1969]. Chaotic Relaxation. *Lin. Alg. Appl.*, 2: pp. 199–222.

A. Cleary [1989]. *Algorithms for Solving Narrowly Banded Linear Systems on Parallel Computers by Direct Methods.* Ph.D. thesis, Applied Mathematics, University of Virginia, Charlottesville, VA.

P. Concus and G. Golub [1976]. A Generalized Conjugate Gradient Method for Nonsymmetric Systems of Linear Equations, in *Lecture Notes in Economics and Mathematical Systems*, R. Glowinski and J. Lions (eds), Springer-Verlag, Berlin.

P. Concus, G. Golub, and G. Meurant [1985]. Block Preconditioning for the Conjugate Gradient Method. *SIAM J. Sci. Stat. Comput.*, 6: pp. 220–252.

P. Concus, G. Golub, and D. O'Leary [1976]. A Generalized Conjugate Gradient Method for the Numerical Solution of Elliptic Partial Differential Equations, in *Sparse Matrix Computations*, J. Bunch and D. Rose (eds.), Academic Press, New York.

R. Courant and D. Hilbert [1953 , 1962]. *Methods of Mathematical Physics.* Volumes 1 and 2, Interscience, New York.

E. Craig [1955]. The N-step Iteration Procedures. *J. Math. Physics*, 34: pp. 64–73.

P. Davis and P. Rabinowitz [1984]. *Methods of Numerical Integration.* Academic Press, New York.

C. de Boor [1978]. *A Practical Guide to Splines.* Springer-Verlag, New York.

T. Dekker and W. Hoffman [1989]. Rehabilitation of the Gauss-Jordan Algorithm. *Numerische Mathematik*, 54: pp. 591–599.

R. Dembo, S. Eisenstat, and T. Steihaug [1982]. Inexact Newton Methods. *SIAM J. Numer. Anal.*, 19: pp. 400–408.

J. Dennis and J. Moré [1977]. Quasi-Newton Methods: Motivation and Theory. *SIAM Rev.*, 19: pp. 46–89.

J. Dennis and R. Schnabel [1983]. *Numerical Methods for Unconstrained Optimization and Nonlinear Equations.* Prentice-Hall, Englewood Cliffs, NJ.

J. Dongarra, J. Bunch, C. Moler, and G. Stewart [1979]. *LINPACK Users' Guide.* SIAM, Philadelphia.

J. Dongarra, J. DuCroz, S. Hammerling, and I. Duff [1990]. A Set of Level 3 Basic Linear Algebra Subprograms. *ACM Trans. Math. Softw.*, 16: pp.1–17.

J. Dongarra, I. Duff, D. Sorensen, and H. van der Vorst [1990]. *Solving Linear Systems on Vector and Shared Memory Computers.* SIAM, Philadelphia.

J. Dongarra, F. Gustavson, and A. Karp [1984]. Implementing Linear Algebra Algorithms for Dense Matrices on a Vector Pipeline Machine. *SIAM Rev.*, 26: pp. 91–112.

J. Dongarra and A. Hinds [1979]. Unrolling Loops in FORTRAN. *Software Pract. Exper.*, 9: pp. 219–229.

J. Dongarra and R. van de Geijn [1992]. *Reduction to Condensed Form for the Eigenvalue Problem on Distributed Memory Architectures.* Oak Ridge National Laboratory Report TM-12006. To appear in Parallel Computing, 1993.

P. Dubois, A. Greenbaum, and G. Rodrigue [1979]. Approximating the Inverse of a Matrix for Use in Iterative Algorithms on Vector Processors. *Computing*, 22: pp. 257–268.

I. Duff, A. Erisman, and J. Reid [1986]. *Direct Methods for Sparse Matrices.* Oxford University Press, Oxford.

I. Duff and G. Meurant [1989]. The Effect of Ordering on Preconditioned Conjugate Gradients. *BIT*, 29: pp. 635–657.

R. Earnshaw and N. Wiseman [1992]. *An Introductory Guide to Scientific Visualization.* Springer-Verlag, New York.

A. Edelman and M. Ohlrich [1991]. Editors note. *SIAM J. Mat. Anal. Appl. 12*, no. 3.

S. Eisenstat [1981]. Efficient Implementation of a Class of Conjugate Gradient Methods. *SIAM J. Sci. Stat. Comput.*, 2: pp. 1–4.

S. Eisenstat, H. Elman, and M. Schultz [1983]. Variational Iterative Methods for Nonsymmetric Systems of Equations. *SIAM J. Numer. Anal.*, 20: pp. 345–357.

S. Eisenstat, M. Heath, C. Henkel, and C. Romine [1988]. Modified Cyclic Algorithms for Solving Triangular Systems on Distributed Memory Multiprocessors. *SIAM J. Sci. Stat. Comput*, 9: pp. 589–600.

H. Elman [1982]. *Iterative Methods for Large, Sparse, Nonsymmetric Systems of Linear Equations.* Ph.D. thesis, Computer Science, Yale University, New Haven, CT.

H. Elman and G. Golub [1990, 91]. Iterative Methods for Cyclically Reduced Non Self Adjoint Linear Systems, I and II. *Math. Comp.*, 54,56: pp. 671–700, 215–242.

R. Elmasri and S. Navathe [1989]. *Fundamentals of Database Systems.* Benjamin/Cummings Publishing Co., Menlo Park, CA.

V. Faber and T. Manteuffel [1984]. Necessary and Sufficient Conditions for the Existence of a Conjugate Gradient Method. *SIAM J. Numer. Anal.*, 21: pp. 352–362.

C. Fischer and R. LeBlanc Jr. [1988]. *Crafting a Compiler.* Benjamin/Cummings Publishing Co., Menlo Park, CA.

R. Fletcher [1976]. Conjugate Gradient Methods for Indefinite Systems, in *Lecture Notes in Mathematics*, G. Watson (ed), Numerical Analysis Dundee Conference.

K. Fong and T. Jordan [1977]. *Some Linear Algebraic Algorithms and Their Performance on the CRAY-1.* Los Alamos National Laboratory Report No. LA-6774.

G. Forsythe and W. Wasow [1960]. *Finite Difference Methods for Partial Differential Equations.* Wiley, New York.

G. Fox, M. Johnson, G. Lyzenga, S. Otto, J. Salmon, and D. Walker [1988]. *Solving Problems on Concurrent Processors.* Prentice-Hall, Englewood Cliffs, New Jersey.

R. Freund, G. Golub, and N. Nachtigal [1992]. Iterative Solution of Linear Systems. *Acta Numerica*, 1: pp. 57–100.

R. Freund, M. Gutknecht, and N. Nachtigal [1993]. An Implementation of the Look-Ahead Lanczos Algorithm for Non-Hermitian Matrices. *SIAM J. Sci. Stat. Comput.*, 14: pp. 137–158.

R. Freund and N. Nachtigal [1991]. QMR: A Quasi-minimal Residual Method for Non-Hermitian Linear Systems. *Numerische Mathematik*, 60: pp. 315–339.

R. Friedhoff and W. Benzon [1989]. *Visualization.* Henry N. Abrams, Inc., New York.

P. Garabedian [1986]. *Partial Differential Equations; Second Edition.* Chelsea Publishing Co., New York.

B. Garbow, J. Boyle, J. Dongarra, and C. Moler [1977]. *Matrix Eigensystem Routines - EISPACK Guide Extension.* Springer-Verlag, New York.

C. W. Gear [1971]. *Numerical Initial Value Problems in Ordinary Differential Equations.* Prentice-Hall, Englewood Cliffs, NJ.

Bibliography

A. Geist and M. Heath [1986]. Matrix Factorization on a Hypercube Multiprocessor. In Heath [1986], pp. 161-180.

A. George and J. Liu [1981]. *Computer Solution of Large Sparse Positive Definite Systems.* Prentice-Hall, Englewood Cliffs, NJ.

G. Golub and D. O'Leary [1989]. Some History of the Conjugate Gradient and Lanczos Algorithms: 1948-1976. *SIAM Rev.*, 31: pp. 50–102.

G. Golub and J. Ortega [1991]. *Scientific Computing and Differential Equations.* Academic Press, New York.

G. Golub and C. Van Loan [1989]. *Matrix Computations, Second Edition.* Johns Hopkins Press, Baltimore.

N. Gould [1991]. On Growth in Gaussian Elimination with Complete Pivoting. *SIAM J. Mat. Anal. Appl.*, 12: pp. 354–361.

A. Greenbaum and Z. Strakos [1992]. Predicting the Behavior of Finite Precision Lanczos and Conjugate Gradient Computations. *SIAM J. Mat. Anal. Appl.*, 13: pp. 121–137.

A. Griewank and C. Corliss (eds)[1991]. *Automatic Differentiation of Algorithms.* SIAM, Philadelphia.

R. Grossman (ed) [1989]. *Symbolic Computation: Application to Scientific Computing.* SIAM, Philadelphia.

J. Gustafson, G. Montry, and R. Benner [1988]. Development of Parallel Methods for a 1024-Processor Hypercube. *SIAM J. Sci. Stat. Comput.*, 9: pp. 609–638.

I. Gustafsson [1978]. A Class of First Order Factorization Methods. *BIT*, 18: pp. 142–156.

M. Gutknecht [1993]. Variants of BiCGSTAB for Matrices with Complex Spectrum. *SIAM J. Sci. Stat. Comput.*, to appear.

R. Haberman [1983]. *Elementary Applied Partial Differential Equations.* Prentice-Hall, Englewood Cliffs, NJ.

W. Hackbush [1985]. *Multigrid Methods with Applications.* Springer-Verlag, New York.

L. Hageman and D. Young [1981]. *Applied Iterative Methods.* Academic Press, New York.

W. Hager [1989]. Updating the Inverse of a Matrix. *SIAM Rev.*, 31: pp. 221–239.

E. Hairer, C. Lubich, and M. Roche [1989]. *The Solution of Differential-Algebraic Systems by Runge-Kutta Methods*. Springer-Verlag, New York.

E. Hairer, S. Norsett, and G. Wanner [1987]. *Solving Ordinary Differential Equations: I. Non-Stiff Problems*. Springer-Verlag, New York.

C. Hall and T. Porsching [1990]. *Numerical Analysis of Partial Differential Equations*. Prentice-Hall, Englewood Cliffs, NJ.

L. Hayes [1978]. *Timing Analysis of Standard Iterative Methods on a Pipeline Computer*. Report CNA-136, Center for Numerical Analysis, University of Texas, Austin, TX.

M. Heath (ed) [1986]. *Hypercube Multiprocessors 1986*. SIAM, Philadelphia.

M. Heath (ed) [1987]. *Hypercube Multiprocessors 1987*. SIAM, Philadelphia.

D. Heller [1976]. Some Aspects of the Cyclic Reduction Algorithm for Block Tridiagonal Linear Systems. *SIAM J. Numer. Anal.*, 13: pp. 484–496.

J. Hennessy and J. Patterson [1990]. *Computer Architecture: A Quantitative Approach*. Morgan Kaufman Publishers, Inc., San Mateo, CA.

P. Henrici [1962]. *Discrete Variable Methods in Ordinary Differential Equations*. Wiley, New York.

M. Hestenes [1956]. The Conjugate Gradient Method for Solving Linear Systems. *Proc. Sixth Symp. Appl. Math*, McGraw-Hill, New York, pp. 83-102.

M. Hestenes and E. Stiefel [1952]. Methods of Conjugate Gradients for Solving Linear Systems. *Journal of Research of the National Bureau of Standards*, 49: pp. 409–436.

N. Higham [1990]. Exploiting Fast Matrix Multiplication within the Level 3 BLAS. *ACM Trans. Math Softw.*, 16: pp. 352–368.

A. Hindmarsh [1983]. ODEPACK, A Systematized Collection of ODE Solvers, in *Scientific Computing*, R. Stepleman (ed.), North Holland, Amsterdam.

R. Hockney [1965]. A Fast Direct Solution of Poisson's Equation Using Fourier Analysis. *J. ACM*, 12: pp. 95–113.

R. Hockney [1970]. The Potential Calculation and Some Applications. *Meth. Comput. Phys.*, 9: pp. 135–211.

R. Hockney and C. Jesshope [1988]. *Parallel Computers 2*. Adam Hilger, Bristol and Philadelphia.

R. Horn and C. Johnson [1985]. *Matrix Analysis.* Cambridge University Press, New York.

R. Horn and C. Johnson [1991]. *Topics in Matrix Analysis.* Cambridge University Press, New York.

A. Householder [1964]. *The Theory of Matrices in Numerical Analysis.* Ginn (Blaisdell), Boston.

I. Ipsen, Y. Saad, and M. Schultz [1986]. Complexity of Dense Linear Systems Solution on a Multiprocessor Ring. *Lin. Alg. Appl.*, 77: pp. 205–239.

E. Isaacson and H. Keller [1966]. *Analysis of Numerical Methods.* Wiley, New York.

C. Johnson [1987]. *Numerical Solution of Partial Differential Equations by the Finite Element Method.* Cambridge University Press, New York.

O. Johnson, C. Micchelli, and G. Paul [1983]. Polynomial Preconditioners for Conjugate Gradient Calculations. *SIAM J. Numer. Anal.*, 20: pp. 362–376.

L. Johnsson [1985]. Solving Narrow Banded Systems on Ensemble Architectures. *ACM Trans. Math. Softw.*, 11: pp. 271–288.

L. Johnsson [1987]. Solving Tridiagonal Systems on Ensemble Architectures. *SIAM J. Sci. Stat. Comput.*, 8: pp. 354–392.

H. Jordan [1986]. Structuring Parallel Algorithms in an MIMD, Shared Memory Environment. *Parallel Comput.*, 3: pp. 93–110.

J. Keener [1988]. *Principles of Applied Mathematics.* Addison-Wesley Publishing Co., Reading, MA.

D. Kershaw [1978]. The Incomplete Choleski-Conjugate Gradient Method for the Iterative Solution of Systems of Linear Equations. *J. Comp. Phys.*, 26: pp. 43–65.

D. Keyes and W. Gropp [1987]. A Comparison of Domain Decomposition Techniques for Elliptic Partial Differential Equations and their Parallel Implementation. *SIAM J. Sci. Stat. Comput.*, 8: pp. s166–s202.

D. Kincaid, T. Oppe, and D. Young [1986]. Vector Computations for Sparse Linear Systems. *SIAM J. Alg. Disc. Meth.*, 7: pp. 99–112.

D. Kuck [1976]. Parallel Processing of Ordinary Programs, *Advances in Computers 15.* Academic Press, New York.

J. Lambiotte [1975]. *The Solution of Linear Systems of Equations on a Vector Computer*. Ph.D. thesis, Applied Mathematics, University of Virginia, Charlottesville, VA.

J. Lambiotte and R. Voigt [1975]. The Solution of Tridiagonal Linear Systems on the CDC STAR-100 Computer. *ACM Trans. Math. Softw.*, 1: pp. 308–329.

P. Lancaster and M. Tismenetsky [1985]. *The Theory of Matrices*. Academic Press, New York.

C. Lanczos [1952]. Solution of Systems of Linear Equations by Minimized Iterations. *Journal of Research of the National Bureau of Standards*, 49: pp.33–53.

D. Lawrie and A. Sameh [1984]. The Computation and Communication Complexity of a Parallel Banded System Solver. *ACM Trans. Math. Softw.*, 10: pp. 185–195.

C. Lawson and R. Hanson [1974]. *Solving Least Squares Problems*. Prentice-Hall, Englewood Cliffs, NJ.

N. Madsen, G. Rodrigue, and J.Karush [1976]. Matrix Multiplication by Diagonals on a Vector/Parallel Processor. *Inf. Proc. Lett.*, 5: pp. 41–45.

L. Mansfield [1991]. Damped Jacobi Preconditioning and Coarse Grid Deflation for Conjugate Gradient Iteration on Parallel Computers. *SIAM J. Sci. Stat. Comput.*, 12: pp. 1314–1323.

T. Manteuffel [1980]. An Incomplete Factorization Technique for Positive Definite Linear Systems. *Math. Comp.*, 34: pp. 473–497.

B. Mattingly, C. Meyer, and J. Ortega [1989]. Orthogonal Reduction on Vector Computers. *SIAM J. Sci. Stat. Comput.*, 10: pp. 372–381.

O. McBryan and E. van de Velde [1985]. Parallel Algorithms for Elliptic Equations. *Commun. Pure Appl. Math*, 38: pp. 769–795.

U. Meier [1985]. A Parallel Partition Method for Solving Banded Systems of Linear Equations. *Parallel Comput.*, 2: pp. 33–43.

J. Meijerink and H. van der Vorst [1977]. An Iterative Solution for Linear Systems of Which the Coefficient Matrix is a Symmetric M-Matrix. *Math. Comp.*, 31: pp. 148–162.

R. Melhem [1987]. Determination of Stripe Structures for Finite Element Matrices. *SIAM J. Numer. Anal.*, 24: pp. 1419–1433.

R. Mendez (ed.) [1990]. *Visualization in Supercomputing.* Springer Verlag, New York.

C. Moler [1972]. Matrix Computations with Fortran and Paging. *Commun. ACM*, 15: pp. 268–270.

C. Moler [1986]. Matrix Computation on Distributed Memory Multiprocessors, in Heath [1986], pp. 181-195.

R. Morison and S. Otto [1987]. The Scattered Decomposition for Finite Elements. *J. Sci. Comput.*, 2: pp. 59–76.

N. Nachtigal, S. Reddy, and L. Trefethen [1992]. How Fast Are Nonsymmetric Matrix Iterations? *SIAM J. Mat. Anal. Appl.*, 13: pp. 778–795.

N. Nachtigal, L. Reichel, and L. Trefethen [1992]. A Hybrid GMRES Algorithm for Nonsymmetric Linear Systems. *SIAM J. Mat. Anal. Appl.*, 13: pp. 796–825.

W. Newman and R. Sproul [1979]. *Principles of Interactive Computer Graphics, Second Edition.* McGraw-Hill, New York.

D. O'Leary [1984]. Ordering Schemes for Parallel Processing of Certain Mesh Problems. *SIAM J. Sci. Stat. Comput.*, 5: pp. 620–632.

D. O'Leary [1987]. Parallel Implementation of the Block Conjugate Gradient Algorithm. *Parallel Comput.*, 5: pp. 127–140.

D. O'Leary and G. Stewart [1985]. Data-Flow Algorithms for Parallel Matrix Computations. *Comm. ACM*, 28: pp. 840–853.

D. O'Leary and R. White [1985]. Multi-splittings of Matrices and Parallel Solution of Linear Systems. *SIAM J. Alg. Disc. Meth.*, 6: pp. 630–649.

J. Ortega [1987]. *Matrix Theory: A Second Course.* Plenum Press, New York.

J. Ortega [1988a]. *Introduction to Parallel and Vector Solution of Linear Systems.* Plenum Press, New York.

J. Ortega [1988b]. The ijk Forms of Factorization Methods I. Vector Computers. *Parallel Comput.*, 7: pp. 135–147.

J. Ortega [1988c]. Efficient Implementations of Certain Iterative Methods. *SIAM J. Sci. Stat. Comput.*, 9: pp. 882–891.

J. Ortega [1990]. *Numerical Analysis: A Second Course.* (Reprint of 1972 original.) SIAM, Philadelphia.

J. Ortega [1991]. Orderings for Conjugate Gradient Preconditionings. *SIAM J. Sci. Stat. Comput.*, 1: pp. 565–582.

J. Ortega and W. Rheinboldt [1970]. *Iterative Solution of Nonlinear Equations in Several Variables*. Academic Press, New York.

J. Ortega and C. Romine [1988]. The ijk Forms of Factorization Methods II. Parallel Computers. *Parallel Comput.*, 7: pp. 149–162.

J. Ortega and R. Voigt [1985]. Solution of Partial Differential Equations on Vector and Parallel Computers. *SIAM Rev.*, 27: pp. 149–240.

C. Paige and M. Saunders [1975]. Solution of Sparse Indefinite Systems of Linear Equations. *SIAM J. Numer. Anal.*, 12: pp. 617–629.

B. Parlett [1980]. *The Symmetric Eigenvalue Problem*. Prentice-Hall, Englewood Cliffs, NJ.

B. Parlett, D. Taylor, and Z. Liu [1985]. A Look-Ahead Lanczos Algorithm for Unsymmetric Matrices. *Math. Comp.*, 44: pp. 105–124.

S. Parter and M. Steuerwalt [1980]. On k-line and $k \times k$ Block Iterative Schemes for a Problem Arising in 3-D Elliptic Difference Equations. *SIAM J. Numer. Anal.*, 17: pp. 823–839.

S. Parter and M. Steuerwalt [1982]. Block Iterative Methods for Elliptic and Parabolic Difference Equations. *SIAM J. Numer. Anal.*, 19: pp. 1173–1195.

G. Peters and J. Wilkinson [1975]. On the Stability of Gauss-Jordan Elimination with Pivoting. *Comm. ACM*, 18: pp. 20–24.

J. Peterson and J. Silberschatz [1985]. *Operating Systems Concepts, Second Edition*. Addison-Wesley Publishing Co., Reading, MA.

E. Poole and J. Ortega [1987]. Multicolor ICCG Methods for Vector Computers. *SIAM J. Numer. Anal.*, 24: pp. 1394–1418.

T. Pratt [1984]. *Programming Languages*. Prentice-Hall, Englewood Cliffs, NJ.

P. Prenter [1975]. *Splines and Variational Methods*. Wiley, New York.

C. Puglisi [1992]. Modification of the Householder Method Based on the Compact WY Representation. *SIAM J. Sci. Stat. Comput.*, 13: pp. 723–726.

G. Rayna [1987]. *Software for Algebraic Computation*. Springer-Verlag, New York.

J. Reid [1971]. On the Method of Conjugate Gradients for the Solution of Large Sparse Systems of Linear Equations. *Proc. Conf. Large Sparse Sets of Linear Equations*, Academic Press, New York.

R. Richtmyer and K. Morton [1967]. *Difference Methods for Initial Value Problems*. Interscience-Wiley, New York.

W. Ronsch [1984]. Stability Aspects in Using Parallel Algorithms. *Parallel Comput.*, 1: pp. 75–98.

S. Rubinow [1975]. *Introduction to Mathematical Biology*. Wiley-Interscience, New York.

Y. Saad [1981]. Krylov Subspace Methods for Solving Large Unsymmetric Linear Systems. *Math. Comp.*, 37: pp. 105–126.

Y. Saad [1982]. The Lanczos Biorthogonalization Algorithm and Other Oblique Projection Methods for Solving Large Unsymmetric Systems. *SIAM J. Numer. Anal.*, 19: pp. 485–506.

Y. Saad [1984]. Practical Use of Some Krylov Subspace Methods for Solving Indefinite and Unsymmetric Linear Systems. *SIAM J. Sci. Stat. Comput.*, 5: pp. 203–228.

Y. Saad [1985]. Practical Use of Polynomial Preconditionings for the Conjugate Gradient Method. *SIAM J. Sci. Stat. Comput.*, 6: pp. 865–882.

Y. Saad [1989]. Krylov Subspace Methods on Supercomputers. *SIAM J. Sci. Stat. Comput.*, 10: pp. 1200–1232.

Y. Saad and M. Schultz [1985]. Conjugate Gradient-Like Algorithms for Solving Nonsymmetric Linear Systems. *Math. Comp.*, 44: pp. 417–424.

Y. Saad and M. Schultz [1986]. GMRES: A Generalized Minimal Residual Algorithm for Solving Nonsymmetric Linear Systems. *SIAM J. Sci. Stat. Comput.*, 7: pp. 856–869.

Y. Saad and M. Schultz [1988]. Topological Properties of Hypercubes. *IEEE Trans. Comput.*, C-37: pp. 867–872.

Y. Saad and M. Schultz [1989a]. Data Communication in Hypercubes. *J. Dist. Parallel Comput.*, 6: pp. 115–135.

Y. Saad and M. Schultz [1989b]. Data Communication in Parallel Architectures. *Parallel Comput.*, 11: pp. 131–150.

W. Schiesser [1991]. *The Numerical Method of Lines: Integration of Partial Differential Equations*. Academic Press, New York.

R. Schreiber and W. Tang [1982]. Vectorizing the Conjugate Gradient Method. *Proc. Symposium CYBER 205 Applications*, Fort Collins, Colorado.

R. Schreiber and C. Van Loan [1989]. A Storage-Efficient WY Representation for Products of Householder Transformations. *SIAM J. Sci. Stat. Comput.*, 10: pp. 52–57.

R. Sethi [1989]. *Programming Languages*. Addison-Wesley, Reading, MA.

R. Skeel [1980]. Iterative Refinement Implies Numerical Stability for Gaussian Elimination. *Math. Comp.*, 35: pp. 817–832.

P. Sonneveld [1989]. CGS, A Fast Lanczos-type Solver for Nonsymmetric Linear Systems. *SIAM J. Sci. Stat. Comput.*, 10: pp. 36–52.

D. Sorensen [1985]. Analysis of Pairwise Pivoting in Gaussian Elimination. *IEEE Trans. Comput.*, C-34: pp. 274–278.

G. W. Stewart [1973]. *Introduction to Matrix Computations*. Academic Press, New York.

G. W. Stewart [1974]. Modifying Pivot Elements in Gaussian Elimination. *Math. Comp.*, 28: pp. 527–542.

G. W. Stewart and J.-G. Sun [1990]. *Matrix Perturbation Theory*. Academic Press, New York.

H. Stone [1973]. An Efficient Parallel Algorithm for the Solution of a Tridiagonal Linear System of Equations. *J. ACM*, 20: pp. 27–38.

H. Stone [1975]. Parallel Tridiagonal Equation Solvers. *ACM Trans. Math. Softw.*, 1: pp. 289–307.

H. Stone [1990]. *High-Performance Computer Architecture, Second Edition*. Addison-Wesley, Reading, MA.

G. Strang and G. Fix [1973]. *An Analysis of the Finite Element Method*. Prentice-Hall, Englewood Cliffs, NJ.

V. Strassen [1969]. Gaussian Elimination Is Not Optimal. *Numerische Mathematik*, 13: pp. 354–356.

J. Strikwerda [1989]. *Finite Difference Schemes and Partial Differential Equations*. Wadsworth Pub. Co., Belmont, CA.

A. Stroud [1971]. *Approximate Calculation of Multiple Integrals*. Prentice-Hall, Englewood Cliffs, NJ.

P. Swarztrauber and R. Sweet [1989]. Vector and Parallel Methods for the Direct Solution of Poisson's Equation. *J. Comp. Appl. Math.*, 27: pp. 241–263.

Symbolics [1987]. *Macsyma User's Guide*. Symbolics, Inc., Cambridge, MA.

J. Traub [1964]. *Iterative Methods for the Solution of Equations*. Prentice-Hall, Englewood Cliffs, NJ.

A. van der Sluis and H. van der Vorst [1986]. The Rate of Convergence of Conjugate Gradients. *Numerische Mathematik*, 48: pp. 543–560.

H. van der Vorst [1982]. A Vectorizable Variant of Some ICCG Methods. *SIAM J. Sci. Stat. Comput.*, 3: pp. 350–356.

H. van der Vorst [1989a]. High Performance Preconditioning. *SIAM J. Sci. Statist. Comput.*, 10: pp. 1174–1185.

H. van der Vorst [1989b]. ICCG and Related Methods for 3D Problems on Vector Computers. *Comput. Phys. Comm.*, 53: pp. 223–235.

H. van der Vorst [1990]. The Convergence Behavior of Preconditioned CG and CG-S in the Presence of Rounding Errors, in *Lecture Notes in Mathematics*, 1457, O. Axelsson and L. Kolotilina (eds), Springer-Verlag, Berlin, pp. 126-136.

H. van der Vorst [1992]. BiCGSTAB: A Fast and Smoothly Converging Variant of Bi-CG for the Solution of Nonsymmetric Linear Systems. *SIAM J. Sci. Stat. Comput.*, 13: pp. 631–644.

C. Van Loan [1991]. *Computational Frameworks for the Fast Fourier Transform*. SIAM, Philadelphia.

R. Varga [1960]. Factorization and Normalized Iterative Methods in Boundary Problems, in *Differential Equations*. R. Langer (ed), University of Wisconsin Press, Madison.

R. Varga [1962]. *Matrix Iterative Analysis*. Prentice-Hall, Englewood Cliffs, New Jersey.

P. Vinsome [1976]. ORTHOMIN, An Iterative Method for Solving Sparse Sets of Simultaneous Linear Equations. In *Proc. Fourth Symposium Reservoir Simulation*, Society of Petroleum Engineers of AIME, pp. 149-159.

G. Wahba [1990]. *Spline Models for Observational Data*. SIAM, Philadelphia.

H. Wang [1981]. A Parallel Method for Tridiagonal Systems. *ACM Trans. Math. Softw.*, 7: pp. 170–183.

W. Ware [1973]. The Ultimate Computer. *IEEE Spectrum*, 10(3): pp. 89–91.

O. Widlund [1978]. A Lanczos Method for a Class of Non-Symmetric Systems of Linear Equations. *SIAM J. Numer. Anal.*, 15: pp. 801–812.

J. Wilkinson [1961]. Error Analysis of Direct Methods of Matrix Inversion. *J. ACM*, 10: pp. 281–330.

J. Wilkinson [1963]. *Rounding Errors in Algebraic Processes*. Prentice-Hall, Englewood Cliffs, NJ.

J. Wilkinson [1965]. *The Algebraic Eigenvalue Problem*. Oxford University Press, New York.

S. Wolfram [1988]. *A System for Doing Mathematics by Computer*. Addison-Wesley, Redwood City, CA.

D. Young [1950]. *Iterative Methods for Solving Partial Difference Equations of Elliptic Type*. Ph.D. thesis, Mathematics, Harvard University.

D. Young [1954]. Iterative Methods for Solving Partial Difference Equations of Elliptic Type. *Trans. Amer. Math. Soc.*, 76: pp. 92–111.

D. Young [1971]. *Iterative Solution of Large Linear Systems*. Academic Press, New York.

D. Young and R. Gregory [1990]. *A Survey of Numerical Mathematics*. Chelsea Publishing Co., New York.

D. Young and K. Jea [1980]. Generalized Conjugate Gradient Acceleration of Nonsymmetrizable Iterative Methods. *Lin. Alg. Appl.*, 34: pp. 159–194.

D. Young and T.-Z. Mai [1990]. The Search for Omega, in *Iterative Methods for Large Linear Systems*, D. Kincaid and L. Hayes (eds.), Academic Press, New York.

Author Index

A

Adams, L., 351, 393, 413
Aho, A., 13, 413
Allgower, E., 130, 413
Amdahl, G., 70, 413
Ames, W., 199, 413
Anderson, E., 14, 87, 230, 301, 413, 414
Andrews, G., 70, 413
Arioli, M., 253, 414
Arnoldi, W., 410, 414
Ascher, U., 186, 188, 414
Ashby, S., 414
Ashcraft, C., 352, 394, 414
Averick, B., 411, 414
Axelrod, T., 70, 414
Axelsson, O., 212, 379, 393, 410, 414

B

Bai, Z., 413
Bailey, D., 70, 88, 414
Bank, R., 393, 414
Barker, V., 212, 414
Bartlett, M., 231
Baudet, G., 351, 414
Becker, E., 212, 415
Benner, R., 419
Benzon, W., 13, 418
Berger, M., 350, 415
Bischof, C., 301, 413, 414
Bokhari, S., 351, 415
Boyle, J., 418
Brandt, A., 415

Brenan, K., 170, 415
Briggs, W., 352, 369, 415
Brown, P., 170, 415
Bunch, J., 14, 230, 416
Businger, P., 253, 415
Butcher, J., 166, 167, 168, 170, 415
Byrne, G., 170, 415

C

Campbell, S., 170, 415
Carey, G., 212, 415
Chan, T., 270, 415
Chandra, R., 410, 415
Char, B., 13, 415
Chazan, D., 351, 415
Cleary, A., 318, 416
Concus, P., 379, 393, 395, 409, 416
Corliss, C., 130, 419
Courant, R., 199, 416
Craig, E., 409, 416

D

Dahlquist, G., 169
Davis, P., 149, 416
De Boor, C., 103, 416
Dekker, T., 292, 416
Dembo, R., 411, 416
Demmel, J., 413, 414
Dennis, J., 130, 378, 416
Dongarra, J., 14, 62, 69, 70, 87, 88,
    230, 300, 301, 395, 413, 416, 417, 418
Douglas, C., 393, 414

Dubois, P., 393, 417
Du Croz, J., 413, 416
Duff, I., 14, 62, 88, 231, 301, 394, 395, 414, 416, 417

E

Earnshaw, R., 13, 417
Edelman, A., 252, 417
Eisenstat, S., 393, 395, 410, 416, 417
Elman, H., 352, 395, 410, 417
Elmasri, R., 13, 418
Erisman, A., 417

F

Faber, V., 410, 418
Fischer, C., 13, 418
Fletcher, R., 409, 418
Fix, G., 212, 426
Fong, K., 300, 418
Forsythe, G., 212, 418
Fox, G., 418
Freidhoff, R., 13, 418
Freund, R., 411, 418

G

Garabedian, P., 199, 418
Garbow, B., 14, 418
Gear, W., 166, 167, 170, 418
Geddes, K., 415
Geist, A., 293, 419
Georg, K., 130, 413
George, A., 231, 419
Golub, G., 46, 104, 115, 230, 252, 253, 269, 270, 318, 352, 379, 380, 393, 395, 409, 411, 416, 417, 418, 419
Gonnet, G., 415
Gould, N., 252, 419
Greenbaum, A., 379, 413, 417, 419

Gregory, R., 103, 428
Griewank, A., 130, 419
Grimes, R., 352, 394. 414
Gropp, W., 395, 421
Grossman, R., 419
Gustafson, J., 70, 419
Gustafsson, I., 394, 419
Gustavson, F., 87, 417
Gutknecht, M., 411, 418, 419

H

Haberman, R., 199, 419
Hackbrush, W., 352, 369, 419
Hageman, L., 330, 379, 394, 419
Hager, W., 230, 419
Hairer, E., 166, 170, 420
Hall, C., 199, 212, 420
Hammarling, S., 413. 416
Hansen, P., 270, 415
Hanson, R., 115, 269, 422
Hayes, L., 352, 420
Heath, M., 293, 417, 419, 420
Helier, D., 318, 420
Henkel, C., 417
Hennesy, J., 13, 420
Henrici, P. 166, 167, 169, 420
Hestenes, M., 378, 379, 393, 409, 420
Higham, N., 88, 420
Hilbert, D., 199, 416
Hindmarsh, A., 167, 170, 415, 420
Hinds, A., 70, 88, 417
Hockney, R., 62, 270, 318, 420
Hoffman, W., 292, 416
Horn, R., 46, 212, 421
Householder, A., 421

I

Ipsen, I., 293, 421
Isaacson, E., 199, 421

Author Index

J

Jea, K., 410, 428
Jesshope, C., 44, 62, 270, 318, 420
Johnson, C., 212, 421
Johnson, C., 46, 212, 421
Johnson, O., 393, 421
Johnson, M., 418
Johnsson, L., 318, 421
Jordan, H., 70, 351, 413, 421
Jordan, T., 300, 418

K

Karp, A., 87, 417
Karush, J., 422
Keener, J., 199, 421
Keller, H., 199, 421
Kershaw, D., 394, 421
Keyes, D., 395, 421
Kincaid, D., 421
Kuck, D., 421

L

Lambiotte, J., 318, 351, 422
Lancaster, P., 46, 422
Lanczos, C., 409, 422
Lawrie, D., 318, 422
Lawson, C., 115, 269, 422
LeBlanc, T., 13, 418
Lee, K., 414
Leong, B., 415
LeVeque, R., 413
Lindskog, G., 379, 414
Liu, J., 231, 419
Liu, Z., 424
Lubich, C., 170, 420
Lyzenga, G., 418

M

Madsen, N., 87, 422
Mai, T-Z., 330, 428
Mansfield, L., 395, 422
Manteuffel, T., 394, 410, 414, 418, 422
Mattheij, R., 414
Mattingly, B., 301, 422
McBryan, O., 422
McKenney, A., 413
Meier, U., 318, 422
Meijerink, J., 393, 422
Melhem, R., 87, 422
Mendez, R., 13, 423
Meurant, G., 394, 395, 416, 417
Meyer, C., 422
Micchelli, C., 421
Miranker, W., 351, 415
Moler, C., 70, 300, 416, 418, 423
Monogan, M., 415
Montry, G., 419
Moré, J., 416
Morison, R., 351, 423
Morrison, W., 230
Morton, K., 199, 212, 425

N

Nachtigal, S., 409, 411, 418, 423
Navathe, S., 13, 418
Newman, W., 13, 423
Norsett, S., 166, 420

O

Oden, J., 212, 415
Ohlrich, M., 252, 417
O'Leary, D., 293, 351, 379, 393, 416, 410, 423
Oppe, T., 421

Ortega, J., 46, 47, 103, 104, 130, 169, 187, 212, 252, 293, 300, 318, 330, 351, 378, 393, 394, 395, 411, 413, 414, 419, 422, 423, 424
Ostrouchov, S., 413
Otto, S., 351, 418, 423

P

Paige, C., 410, 424
Parlett, B., 47, 411, 424
Parter, S., 351, 424
Patterson, J., 13, 420
Paul, G., 421
Peters, G., 292, 424
Peterson, J., 13, 424
Petzold, C., 170, 415
Poole, E., 394, 424
Porsching, T., 199, 212, 420
Pratt, T., 13, 424
Prenter, P., 104, 424
Puglisi, C., 301, 424

R

Rabinowitz, P., 149, 416
Rayna, G., 13, 424
Reddy, S., 409, 423
Reichel, L., 411, 423
Reid, J., 378, 417, 425
Rheinboldt, W., 130, 187, 424
Richtmyer, R., 199, 212, 425
Roche, M., 170, 420
Rodrigue, G., 417, 422
Romine, C., 417, 424
Rönsch, W., 70, 425
Rubinov, S., 166, 425
Runge, C., 104
Russell, R., 414

S

Saad, Y., 62, 87, 393, 410, 414, 421, 425
Salmon, J., 418
Sameh, A., 318, 422
Saunders, M., 410, 424
Saylor, P., 414
Schiesser, W., 199, 425
Schnabel, R., 130, 378, 416
Schneider, F., 70, 413
Schreiber, R., 301, 394, 426
Schultz, M., 62, 410, 417, 421, 425
Sethi, R., 13, 413, 426
Sherman, J., 230
Silberschatz, J., 13, 424
Simon, H., 414
Skeel, R., 253, 426
Sonneveld, P., 411, 426
Sorenson, D., 293, 413, 416, 426
Sproul, R., 13, 423
Steihaug, T., 416
Steurwalt, M., 351, 424
Stewart, G., 47, 253, 269, 293, 416, 423, 426
Stiefel, E., 378, 379, 409, 420
Stone, H., 62, 318, 426
Strakos, Z., 379, 419
Strang, G., 212, 426
Strassen, V., 70, 88, 426
Strikwerda, J., 199, 212, 426
Stroud, A., 149, 426
Sun, J.-G.,253, 426
Swarztrauber, P., 270, 318, 427
Sweet, R., 270, 318, 427

T

Tang, W., 394, 426
Taylor, D., 424
Timenetsky, M., 46, 422
Traub, J., 130, 427
Trefethen, L., 409, 411, 423

Author Index

U

Ullman, J., 13, 413

V

Van de Geijn, R., 301, 417
Van de Velde, E., 422
Van der Sluis, A., 379, 427
Van der Vorst, H., 352, 379, 393, 411, 416, 422, 427
Van Loan, C., 46, 115, 230, 252, 253, 269, 270, 301, 318, 380, 415, 419, 426, 427
Varga, R., 47, 212, 330, 351, 393, 427
Vinsome, P., 409, 427
Voigt, R., 318, 422, 424

W

Wahba, G., 427
Walker, D., 418
Wang, H., 318, 427
Wanner, G., 166, 420
Ware, W., 70, 428
Wasow, W., 212, 418
Watt, S., 415
White, R., 351, 423
Widlund, O., 409, 428
Wilkinson, J., 47, 252, 292, 424, 428
Wiseman, N., 13, 417
Wolfram, S., 13, 428
Woodbury, M., 230

Y

Young, D., 103, 330, 352, 379, 394, 410, 413, 419, 421, 428

Subject Index

A

Adams-Bashforth methods, 155*ff*
Adams-Moulton methods, 157*ff*
Adaptive method, 329
Adjoint equation, 188
Affine subspace, 376
Affine transformation, 355
Alternating direction method, 210
Amdahl's law, 66
Annihilation pattern, 293
Arnoldi process, 403*ff*
Arrowhead matrix, 228
  block, 229
Asynchronous method, 334, 343
Automatic differentiation, 130
Axpy, 79

B

Back substitution, 217
Backward error analysis, 8, 239
Bairstow's method, 134
Balancing, 241
Banded matrix, 84, 220*ff*, 242, 302*ff*
  symmetrically, 84, 87
Bank conflict, 51
Barrier, 70
Basis, 20
BiCGSTAB algorithm, 411
Biconjugate gradient method, 401
Bidiagonal matrix, 221
Binary tree, 63

Biorthogonal vectors, 401
Bisection method, 123*ff*
Bit-reversal, 268
BLAS, 79, 88, 301
Block methods, 77*ff*, 297*ff*, 337*ff*
Block storage, 283
Boundary conditions, 173
  Dirichlet, 180, 203
  Neumann, 180, 203
  periodic, 181
Boundary points, 173
Boundary value problem, 173
  two-point, 173
  nonlinear, 173, 182
Breakdown, 411
Bunch-Kaufman algorithm, 270
Busy waiting, 70

C

C, 12
Cache memory, 77
Catastrophic cancellation, 7
CDC STAR-100, 50
CGNE method, 397*ff*
CGNR method, 397*ff*
CGS algorithm, 411
Chaining, 52
Chaotic relaxation, 351
Characteristic polynomial, 28, 146, 168
Checkerboard ordering, 340

435

Cholesky factorization, 257*ff*, 279*ff*
  incomplete, 388*ff*
    no-fill, 389
    root-free, 271
Cluster, 60
Coarse grid, 354*ff*
Cofactor expansion, 23
Column sweep algorithm, 277, 304
Common memory, 54
Communication, 64
Communication length, 57-58
Companion matrix, 38
Complete pivoting, 252
Completely connected, 55
Computational complexity, 8
Computational engineering, 1
Computational science, 1
Compute-ahead, 274
Concave function, 122
Condition number, 249*ff*
Congruence transformation, 381
Conjugate directions, 373
Conjugate gradient method, 374*ff*
  block, 379
  preconditioned, 382
Conjugate residual method, 399
  generalized, 399*ff*
Conjugate transpose, 17
Connection Machine, 54
Consistency, 167, 194
Consistently ordered, 328
Contention, 55
  bus, 57
Contiguous, 51
Continuation, 131
Convergence error, 8
Convergence tests, 335
Convex function, 121
  strictly, 121
Craig's method, 397
Crank-Nicolson method, 197
CRAY, 50
Critical section, 64
Crossbar, 55, 56
Cross-over point, 53
Crout form, 219
CYBER, 50
Cycle machine, 50
Cyclic reduction, 306*ff*

D

Data flow, 68
Data management, 12
Davidenko's method, 131
Daxpy, 79
Deferred correction, 187
Determinant, 22*ff*, 246
Diagonal matrix, 17
  block, 17
Diagonal ordering, 348, 388
Diagonal scaling, 382
Diagonally dominant, 46, 178, 242, 252
  column, 46
  irreducibly, 46, 178
  row, 46, 177
  strictly, 46
Diagonally sparse matrix, 85, 87
Diameter, 57
Differential-algebraic equation, 170
Diffusion equation, 191
Direct-connect routing, 57
Direct method, 215
Direction vector, 371
Discretization error, 8, 141, 152*ff*, 196
  global, 141, 175
  local, 141, 166, 175, 193
Distributed memory, 54
Divide and conquer, 63
Domain decomposition, 278, 312, 377*ff*
Doolittle form, 219
Dot product, 17
Double-precision, 7

E

Edge contention, 58
Edge values, 334, 367
Efficiency, 8, 65
Elliptic equation, 199
Eigenfunction, 360
Eigensystem, 28
Eigenvalue, 28*ff*
  multiplicity, 29
  simple, 29
Eigenvector, 28*ff*
  left, 29

Subject Index 437

Eisenstat trick, 388, 393
EISPACK, 10
Elementary reflection matrix, 261
Equilibration, 241
Euler's method, 151$f\!f$
  backward, 164
Explicit method, 158, 195

F

Fan-in, 63
Far read, 61
Fast Fourier Transform (FFT), 266
Fast Poisson solvers, 265$f\!f$
Fast Sine Transform, 268
Fill, 223
  Poisson matrix, 225$f\!f$
Fine grid, 353$f\!f$
Finite difference, 151, 174, 186, 192, 204
  one-sided, 178
Finite element method, 212
First-order method, 153
Fork, 64
Forward reduction, 216
Frobenius matrix, 38
Full orthogonalization (FOM), 404
Full weighting restriction, 362
Fundamental solution, 168

G

Galerkin technique, 363
Gather, 51
Gauss-Jordan, 292
Gauss transform, 313
Gauss-Seidel method, 322$f\!f$, 340$f\!f$, 357$f\!f$, 372
  block, 347
  line, 347
  symmetric, 385
Gaussian elimination, 215$f\!f$, 310
  two-way, 319

Gaussian quadrature, 145
GCG(k) method, 400
Gerschgorin's theorem, 43
Givens reduction, 259$f\!f$, 281$f\!f$, 290$f\!f$
Givens transformation, 259
GMRES method, 404$f\!f$
Grammian matrix, 110
Granularity, 63
Graph, 25
  directed, 25, 179
  strongly connected, 25
Growth factor, 252

H

Hadamard matrix, 26
Hadamard product, 19
Hankel matrix, 108, 147, 230
Heat equation, 191$f\!f$, 199, 202, 209
Helmholtz equation, 331
Hermitian matrix, 18
Hessenberg matrix, 254
Heun's method, 154
Higher-order methods, 154
Higher-precision, 7
Hilbert matrix, 245
Householder-John theorem, 330
Householder reduction, 261$f\!f$, 280$f\!f$, 288$f\!f$
Householder transformation, 261
Hyperbolic equation, 199
Hypercube, 58$f\!f$, 59

I

ICCG method, 390
Idempotent matrix, 27
Identity matrix, 17
IEEE standard, 11
ijk forms, 76, 88, 295$f\!f$, 302$f\!f$
Ill-conditioned, 8, 125, 161, 243$f\!f$
Illiac IV, 54

Ill-posed, 8
ILU factorization, 406
Implicit method, 158, 195
ICCG method, 390
Incomplete orthogonalization (IOM), 404
Indefinite matrix, 32
Initial value problem, 150
Injection, 357
Inner product, 19, 41
  algorithm, 71$f\!f$, 82, 277, 287, 304
Inner iteration, 408
Interchanges, 236$f\!f$, 305
Interconnection schemes, 55
Interleaved storage, 283
Interpolation, 93
  error, 98
  Hermite, 105
Interpolating polynomial, 93, 355
  representation, 97
  uniqueness, 96
Inverse matrix, 21, 222
Irreducible matrix, 25
Iterative process, 8, 215
  subsidiary, 384
Iterative refinement, 253

J

Jacobi transformation, 259
Jacobi's method, 321$f\!f$, 372, 384
  block, 338
  damped, 361
  line, 338
Jacobian matrix, 128, 183
Jacobian transformation, 259
Jagged diagonal, 87
Join, 64
Jordan canonical form, 35$f\!f$

K

Kronecker product, 209
Kronecker sum, 209
Krylov subspace, 375$f\!f$

L

Lagrange polynomials, 94
Lanczos method, 379
Lanczos vectors, 401
LAPACK, 230, 250
Laplace's equation, 199, 323
Lax equivalence theorem, 195
Least squares approximation, 106$f\!f$, 264$f\!f$
Left-looking, 295
Linear array, 56-57
Linear combination algorithm, 72, 75, 81
  dual, 75
Linearization, 117, 126
Linearly independent vectors, 20
LINPACK, 10, 230, 250
Load balancing, 62$f\!f$
  dynamic, 62
  static, 62
Look-ahead, 411
Loop unrolling, 72, 88
  depth of, 74
Lotka-Volterra equations, 151
LU decomposition, 249
LU factorization, 217$f\!f$, 241, 275$f\!f$, 302$f\!f$
  incomplete, 406$f\!f$

M

MACSYMA, 13
Maintainability, 10
MAPLE, 13
Math Works, 14
MATHEMATICA, 13
MATLAB, 10
Matrix, 15$f\!f$
  dimension, 15
  H-, 394
  M-, 394
  rectangular, 15
  square, 15
Matrix-matrix multiplication, 74$f\!f$, 83$f\!f$
  fast, 80
Matrix polynomial, 30
Matrix-vector multiplication, 71

Subject Index 439

Megaflop, 53
Memory delay, 61
Memory-to-memory, 50
Mesh connection, 56-58
Message passing, 55
Method of lines, 198
Midpoint rule, 138
  composite, 140
  error, 143
Milne's method, 172
MIMD, 54
Minimization methods, 371ff
Minimum degree algorithm, 231
Minimum residual algorithm, 401, 410
Modeling, 3ff
Moments, 108, 146, 147
Multicoloring, 343ff
Multigrid method, 353ff
  two-grid, 363
Multiplication by diagonals, 84ff, 87
Multisplitting, 351
Multistep methods, 155ff
  linear, 162
Murnaghan-Wintner theorem, 36

N

Natural ordering, 206
Near read, 61
Negative definite matrix, 32
Nested dissection, 231
Nested iteration, 354
NETLIB, 14
Network
  bus, 57
  ring, 57
  switching, 56-57
Newton-Cotes formulas, 139
Newton form of interpolation, 95
Newton's method, 118ff, 183
  convergence, 119ff, 129
  global, 121
  inexact, 408
  simplified, 130
  systems, 127ff
  truncated, 408

Neumann expansion, 48
Nonlinear problem, 407
Nonsingular matrix, 23, 29, 31
Norm, 39ff
  elliptic, 40
  Euclidean, 40
  ℓ_p, 40
  matrix, 42
Norm equivalence theorem, 42
Normal equations, 108, 397
Numerical analysis, 3
Numerical integration, 137ff

O

Oblique projection, 410
Odd-even ordering, 310
Odd-even reduction, 308
ODEPACK, 167
One-way dissection, 228
Orthodir, 410
Orthogonal basis, 41
Orthogonal matrix, 21
Orthogonal polynomials, 110ff
Orthogonal vectors, 19, 41
  A-, 373
Orthomin, 400
Orthonormal vectors, 19, 41
Orthores, 410
Ostrowski-Reich theorem, 330
Ostrowski's theorem, 327
Outer iteration, 408
Outer product, 19
  algorithm, 75
  dual, 76

P

Page, 61
Pairwise pivoting, 293
Panel storage, 301
Parabolic equation, 191

Parallel computer, 11, 49, 81
Parallel computation, 144, 165
Partial pivoting, 240
Peaceman-Rachford method, 210
Peclet number, 178
Perfectly conditioned, 248
Permutation matrix, 21
Piecewise polynomial, 98$f\!f$
Pipelining, 49, 285
Plane rotation matrix, 259
Poisson's equation, 204
  generalized, 336
Polar decomposition, 39
Polynomial preconditioning, 393
Pool of tasks, 62
Portability, 10
Positive definite matrix, 24, 242, 252, 315
Positive real matrix, 255
Precedence graph, 68
Preconditioning, 381$f\!f$, 406$f\!f$
  left, 406
  right, 406
Predator-prey equations, 151, 166
Predictor-corrector method, 158
Principal submatrix, 16, 236
  leading, 16
Profile, 236
Prolongation, 357
Property A, 328

Rate of convergence, 324
Rectangle rule, 138
  composite, 140
  error, 143
Recursion, three-term, 112
Red-black ordering, 340, 369, 388
  by lines, 348
REDUCE, 13
Reduced system, 394
Reducible matrix, 24
Register-to-register, 50
Regula Falsi, 124
Relaxation methods, 321$f\!f$
  univariate, 372
Reliability, 10
Reordering, 227$f\!f$
Residual vector, 246
Restart, 400
Restoration time, 51
Reynolds number, 178
Result rate, 53
Richardson extrapolation, 148, 167, 187
Richardson's method, 372
Right-looking, 295
Robustness, 10
Romberg integration, 148
Rounding errors, 6$f\!f$, 143
Row-wise ordering, 206
RSCG method, 394
Runge-Kutta method, 154
Runge-Kutta-Fehlberg method, 168
Rutishauser form, 379

Q

QMR algorithm, 411
QR factorization, 259$f\!f$, 280$f\!f$
Quadrature rules, 138

R

Rank, 20
  full, 20
Rank revealing, 270

S

Sameh-Kuck pattern, 294
Saxpy, 79
Scatter, 51
Scattered decomposition, 351
Scattering, 293
Schur complement, 313, 394
Schur form, real 36
Schur product, 19
Schur's theorem, 36

Subject Index 441

Scientific computing, 1*ff*
Secant method, 122*ff*
Self-adjoint, 187
Semibandwidth, 84
Semidefinite matrix, 32
Send-ahead, 284
Separable, 270
Separator set, 228, 313
SGS, see Gauss-Seidel, symmetric
Shared memory, 54
Sherman-Morrison formula, 224, 230
Sherman-Morrison-Woodbury formula, 225, 230
Shifting strategy, 394
Shooting method, 184
   multiple, 188
SIMD, 54
Similar matrices, 33*ff*
Similarity transformation, 33
Simpson's rule, 139
   composite, 140
   error, 143
Single precision, 7
Singular value, 37
Singular value decomposition, 36
Skew-symmetric matrix, 32
SOR method, 326*ff*
   block, 347
   line, 347
Sparse matrix, 206
Spectral radius, 37
Spectrum, 28
Speed-up, 65
   scaled, 67, 70
Spin locks, 70
Spinning, 70
Splines, 100*ff*
   cubic, 101*ff*
   B-, 102
SSOR, 385
Stable method, 162
   A-, 170
   stiffly -, 169
   strongly -, 162
   unconditionally -, 196
   weakly -, 163
Stability, 194
Start-up time, 53

Stationary method, 324
Stencil, 197
Steepest descent method, 372
Stiff equations, 163
Strassen's method, 80, 88
Stride, 51
Stripes, 87
Stripmining, 73
Submatrix, 16
   leading principle, 16
   principle, 16
Subspace, 20
Successive overrelaxation, See SOR
Symbolic computation, 13
Symmetric differencing, 187
Symmetric matrix, 18, 31*ff*
Synchronization, 64, 70
Synchronous method, 334
Systems of equations, 158, 185

T

Taylor series, 91
   error, 92
Thinking Machine, Inc., 54
Threshold, 390
Threshold pivoting, 253
Time splitting method, 210
Toeplitz matrix, 230
Torus assignment, 293
Torus ordering, 352
Trace, 27
Trapezoidal rule, 139, 164, 185
   composite, 140
   error, 143
Transpose, 17
Tree operation, 63
Triangular matrix, 17
   block, 17
Triangular reduction, 216
Triangular systems, 217, 277, 285
Tridiagonal matrix, 84
   block, 207

Truncated series, 384
Truncation, 400
Truncation error, 8, 141

U

Unit ball, 40
Unit sphere, 40
Unitary matrix, 21
UNIX, 12
Unstable, 161
  numerically, 237
Updating, 275
  delayed, 296
  immediate, 296
Upwind differences, 178

V

Validation, 4*ff*
Vandermonde matrix, 94. 105, 109, 140, 230
V-cycle, 358

Vector, 15*ff*
  column, 15
  machine, 50
  row, 15
Vector computation, 144, 165
Vector computer, 11, 31, 69
Vector register, 50
Vector sum algorithm, 277
Visualization, 12
VODE, 170
Volume-to-surface effect, 334

W

Ware's Law, 66
Wave equation, 199
Wave front, 349
W-cycle, 358
Work unit, 366
Wormhole routing, 57
Wrapped storage, 283, 293

Z

Zaxpy, 79
Zebra ordering, 347